METALS SPECIATION, SEPARATION, AND RECOVERY

Volume II

Edited by

JAMES W. PATTERSON
Director, Industrial Waste Elimination Research Center,
IIT Center, Chicago, Illinois

ROBERTO PASSINO
Director, IRSA
Italian National Research Council
Rome, Italy

Proceedings of the Second International Symposium on
Metals Speciation, Separation, and Recovery
Rome, Italy
May 14–19, 1989

Sponsored by
Water Research Institute
Italian National Research Council
and
Industrial Waste Elimination Research Center
Illinois Institute of Technology

Library of Congress Cataloging-in-Publication Data

International Symposium on Metals Speciation, Separation, and Recovery (2nd : 1989 : Rome, Italy)
    Metals speciation, separation, and recovery / edited by James W. Patterson, Roberto Passino.
       p. cm.
    "Proceedings of the Second International Symposium on Metals Speciation, Separation, and Recovery, Rome, Italy, May 14-19, 1989: sponsored by Water Research Institute, Italian National Research Council, and Industrial Waste Elimination Research Center, Illinois Institute of Technology."
    Includes bibliographical references.
    1. Metals--Environmental aspects--Congresses. 2. Metals-Speciation--Environmental aspects--Congresses. 3. Metallurgy--Congresses. I. Patterson, James William, 1940- . II. Passino, Roberto. III. Industrial Waste Elimination Research Center (U.S.) IV. Istituto di ricerca sulle acque (Italy) V. Title.
TD196.M4I577   1990
628.5'2--dc20   89-13874
ISBN 0-87371-268-4

COPYRIGHT © 1990 BY LEWIS PUBLISHERS, INC.
ALL RIGHTS RESERVED

Neither this book nor any part may be reproduced or transmitted in any form or by any means, electronic or mechanical, including photocopying, microfilming, and recording, or by any information storage and retrieval system, without permission in writing from the publisher.

LEWIS PUBLISHERS, INC.
121 South Main Street, Chelsea, Michigan 48118

PRINTED IN THE UNITED STATES OF AMERICA

PREFACE

This volume contains the proceedings of the Second International Symposium on Metals Speciation, Separation, and Recovery, held in Rome, Italy, May 14-19, 1989. Both this and the first symposium, held in Chicago, Illinois in 1986, grew from a belief shared by many environmental engineers and scientists that there is a need for a searching reassessment of current policies for and methods of managing toxic metal pollutants.

These proceedings contain the two keynote and 24 scientific and technical papers presented at the symposium, together with a written discussion of each paper. By intent, attendance at the symposium was limited to approximately 100 engineers and scientists having research and practical interests directly related to the symposium subject. In consequence of the focused interests of the participants, and the relevance of the symposium subject to environmental protection, discussion and debate were vigorous. Following the presentation of papers and invited discussions, the symposium participants divided into four theme groups to explore the state of the art, and define research needs. A summary of this effort is included in these proceedings.

In recognition of the wide audience concerned with metals speciation, separation, and recovery, it was decided to publish these proceedings, as were the proceedings of the First Symposium. The editors would like to express their appreciation to the authors of the papers and written discussions and to the other symposium participants.

ACKNOWLEDGMENTS

The Second International Symposium on Metals Speciation, Separation, and Recovery was sponsored by the Industrial Waste Elimination Research Center of the Illinois Institute of Technology, and the Water Research Institute of the National Research Council of Italy. The Water Research Institute was established in 1968 to carry out research activities in the fields of hydrology, hydrogeology, water supply, and water pollution control. The Industrial Waste Elimination Research Center (IWERC), established in 1980, is an Environmental Research Center supported, in part, through the U.S. Environmental Protection Agency (EPA) Office of Exploratory Research. The center's research is directed toward developing and refining technologies to minimize or eliminate the generation of industrial pollutants.

Appreciation is expressed to Mrs. Gloria Boyda of IIT and to Mrs. Eileen Schmitz of CAPS, Ltd., who worked so long and diligently to organize the symposium and gather the manuscripts. Mr. Dario Marani was invaluable in assisting in the planning and implementation of the Symposium.

Particular appreciation is expressed to Mr. Lou W. Lefke, associate director of the U.S. EPA Risk Reduction Engineering Laboratory; Mr. Robert A. Papetti and Ms. Karen Morehouse of the Environmental Research Centers Program, U.S. EPA Office of Exploratory Research; and the members of the IWERC Scientific Advisory Committee and Industry Advisory Council.

CONTENTS

KEYNOTE PAPERS

Cycles, Fluxes and Speciation of Trace Metals in the Environment 3
    Roy M. Harrison

Metals Control Technology; Past, Present, and Future 27
    James W. Patterson

PART I: CHEMISTRY OF TOXIC METALS

1. Metals Speciation in Sludges from Wastewaters Treatment by Bulk and Surface (XPS) Analysis 45
    Elio Desimoni, A. Marcone, G. Tiravanti, and P. G. Zambonin

    1. Abstract 45
    2. Introduction 45
    3. X-Ray Photoelectron Spectroscopy 47
    4. Samples Treatment and Experimental Techniques 49
    5. Results and Discussion 51
        5.1 XPS Analysis of Standards 51
        5.2 Bulk Chemistry of Starch Xanthate/Metal Ions Systems 53
        5.3 XPS Analysis of SX-Na, SX-Cd, and SX-Hg Specimens 59
    6. Conclusions 65
    Acknowledgments 65
    References 65

Discussion: *Paolo Cescon* 69

2. Chemical Equilibrium Analysis of Lead and Beryllium Speciation in Hazardous Waste Incinerators 73
    Alexander P. Mathews

    1. Introduction 73
    2. Metal Transformations in Incineration Systems 74
    3. Metal Emissions from Incinerators 76
    4. Modeling Approach 77
        4.1 Thermodynamic Analysis 77

5. Model Simulations and Analysis 79
   5.1 Effect of Incinerator Operating Conditions on Be Speciation 79
   5.2 Effect of Incinerator Operating Conditions on Pb Speciation 81
6. Conclusions 82
References 83

Discussion: *Sabino A. Bufo* 85

3. Nucleation and Crystal Growth Studies on Precipitation of Cadmium Hydroxide from Aqueous Solutions 91
J. W. Patterson, B. Luo, D. Marani, and R. Passino

1. Abstract 91
2. Introduction 91
3. Background 93
   3.1 Solubility Phenomena 93
   3.2 Nucleation and Crystal Growth Theories 95
4. Experimental 97
5. Results and Discussion 100
   5.1 Metal Precipitation in the Absence of an Inhibiting Agent 100
   5.2 Metal Precipitation in the Presence of Citric Acid 106
6. Conclusions 110
Acknowledgments 110
References 110

Discussion: *Roger A. Minear* 114

4. Effect of Speciation on the Rates of Oxidation of Metals 125
Frank J. Millero

1. Introduction 125
2. Background 125
3. Effect of pH on the Rates of Oxidation 126
4. Effect of the Major Ionic Components of Natural Waters on the Rates of Oxidation 131
Acknowledgments 140
References 140

Discussion: *Giuseppe Macchi* 142

PART II: PRECIPITATION PHENOMENA

5. **Experiments on the Simultaneous Oxidizing Extraction Process of Fe(II)**   149
   J. Mario Diaz, A. E. Fernandez, A. T. Aguayo, and J. Viguri

   1. Summary   149
   2. Introduction   149
   3. Experimental Methods   150
   4. Experimental Results and Discussion   152
      - 4.1 Iron Oxidation   153
         - Regimes Analysis   153
         - Kinetic Study   155
         - Precipitation Limits   156
      - 4.2 Liquid Extraction   157
         - Extraction Equilibrium   157
         - Extraction Kinetics   159
         - Extractive Oxidation   161
   5. Symbols   162
   References   163

   Discussion: *A. Lopez and D. Petruzzelli*   165

6. **Hydrolysis, Precipitation and Aging of Copper(II) in the Presence of Nitrate**   169
   J. W. Patterson, R. E. Boice, C. Petropoulou, D. Marani, and G. Macchi

   1. Introduction   169
   2. Background   170
      - 2.1 Phenomenology of Cu(II) Hydrolysis-Precipitation   170
      - 2.2 Solubility of Cupric Hydroxide-Oxide Precipitates   171
      - 2.3 Basic Cupric Salts   171
   3. Experimental   174
      - 3.1 Batch Tests   174
      - 3.2 Continuous Stirred Tank Reactor (CSTR) Tests   175
      - 3.3 Analytical Measurements   175
   4. Results and Discussion   177
      - 4.1 Cupric Hydroxide Precipitation and Aging   177
      - 4.2 Effect of Reagents Mixing Procedure   183
      - 4.3 Thermodynamic Considerations   186
      - 4.4 Basic Salt Precipitation   189

5. Summary and Conclusions	193
Acknowledgments	194
References	194

Discussion: *Luigi Campanella* — 198

7. Role of Copper Complexation on Treatment Efficiency and Design of Two Industrial Wastestreams
Kostas Saranteas and I. W. Wei — 203

1. Abstract	203
2. Introduction	203
3. Some Theoretical Considerations:	204
3.1 Cupric Ion Electrode	204
3.2 Chemical Precipitation	205
3.3 Activated Carbon Treatment	205
3.4 Experimental Procedure	206
4. Results and Discussion	207
4.1 Free Copper Analysis	207
4.2 Chemical Precipitation	208
Hydroxide Precipitation	208
Sulfide Precipitation	211
4.3 Activated Carbon	212
4.4 Combined Waste Treatment	220
5. Conclusions	222
References	224

Discussion: *Alan Bowers and Soon H. Cho* — 226

PART III: METAL SPECIATION AND COMPLEXATION IN NATURAL SYSTEMS

8. Speciation of Aluminum in Geothermal Brines: Comparison of Different Methodologies
Marco Achilli, G. Ciceri, R. Ferraroli, G. Culivicchi, and S. Pieri — 237

1. Summary	237
2. Introduction	237
3. Experimental	239
3.1 Apparatus	239
3.2 Reagents	239
3.3 Preliminary Studies	241
3.4 Pyrocathechol Violet (PCV) Colorimetric Method	241

		3.5	Adsorption on Alumina	242

    3.5 Adsorption on Alumina   242
    3.6 Filtration and Ultrafiltration   242
    3.7 8-HQ Complexation-MIBK Extraction
         Method   242
    3.8 Ion Exchange-PCV Complexation Method   243
    3.9 Chelation with Supported 8-HQ   243
 4. Results and Discussion   244
    4.1 Hydroxy-Al Solutions of Different
         [OH$^-$]/[Al] Ratio   244
    4.2 Pyrocatechol Violet (PCV) Colorimetric
         Method   246
         Preliminary Tests on Al-PCV Complex   246
         Results from PCV Colorimetric Method
         Application   248
    4.3 Adsorption on Alumina   248
    4.4 Filtration and Ultrafiltration   249
    4.5 8-HQ Complexation-MIBK Extraction
         Method   250
    4.6 Ion Exchange-Complexation Method   252
    4.7 Chelation with Supported 8-HQ   254
    4.8 Proposed Method   256
 5. Conclusions   256
    References   258

Discussion: *Frank J. Millero*   261

9. Speciation of Tin in Sediments of Arcachon Bay (France) — 263
   M. Astruc, R. Lavigne, R. Pinel, F. Leguille,
   V. Desauziers, P. Quevauviller, and O. Donard

 1. Introduction   263
 2. Material and Methods   265
    2.1 Sampling   265
    2.2 Analytical Techniques   265
 3. Results and Discussion   267
    3.1 Speciation of Tin in the Sediment   267
    3.2 Total Tin   267
    3.3 Total Recoverable Inorganic Tin (TRIT)   267
    3.4 Butylated Species   267
    3.5 Interstitial Waters   270
 4. Conclusions   272
    References   273

Discussion: *Maurizio Pettine and A. Puddu*   275

10. Metal Complexation by Water-Soluble Organic Substances in Forest Soils   283
A. T. Kuiters and W. Mulder

   1. Abstract   283
   2. Introduction   284
   3. Metal-Complexing Capacity of Leaf-Litter Leachates   285
      3.1 Leaching Procedure   285
      3.2 The Size-Exclusion Chromatographic Technique   285
      3.3 The Ion-Exchange Technique   287
      3.4 Analytical Procedures   287
      3.5 Results   287
   4. Metal-Solubilization from Forest Soils by the Litter Leachates   291
      4.1 Batch Equilibrium Experiments   291
      4.2 Results   292
   5. Discussion   294
   Acknowledgments   295
   References   296

Discussion: *Kostantinos Fytianos*   298

11. Theoretical and Experimental Drawbacks in Heavy Metal Speciation in Natural Waters   301
Paolo Papoff, M. Betti, and R. Fuocco

   1. Introduction   301
   2. The Biological Problem   302
      2.1 Short Comments on the Biological Postulates   305
      2.2 Conclusion Concerning the Biological Problem   307
   3. The Chemical Problem   308
      3.1 The Determination of the Total Concentration C_M 310
      3.2 Bias Concerning C_M as Results from Interlaboratory Exercise   312
      3.3 Chemical Speciation of Single Species   314
      3.4 Chemical Speciation of Classes of Compounds   315
      3.5 Conclusions Concerning Speciation by Electrochemical Measurements   316
   4. Conclusions   318
   Acknowledgments   319
   References   319

Discussion: *Conny Haraldsson*   324

PART IV: SORPTION ONTO SURFACES

12. **Proton and Metal Ion Binding on Humic Substances**   329
    *J. C. M. DeWit, W. H. Van Riemsdijk, and
    L. K. Koopal*

    1. Abstract   329
    2. Introduction   329
    3. The Mono-Component Proton Adsorption Isotherm   331
        3.1 Theoretical Framework   331
        3.2 Curvature and Ionic Strength Effects   334
        3.3 Parameter Sensitivity   335
    4. Calculation of $\Theta_{i,H}(pH_s)$ Curves from Experimental Proton Adsorption Data for Humic and Fulvic Acids   336
    5. Metal-Ion Adsorption   341
        5.1 Theoretical Framework   341
        5.2 Model Calculations for Metal Ion Adsorption   343
        5.3 Monodentate Metal Ion Adsorption   343
        5.4 Bidentate Metal Ion Adsorption on Salicylic Acid and Phthalic Acid-Like Surface Groups   345
    6. Conclusions   349
    Acknowledgments   350
    References   350

Discussion: *Virginia L. Cunningham*   354

13. **Polyelectrolytic Metal Ions Sequestrants**   359
    *L. Campanella, V. Crescenzi, M. Dentini,
    C. Fabiani, F. Mazzei, and A. I. Nero Scheffino*

    1. Introduction   359
    2. Results and Discussion   360
        2.1 Potentiometric and Equilibrium Dialysis Data   360
        2.2 UV Spectral Data   363
        2.3 Calorimetric Data   365
        2.4 Membrane Filtration Data   367
    3. Concluding Remarks   369
    4. Experimental   370
        4.1 Samples   370
        4.2 Instrumentation   370
    Acknowledgments   372
    References   372

Discussion: *Arup K. Sengupta*   373

14. Chemical Speciation of Heavy Metals in Soils Following   377
    Land Application of Conditioned Biological Sludges and
    Raw Pig Manure
    M. Baldi, M. C. Negri, and A. G. Capodaglio

    1. Introduction   377
    2. Preliminary Investigation   378
    3. Heavy Metals in Agricultural Soil   379
    4. Materials and Methods   381
    5. Results and Discussion   386
    6. Conclusions   390
    References   390

Discussion: *Gianniantonio Petruzzelli*   393

PART V: ION SEPARATION

15. A Process to Improve the Regenerated Effluent   399
    Concentration of Ion Exchange
    Xiang Liangkui and Peng Dangcong

    1. Introduction   399
    2. Background   399
    3. Limit Regenerated Effluent Concentration Theory   400
    4. Step by Step Regeneration Process   402
    5. Experimental Results   403
    6. Conclusions   405
    Acknowledgments   405
    References   405

Discussion: *Lorenzo Liberti and D. Petruzzelli*   406

16. Heavy Metal Removal Using Natural Zeolite   417
    M. D. Loizidou

    1. Introduction   417
    2. Theoretical Aspects of the Ion Exchange Process   418
    3. Experimental Procedure   420
        3.1 Zeolite Treatment and Analyses   420
        3.2 Ion Exchange Experiments   421
    4. Results   422
    5. Discussion   422
        5.1 Exchange Capacity   422
        5.2 Ion Exchanges   423
            Single Ion in Solution   423
            Mixed Ion Solutions   426
            Regeneration of the Zeolite   428
            Influence of the Anions   428

6. Concluding Remarks	432
References	432

Discussion: *Robert H. Rosset*   434

17. Metal Ion Separations from Hazardous Waste Streams by Impregnated Ceramic Membranes   437
J. Yi, R. Ferreira, and L. L. Tavlarides

1. Abstract	437
2. Introduction	438
3. Coupled Transport of Metal Ions in Ceramic Membranes	441
4. Experimental Apparatus and Technique	444
5. Preliminary Results	446
6. Conclusion and Future Work	449
Acknowledgments	449
Notation	449
References	450

Discussion: *Peter C. Nwogu*   452

PART VI: SOILS CONTAMINATION AND DECONTAMINATION

18. Partitioning of Heavy Metals into Selective Chemical Fractions in Sediments from Rivers in Northern Greece   463
V. Samanidou and K. Fytianos

1. Abstract	463
2. Introduction	463
3. Materials and Methods	464
4. Results and Discussion	464
5. Conclusions	469
References	470

Discussion: *Alexander P. Mathews*   473

19. Chemical Decontamination of Dredged Materials, Sludges, Combustion Residues, Soils, and Other Materials Contaminated with Heavy Metals   477
G. Müller

1. Introduction	477
2. Treatment Techniques	478
3. Chemical Decontamination: A Concept for the Final Disposal of Materials Contaminated with Heavy Metals	481

	3.1 Extraction by Acid Treatment - Selection of Suitable Acids	483
	3.2 Separation of the Decontaminated Solids from the Solvents	483
	3.3 Precipitation of Heavy Metals as Hydroxides	483
	3.4 Subsequent Precipitation of Remaining Minor Concentrations of Cadmium and Other Heavy Metals	484
4.	Advantages of Chemical Decontamination-Practical Realization	486
References		488

Discussion: *Raymond W. Regan, Sr.* 490

20. **Results of Bench-Scale Research Efforts to Wash Contaminated Soils at Battery-Recycling Facilities** 497
    J. L. Hessling, M. P. Esposito, R. P. Traver, and R. H. Snow

1.	Introduction and Background	497
2.	Summary of Findings	499
3.	Site Profiles	499
	3.1 Site A	500
	3.2 Site B	500
	3.3 Site C	502
	3.4 Site D	502
	3.5 Site E	502
	3.6 Site F	503
	3.7 SSM	503
4.	Soil Characterization	504
5.	Experimental Soil-Washing Procedures	505
6.	Results and Discussion	506
7.	Conclusions	510
Acknowledgments		511

Discussion: *Giovanni Tiravanti* 512

PART VII: WASTE REDUCTION AND RECOVERY CASE STUDIES

21. **Three Case Studies of Waste Minimization Through Use of Metal Recovery Processes** 517
    M. L. Apel, J. Bridges, M. F. Szabo, and S. H. Ambekar

1.	Introduction	517
2.	Background	518
3.	Case Study No. 1	519

3.1	Background	519
3.2	Printed Circuit Board Production Process	519
3.3	Engineering Design: Spray Rinse System	520
3.4	Results and Discussion	522
	Rinse Efficiency	522
4. Materials Balance		523
5. Economic Analysis		526
6. Conclusions		527
7. Case Study No. 2		527
7.1	Background	527
7.2	Experimental Equipment and Procedures	529
7.3	Results and Discussion	530
7.4	Conclusions	533
8. Case Study No. 3		533
8.1	Background	533
8.2	Experimental Procedures and Equipment	535
8.3	Results and Discussion	536
	Ammonium Persulfate and Ammonium Carbonate Leaching	536
	Sodium Sulfide Precipitation	538
8.4	Conclusions	538
References		539

Discussion: *W. Wesley Eckenfelder, Jr.*   542

22. Metro Recovery Systems – A Centralized Metals Recovery and Treatment Facility in Twin Cities, U.S.A.   547
J. J. Chen

1. Introduction		547
2. Project Evaluations		548
2.1	Technical	548
	Types of Resins Selected	549
	Cross Contamination of Resins	549
	Fouling of Resin	550
	Ion Exchange Effluent Monitoring	550
	Water Reuse	550
	Dead Rinse	551
2.2	Economics	552
2.3	Market Analysis	555
	Tier I – Contract Companies	555
	Tier II – Other Electroplating and Printed Circuit Companies	556
	Tier III – Other Industries Producing Inorganic Waste	557
	Tier IV – Future Markets	558

	3. The Central Treatment and Recovery Facility	558
	3.1 The In-Plant Equipment	558
	3.2 The CTRF	559
	3.3 The Current Status	561
	4. Summary and Conclusions	563
	Acknowledgments	564
	References	564

Discussion: *Enrico Rolle*   565

23. Recovery of Metals from Waste Streams by Hydrometallurgical Processes   567
C. L. van Deelen

	1. Summary	567
	2. Introduction	567
	3. Metal Recovery by Hydrometallurgical Techniques	568
	4. Recovery of Metals from Spent Hydrodesulphurizing Catalyst	569
	4.1 Data on Hydrodesulphurizing Catalyst	569
	4.2 Existing Processes for Recovery of Metals	570
	4.3 TNO Metal Recovery Process	572
	5. Recovery of Metals from Spent Batteries	575
	5.1 General	575
	5.2 Existing Processes for Recycling of Spent Batteries	577
	5.3 TNO Processes for Recovery of Metal from Spent Batteries	578
	Processing of Nickel-Cadmium Batteries	578
	Processing of Alkaline Batteries	580
	6. Conclusions	582
	References	582

Discussion: *Domenico Petruzzelli and A. Lopez*   584

24. Chromium Recovery from Tannery Sludge by Incineration and Acid Extraction   587
M. Beccari, L. Campanella, E. Cardarelli, M. Majone, and E. Rolle

	1. Introduction	587
	2. Definition of Objectives	590
	3. Experimental	590
	3.1 Sludge Characterization	590
	3.2 Incineration Procedure	591
	3.3 Ash Characterization and Extraction Procedure	592

4.	Experimental Results	592
	4.1 Incineration Tests	593
	4.2 Extraction Tests	597
	4.3 Evaluation of Environmental Impact	598
5.	Conclusions	600
	References	601

Discussion: *Chriso Petropoulou*   603

RESEARCH NEEDS WORKSHOP: Research Needs in Metals Speciation, Separation, and Recovery 611
J. W. Patterson and C. Petropoulou

Group A:	Analysis and Speciation Chemistry	611
Group B:	Solid-Liquid Interface Chemistry	613
Group C:	Selective Separation Technology	615
Group D:	Metal Sludges Management	619

References   621

Index   623

METALS SPECIATION, SEPARATION, AND RECOVERY

Volume II

KEYNOTE PAPERS

CYCLES, FLUXES AND SPECIATION OF TRACE METALS IN THE ENVIRONMENT

Roy M. Harrison
Institute of Aerosol Science, University of Essex, England

INTRODUCTION

Looking through the published literature on metals in the environment, one is immediately struck by the predominance of papers on process, pathway and mechanism studies. thus for each metal, processes such as atmospheric wet and dry deposition, aquatic speciation, partitioning, precipitation and accumulation in sediments are generally rather well understood in terms of the process efficiency and mechanism, and there is an increasing body of knowledge on chemical speciation. It is undeniable that a deep appreciation of such processes, pathways and mechanisms is an essential pre-requisite to a full understanding of metal behavior in the environment.

For the pure geochemist, the pursuit of an understanding of such aspects of environmental behavior is itself an admirable exercise. However, the majority of workers in the fields of environmental science and engineering see these studies as only one part of a more major aim, which is to provide effective mechanisms of control of trace metals in the environment at points where damage to humans, or to the environment itself may arise. a second major component of such a broad aim is the development and validation of engineering control procedures, a subject which will be a major theme of this Symposium; i.e. metals separation and recovery. The other major theme of the meeting relates to metal speciation, a subject of clear relevance to removal and recovery, as well as to environmental process and pathway studies and to toxicity.

However, if we are to develop a full comprehension of the knowledge necessary for our main aim of control of trace metals in the environment, one crucial piece of the jigsaw is missing. We need a greatly improved knowledge of metal fluxes through the various environmental pathways if we are to formulate the most effective and perhaps more importantly the most cost-effective means of control of metals in critical areas of the environment.

4 INTRODUCTION

Biogeochemical Cycles of Trace Metal Species

A major focus of interest in recent years has been the behavior of alkyllead compounds in the environment. These are compounds for which there are excellent speciation techniques available [1, 2] so few doubts can exist about actual chemical forms. The majority of emissions are in the form of tetraalkyllead (PbR_4; $R = CH_3$ or C_2H_5) and environmental degradation proceeds *via* trialkyllead (PbR_3^+) and dialkyllead PbR_2^{2+}) to inorganic Pb(II). There is some evidence [3, 4] of a very inefficient environmental alkylation of Pb^{2+} to form volatile alkyllead species which may affect atmospheric concentrations of these compounds away from areas of substantial pollutant emissions.

In 1985 we proposed the environmental cycle of alkyllead [5] shown in Figure 1. This shows the assumed major processes, as inferred from chemical analysis of environmental samples. To assign reliable numerical values of fluxes or reservoirs is currently beyond our capability. We have approximate knowledge of the source strength term for emissions to atmosphere, but even that is poorly quantified and recent work shows that our knowledge of the speciation of vehicular emissions is incomplete [6].

Assignment of reliable values to fluxes and reservoir burdens can only be carried out on the basis of extensive measurement and process studies. Most research programmes are too small, either in funding or in outlook to provide the necessary data. The need for such data, is, however, so acute that new research studies will need to be focussed towards this objective. One such study, which exemplifies the problems involved is research into trace metal pollution of the North Sea. This will be described briefly in the sections following.

POLLUTANT INPUTS TO THE NORTH SEA

The North Sea (Figure 2) is a relatively enclosed body of water, with major inputs of pollutants from all of the European countries. The reason for indicting all countries is that atmospheric deposition plays a major role as a pollutant input, which is derived to a greater or lesser extent from all European countries (and to a slight degree from countries outside Europe). Critchley [7] has estimated a budget for inputs of selected trace metals to the North Sea (Table 1). These estimates are obviously subject to temporal change due to genuine changes in source strength, for example those western European countries. However, for a given point in time,

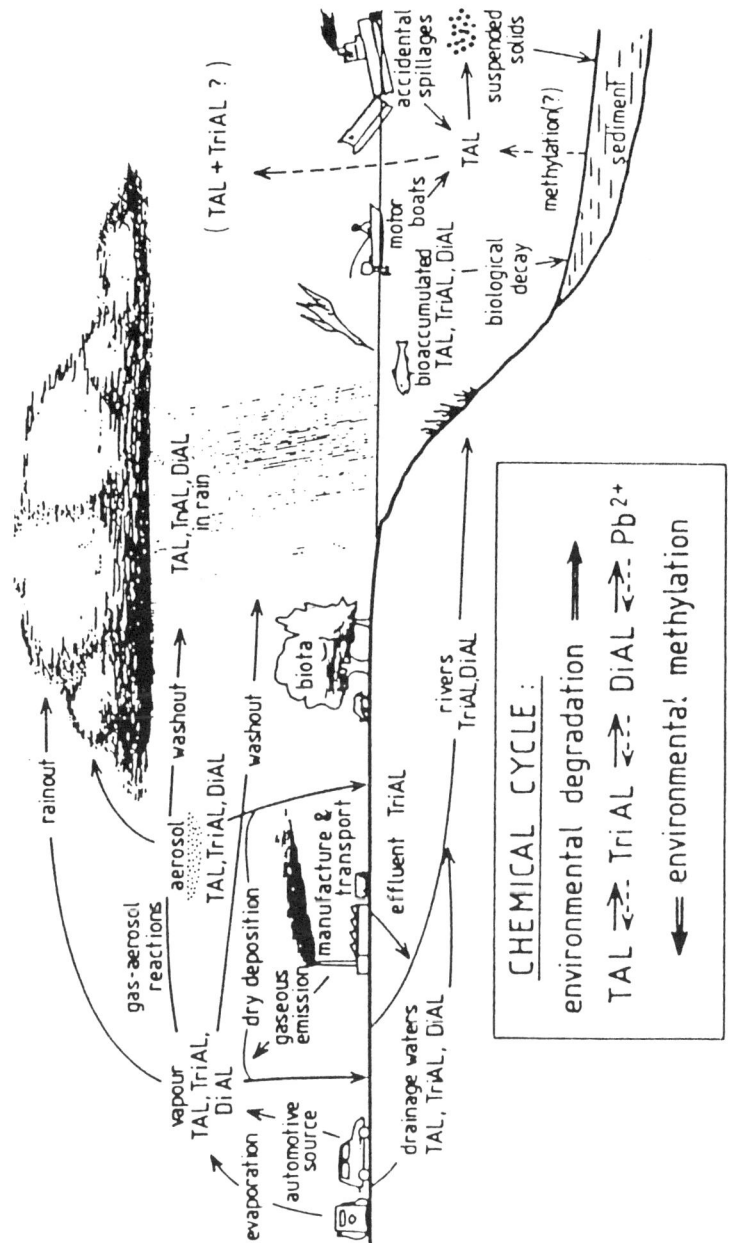

FIGURE 1. *The Biogeochemical Cycle of Alkyllead Compounds.*

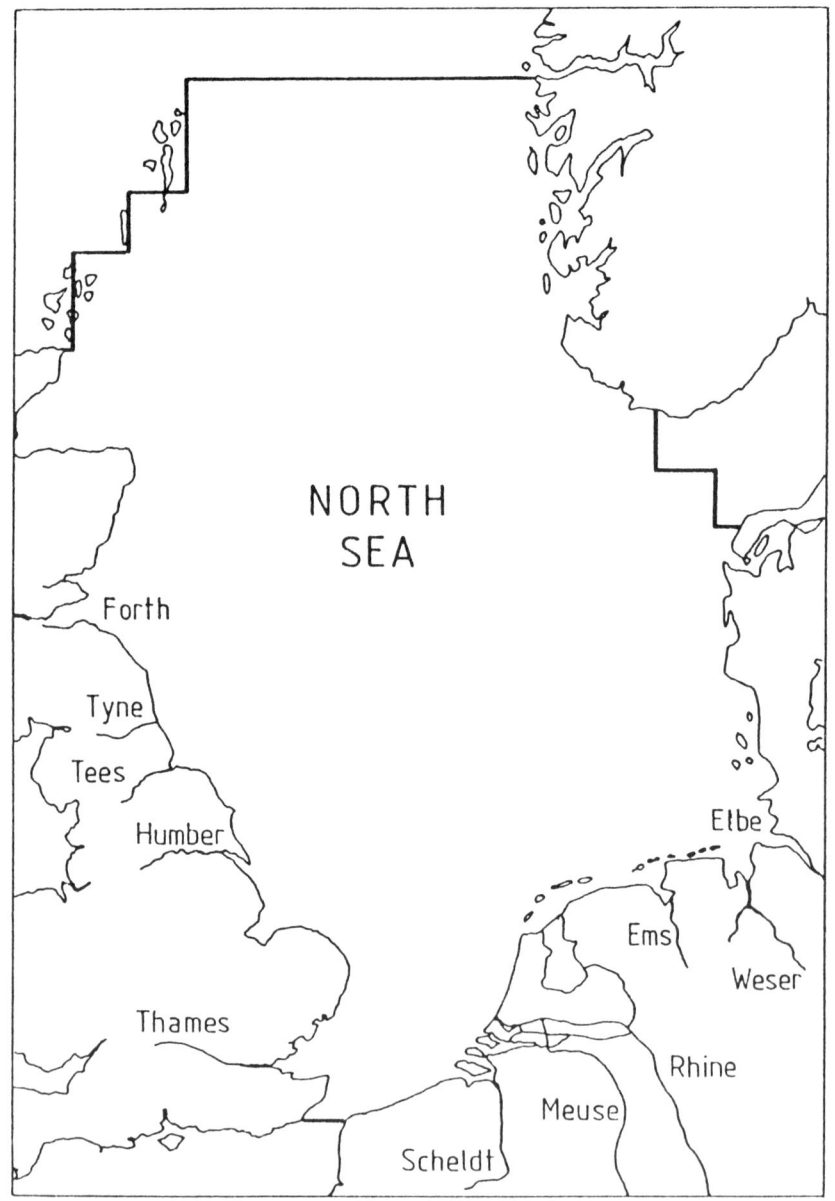

FIGURE 2. The North Sea, Showing Accepted Boundaries and Major River Inputs.

these estimates are subject to massive uncertainties which arise primarily from the scarcity of research into reliable quantification of fluxes.

Before discussing some of the problems inherent in flux determination, two points need to be raised. Firstly, it is clear from Table 1 that for the North Sea as a whole, atmospheric deposition and river inputs are by far the major sources of trace metals. This point will be reflected in the emphasis of subsequent discussions. Secondly, it must be borne in mind that these inputs are not spatially uniform, and as the North Sea is a poorly mixed body, spatial variations in metal inputs will be reflected in considerable local variation in metal content of waters and sediments.

Problems in Determination of Atmospheric Fluxes to the Sea

There have been two main approaches to the estimation of atmospheric input fluxes. The first, and most obvious, is to use deposit gauges to collect total deposition for analysis. This approach has been used in two surveys by Cambray and coworkers [8, 9]. It suffers from two major disadvantages. The first is that a deposit gauge cannot realistically parallel the behavior of a seawater surface as a collector of dry deposition and thus deposition velocities, and hence measured fluxes may differ between gauge measurements and the adjacent sea surface. Secondly, when such samples are placed on platforms over the sea, they collect a great deal of sea spray. This spray contains the trace metals of interest, but representing only a recycled sea-derived fraction which should not be included in the estimate of atmospheric deposition. Subtraction of this component based upon use of bulk seawater metal/sodium ratios is highly controversial as some workers believe that there is a major "maritime effect" in which sea spray is highly enriched in trace metals relative to bulk seawater [10] whilst other workers deny the existence of such an effect. There is thus, no agreed method of correcting data for recycled spray. The tendency has thus been to use sites a few kilometers inland at which sea spray inputs are negligible. Modeling studies, however, show rather strong gradients of atmospheric concentrations over the North Sea [11] and thus use of only data from coastal sites will lead to and overestimation of metal inputs.

The second approach is to measure airborne concentrations over the sea. This may be performed using static platforms [10], aircraft [12] or ship-borne samplers as in our own current work. Much data has also been derived from coastal sites, although this is subject to the same problems of overestimation as the deposition data from such sites. Collection of air concentration data is to some extent limited by weather conditions, as sea spray creates major problems in high wind conditions and low altitude sampling has to be curtailed. From air concentration data, deposition fluxes are calculated *via* the deposition velocity:

Flux (μg m^{-2} s^{-1}) = Atmospheric conc. x deposition velocity
(μg m^{-3})   (m s^{-1})

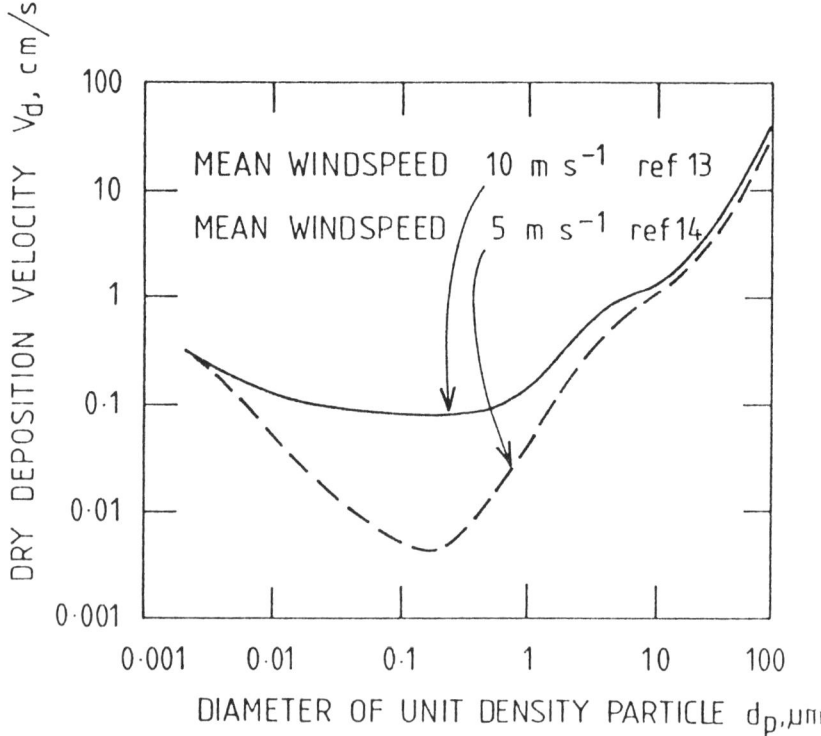

FIGURE 3. Calculated Deposition Velocities to Water Surfaces as a Function of Particle Size.

Deposition velocities are a very strong function of particle size (see Figure 3) and there is a real problem in selecting an appropriate deposition velocity in the absence of full knowledge of a size distribution. Our own present work collects size-fractionated samples, thus allowing a better estimation of dry deposition fluxes by calculation for each size fraction separately. This however, requires longer sampling intervals, this reducing temporal/spatial resolution on a mobile platform. Figure 3 shows how deposition velocities increase rapidly for larger particles and Buat-Menard and coworkers [15] have shown in measurements over the Mediterranean Sea that although large particles ($>5\mu m$) represent only a minor part of the aerosol mass, they may make a major contribution to the atmospheric deposition flux. This creates two problems: firstly, conventional air samplers do not efficiently sample particles of $>10\mu m$ and thus tend to underestimate their concentration. We are thus developing samplers with better inlet characteristics. Secondly, the large particles tend to contain a large component of recycled sea-surface material and quantitative subtraction of this provides a real problem, as with deposition measurements.

Estimation of wet deposition may be from land-based deposition samples from wet-only collectors, or from total collectors after subtraction of the calculated dry deposition component. This must then be scaled by the ratio of rainfall depth over the sea to that at the near-coastal site. This is problematic as the data base for rainfall at sea is very sparse and most work for the North Sea has assumed rainfall of ca. 50% of that over adjacent land, based upon one set of data from Cambray *et al.* [8]. It is also possible to estimate wet fluxes from atmospheric concentrations using the scavenging ratio:

$$\text{Conc. in rain} = \text{Conc. in air} \times \text{Scavenging ratio}$$
$$(\text{mg kg}^{-1}) \qquad\qquad (\text{mg kg}^{-1})$$

Scavenging ratios relating to the above units of concentration are typically ~500-2000, but vary according to the element in a manner apparently related to particle size [10, 15]. If calculated deposition fields are to reflect reduced atmospheric concentrations over the central sea relative to coastal areas, an approach based upon use of scavenging ratios is essential.

One has to take into account the compounded uncertainties in: (i) the atmospheric concentration field (perhaps ±30%); (ii) the deposition velocity (±80%); (iii) the rainfall field over the sea (±25%, perhaps); (iv) the scavenging ratio (±40%); (v) the correction for re-cycled material (?%). One is left with the feeling that local estimates of atmospheric inputs of metals are really order of magnitude figures rather than quantitative fluxes. This is less true for estimates to the whole sea, as underestimation of a deposition velocity, or scavenging ratio leads to reduced deposition estimates close to source which are to a large extent compensation by relatively greater deposition further afield. Research on mechanisms and processes is clearly highly pertinent to reducing these uncertainties, but it needs to be focussed upon the problems of flux determination.

Estimation of Riverine Fluxes of Metals

For most metals, river inputs represent the major estimated source of input to the North Sea (Table 1). These fluxes are again very difficult to quantify reliably. The instantaneous load in a river is given by the product of concentration and volumetric discharge (flow).

$$\text{Load}_{(Inst)} = \text{Concentration}_{(Inst)} \times \text{discharge}_{(Inst)}$$
$$(g\ s^{-1}) \qquad (g\ m^{-3}) \qquad (m^3\ s^{-1})$$

There are, however, very real problems in obtaining loads over long periods appropriate to the pollution of major seas. The availability of continuous metal concentration data for rivers is essentially zero, and in most instances only periodic (perhaps weekly) grab sample information exists, although the records of discharge may be continuous. Many data sets have been obtained using analytical methods of inadequate sensitivity and include many values below detection limit; this creates obvious problems. There are less obvious, but still major problems in estimating metal fluxes from grab sample data with instantaneous, and in some cases also continuous flow measurements. Walling and Webb [17] have reviewed the procedures which have been used for estimating long-term fluxes from grab sample information, concluding that some methods introduce a bias into the result.

In collaboration with the U.K. Water Research Center (WRc), we have operated a continuous flow-proportional water sampler in a major U.K. river. The sampler, designed by WRc, collects up to 2000 5ml subsamples of water over a one-week period at a frequency which is adjusted continuously in response to discharge as measured by an ultrasonic flow gauge. It thus produces a one-week flow-proportional sample, and the weekly load is estimated from

Weekly load = Conc. (Flow prop) x Ave. Flow x Sampling Period
$$\quad (g) \qquad\qquad (g\ m^{-3}) \qquad\quad (m^3\ s^{-1}) \qquad (s)$$

TABLE 1. ***Summary of Trace Metal Inputs to the North Sea (After Critchley; ref 7).***

Input route	% contribution to input to North Sea						
	Cd	Cr	Cu	Hg	Ni	Pb	Zn
Atmospheric deposition	28	25	25	13	44	58	35
rivers	65	65	33	66	43	31	44
direct discharges to coastal waters and estuaries	5	7	5	13	8	7	18
sewage sludge dumpling	2	2	2	7	2	1	1
Industrial wastes dumping	<1	1	3	1	3	3	2
Total (%)	100	100	100	100	100	100	100
Total input (t/y)	430	2830	6020	45	3740	7840	31100
Estimated load redistributed as dredgings (t/y)	62	2520	1730	35	540	3750	15800

Note: the area of the North Sea was taken as 530 000 km^2.

In conjunction with the collection of a weekly integrated sample, an instantaneous grab sample was collected each week, and the flow at that time recorded. Three methods were then used to calculate average weekly loads over a 40 week period.

Method 1

$$\text{Load (1)} = \frac{1}{n}\sum_{i=1}^{n} C_i Q_i$$

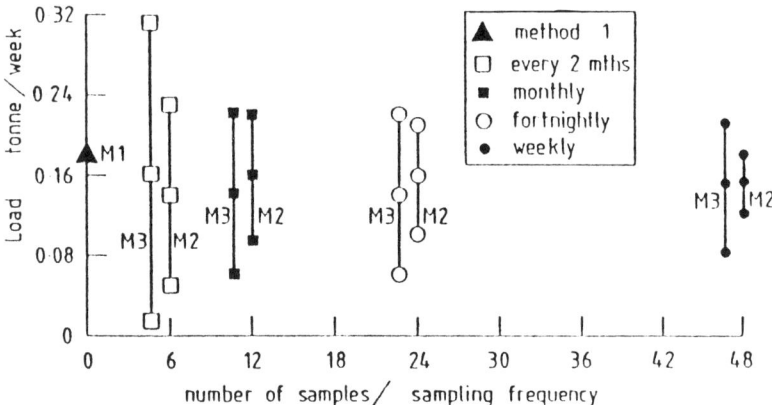

FIGURE 4. Calculated Loads of Lead with 95% Confidence Limits, Calculated by Methods 1, 2 and 3.

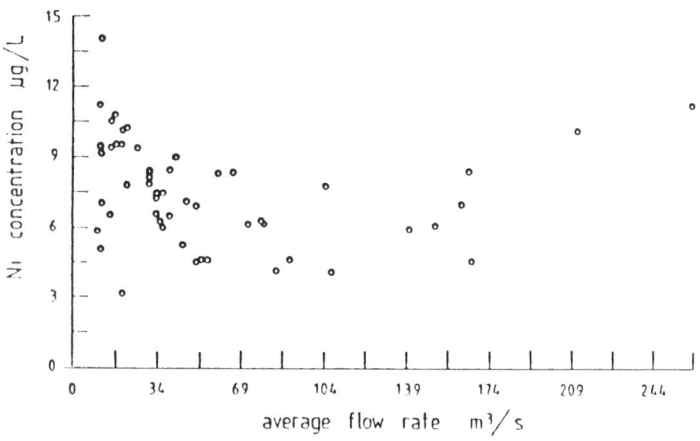

FIGURE 5. Variation in Nickel Concentration in Flow-Proportional Composite Weekly Sample as a Function of Average River Discharge.

where n is the number of weekly samples, C_i and Q_i are respectively the flow proportional composite sample concentration and the average discharge over the ith weekly interval.

Method 2

$$\text{Load (2)} = \left[\frac{1}{n} \sum_{i=1}^{n} c_i q_i \right] \frac{\bar{Q}}{\frac{1}{n} \sum_{i=1}^{n} q_i}$$

in which c_i and q_i are respectively the instantaneous concentration and discharge at time t_i and $\bar{Q}$ is the average discharge over the whole period, derived from the continuous ultrasonic gauge. This method thus contains an adjustment for the difference between $\bar{Q}$, the true average discharge, and the arithmetic average of the instantaneous discharges, represented by the denominator in the second term of the equation.

Method 3

$$\text{Load (3)} = \frac{1}{n} \sum_{i=1}^{n} c_i q_i$$

This is thus the simple arithmetic average of instantaneous loads.

Using these formulas, loads were estimated for a number of analytes, assuming different sampling frequencies: weekly, fortnightly, monthly and bimonthly. Figure 4 exemplifies the results obtained. Since Load (1) is based upon flow-proportional composite samples and average discharges, it is taken as the true load. As may be seen, the averages derived from grab samples are often close to the load (1) value, although the error bars, representing 95% confidence intervals become larger as the sampling frequency is decreased. In general, method 2 gives a better (narrower) confidence interval than method 3.

This exercise serves to demonstrate the uncertainties inherent in estimating riverine fluxes from the commonly available grab sample data. The sampler used in this work was installed just above the tidal reaches of the river. Measurement of fluxes within or below the estuary is fraught with problems due to massive variations in flow and the complications introduced by flow reversal. Thus, most available data are for the lower reaches of the river and take no account of sedimentation processes affecting particulate-associated metals within the estuary, or possible re-suspension of bedload during conditions of high flow.

14 INTRODUCTION

The operation of a flow-proportional continuous sampler can, as well as providing accurate flux information, provide interesting insights into metal behavior in the river. Figure 5 shows the dependence of weekly composite average concentration and average flow for nickel. The U-shaped behavior observed was typical of trace metals generally, but was in marked contrast to the behavior of dissolved phosphate and nitrate. At flow rates below 80-100 m^3 s^{-1} the prime effect is one of decreasing concentration with discharge, reflecting a dilution process. Above this flow, however, metal concentrations increase again, presumably because of resuspension of bottom sediments under more energetic conditions.

METAL BUDGET FOR A MAJOR HIGHWAY

Whilst it is relatively easy to monitor fluxes of metals from a point source discharge into a water course, estimation of fluxes from a more diffuse source such as a highway is far more difficult. We have done so for the M6 motorway in England at a time when it carried around 30,000 gasoline vehicles per day [18]. Various measurements were taken and estimates made, as follows:

(i) An automated sampler was used to collect sequential water samples and measure volumetric discharge through a number of storm events. Whilst the predominantly dissolved metals showed a "first flush" effect of high concentration early in the storm event, other metals showed a more complex temporal variation related to storm intensity and the flushing of large-grained sediment through the drainage system [19]. Using average concentrations and an integrated volumetric discharge, annual fluxes of metals in runoff water were estimated (see Table 2).

(ii) By placing total deposit gauges at various distances from the roadside [20], estimates of integrated deposition to roadside verges were obtained (Table 2).

(iii) Soil cores were taken from the roadside verges at different distances from the road and used to estimate an integrated deposition to roadside soil since the opening of the road.

(iv) Using statistics of vehicle density since the opening of the road in 1960, an integrated lead emission was estimated [18]. Results are shown in Table 2.

These results, which are broadly consistent with those of other published studies demonstrate that for lead, the only metal for which a source strength can be estimated, road runoff and local atmospheric deposition account for only around 10% of emissions.

The remainder is presumed to be transported via the atmosphere over much larger distances. This may not be true of cadmium and copper, which appear to be produced as rather larger particles, less capable of a prolonged atmospheric existence [20, 21]. Drainage water from the road entered local water courses causing appreciable pollution as mean concentrations were: lead, 353 μg l^{-1}; cadmium, 3.4 μg l^{-1} and copper 87μg l^{-1}. Knowledge of fluxes per meter of highway per vehicle provides useful information for larger-scale flux estimations.

TABLE 2.   Metal Fluxes Derived from Measurements on M6 Motorway (ref. 18) Annual Fluxes: (g m^{-1} y^{-1}):

	Pb	Cd	Cu
(a) Removal in drainage water	14 (5.1)*	0.18	3.6
(b) Deposition to verges	8.3 (2.9)*	0.07	0.29
(c) Vehicular source strength	281	U$^+$	U

Fluxes since opening of road: (gm^{-1})

(a) Accumulation in verges	422 (7.5)*	8.3	99
(b) Vehicular souce strength	5602	U	U

* % of source strength
U = unknown

METAL FLUXES   TO SOILS AND PLANTS

Critchley [7] has estimated metal fluxes from various sources to U.K. agricultural land (Table 3). The same caveat applies to these data as to the North Sea inventory; whilst atmospheric deposition may overall be the largest input (other than for mercury), locally, other sources such as sewage sludges may be considerably more important.

Measurements of atmospheric deposition over land are rather easier to make than over the sea, but there are nonetheless controversial aspects to the estimation of metal deposition. Surrogate surfaces (i.e. deposit gauges) cannot reliably simulate the behavior of natural surfaces, and estimates based upon measurement of an atmospheric concentration and multiplication by a deposition velocity are subject to much the same reservations are estimates made in the marine environment (excepting the presence of a "maritime effect").

16 INTRODUCTION

If it were considered desirable to limit inputs of metals into agricultural systems, then Table 3 might provide some basis for a control strategy. However, limiting actual input into growing plants is quite another matter. These may come either from the soil, or from the atmosphere by direct deposition. Apportionment of the contributions arising from the soil and atmospheric sources is experimentally difficult, but has been achieved in our research group by use of three methods, all of which give results in broad agreement:

(a) addition of isotopic tracers to soils. Both ^{210}Pb and ^{109}Cd have been added to soils as a tracer of uptake from this source. It appears that the tracer is of similar availability to the stable element in a weathered soil [21, 23].

TABLE 3. Summary of Trace Metal Inputs to UK Agricultural Land (after Critchley; ref. 7).

Input route	% contribution to input to UK agricultural land								
	Cd	Cr	Cu	Hg	Ni	Pb	Zn	As	Se
Atmospheric deposition	50	45	77	37	93	92	90	89	91
Sewage sludge	14	29	6	5	7	5	5	3	2
Inorganic phosphatic fertilizers	36	25	1	1	<1	3	1	8	7
Other sources*	0	0	16	57	0	0	4	0	0
Total (%)	100	100	100	100	100	100	100	100	100
Total input (t/y)	110	1020	4780	49	990	5000	15600	320	100

*Pig slurry (copper); seed dressings (mercury); pig slurry and poultry waste (zinc).

(b) Use of a dual plant growth cabinet in which plants are grown in two chambers identical except for the fact that one is ventilated with metal-free filtered air, and the other with polluted ambient air. By this mechanism, the soil-derived contribution may be identified from the set of plants grown in filtered air, whilst the other plants contain both soil and atmosphere-derived trace metals [24].

(c) Plants have been grown on an identical soil in locations widely differing air metal concentrations. Moss bag collectors are used at the same locations to provide a concurrent measure of atmosphere deposition metals. By plotting plant metal content versus relative (moss bag) atmospheric deposition and extrapolation to zero deposition, an intercept representing the soil-derived metal in the plant is obtained [25].

A summary of the data obtained appears in Figures 6 and 7. These shows good general agreement between results for a given plant type and part irrespective of the measurement technique used. The value of these transfer factors is the predictive capability which they provide. Concentration factors vary greatly between elements, with a general order Zn > Cd > Ni > Cr > Pb irrespective of the plant type, reflecting the relative availability of these metals for uptake. Values of AAF, however, are far more a function of plant part than of chemical elemental consistent with dry deposition of particles to plant surfaces of differing physical characteristics (e.g. surface roughness).

The majority of earlier studies of plant uptake of trace metals have entirely ignored direct atmospheric deposition as a pathway. When, however, this is separately quantified, it is found to be quite appreciable (Table 4). These results were obtained in rural parts of the United Kingdom where both soil and atmospheric metal levels are typical of background agricultural locations throughout Europe.

GLOBAL FLUXES OF METALS

Niragu and Pacyna [26] have provided a rather sophisticated global inventory of trace metals. By estimating emission factors for different processes, together with global production/consumption of the commodity associated with that process, they have been able to estimate inputs of metals to air, aquatic ecosystems and soils. In Table 5, details of the inventory are presented for the element Cd.

Whilst the production of such inventories is enormously important and provides the first step in a global flux model, much

TABLE 4. Range of Atmospheric Contribution (%) Estimated from Field-Grown Plants (from ref. 25).

Plant	Metal				
	Cd	Cr	Ni	Pb	Zn
Pea leaves	6-94	0-39	0-69	12-94	42-51
pods	10-99	12-72	5-45	72-89	54-56
fruits	0-80	0-29	0-69	0-30	11-57
Radish leaves	40-80	0-44	77-80	0-78	0-62
Storage roots	25-65	0-54	0	11-57	0-44

To assist in the interpretation of results, two transfer factors were defined as follows:

$$\text{Concentration Factor} = \frac{\text{Soil-derived metal in plant } (\mu g\ g^{-1}\ \text{dry wt})}{\text{metal in soil } (\mu g\ g^{-1}\ \text{dry wt})}$$

$$\text{Air Accumulation Factor} = \frac{\text{Metal in plant derived from atmospheric deposition } (\mu g\ g^{-1}\ \text{dry wt})}{\text{Metal in air concn } (\mu g\ m^{-3})}$$

development is required in two contexts. Firstly, global fluxes obscure a great deal of regional variation which can best be accommodated though regional-scale flux models. Secondly, and

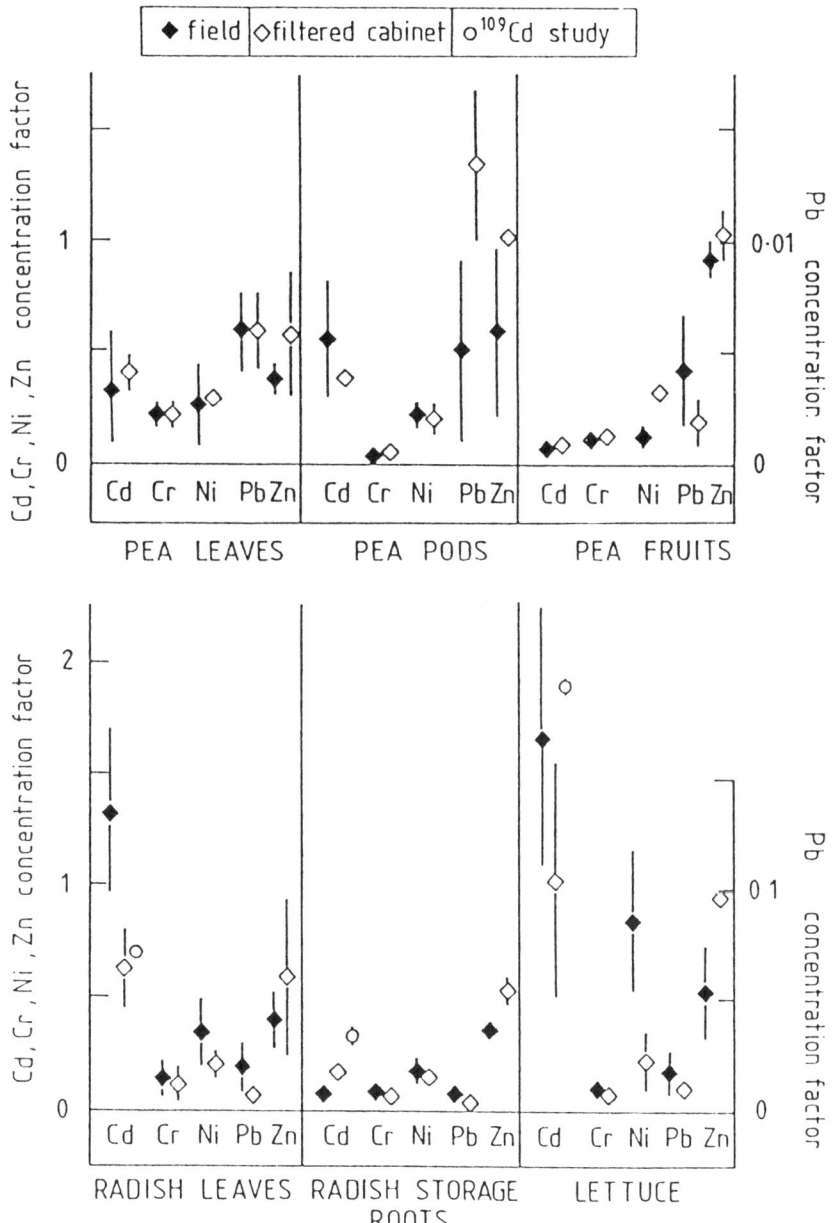

FIGURE 6. Concentration Factors (CF) for Trace Metals, as Defined in text.

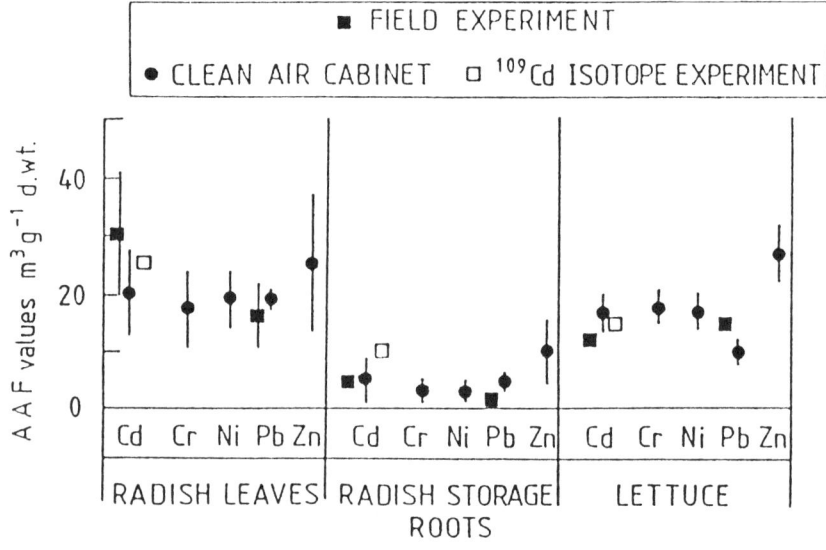
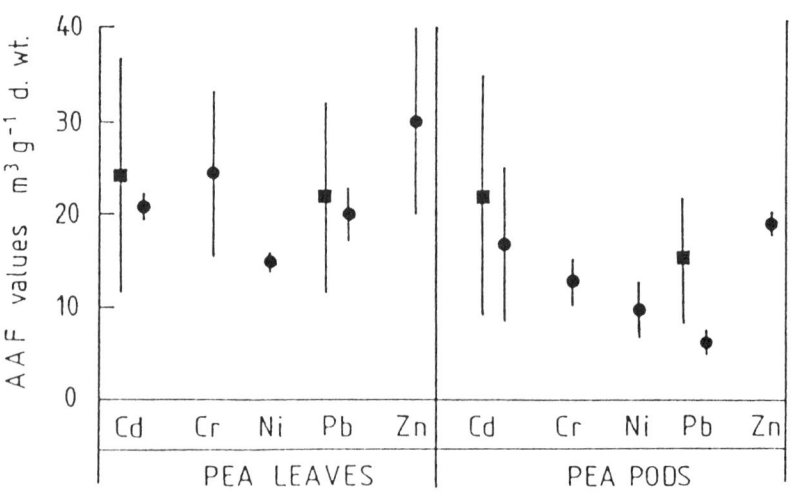

FIGURE 7. Air Accumulation Factors AAF for Trace Metals, as Defined in Text.

perhaps more importantly, the input inventory tells only a part of the story. We need to know the internal pathways of a system and the outputs into other environmental compartments. Thus, for example, for the North Sea, we need to know not only the riverine flux of metal into an estuary, we need the flux outwards into the North Sea. We then need to know transfer fluxes between metal and bottom sediment (in both directions) and with biota, and fluxes in waters advected in and out of the North Sea boundaries. Only then will we have knowledge approaching that of a true budget. Combined with knowledge of chemical speciation, such a budget would provide a truly powerful tool in decision making for pollution control.

TABLE 5. Global Emission Fluxes for Cd (after Nriagu and Pacyna; ref 26).

Atmosphere Source Category	Global production/Consumption (10^9 kg y^{-1})	Cd Emission (10^3 kg y^{-1})
Coal combustion		
-electric utilities	(15.5×10^9 MJ)	77-387
-industry and domestic	999	99-495
Oil combustion		
-electric utilities	(5.8×10^9 MJ)	23-174
-industry and domestic	358	18-72
Pyrometallurgical non-ferrous metal production		
-mining		0.6-3
-Pb production	3.9	39-195
-Cu-Ni production	8.5	1,700-3,400
-Zn-Cd production	4.6	920-4,600

Source	(col 2)	Cd
Secondary non-ferrous		2.3-3.6
metal production		28-284
Steel and iron mfg	710	
Refuse incineration		56-1,400
-municipal	140	3-36
-sewage sludge	3	68-274
Phosphate fertilizers	137	8.9-534
Cement production	890	60-180
Wood combustion	600	
Mobile sources	647(gasoline)	3,100-12,00
Total emissions		7,570
Median value		

Water Source Category	Annual global discharge ($10^9 m^3$)	Cd Discharge (10^6 kg y^{-1})
Domestic wastewater	90	0.18-1.8
-Central	60	0.3-1.2
-Non Central	6	0.01-0.24
Stream electric		
Base metal mining and dressing	0.5	0.03
Smelting and refining	7	
-Iron and steel	2	0.01-3.6
-Non-ferrous metals		
Manufacturing processes	25	0.5-1.8
-Metals	5	0.1-2.5
-Chemicals	3	-
-Pulp and paper	0.3	-
-Petroleum products		0.9-3.6
Atmospheric fallout	(6×10^9 kg)	
Dumping of sewage sludge		0.08-1.3
Total input, water		2.1-17
Median value		9.4

Soils

Source Category	Annual global discharge ($\times 10^{12}$ kg)	Cd Input (10^6 kg y^{-1})
Agric and food wastes	15	0-3.0
Animal wastes, manure	2	0.2-1.2
Logging & other wood wastes	11	0-2.2
Urban refuse	440	0.88-7.5
Municipal sewage sludge	20	0.02-0.34
Miscellaneous organic wastes including excreta	210	0-0.01
Solid wastes, metal mfg	380	0-0.08
Coal fly ash and bottom fly ash	3,720	1.5-13
Fertilizer	166	0.03-0.25
Peat (agricultural and fuel uses)	375	0-0.11
Wastage of commercial products		0.78-1.6
Atmospheric fallout		2.2-8.4
Total Input, soils		5.6-38
Median value		22
Mine tailings		2.7-4.1
Smelter slags and wastes		1.6-3.2
Total discharge on land		9.9-45

REFERENCES

1. Radojevic M. and R.M. Harrison. "Concentrations and Pathways of Organolead Compounds in the Environment: A Review," *Sci. Tot. Environ.*, 59:157-180 (1987).

2. Harrison R.M. and S. Rapsomanikis. "Environmental Analysis Using Chromatography Interfaced with Atomic spectroscopy," Ellis Horwood, Chichester (1989).

3. Hewitt C.N. and R. M. Harrison. "Atmospheric Concentrations and Chemistry of Alkyllead Compounds and Environmental Alkyllation of Lead," *Environ. Sci. Technol.*, 21:260-266 (1987).

4. Hewitt C.N. and M. Rashed. "Organic Lead Compounds in Vehicle Exhaust," *Appl. Organomet. Chem.*, 2:95-100 (1988).

5. Harrison R.M. and A.G. Allen. "Environmental Sources and Sinks of Alkyllead Compounds," *Appl. Organomet. Chem.*, 3:49-58 (1989).

6. Harrison, R.M., C.N. Hewitt, and M. Radojevic, "Environmental Pathways of Akyllead Compounds," *Proc. Int. Conf. Heavy Metals in the Environment*, Athens, Greece, 1:82-84 (1985).

7. Critchley R.F. "An Assessment of Trace Metal Inputs and Pathways to the Marine and Terrestrial Environments," *Proc. Int. Conf. Heavy Metals in the Environment*, Heidelberg, F.R.G., pp. 1108-1111 (1983).

8. Cambray R.S., D.F. Jeffries, and G. Topping. "An Estimate of the Input of Atmospheric Trace Metals into the North Sea and the Clyde Sea," *AERE Report R7733*, HMSO, London (1975).

9. Cambray R.S., D.F. Jeffries, and G. Topping. "The Atmospheric Input of Trace Elements to the North Sea," *Marine Science Communications*, 5:175-194 (1979).

10. Peirson D.H., P.A. Cawse and R.S. Cambray, "Chemical Uniformity of Airborne Particulate Matter, and a Maritime Effect," *Nature*, 251:675-679 (1974).

11. Petersen G., H. Weber and H. Grassl. "Modelling the Atmospheric Transport of Trace Metals from Europe to the North Sea and the Baltic Sea," *Control and Fate of Atmospheric Trace Metals*, J.M. Pacyna and B. Ottar (Eds.), Kluwer, Dordrecht, 57-83 (1989).

12. Otten P., C. Rojas, L. Wouters, and R. Van Grieken. "Atmospheric Deposition of Heavy Metals (Cd, Cu, Pb and Zn) into the North Sea," University of Antwerp Report, (1989).

13. Buat-Ménard P., personal communication.

14. Slinn, S.A. and W.G.N. Slinn. "Predictions for Particle Deposition on Natural Waters," *Atmos. Environ.*, 14:1013-1016 (1980).

15. Williams, R.M. "A Model for the Dry Deposition of Particles to Water Surfaces," *Atmos. Environ.*, 16:1933-1938 (1982).

16. Jaffrezo, J.L. and J. L. Colin. "Rain-Aerosol Coupling in Urban Area: Scavenging Ratio Measurements and Identification of Some TRansfer Processes," *Atmos. Environ.*, 22:463-465 (1988).

17. Walling, C.E. and B.W. Webb. "Estimating the Discharge of Contaminants to Coastal Wasters by Rivers: Some Cautionary Comments," *Marine Pollution Bulletin*, 16:488-492 (1985).

18. Harrison, R.M., W.R. Johnston, J.C. Ralph, and S.J. Wilson. "The Budget of Lead, Copper and Cadmium for a Major Highway," *Sci. Tot. Environ.*, 46:137-145 (1985).

19. Harrison, R.M. and S.J. Wilson. "The Chemical Composition of Highway Drainage Waters, I. Major Ions and Selected Trace Metals," *Sci. Tot. Environ.*, 43:63-77 (1985).

20. Harrison, R.M. and W.R. Johnston. "Deposition Fluxes of Lead, Cadmium, Copper and Polynuclear Aromatic Hydrocarbon (PAH) on the Verges of a Major Highway," *Sci. Tot. Environ.*, 46:121-135 (1985).

21. Harrison, R.M. and C.R. Williams. "Airborne Cadmium, Lead and Zinc at Urban and Rural Sites in North-West England," *Atmos. Environ.*, 16:2669-2681 (1982).

22. Harrison, R.M. and W.R. Johnston. "Experimental Investigations on the Relative Contribution of Atmosphere and Soils to the Lead Content of Crops," *Pollutant Transport and Fate in Ecosystems*, P.J. Coughtrey, M.H. Martin and M.H. Unsworth (eds), British Ecol. Soc. Spec. Publ. No. 6, Blackwell, Oxford, pp. 239-247 (1987).

23. Harrison, R.M. and M.B. Chirgawi. "The Assessment of Air and Soil as Contributors of Some Trace Metals to Vegetable Plants. II. Translocation of Atmospheric and Laboratory-Generated Cadmium Aerosols to and within Vegetable Plants, *Sci. Tot. Environ.*, in press.

24. Harrison, R.M. and M.B. Chirgawi. "The Assessment of Air and Soil as Filtered Air Growth Cabinet, *Sci. Tot. Environ.*, in press.

25. Harrison, R.M. and M.B. Chirgawi. "The Assessment of Air and Soil as Contributors of Some Trace Metals to Vegetable Plants. II. Experiments with Field-Grown Plants," *Sci. Tot. Environ.*, in press.

26. Niragu, J.O. and J.M. Pacyna. "Quantitative Assessment of Worldwide Contamination of Air, Water and Soils by Trace Metals," *Nature,* 333:134-139 (1988).

METALS CONTROL TECHNOLOGY; PAST, PRESENT AND FUTURE

James W. Patterson
Illinois Institute of Technology, Chicago, Illinois

INTRODUCTION

The challenges of environmental protection have escalated rapidly over the past two decades as we increasingly recognize the interlinkages of our environment, and the sometimes subtle but dangerous consequences of our release of industrial pollutants into the ecosphere. We can broadly categorize industrial pollutants into organic compounds and inorganics. At the most fundamental level, we conclude that the most protective method to avoid adverse environmental impacts for industrial organic pollutants is to destroy those organic compounds, by biological or thermal oxidation, converting them to carbon dioxide and associated oxidation products. When we choose other options for these industrial organics, such as landfill or atmospheric release, we merely shift a localized to a dispersed problem, and amplify the adverse effects and the consequent costs, for remediation.

For inorganics, destruction is not an option. Environmental management technology for inorganics has fundamentally involved either dispersing the pollutant into the environment, or shifting its physical form preparatory to storage. Dispersal of the inorganics has the expected consequences, comparable to the discharge of toxic industrial organics. However, for the inorganics, there is not even the prospect of "assimilative capacity." There is no such capacity on the land, in the water or in the atmosphere for a toxic inorganic such as mercury or lead. The other option, a shift in physical form, is best personified by the precipitation treatment of an industrial wastewater, thereby producing a sludge. Landfilling of that sludge will surely, and inevitably, lead to leaching of that landfilled material into the groundwater.

As a consequences, for inorganic industrial pollutants, there is increasing pressure on industry, and within the technological community, to develop new and safer options for management of these inorganics. My topic today is metals control technology. I wish to address it within the context of this technological challenge to develop new and safer management options, and I will place this in perspective by considering past and present practices.

PAST PRACTICES

Most metal-bearing wastes originate as wastewaters(Patterson, 1985). In considering past practices, I decided to review two texts which I studied in the 1960's when I was a graduate student in environmental engineering. The two texts were both concerned solely with industrial pollution control. They were certainly not the earliest texts published on that subject but, at the time of my graduate studies, were widely used as textbooks and references. The author of each is internationally renown as an academic leader in the field of industrial pollution control.

In musing on the history of metals control technology, I decided that it would be interesting to refer back to these two now dusty references, which I so assiduously studied over two decades ago. I reviewed the relevant sections of those two texts, and found that indeed they did address metals pollution control. A number of metals control technologies were described (Table 1). Further, it struck me that the list of technologies was not substantially different from the conventional treatment technologies which I had reviewed (Table 2) in my 1986 perspective address at the First International Symposia on Metals Speciation, Separation and Recovery, in Chicago, Illinois.

Table 1. Metals Control Technologies (Nemerow, 1963; Eckenfelder, 1966).

Precipitation
Oxidation/Precipitation
Reduction/Precipitation
Coagulation/Coprecipitation
Ion Exchange
Adsorption

Upon reflection, I realized that from the perspective of end-of-pipe treatment, there has been little incentive to evolve new and innovative metals control technologies, and thus little reason to have seen significant advances over the past two or more decades. The end-of-pipe metals control technologies perform quite well in the context of purging the industrial wastewater of metals prior to effluent discharge. Evolution in metals control technology has been instigated not by a failure of the existing technologies to achieve good effluent quality, but by our escalating concerns for the subsequent, sludge management, phases of metals pollution control. That is, the motivation to find alternative methods of metals management has not evolved primarily from a need to achieve a higher quality of effluent (although that can be achieved), but from a recognition that the treatment residue, the sludge, is simply not disposable in a safe environmental manner. This is recognized not only by environmental engineers and scientists, but by the industrial generators of the sludges. These industrial generators are increasingly paying the very expensive price of the clean-up of past sludge disposal sites, where leaching has released metals into the groundwater.

Table 2. Principal Technologies for Metals Separation (Patterson, 1987).

A. Conventional Treatment Technologies (Ranked)
   Precipitation (incl. primary coprecipitation)
   Oxidation/Precipitation
   Reduction/Precipitation
   Concentration/Precipitation
   Secondary Coprecipitation

B. Recognized Recovery Technologies (Unranked)
   Evaporative Recovery
   Ion Exchange
   Membrane Separation
   Reductive Electrolysis

C. Emerging Recovery Technologies (Unranked)
   Differential Precipitation
   Extractive Metallurgy
   Selective Adsorption

CURRENT STATUS

In the past few years, the concepts of waste avoidance and waste minimization have become quite popular. In fact, the U.S. EPA has recently established a new Office of Pollution Prevention. Waste minimization strategies include (Table 3) source reduction or avoidance, waste separation and concentration for volume reduction, waste exchange for alternative beneficial use, and pollutant recycle, recovery and reuse.

Table 3. Waste Minimization Strategies (USEPA, 1987)

Source Reduction
Waste Separation and Concentration
Waste Exchange
Recycle, Recovery and Reuse

Source avoidance or reduction can involve a number of options (Table 4). One particularly effective option is raw material substitution. For example, ICI Colours and Chemicals Division has developed a new product to replace chromium in leather tanning (Anon, 1988a). Material substitution options can effectively reduce the mass of pollutants generated. In contrast, waste segregation and concentration while beneficial, is only capable of isolating the mass of pollutants, into a smaller and more homogeneous volume. This may offer the opportunity to extract and recover pollutants from that more concentrated and homogeneous waste stream however, and is often one element in successful waste avoidance initiatives.

Table 4. Options for Source Reduction or Avoidance

Raw Material Substitution
Raw Material Prepurification
Operating Modifications
Equipment Modifications
"Clean" Manufacturing Technologies

Rockwell International has reported that an automated system to isolate individual waste streams from a manufacturing facility performing metal plating and producing printed circuit boards allows the utilization of five separate metals recovery systems (Anon, 1989a). These include two systems for copper, and one each for lead, nickel and zinc. Sludge generation from wastewater treatment has been reduced 70 by percent and is targeted for 90 percent reduction. Economic benefits include reduction in sludge disposal cost, and savings in raw materials purchases.

Waste exchange, the third element of waste minimization, appears to have limited relevance for metal pollutants. The final option, recycle, recovery and reuse, represents the greatest potential but has limited current application for non-precious metals. As demonstrated in Table 5, only five percent or less of the metal-bearing hazardous wastes currently generated in the United States is processed for metals recovery.

Table 5. U.S. Metal-Bearing Hazardous Waste Generated and Recovered.

Category	Million Metric Tons/Year
Generated (USEPA, 1987)	8 - 11
Recovered (Krieger, 1989)	0.4 - 0

Conceptually, one may subdivide existing and prospective metals control technologies into two groups; metals (incl. salts) concentration, and selective extraction (Table 6). The former category tends to be non-selective, yielding complex mixtures of sludges or solutions (e.g. ion exchange regenerant). The latter category represents technologies which are often only marginally more selective, or work only for very pure or specialized waste streams.

Among the concentration technologies, precipitation/ coprecipitation dominates, followed by ion exchange and membrane processes (reverse osmosis and electrodialysis). Clifford, et al. (1986) have summarized the advantages and disadvantages of these processes. From the narrow viewpoint of effluent quality only, both precipitation and ion exchange are quite effective, as are the membrane processes (albeit more limited in applicability to specific waste streams). From the viewpoint of metals recovery, and reuse, these concentration technologies, and the others listed in Table 6, are of limited value. In general, where the original waste stream is very pure, almost any concentration technique can lead to successful extraction and recovery. If the original waste stream is complex or contains an array of impurities, concentration technlogies will not provide an effective mechanism for selective recovery for reuse.

Precipitation. The most common technology for metals control is precipitation, including primary coprecipitation (Patterson, 1987). In its most frequent application, a highly disordered solids matrix results from rapidly induced supersaturation, typically achieved by pH adjustment. The technology is commonly termed "hydroxide" or

Table 6. Categories of Metals Controls Technologies

Concentration Technologies	Selective Extraction Technologies
Precipitation	Differential Precipitation
Coprecipitation	Reductive Electrolysis
Ion Exchange	Selective Adsorption
Membrane Technology	Differential Solids Extraction
Evaporation	
Liquid Extraction	

"alkaline" precipitation. Depending upon thermodynamic and kinetic factors however, as well as the matrix of ions present, a wide spectrum of both unstable and stable solid phases can initially form. The rapid induction of supersaturation, and resultant solid phase disorder, produces large volumes of gelatinous solids with extremely poor solids handling characteristics. The solids are typically not geochemically stable, and may exhibit property shifts (either desirable or undesirable) over periods of hours, days, or longer. More targeted precipitation techniques, employing for example sulfide and carbonate, are occasionally attempted. Such "selective" or "differential" precipitation will be considered separately under the category of "Selective Extraction Technologies."

Ion Exchange. Many proprietary ion exchange resins are commercially available. While the commercial product literature seeks to reinforce concepts of selectivity, the technical literature convincingly demonstrates the relative non-selectivity of ion exchange. Even the most successful selectivity application, for chromic acid recovery, is operationally complex and limited in the strength of chromic acid recoverable without oxidative damage to the resin matrix. As with precipitation, and from the perspective of effluent quality alone, ion exchange can be effective (Woodward, et al., 1988). Disadvantages, even for this application, include (Clifford, et al., 1986):
- Potential for chromatographic effluent peaking
- Spent regenerant disposal is required
- Effluent quality varies as a function of background ions
- Technology not feasible at high TDS

From the perspective of selective metal recovery, ion exchange appears applicable only where a relatively pure waste stream is to be processed. It may not even be the most appropriate concentration technology for this very narrow circumstance.

Membrane Processes. Although reverse osmosis and, less frequently, electrodialysis, are the membrane processes most often discussed for metals control, ultrafiltration has also been used to an extent. The General Electric Company has developed a process whereby metal-bearing wastewater is separated via an ultrafilter from a liquid ion exchange material, with metal ions transported via the 0.05 μm ultra-filter pores from the wastewater to the liquid ion exchange material (Anon, 1984a). In a second application, ultrafiltration is employed for micellar-enhanced recovery of chromate ion from wastewaters (Christian, et al., 1988). The process is an extension of a micellar-enhanced technique to remove soluble organics from wastewater. In the chromate recovery application, a UF membrane with an effective molecular weight cutoff of 5,000 Daltons was employed. The membrane demonstrated about 95 percent rejection of the micelle-bound chromate ion.

Electrodialysis is most commonly used for water purification from brackish water supplies. One full-scale application has been reported for metals control however, at the Washington Steel Mill in Pennsylvania (Anon, 1988b). The acid resistance membrane concentrates nickel and chromium for recycle to the steel mill, while purifying hydrofluoric and nitric acids for reuse.

The most common membrane process is reverse osmosis (Patterson, 1985; Walker, et al., 1988). The delicate nature of the RO membranes is well recognized. For neutral, prepurified wastewaters, reverse osmosis can be an effective effluent polishing device. The brine requires subsequent treatment. The only well-documented application for metals recovery is for recovery of nickel from certain very clean nickel plating wastewaters (Crampton and Wilmoth, 1982). Clifford, et al., (1986) have cited the high capital cost, membrane fouling, and high level of pretreatment required as disadvantages of the membrane processes. As in the instance of concentration technologies previously discussed, reverse osmosis is a viable pathway to recovery only when applied for relatively pure waste streams.

Evaporation. Concentration by evaporation is straight-forward, excepting that all non-volatile impurities are retained in the residual brine stream. The technology can be energy intensive, and enhanced energy efficiency is obtained only by application of more sophisticated, costly modifications. Innovations such as vapor compression evaporation have somewhat improved the economics of this technology (Anthony, 1989). The non-selective nature of the concentration process renders it generally unattractive as a metals recovery option except, as for other technologies, for relatively pure waste streams. Such instances offer only trivial technological challenges, in any event.

Liquid Extraction. This promising technology has had very limited success to date. It encompasses a variety of approaches, termed "solvent extraction," "liquid ion exchange," "liquid membranes," and simply "liquid-liquid extraction." It has been suggested that among a spectrum of commercial applications, recovery of metals from low grade ores, and from wastewaters, are most promising (Anon, 1984b).

Liquid ion exchange resin is composed of an organic liquid which contains a dissolved, water-insoluble active compound (Knocke, et al., 1980). Acid based liquid ion exchange resins are commercially available (Peterson, et al., 1981). In application, the resins are dissolved in a solvent such as kerosene. The resin is normally regenerated with acid, and the process is thus similar to solid phase ion exchange. It is claimed that the liquid products are more selective, however. Experimental studies using a synthesized plating waste, demonstrated co-extraction of cadmium and nickel (Peterson, et al., 1981). Thus, as with solid phase resin ion exchange, selectivity may not be high. Liquid ion exchange material has been used in conjunction with ultrafiltration, to promote solvent-water phase separation (Anon, 1984a).

Ion exchange functionality is not a mandatory component of liquid extraction. Organic liquid membrane encapsulation of an extractant is another approach. Generally, liquid extraction involves a sequence of at least three operations (Robbins, 1980):
1. Liquid-liquid counter-current contact
2. Solvent or carrier separation from the aqueous phase
3. Raffinate purification or regeneration.

A commercial process for liquid extraction of metals in wastewaters is available (McIlvance, 1989).

FUTURE DIRECTIONS

It appears to me that the most promising areas for future development are within the category of selective extraction technologies. However, such development and application will require a far more sophisticated understanding and utilization of chemical speciation and kinetic controls than is represented in our current state of practice. I will briefly comment on these technologies, and their more intriguing aspects.

Differential Precipitation. As discussed above, precipitation is generally effective for effluent purification, but the sludges generated are not amenable to effective management, nor metals recovery. Targeted differential precipitation can involve a number of strategies:

. Enforced formation of a desired salt (e.g. sufide or carbonate precipitation

. Sequential, step-wise precipitation of relatively pure solid phases

. Formation of solid phases having desirable characteristics relating for example to particle size, density, or surface charge.

Examples of the first type, forced formation of a desired salt, are numerous (Table 7). The strategy normally is applied in order to reduce target pollutant solubility in the wastewater, and thereby achieve better effluent quality. Where multiple metals are present, the resultant sludge takes on the characteristics of the least manageable components, and from the standpoints of sludge handling and metals recovery no advantages are achieved (Horn, 1983; Jenista; 1984). Further, the forced salt formation processes are particularly sensitive to reaction stoichiometry, and secondary ion scavenging of the precipitant can interfere with process efficiency.

Table 7. Examples of Selective Salt Precipitation

Pollutant	Salt
Arsenate	Ferric arsenate (Anon, 1988c)
Chromate	Barium chromate (Brooks, 1986)
Cadmium, Lead	Metal carbonate (Patterson, et al. 1977, Hsu et al., 1981)
Copper	Copper oxide, copper hydroxynitrate (Patterson et al., 1989)
Chromium, Copper, Lead, Zinc	Metal thioacetamide, metal dibromo-oxime (Brooks, 1986)
Many Metals	Metal sulfide (numerous literature citations)

The second option, of sequential stepwise precipitation, is more promising. The technique involves the use of kinetic or other solubility controls to sequentially form and separate relatively pure salts, or well-defined salt mixtures. The technique has been utilized to sequentially precipitate and recover copper and then silver from a silver plating rinsewater (Gould, 1984). Jenke and Diebold (1983)

36 INTRODUCTION

have modeled and then tested a selective sequential titration scheme to recover metals from acid mine drainage. Unfortunately, the modeling effort relied solely upon thermodynamics and assumed equilibrium. As has been reported by Karra, et al. (1984) and by Wajun, et al. (1985), kinetic features of selective precipitation frequently control real-time results. Cho and Kim (1987) have convincingly demonstrated on a mixed cadmium-copper waste that manipulation of kinetic behavior and solubility features allows highly efficient selective precipitation to occur.

The third option seeks to control the discrete physical and adsorbed characteristics of the solid phase formed. Freshly precipitated salts typically have an electrical surface charge which results in adsorption of counterions from the wastewater. The solid phase is thereby made less chemically "pure," and less attractive for recovery. Surface charge management can suppress (or enhance) this adsorptive coprecipitation. In a like manner, the strategy may seek to achieve larger, or perhaps more dense (crystalline) solid salts, even if achieved within a matrix of other gelatineous precipitates. Physical separation techniques such as sedimentation or centrifugation can then separate the intermingled solid phases.

Reductive Electrolysis. Copper recovery by electrolytic reduction is a well-established technology, for spent printed circuit board and similar baths. The technique is varied, ranging from application of imposed electrolytic potential to plate out the copper (electrowinning), to spontaneous electrolysis employing a sacrificial anode such as scrap iron (cementation). In one interesting variation, alkaline precipitated copper slurry is treated in an electrolytic cell for copper recovery and reuse (Anon, 1989b). A similar induced electrolytic technique has been described for the recovery of lead previously leached from contaminated soils (PEI Associates, 1986). The spontaneous cementation process, using sacrificial metallic aluminum, is reported to effectively recover copper, lead and zinc (Meyers, 1988). Nickel, which is much closer in the electromotive series to aluminum than the other three metals, was not reduced. Properly applied, electrowinning and cementation seem to be promising technologies for selective metals extraction for recovery.

Selective Adsorption. This technology, employing biological materials, mineral oxides, or polymeric resins, was the topic of much discussion at the 1986 First Symposium on Metals Speciation, Separation, and Recovery. One interesting recent trend in this technology is to attempt to coat coarse or porous media such as sand with an adsorbing surface. In one instance, ferrihydrite is coated onto sand (Edwards and Benjamin, 1988). In another, it is proposed to coat the sand grains with crown ethers (Anon, 1988d). This latter application is specifically proposed for lead removal. Progress in developing the selective adsorption technology has been generally disappointing, however. As in 1986, I caution that the key to successful application will require demonstration of the technology's ability to <u>selectively</u> adsorb specific metals from complex wastewater matrices, followed by elution or other recovery of a metal-rich concentrate. If the technology can only be applied effectively to relatively pure waste streams, then other concentration technologies will likely prevail.

Differential Solids Extraction. Metallurgical and other selective solids extraction techniques appear to offer fruitful opportunities to recover metals from wastewater treatment sludges (Brooks, 1984). Application will be enhanced by the generation of relatively "pure" sludges such as may result from differential precipitation. In 1986, I described three different categories of extractive metallurgy (Patterson, 1987):
- Pyrometallurgy, involving thermal techniques
- Hydrometallurgy, using leaching techniques
- Biohydrometallurgy, applying microbes to enhance leaching

To this list can be added a possible fourth, of electrolytic reduction from metal hydroxide slurries (Anon, 1989b). The potential of this innovation is not yet well-established. Differential thermal extraction of lead from mine-waste contaminated soils and dusts has been reported by Moon and Thornton (1988). The process sequentially stripped lead halides followed by lead sulfide. Hydro/leaching techniques continue to be developed, and are considered to be cleaner approaches than the pyrometallurgical technologies, due to the air pollutants associated with the latter processes. Table 8 lists example leachants and metals recovery applications. Biohydrometallurgy is still in an early stage of development. One potential limitation on its application to metal recovery from wastewater treatment sludge is that the biological route normally requires acidic conditions (Smith, et al. 1988), while treatment sludges are typically quite alkaline.

CONCLUSIONS

With respect to inorganic toxic pollutants, we must accept that there are no available destructive technologies, nor natural environmental assimilative capacity as are options for most industrial organic compounds. Thus, our management strategies must be formulated within a context of those options available to us for safe environmental management. Current metals control technologies have evolved from a need to achieve acceptable effluent quality. Existing treatment technologies achieve this objective efficiently, and economically. The motivations for change in metals control technology thus derive not from lack of satisfaction with achieving their original objective, but from the consequence of that achievement---sludge management.

Our array of opportunities involve three basic strategies, to eliminate or reduce the masses of metal sludges now disposed into the environment:

- Source avoidance or reduction
- Direct metals concentration and selective recovery for reuse, from the waste stream
- Selective metallurgical extraction of metals for recovery and reuse, from sludges.

In my opinion, our most fruitful opportunities for metal recovery, and future avoidance of environmental pollution, must be representative of one of these above three strategies.

Table 8. Hydrometallurgical Leachants

Leachants	Application
Ammonium Carbonate/ Ammonium Bisulfide	Lead extraction (Anon, 1984c)
Ammonium Carbonate	Copper extraction (Leak, 1988)
Ferric Sulfate	Copper extraction (Anon, 1989c)
Acids	Numerous (e.g. Tanning sludge, (Majone, 1986)

REFERENCES

1. Anonymous. "GE Continuous Process Extracts Metals from Wastewaters," *Chemical & Engineering News*, (June 25, 1984a). p. 33.

2. Anonymous. "Liquid Membrane Technology," *Breakthrough, The Newsletter of Innovation,* 11:3. (1984b).

3. Anonymous. "Lead-Recovery Route Eliminates Smelting," *Chemical Engineering*, (August 20, 1984c). p. 42.

4. Anonymous. "Chrome Cure," *Water Quality International*, No. 4, (1988a). p. 27.

5. Anonymous. "An Acid-Waste Recovery Route Seeks New Frontiers," *Chemical Engineering*, (December 19, 1988b). p. 23.

6. Anonymous. "A Route for Cutting Wastewater Outfall by 80-90% in Arsenic Operations," *Chemical Engineering*, (May 9, 1988c). p. 23.

7. Anonymous. "Toxic Trap," *Water Quality International*, No. 4, (1988d). p. 27.

8. Anonymous. "Automated Waste Recovery Cuts Materials and Disposal Costs," *Chemical Engineering*, (January, 1989a). p. 50.

9. Anonymous. "Electrochemical Copper-Removal from Dye-Making Wastes Boosts Savings," *Chemical Engineering*, (April, 1989b). p. 17.

10. Anonymous. "A Hydrometallurgical Copper-Recovery Process Could Replace Smelting," *Chemical Engineering*, (March, 1989c). p. 21.

11. Anthony, D. "Evaporate and Crystallize Waste Brines," *Chemical Engineering*, (April, 1989d). p. 138.

12. Brooks, C.S. "Metal Recovery from Waste Sludges," *Proceedings Thirty-Ninth Industrial Waste Conference*, Purdue University, (1984).

13. Brooks, C.S. "Metal Recovery from Industrial Wastes," *Journal of Metals*, (July, 1986). p. 50.

14. Cho, S-H. and Y-K. Kim. "Selective Precipitation of a Single Metal from the Mixed Metal Solution Part II. Precipitation Kinetics of Copper and Selective Precipitation," *J. Korean Society of Environmental Engineers*, 9:2:33, (1987).

15. Christian, S.D., S. NiBhat, E.E. Tucker, J.F. Scamehorn, and D.A. El-Sayed. "Miceller-Enhanced Ultrafiltration of Chromate Anion from Aqueous Streams," *AIChE Journal*, 34:2:189, (1988).

16. Clifford, D., S. Subramonian, and T. Song. "Recovering Dissolved Inorganic Contaminants from Wastes," *Environmental Science & Technology*, 20:11:1072, (1986).

17. Crampton, D. and P. Welmoth. "Reverse Osmosis in the Metal Finishing Industry," *Metal Finishing,* (March, 1982).

18. Eckenfelder, W. Wesley Jr. *Industrial Water Pollution Control,* (McGraw-Hill Book Company, New York, 1966).

19. Edwards, M. and M.M. Benjamin. "Adsorptive Filtration Using Coated Sand: A New Approach for Treatment of Metal-Bearing Wastes," presented at *WPCF Annual Conference,* Dallas, Texas, (1988).

20. Gould, J. "There's No Gold in This Wastewater," *Pollution Engineering,* (July, 1984). p. 10.

21. Horn, G. "The Effect of Carbonate on the Solubility of Cadmium and Copper," M.S. Thesis, Illinois Institute of Technology, Chicago, Illinois, (1983).

22. Hsu,D.Y., M.D.R. Riddell, and B. Bonamico. "Soda Ash Improves Lead Removal in Lime Precipitation Process," *Proceedings Thirty-Sixth Industrial Waste Conference,* Purdue University, (1981).

23. Jenista, J. "The Effect of Carbonate on the Solubility of Zinc and Copper," M.S. Thesis, Illinois Institute of Technology, Chicago, Illinois, (1984).

24. Jenke, D.R. and F.E. Diebold. "Recovery of Valuable Metals from Acid Mine Drainage by Selective Titration," *Water Research,* 17:11:1585, (1983).

25. Karra, S.B., C.N. Haas, V. Tare, and H.E. Allen. "Kinetic Limitations in the Selective Precipitation Treatment of Electronic Wastes," *Air, Water and Soil Pollution,* (1984).

26. Knocke, W.R., T. Clevender, M.M. Ghosh and J.T. Novak. "Recovery of Metals from Electroplating Wastes," *Proceedings of the Thirty-Fifth Industrial Waste Conference,* Purdue University, (1980).

27. Krieger, J. "Hazardous Waste Management Database Starts to Take Shape," *Chemical & Engineering News,* (February 18, 1989).

28. Leak, V.G. "Metal Removal/Recovery Using Non-Electrolytic Metal Recovery," report to *U.S. EPA Hazardous Waste Engineering Research Laboratory,* (June, 1988).

29. Majone, M. "Aluminum and Iron Separation from Chromium Solutions by Precipitation With Cupferron," *Envir. Technology Letters,* 7:531, (1986).

30. McIlvane, R. "Pollution Control Developments in Europe," *Pollution Engineering,* (February, 1989). p. 84.

31. Meyers, S.C. "Recovery of Metals Using Aluminum Displacement," report to *U.S. EPA Hazardous Waste Engineering Research Laboratory,* (June, 1988).

32. Moon, C.H. and I. Thornton. "Differential Thermal Extraction of Lead Species in Mine-Waste Contaminated Soils and Dusts," *Environ. Technology Letters,* 9:1367, (1988).

33. Nemerow, Nelson L. *Theories and Practices of Industrial Waste Treatment,* (Addison-Wesley Publishing Company, Inc., Reading, Massachusetts, 1963).

34. Patterson, J.W. *Industrial Wastewater Treatment Technology,* 2nd Edition, (Butterworth Publishers, Stoneham, Massachusetts, 1985).

35. Patterson, J.W. "Metals Separation and Recovery," in *Metals Speciation, Separation and Recovery,* (Lewis Publishers, Chelsea, Michigan, 1987).

36. Patterson, J.W., R.E. Boice, C. Petropoulou, D. Marani, and G. Macchi. "Hydrolysis, Precipitation and Aging of Copper(II) in the Presence of Nitrate," presented at *2nd International Symposium on Metals Speciation, Separation and Recovery,* (Rome, Italy, May, 1989).

37. Patterson, J.W., H.E. Allen, and J.J. Scala, "Carbonate Precipitation of Heavy Metals Pollutants," *Journal, WPCF,* 49:2397.

38. PEI Associates. "Electromembrane Process for Recovery of Lead from Contaminated Soils," report to the *National Science Foundation, Small Business Innovation Research Program,* (July, 1986).

39. Petersen, J.J., N.C. Burbank, Jr., and G.C. Amy. "Liquid Ion Exchange Pretreatment for Removal of Heavy Metals from Plating Rinsewater," *Proceedings of the Thirty-Sixth Industrial Waste Conference,* Purdue University, (1981).

40. Robbins, L.A. "Liquid-Liquid Extraction: A Pretreatment Process for Wastewater," *Chemical Engineering Progress*, (October, 1980). p. 58.

41. Smith, J.R., R.G. Luthy, and A.C. Middleton. "Microbial Ferrous Iron Oxidation in Acidic Solution," *Journal, WPCF*, 60:4:518, (1988).

42. U.S. EPA. "The Hazardous Waste System," report of the *EPA Office of Solid Waste and Emergency Response*, Washington, D.C., (1987).

43. Wajon, J.E., G-E. Ho, and P.J. Murphy. "Rate of Precipitation of Ferrous Iron and Formation of Mixed Iron-Calcium Carbonates by Naturally Occurring Carbonate Materials," *Water Research*, 19:7:831, (1985).

44. Walker, J.F., C.H. Brown Jr., and J.H. Wilson. "Treatment of Chromium Contaminated Plating Shop Rinse Streams by Reverse Osmosis," presented at *WPCF Annual Conference*, Dallas, Texas, (1988).

45. Woodward, T.L., C.M. Klevens, and R.C. Ganley. "Ion Exchange Treatment of Industrial Metal Rinsewaters to Meet Categorical Pretreatment Standards," presented at *WPCF Annual Conference*, Dallas, Texas, (1988).

PART I: CHEMISTRY OF TOXIC METALS

METAL SPECIATION IN SLUDGES FROM WASTEWATERS TREATMENT BY BULK AND SURFACE (XPS) ANALYSIS

E. Desimoni
Institute of Chemistry, University of Basilicata, Potenza, Italy

A. Marcone, G. Tiravanti
Water Research Institute, I.R.S.A. - C.N.R., Bari, Italy

P.G. Zambonin
Department of Chemistry, University of Bari, Italy

1. ABSTRACT

An investigation was performed by combining bulk and surface analysis techniques in order to elucidate micromechanisms of toxic metals uptake from wastewaters by an advanced method based on the use of semisynthetic polymers (e.g. starch xanthate) as precipitation of silver can be interpreted on the basis of a simple model involving the precipitation of a mixture of metal sulphides and xanthates whose composition is controlled by the pH of the watewater specimen. On the contrary, the analysis of the data relevant to the precipitation of Hg(II) and Cd(II) suggests that side reactions can play an important role in the precipitation process. The key for a reasonable interpretation of the behaviour of divalent metals and, in particular, of peculiar features of their precipitation pathways seems offered by XPS surface speciation followed by the critical comparison of bulk and surface data.

2. INTRODUCTION

Many industries, such as those processing non-ferrous metals, pigments, storage batteries, metal finishing and plating discharge heavy metals in their waste streams. Precipitation, ion exchange, adsorption and reverse osmosis techniques have been successfully applied [1] to scavenge these metal ions, some of which are extremely toxic even at trace levels.

Interest has recently arose in advance wastewater treatments making use of agricultrural by-products such as modified starch and cellulose. Promising results have been obtained by using the water-soluble starch xanthate (S44X) or cellulose xanthate (CX) as precipitating agents [2,3]. The MEXICO (Metals Extraction by Xanthate

Insolubilization and Chemical Oxidation) process [4] proved to be efficient in removing Cd, Cu, Ag, Pb, Ni, Cr(III) and Hg(II) from acqueous solutions within a wide pH range (e.g. 3-11). The produced sludges are post-treated with an oxidizing NaClO solution, so that the recovery of metals in concentrated solutions can easily be achieved [4].

Even if the MEXICO process has been developed and tested on pilot scale for mercury removal and recovery, nevertheless very little systematic research has been carried out in understanding the mechanism of heavy metal precipitation. Generally the available information refers to the liquid phase, without taking into account the precipitate itself. For example, it is known [2] that different amounts of reagent are needed at different pH values to precipitate the same amounts of metal by the SX precipitation process.

An investigation was recently initiated [5] to characterize solid sludges of the SX process by X-ray Photoelectron Spectroscopy (XPS). The aim is to evaluate the reproducibility of the polymer structure, to elucidate the actual metals oxidation state, their chemical environment and the distribution of metal ions and active polymer sites within the particles of the polymeric sludge. Likely the more this information is relevant to the top surface layers of the reacting polymer particles, the more they are related to the most active sites of the particles themselves [6]. For general information on XPS capabilities in the field of polymers see for example references 7-9. The reliability of XPS technique in collecting real 'surface information' from the system under study was previously [5] ascertained by observing the morphology of SX/Cd sludges by Scanning Electron Microscopy (SEM). The size of the polymer grains proved enormously larger than the actual thickness sampled by XPS. The surface sensitivity of XPS, when analyzing sludges characterized by poorly defined surfaces, was confirmed a posteriori via the comparison of the results of bulk and surface analysis (see reference 5 and following paragraphs).

The XPS measurements are paralleled by potentiometric titrations aimed to obtain the experimental values of the xanthate/metal molar reaction ratio; uni- and di-valent metal ions are tested to evaluate possible effects of the ionic charge. The bulk xanthate/metal molar ratio in the polymeric sludge is measured by standard methods [5].

Cadmium was selected for the first approach since it represents a good compromise between environmental concern and spectral complexity. The XPS sprectra relevant ot the O1s, C1s, and S2p regions of the polymer before the precipitation (SX-Na form) were noticeably larger than those observed after the metal uptake (SX-Cd form). For example, the S2p region likely contained all the possible contributions from sulphide to sulphate. This was explained in terms of polymer stabilization, likely via cross-linking of adjacent polysaccharide chains and/or by dissolution, during the precipitation step, of possible contaminants of the polymer solution of SX which contribute to the XPS signal of the SX-Na form.

The surface Cd/S elemental ratio was found higher than the bulk one as well as that of the maximum theoretical value for SX-Cd. Moreover, when minimizing the thickness of the explored surface layer by variable angle XPS techniques, the Cd/S elemental ratio was found to increase over the unit (vide infra, Figure 10).

These results were interpreted by considering the precipitation of CdS and cadmium xanthate in the bulk and by cadmium oxygenated species, likely cadmium hydroxide, over the outermost surface monolayers (at the lowest possible sampling depth).

Generally speaking the study appears promising and more complex systems are being considered. In particular, this paper describes investigations relevant to the recovery of mercury.

The results on the SX-Hg system are here critically compared with those relevant to the SX-Ag and SX-Cd [5] and all the matter is discussed by paying attention to the effects of side-reactions and pH on the metal/polymer reaction ratio.

Literature is quite poor on XPS information [10-20]) of mercury compounds so that, in order to collect a proper set of spectroscopic parameters necessary for analyzing the complex spectra relevant to the mercury containing sludges, standard specimens of mercury sulphides were preliminarily analyzed.

3. X-RAY PHOTOELECTRON SPECTROSCOPY

Electrons of inner shells can be ejected by the atoms of a solid surface irradiated by X-ray photons of sufficient energy, hv. The kinetic energy (KE) of the photoelectrons is given by the equation

$$KE = hv - BE - (\emptyset + S) \qquad (1)$$

in which BE is the binding energy of the photoemitting level, $\emptyset$ is the work function of the spectrometer and S the eventual surface charge of non-conducting samples. Equation 1 shows that XPS is capable of providing elemental analysis since no two atoms in the periodic table exhibit the same set of electron energies. Moreover, since a given atom is characterized by different energy levels as a function of its oxidation state and of its chemical environment, XPS can by exploited for collecting speciation information. This means that XPS can in principle distinguish between different oxidation states as well as between the different chemical forms of the same oxidation state. Differences in BE of a given level of a given atom are named 'chemical shifts' and range between 0.1 and 10 eV.

XPS allows performance of at least semi-quantitative analysis since the photoelectron current depends on the density and distribution of the photoemitting atoms.

The photoemission process leaves a hole in a core level. Regardless of how this hole is produced, de-excitation must occur by one of two competing processes, X-ray fluorescence or Auger emission. The first consists of filling the hole by an outer electron whose excess energy is emitted as an X-ray photon. In the second case, this excess energy is imparted to another electron which is emitted too. The prevalence of X-ray fluorescence over Auger emission decay depends on the atomic number and on the BE of the initial hole. In practice Auger decay predominates for core levels of BE lower than 2 KeV.

XPS peaks are named according to photoemitting levels, e.g. O1s, Fe2p3/2, W4f7/2, W4f5/2 etc. Auger peaks are named according to the levels involved in the emission; thus KLL means the initial hole was in level K and an L electron filled the vacancy, the energy difference being given up to eject a second L electron. The difference in nomenclature is convenient to avoid confusion in labelling peaks.

XPS is a surface analysis technique since photoelectrons must leave the sample without inelastic collisions in order to be detectable as peaks. If they lose unquantified fractions of their KE they are counted as background, together with all the electrons which, originated at various depths, lose part of their energy during transport out to the solid surface. In practice they diffsue to and escape from the specimen surface if their parent atom lies within the sampling depth (SD), which ranges between 10-100 angstrom. SD can be varied, within certain limits, by rotating the specimen in order to vary the angle at which photoelectrons leave the surface (variable take-off angle XPS or VAXPS), or by changing the X-ray source, since the SD depends on the KE of the photoelectron.

It is possible to perform depth profile analyses by alternating cycles of argon ion sputtering and analysis. In this way, each analysis step collects information from progressively deeper layers of the specimen. However it must be remembered that the ion bombardment frequently induces alterations of surface composition and morphology, and that depth quantifications require quite complex calibrations.

XPS measurements are carried out under vacuum since photoelectrons must be able to enter the energy analyzer without any scattering. To this aim a pressure of 10 exp(-5) mbar is in theory sufficient. However, under this pressure a thin gaseous film is still strongly adsorbed onto the whole inner surface of the spectrometer and then onto the surface of the specimen. In order to collect information relevant to the 'clean' specimen surface, this surface contamination must be minimized by raising the temperature of the whole machine (before sample insertion) to about 200°C for a period of several hours. The so-called 'bakeout' allows to work, if necessary, at pressures as low as 10 exp(-11) mbar.

The soft X-ray sources capable of emitting Kα lines of reasonably narrow width are Mg (1253.6 eV, FW=0.7 eV), Al (1456.6 eV, FW=0.85 eV) and Si (1739.5 eV, FW=1.0 eV). Usually Mg and Al anodes are both available on commercial instruments. Satellite lines (high energy lines) and the so-called Bremsstrahlung (or 'braking radiation') continuum are also present in the source emission spectrum. A partial elimination of satellites and Bramsstrahlung emission can be obtained by monochromators or by placing a very thin aluminum window behind the X-ray gun exit aperture.

In XPS, spectra of the whole KE range (wide or survey scan) contain information about the qualitative elemental composition of the specimen since each peak or doublet can be assigned to its parent atom on the basis of tabulated BE or KE values.

A detailed analysis of regions surrounding each photoelectron peak (detailed scans) allows quantitative analysis and chemical shift evaluation. For deeper information see for example reference 11.

4. SAMPLES TREATMENT AND EXPERIMENTAL TECHNIQUES

Most of the information on starch xanthate preparation, purification and characterization, SX-metal precipitation, potentiometric titrations and XPS measurements was described previously [1-5]. The likely structure of SX is shown in Figure 1.

HgS (red, rhombohedral, cinnabar, 99.5%) was obtained from Aldrich Chemical Company (USA).

All the standard specimens were bombarded with argon ions to test their stability under sputtering.

To minimize the unavoidable contaminations encountered in a real industrial process, synthetic wastewaters were prepared in the laboratory by dissolving a proper amount of mercury nitrate into 0.1 M NaNO3. Usual feed concentrations in the course of pilot plant tests ranged between 7-100 ppm [4] but, in this work, a metal concentration of the order of 1000 ppm was used to obtain a reasonable amount of sludge for spectroscopic analysis.

Mercury, cadmium and silver contents in the SX/M specimens were determined by standard flame atomic absorption methods after digestion in concentrated nitric acid [5, 21].

For XPS measurements the specimens were pressed to form thin pellets. Alternatively they were stuck onto a conductive-adhesive coated copper-tape (3M U.K.). Starch xanthate specimens (sodium, cadmium and mercury forms) were loaded on the sample rod under nitrogen atmosphere to avoid possible [22] hydrolysis phenomena. Each specimen was analysed at least in duplicate.

The spectrometer was a Leybold-Heraeus LHS10. Spectra were excited on using unmonochromatized Mg Kα radiation (1253.6 eV) usually operated at 14 KV and 20 mA. When analyzing SX specimens the source was operated at 14 KV and 10 mA to minimize possible specimen degradation by heating, secondary electron bombardment and/or X-ray radiation [23]; for the same reason sometimes specimens were cooled by circulating liquid nitrogen through the sample rod.

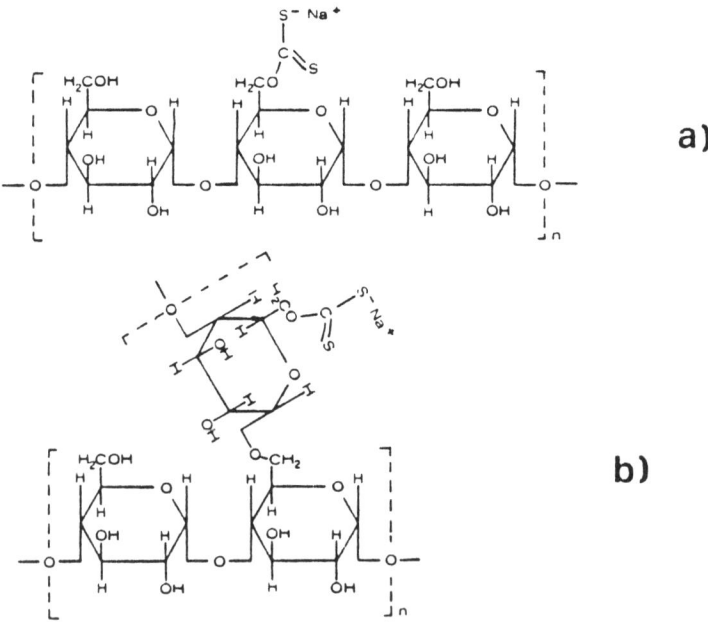

FIGURE 1. A Likely Structure of the Starch Xanthate Polymer: a) Amylose (15-30); b) Amylopectin (70-85%).

The pressure in the analysis chamber was always lower than 5.10 exp(-8) mbar. Binding Energies (BE) are referred to Au 4f 7/2 level (84.0 eV). FAT mode was always used for detail spectra with a pass energy of 50 eV.

Data acquisition and analysis was performed by using a home-made data station [24]. The processing software allows to perform smoothing, 1st and 2nd derivatives, satellites and linear/structured background subtraction, symmetric/asymmetric peak stripping, etc. Spectral synthesis can be performed by adding all the peak/doublets to the background+satellites contribution and by evaluating the goodness of fit by the Chi-square test.

The increase of the background at the highest KE observed in some Hg4f region spectra is due to the satellites contribution. The error window at the bottom of the fits displays the difference between experimental (dots) and synthetized (continuous curve) channel points expanded by the specified magnification factor.

Sensitivity factors derived by Wagner et al. [25] were used for quantification of spectra recorded by LHS10 spectrometer since results previously obtained [26] showed the transmission analyser function varying with the inverse of the photoelectron kinetic energy (as required by the Wagner data).

Kinetic energy (KE) in all the figures was not corrected for surface charging.

5. RESULTS AND DISCUSSION

5.1 XPS Analysis of Standards

The analysis of the results relevant to mercury sulphides evidences a certain degree of surface contamination and/or degradation of the specimens and a probable influence of UHV and/or X-ray irradiation on the stability of both polymorphous forms (see reference 27 for a review on radiation damage during surface analysis). The situation is made even more complex by possible transformations of one form into the other: according to the basic chemistry the red HgS can subliminate into the black under vacuum and the black changes in red by grinding or heating.

Figure 2 shows the Hg4f region of different red HgS specimens. Curves A, B and D are relevant to 'as received' standards while curve C was obtained by bombarding with Ar ions (2' at 3.0 KV) the specimen of curve A. As can be seen, the Hg4f doublet sometimes appears skewed towards lower KEs, thus suggesting the presence of 'oxidized' surface species (compare curves A and B). Noticeably curve B is very similar to the one obtained

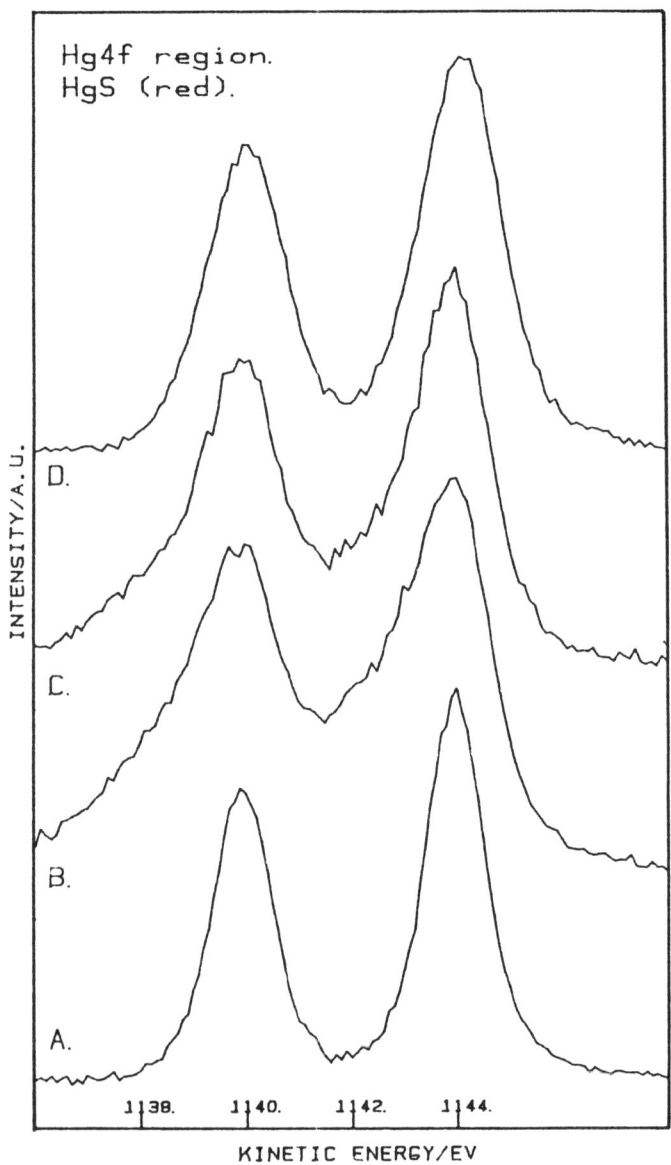

FIGURE 2. Hg4f Regions of Red HgS. A, B and D: Hg4f Regions of 'As Received' Specimens. C: Spectrum of the Same Specimen of Curve A But After 2' Sputtering With Ar+ Ions At 3.0 KV.

under sputtering (curve C) thus suggesting that ion bombardment can induce chemical degradation. The doublet can also appear rather broadened (compare curves A and D). It can be here anticipated that, fortunately, such unusual effects are not so evident when analysing SX-M systems.

A detailed analysis of the region A of Figure 2 is reported in Figure 3. Two doublets are evidenced at both sides of the main Hg4f doublet. The one appearing at higher kinetic energies can be likely ascribed to Hg(I) on the basis of the data of Table I, which reports the only information available on Hg(I) (as calomel) [15]. The chemical shift between Hg(II) and Hg(I) calculated from the Table, i.e. 1.4 eV, is in fact in good agreement with one observed between the two rightmost doublets in Figure 2 (+1.3+/-0.1 eV). Hg(I) can be likely produced by degradation under X-ray irradiation, as previously observed for other species, such as for example, dichromate, which is reduced to Cr(III) [28].

In the absence of mercury species more oxidized than Hg(II), the other doublet sometimes appearing at lower KE values (-1.1 +/- 0.1 eV vs. Hg(II)) can be tentatively ascribed to some surface contamination or to hydrated species or, alternatively, to differential charging. Noticeably the Hg/S elemental ratio obtained after subtraction of the extra contribution of the two side doublets is 1.0 +/-0.1, as required by the stoichiometry.

Figure 4 reports the S2p region of the same red HgS specimen.

Similar results were obtained by analyzing black HgS (see Figures 5 and 6).

The presence of more or less oxidized surface impurities is difficult to avoid and can perhaps explain the paucity of XPS spectral data relevant to mercury compounds.

The results of the analysis of standard mercury sulphides are summarized in Table II, where they are compared with previous literature information.

5.2 Bulk Chemistry of Starch Xanthate/Metal Ions Systems

As previously mentioned, the precipitation efficiency of the SX process in terms of residual metal concentration in solution is strongly influenced by the Metal/SX molar ratio, and by the pH of the solution. In order to understand the behaviour of such a complex system, a simple precipitation model can be postulated.

SX reacts with uni-divalent metals according to the following scheme [22].

$$M^{z+} + z\ R\text{-}O\text{-}CS_2^- = (R\text{-}O\text{-}CS_2)_z M \qquad (2)$$

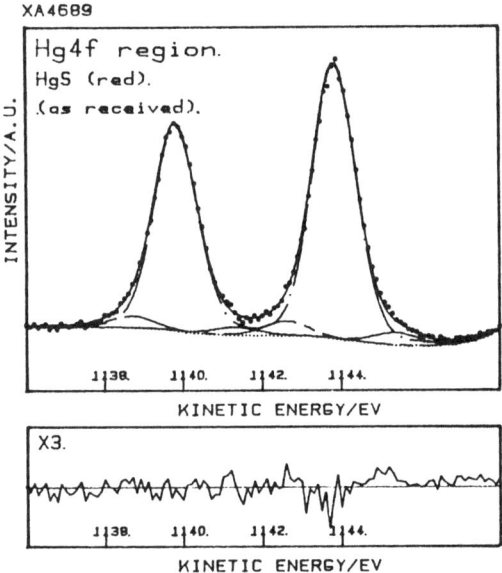

FIGURE 3. *Hg4f Doublet of Red HgS.*

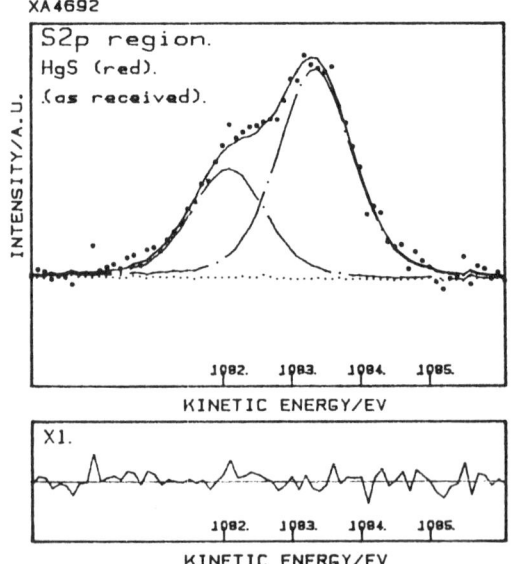

FIGURE 4. *S2p Doublet of the Specimen in Figure 2.*

TABLE I. *Binding energies of Hg/Cl Compounds (from ref. 15).*

	Hg(0)	Hg(I)	Hg(II)
7H4gf7/2Hg	99.9+/-0.1	100.1+/-0.2	101.5+/-0.2
4f5/2	104.01+/-0.1	104.4+/-0.2	105.7+/-0.1

Values were referred to Au4f7/2 = 83.4 eV in the original work and were corrected in this table to be referred to Au4f7/2 = 84.0 eV.

TABLE II. *Spectral Parameters of Hg and HgS.*

METALLIC MERCURY			
References	14	16	17
Hg4f7/2	98.5+/-.5	99.9	98.5
FWHM	102.2+/-.5	0.74	102.2
Hg4f5/2		104.0	
FWHM		0.73	
Ref level	Fermi	Fermi	Fermi
MERCURY SULPHIDE			
References	14	tw	tw
Species	HgS	HgS(b)	HgS(r)
Hg4f7/2	100.55	100.2	100.7
FWHM		1.4	1.4
Hg4f5/2	104.61	104.3	104.8
FWHM		1.4	1.4
R		1.35	1.35
GL		0.65	0.70
S2p3/2		161.6	161.4
FWHM	161.44	1.3	1.3
S2p1/2		162.9	162.7
FWHM	162.63	1.3	1.3
R		2.0	1.9
GL		0.7	0.7
Ref level	Fermi	Au4f7/2 84.0 eV	Au4f7/2 84.0 eV

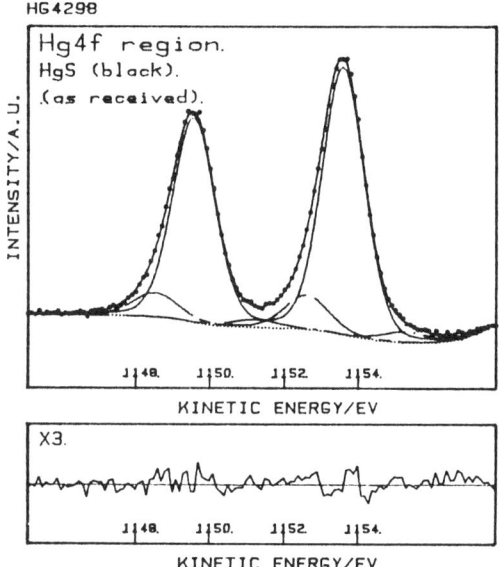

FIGURE 5. *Hg4f Doublet of Black HgS.*

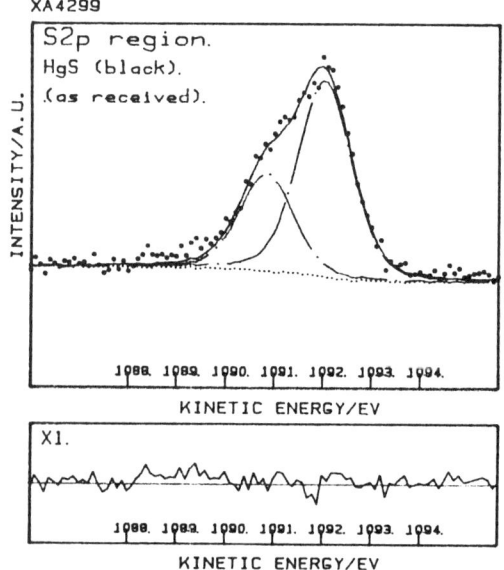

FIGURE 6. *S2p Doublet of the Specimen of Figure 4.*

In alkaline media SX decomposes [22] to give mainly sulphide, carbonates and starch. A general reaction can then be written [5]

$$4/z\ M^{z+} + R\text{-}O\text{-}CS_2^- + nOH^-$$

$$= 2\ M_{2/z}S + R\text{-}OH + H_{5-n}CO_3^{3-n} + (n-3)\ H_2O \qquad (3)$$

in which n ($3 \leq n \leq 5$) varies with the pH value. Depending on pH and reaction kinetics, metal sulphides can precipitate together with metal xanthates in different ratios. It follows that, according to the given model, the maximum theoretical molar ratios between metal and xanthate, M/SX, are 4 and 2 for uni-valent and di-valent metal ions, respectively.

If the precipitation process involves only reactions 2 and 3, it can be shown [5] that it is possible to calculate the amounts of metal precipitated as xanthate or sulphide, under different experimental conditions as a function of the total alkalinity added to the system to maintain constant the pH value during the titration of the metal-containing wastewater with starch xanthate. If SX is the number of moles of xanthic groups, R(OH) the moles of added hydroxide ions and R(M) the moles of metal ions, respectively, it can be shown [5] that R(M) (which coincides with the M/SX molar ratio since referred to SX = 1) can be evaluated by the simple relationship.

$$R(M) = 1/z\ (1 + 3\ R(OH)/n) \qquad (4)$$

The experimental R(M) values obtained in the course of the potentiometric titrations for silver, cadmium and mercury are reported in Table III, together with the theoretical ones calculated on the basis of Equation 4. EtX in the table indicates ethyl xanthate specimens used as standards in a previous work [5].

At pH values higher than 7, when bicarbonate ions prevails, a value of n = 4 was used for calculations relevant to silver. At pH values ranging between 5 and 6, a slightly better fit between experimental and calculated R(M) values can be obtained with n = 3.5, likely because reaction 3 produces H2CO3 (H2O, CO2) and HCO3- in the approximate 1:1 ratio.

Differently from those of silver, the R(M) values calculated for cadmium (1.21 at pH 6.0 and 1.01 at pH 8.5) are noticeably higher than the experimental ones (0.42 and 0.49 respectively: see Table III). This effect was explained [5] on considering that reaction 3 could be far from "theoretical completeness" because of kinetic or steric effects. In that case, the contribution of reaction

58 CHEMISTRY OF TOXIC METALS

TABLE III. Experimetal and Calculated Molar Ratios for Silver, Cadmium and Mercury Extraction at Different, Constant pH Values in Solution.

	n	Metal	pH	SX	R(OH)	R(M)exp	R(M)calc
EtX	3.5	Ag	5.5	1	0.15	0.13	1.13
EtX	4.0	Ag	9.0	1	3.27	3.63	3.45
EtX	4.0	Ag	9.0	1	3.82	3.92	3.86
SX	3.5	Ag	6.0	1	1.85	2.53	2.59
SX	4.0	Ag	9.0	1	4.20	3.77	4.15
SX	3.5	Cd	6.0	1	1.65	0.42	1.21
SX	4.0	Cd	8.5	1	1.36	0.49	1.01
SX	3.5	Hg	5.0	1	0.96	1.00	0.91

TABLE IV. Elemental Composition of Specimens Containing Xanthic Groups.

	S	C	O	Na	Cd	Hg	Notes
EtX/Na	1	1.5	0.5	0.5			th, bulk
	1	4.4	0.6	1.7			exp, surf
SX/Na	1	1.6	1.0	0.7			exp, surf
	1	14.0	8.7	8.0			exp, surf
	1	32.0	17.0	8.4			exp, surf
	1	81.5	46.4	21.4			exp, surf
	1	9.5	8.0	0.5			th, bulk
SX/Cd	1	37.5	26.4		0.65		exp, surf
		+/-0.5	+/-1.5		+/-0.06		
	1				0.19		exp, bulk
					+/-0.03		
	1	9.5	8.0		0.25		th,bulk
SX/Hg	1					1.35	exp, bulk
						+/-0.05	
	1					0.63	exp, surf
						+/-0.06	
	1	9.5	8.0			0.25	th, bulk

Data of SX/Cd are relevant to 5 different specimens. Data of SX/Hg are relevant to 4 different specimens.

2 could be higher than predictable by the simple model based on reactions 2 + 3.

In effect, the experimental values for cadmium can be consistent either with a nearly complete precipitation as cadmium xanthate or with the precipitation of mixed sulphides and xanthates. In any case, some sulphur-containing groups do not react with cadmium ions and this seems confirmed by the results of bulk analysis. In the case of cadmium, in fact, the bulk value of the metal/sulphur elemental ratio M/S, is 0.19 (see Table IV) which is lower than the maximum theoretical one (0.25 in the absence of sulphides) and of the one expected on the basis of the data of Table III (0.42/2 and 0.49/2 to transform M/SX into M/S values).

For what concerns the SX-Hg system, on the contrary, the experimental M/S value (i.e. R(M) = 1.0 in Table III) is slightly higher than the calculated one (i.e. 0.91).

Metallic mercury was also present in the bulk polymeric sludge, as evidenced by analyzing "cold vapor" electrothermal atomic spectroscopy, a sample of nitrogen gas flowed through the sludge. A likely reaction pathway, not considered by the simple model which leads to Equation 4, can be written as follows:

$$2\ R\text{-}O\text{-}CS_2^- + Hg^{2+} = R\text{-}O\text{-}CS\text{-}S\text{-}S\text{-}SC\text{-}O\text{-}R + Hg \quad (5)$$

Reaction 5, however, cannot explain the observed difference between experimental and calculated M/S ratios, since two xanthic groups are necessary for any Hg(0) atom as in reaction 2. Even on considering that the metal can be at least partially lost during bulk analysis, because of its volatility and/or solubility in water [29], the experimental R(M) value should be lower, not higher than the theoretical one. More information was necessary to rationalize these findings. A better understanding of the overall behaviour of the SX-M systems could be achieved on the basis of the analytical speciation of the sludges by XPS.

5.3 XPS Analysis of SX-Na, SX-Cd and SX-Hg Specimens

SX specimens were analyzed by XPS techniques before and after the precipitation of metal ions from the synthetic wastewater.

As observed when studying the SX-Cd system, even in the case of SX-Hg system, the spectra relevant to the C1s, O1s and S2p levels of the mercury form of the sludge (recorded after the uptake of the metal) appear noticeably narrower than those of the sodium form (recorded before the precipitation). Even in the case of mercury, the elemental composition of the mercury form exhibits a much better internal reproducibility than the sodium form (see Table IV).

Examples of the three regions are reported in Figure 7. As previously observed [5], the broadening of Cls and S2p regions of SX-Na specimens (see relevant curves a) indicates the presence of almost all the possible oxidation states of carbon and sulphur [30,31].

Again, the result can be interpreted in terms of polymer "stabilization" by divalent metal ions (cross-linking of adjacent polymeric chains), and/or by dissolution of possible impurities during the precipitation step. In addition, these findings suggest that specimen degradation by hydrolysis [22] in the course of the precipitation process, should be negligible since any degradation should lead to signal enlargements after the precipitation.

A major difference between the SX-Cd and SX-Hg systems is the possible presence of reduced mercury species, together with mercury (II). See Figure 8, which reports an example of the Hg4f region relevant to a SX-Hg specimen maintained at ambient temperature. The little doublet appearing at the highest KE values can be attributed to Hg(0) upon considering that the metal was previously detected into the bulk polymeric sludge (this suggests that Hg(0) could not be ascribed to simple degradation under x-ray irradiation or to surface effects), and that the observed chemical shift (1.7 +/-0.1 eV) seems rather higher than that observed in this work between Hg(II) and Hg(I) (1.4 eV). However, a certain small production of Hg(I) cannot be excluded, for example, according to the following reaction:

$$2 \ R\text{-}O\text{-}CS_2^- + Hg^{2+} = R\text{-}O\text{-}CS\text{-}S\text{-}S\text{-}CS\text{-}O\text{-}R + Hg_2^+ \quad (6)$$

To verify that Hg(0) was not produced by heating under x-ray irradiation, specimens of the SX-Hg sludge maintained at -150° C, were analyzed too (see Figure 9). The result was practically the same as that of Figure 7, thus suggesting that the presence of Hg(0) in the polymer particles does not increase with temperature.

The comparison of the XPS results relevant to the SX-Cd and SX-Hg systems shows other differences in their behaviour.

While in the case of cadmium, the experimental surface and bulk Cd/S ratios were 0.65 and 0.19 respectively, the surface and bulk ratios found for mercury are 1.35 and 0.63 (see Table IV).

The R experimental values of SX-Cd and SX-Hg are reported in Figure 10, as a function of the take-off angle. In the case of cadmium, a value of R = 1.4 was observed by reducing the SD of 50%, that is to say analyzing the first 1-2 surface monolayers. The curve relevant to mercury shows that the R value (always larger than 1), does not vary appreciably with the take-off angle. The

METALS SPECIATION IN SLUDGES   61

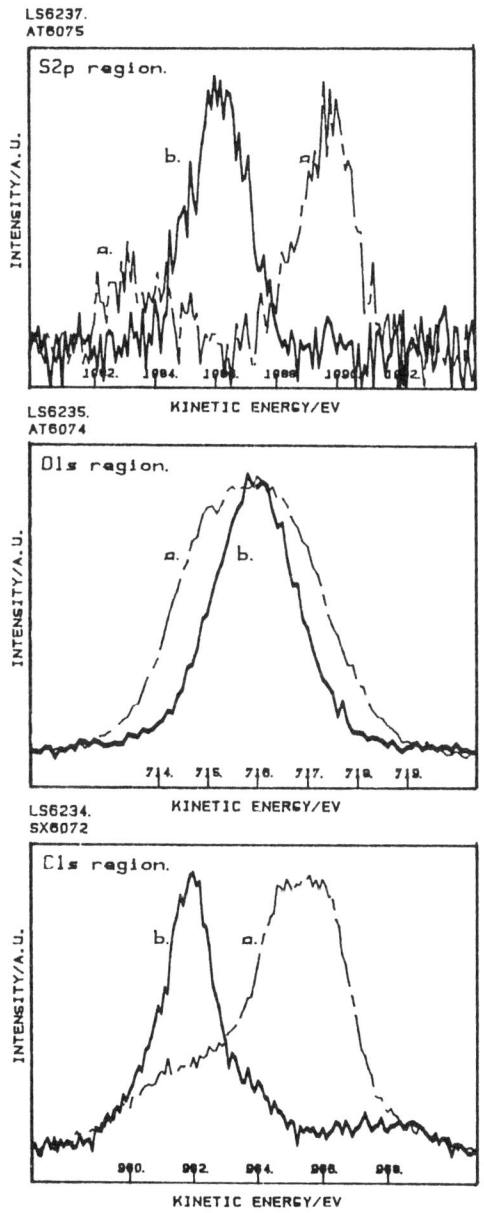

FIGURE 7.   *S2p, P1s and C1s Regions Relevant to SX-Na (A) and SX-Hg (B) Specimens. The KE axis is arbitrarily shifted to facilitate the comparison.*

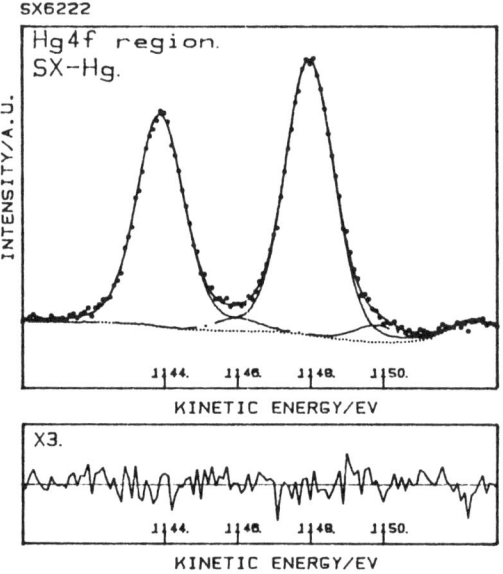

FIGURE 8. *Hg4f Region of a SX-Hg Specimen Maintained at Ambient Temperature.*

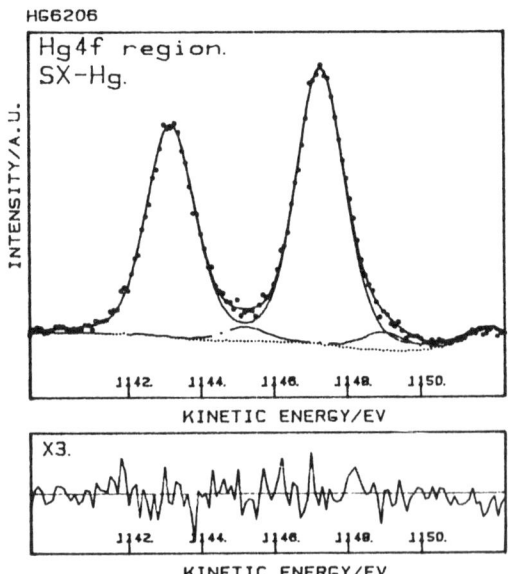

FIGURE 9. *Hg4f Region of a SX-Hg Specimen Maintained at −150°C.*

curve is dotted since the Hg4f signal contains contributions of mercury species different from HgS and mercury xanthate, since at take-off angle lower than 60 degrees (e.g. within the first 1-2 monolayers) the Hg4f signal can still be recorded while the S2p signal cannot be evidenced anymore. HgO and the above mentioned traces of metallic mercury are the most likely surface species. However, a quantitative estimate of the HgO content cannot be obtained from XPS data, since the O1s region contains a very large and masking contribution of starch alcoholic groups. The presence of HgO is also in agreement with the basic chemistry of mercury, since the polymer added to the wastewater has a higher hydroxide content (from the synthesis).

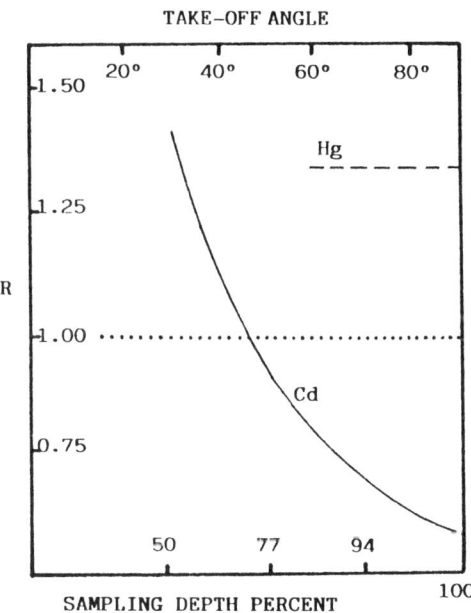

FIGURE 10. *Plot of the Cd/S Elemental Atomic Ratio as a Function of the Take-Off Angle. SD is expressed as percent of the value at 90 take-off.*

In conclusion, for what concerns the top surface layers of the polymer, CdS and cadmium xanthate are both present, together with a certain amount of cadmium hydroxide, while in the case of mercury, mainly HgO and traces of Hg(0) are present.

The Hg/S bulk value (0.63) is slightly higher than the one expected on the basis of the results relevant to potentiometric titrations (1.0/2 from Table III to change Hs/SX into Hg/S). This can be explained on considering that even the mercury analyzed in the bulk of the sludge contains species such as HgO and metallic mercury, other than HgS and mercury xanthate.

As with cadmium, the bulk concentration of mercury is lower than the one present onto the surface of the semisynthetic polymer particles. If surface contamination by HgO and by the metal could be neglected, the bulk sulphide percent could be calculated [5] from the XPS Cd/S elemental ratio R (0.25 R 1) by the equation:

$$\% S^{--} = \frac{4R - 1}{3} \cdot 100 \qquad (7)$$

The average content should be about 55% which, even within the experimental error, indicates a high content of mercury sulphide in the whole mass of the precipitate.

These results can be tentatively rationalized upon considering that the xanthate precipitation of mercury occurs rapidly [32], as verified experimentally, while that of cadmium takes quite a long time. Most likely when the alkaline SX is added to the solution of the metal (maintained at constant pH as indicated in Table IV), cadmium first precipitates as hydroxide onto the surface of the polymer particles. It must be taken into account that, because of the alkalinity of the added xanthate solution, local pH can be easily higher than 8, so that cadmium hydroxide can precipitate onto the surface of the polymer particles. Successively cadmium hydroxide can be partially dissolved, while cadmium ions diffuse towards the inside and hydroxide ions diffuse towards the outside of the polymer particles in such a way that simultaneous in-depth precipitation and dissolution occur. At the end of the process, CdS and cadmium xanthate are mainly present at different concentration levels onto the whole polymer particles. The precipitation of cadmium hydroxide is then responsible for a slower in-depth diffusion of cadmium ions and for the relatively low experimental values of the bulk Cd/S ratio. As mentioned, the presence of a surface layer of residual cadmium hydroxide can explain the experimental Cd/S ratio (i.e. 1.4) higher than the maximum "theoretical" one (R = 1) at the lowest SD.

In the case of the rapid mercury precipitation, on the contrary, reaction 3 is so forced to the right that additional NaOH had to be added to the wastewater to maintain a constant pH during the precipitation, even if the SX solution itself is conditioned at very basic pH values. This suggests that HgO found in the sludge likely co-precipitate with sulphide, and that surface HgO do not retard reaction 3.

6. CONCLUSIONS

Surface analysis techniques seem to represent a valid aid in interpreting the complex precipitation processes from wastewaters. X-ray Photoelectron Spectroscopy in particular, allows to investigate the different speciation chemistry of the metals, even when working on complex substrates such as polymeric sludges. The results obtained in this study, integrated with information on the bulk chemistry, allow to elucidate the precipitation micromechanisms by focusing the attention at the outermost surface layers, e.g. at the most active sites of the polymeric sludge. By this approach, it was possible to evidence the presence and the non-homogeneous distribution of species hypothesized on the basis of bulk elemental ratios and to give a reasonable interpretation of the different precipitation kinetics of the studied toxics metals, in terms of the different role played by surface oxygenated species unexpectedly on the basis of bulk analysis.

ACKNOWLEDGMENTS

Work carried out with the financial assistance of Ministero della Pubblica Istruzione (M.P.I. Rome, Italy) and Consiglio Nazionale delle Ricerche (C.N.R. Rome, Italy). Thanks are due to Mr. G. Sonnante for skilled help in the preparation of xanthates and precipitation processes, and to Mr. A. Tambone for his valuable assistance in some XPS measurements.

REFERENCES

1. Tiravanti, G. and D. Marani. "Technologie di Depurazione Impiegate per la Rimozione Dei Singoli Metalli," *Quaderno IRSA*, n. 71:229-311, (1986). (In Italian)

2. Macchi, G., D. Marani, M. Majone, and M.R. Coretti. "Optimization of Mercury Removal from Chlorakali in Industrial Wastewaters by Starch Xanthate," *Environ. Technol. Letters*, 6:369-380 (1985).

3. Tiravanti, G., G. Macchi, D. Marani, M. Pagano, M. Santori, and R. Passino. "Pilot Scale Treatment of Industrial Effluents Containing Heavy Metals by the MEXICO Precipitation-Oxidation Process," in *Proceedings of the International Conference on Heavy Metals in the Environment*, (CEP Publisher, Athens, 1985) pp.10-14

4. Tiravanti, G., A.C. DiPinto, G. Macchi, D. Marani, M. Santori, and Y. Wang. "Heavy Metals Removal: Pilot Scale Research on the Advanced Mexico Precipitation Process," in *Proceedings of the International Symposium on Metal Speciation, Separation and Recovery*, Chicago, (Lewis Publishers, Inc., Chelsea, MI, 1987) pp. 665-682.

5. Desimoni, E., A.M. Salvi, G. Tiravanti, P.G. Zambonin. "Analytical Characterization of Semisynthetic Polymers Used in Removing Heavy Metals from Wastewaters by X-Ray Photoelectron Spectroscopy," in *Euroanalysis VI-Reviews on Analytical Chemistry*, E. Roth, Ed. (Les Ulis Cedex, France: Les Editions de Physique, 1988) pp. 253-269

6. Farmer, M.E., R.W. Linton. "Correlative Surface Analysis Studies of Environmental Particles," *Environ. Sci. & Technol.*, 18:319-326 (1984).

7. Holm, R. and S. Storp. "Surface and Interface Analysis in Polymer Technology: A Review," *Surf. Interface Anal.*, 2:96-106 [1980].

8. Clark, D.T. "The Modification, Degradation and Synthesis of Polymer Surfaces Studied by Means of ESCA," in *Physicochemical Aspects of Polymer Surfaces*, Vol. 1, K.L. Mittal, Ed., (Plenum Press, New York, 1983). pp. 3-32.

9. Briggs, D. "Application of XPS in Polymer Technology," in *Practical Surface Analysis by Auger and X-Ray Photoelectron Spectroscopy*, D. Briggs, M.P. Seah, Eds., (J. Wiley and Sons Ltd., New York, 1983). p.359.

10. Turner, N.H. "Surface Analysis: X-Ray Photoelectron Spectroscopy and Auger Electron Spectroscopy," *Anal. Chem.*, 60:377R-387R (1988).

11. *Practical Surface Analysis by Auger and X-Ray Photoelectron Spectroscopy*, D. Briggs and M.P. Seah, Eds. (J. Wiley and Sons Ltd., New York, 1983).

12. Sakashita, M., H.H. Strehblow, and M. Bettini. "Anodic Oxide Films and Electrochemical Reactions on HgTe," *J. Electrochem. Soc.*, 129:739-746 (1982).

13. Sun, T. S., S.P. Buchner, and N.E. Beyer. "Oxide and Interface Properties of Anodic Films on Hg(1-x)Cd(x)Te," *J. Vacuum Sci. and Technol.*, 17:1067-1073 (1980).

14. Vesely, C. J. and D.W. Langer. "Electronic Core Levels of the IIb-VIa Compounds," *Phys. Rev. B*, 4:451-462 (1971).

15. Seals, R. D., R. Alexander, L.T. Taylor and J.G. Dillard. "Core Electron Binding Energy Study of Group IIb-VIIa Compounds," *Inorg. Chem.*, 12:2485-2487 (1973).

16. Svensson, S., N. Martensson, E. Basilier, P.A. Malmquist, U. Gelius, and K. Siegbahn. "Core and Valence Orbitals in Solid and Gaseous Mercury by Means of ESCA," *J. Electron Spectrosc. Related Phenom.*, 9:51-65 (1976).

17. Bearden, J. A. and A.F. Burr. "Re-Evaluation of X-Ray Atomic Energy Levels," *Rev. Mod. Phys.*, 19:125-142 (1967).

18. Brown, J. R., G.M. Bancroft, and W.S. Fyfe. "Mercury Removal From Water by Iron Sulphide Minerals. An Electron Spectroscopy for Chemical Analysis (ESCA) Study," *Environ. Sci. Technol.*, 13:1142-1144 (1979).

19. Adams, F. and J. DeWaele. "Surface Analysis in Atmospheric Environmental Studies," *Surf. Interface Anal.*, 12:551-564 (1988).

20. Cecile, J.L. "Application of XPS in the Study of Sulphide Mineral Floatation-A Review," *Dev. Miner. Process*, 6: 61-80 (1985).

21. "Manual of Methods in Aquatic Environment Research, Part III," FAO Fisheries Technical Paper N. 137. 1-124 (1975).

22. Ramachandra, S. *Xanthates and Related Compounds*, (M. Dekker, Inc., New York, 1971).

23. Johansson, L. S., J. Juhanoja and K. Laajaletho, and E. Suoninem. "XPS Studies of Xanthate Adsorption on Metal and Sulphides," *Surf. Interface Anal.*, 9:501-505 (1986).

24. Desimoni, E. and C. Malitesta. "Interfacing and LHS 10 Spectrometer to a Microcomputer: Data Acquisition and Analysis," *Computer Enhanced Spectroscopy*, 3:107-112 (1986).

25. Wagner, C. D., L.E. Davis, M.V. Zeller, J.A. Taylor, R. M. Raymond, and L.H. Gale. "Empirical Atomic Sensitivity Factors for Quantitative Analysis by Electron Spectroscopy for Chemical Analysis," *Surf. Inter. Anal.*, 2:211-225 (1981).

26. Sabbatini, L., C. Malitesta, E. Desimoni, and P.G. Zambonin. "Determination of the Transmission Efficiency in FAT For a Hemispherical Electron Analyzer of the Series LHS 10: Implications in Quantitative Analysis of Surface Species," *Annal. Chim.* (Rome), 74:341-338 (1984).

27. Storp, S. "Radiation Damage During Surface Analysis," *Spectrochim. Acta.*, 40B:745-756 (1985).

28. Desimoni, E., C. Malitesta, P.G. Zambonin and J.C. Riviere. "An X-ray Photoelectron Spectroscopic Study of Some Chromium-Oxygen Systems," *Surf. Interface Anal.*, 13:173-179 (1988).

29. Choi, S.S. and D.G. Tuck. "A Neutron Activation Study of the Solubility of Mercury in Water," *J. Chem. Soc.*, 84:4080 (1962).

30. Lindberg, B. J., K. Hamrin, G. Johansson, U. Gelius, A. Fahlman, C. Nordling, and K. Siegbahn. "Molecular Spectroscopy by Means of ESCA. III. Sulphur Compounds," Uppsala University, Uppsala, UUIP-638, (March 1970).

31. Gelius, U., P.F. Heden, J. Hedman, B.J. Lindberg, R. Manne, R. Nordberg, C. Nording, and K. Siegbahn. "Molecular Spectroscopy by Means of ESCA. III. Carbon Compounds," Uppsala University, Uppsala, UUIP-714, (July 1970).

32. Marani, D., M. Mezzana, R. Passino, and G. Tiravanti. "Treatment of Industrial Effluents for Heavy Metal Removal Using the Water Soluble SX Process," in *Proceedings of the International Conference on Heavy Metals in the Environment*," (Amsterdam, CEP Publisher, 1981). pp. 92-95

DISCUSSION OF:
METALS SPECIATION IN SLUDGES FROM WASTEWATERS TREATMENT BY BULK AND SURFACE XPS ANALYSIS

P. Cescon
Department of Environmental Sciences, University of Venice, Venice, Italy

Knowledge of the chemical state of trace metals in natural waters is important for understanding their reactivity, transport, and toxicity. The potential toxicity of various metals is controlled to a large extent by their physico-chemical forms [1, 2]. Study of heavy metals speciation in natural systems is therefore a considerable interest [3, 4]. Extension of this knowledge to industrial waste and sludge is very important for environmental impact assessment and to set up methodologies for the recovery of toxic elements.

The work under consideration studies the metals abatement by insolubilization with xanthate (MEXICO process) developed in a pilot plant and represents a significant base research following the applicative process. Speciation of Cd and Hg in the sludge obtained by precipitation as xanthate is considered. To explain the abatement mechanism, models for the metal-xanthate reaction are introduced and validated by determination of the Metal/xanthate ratio (M/SX).

X-Ray Photoelectron Spectroscopy is a widely used surface analytical technique when information on elemental composition has to be paralleled by information on the chemical status of the atoms.

Introduction of XPS to study SX-Cd and SX-Hg specimens gives a new methodology for chemical characterization of complex environmental matrices. In this work XPS has been applied for the first time to sludge obtained from wastewater precipitation processes. As suggested by the authors, this approach can give access to information otherwise unavailable, such as the chemical status of toxic metals in the sludge and the in-depth structure of the sludge particles. In addition, it seems to offer a way of identifying surface species, which, unexpected and undetectable on the basis of bulk analysis only, may play an important role in the precipitation micromechanism. Moreover the proposed methodology gives interesting information on the spatial distribution and precipitation mechanism of species.

The researcher must however be acquainted with the limits sometimes accompanying these measurements, limits which are inherent in some previous features of the technique.

A possible problem to be faced when examining complex multi-metal specimens from real processes is the mutual interference of the relevant spectroscopic signals. Analysis in this case is likely to suffer from multielemental spectral structures.

Even is the real value as "surface analytical technique" seems to be confirmed by the differences observed between bulk and surface molar ratios, a drawback of this technique could be the relatively high limit of detection (0.1 - 1.0 % atomic concentration). Elements present at low concentration levels could be 'lost' by the analyst.

From the results we have seen that sometimes, especially when the chemical shifts involved are rather low, curve fitting procedures do not lead to univocal results. Consider the difficult discrimination between Hg (1) and Hg metal.

For example the proposed model for starch xantate/mercury ions systems can be efficaciously integrated with the following reaction:

$$2\ ROCS_2^- + 2\ Hg^{2+} = (ROCS_2)_2 + Hg_2^{2+}$$

which presents the theoretical molar ratios values:

$$Hg/SX = 1$$

$$Hg/S = 0.5$$

close to experimental data:

$$Hg/SX = 1.0$$

$$Hg/S = 0.63$$

Experimental results do not show the presence of Hg (1), so the model can be confirmed only after further studies.

Finally, additional caution must be used in evaluation the goodness of the fits when the number of peaks or doublets increases. Is it possible that in such situation different fittings can be characterized by equivalent statistics?

In the future, for the interpretation of results obtained on real matrices the formation constants of species should be determined. Moreover, the problems originated by surface contamination and by degradation of materials due to irradiation should be examined carefully.

Nevertheless this new approach seems to open the way to a deeper understanding of the precipitation micromechanism and this lecture represents an original way of studying metal speciation in evaluating the speciation chemistry of solid sludges.

REFERENCES

1. Salomons, W. and U. *Forstner. Metals in Hydrocycle.* (Springer Verlag, Berlin, 1984).

2. Florence, T.M. and G.E. Batley. "Chemical Speciation in Natural Waters." *CRC Crit. Rev. Anal. Chem.*, 9:219-296 (1980).

3. Coale, K.H. and K.W. Bruland. "Copper Complexation in the Northeast Pacific." *Limnol. Oceanogr.*, 33:1084-1101 (1988).

4. Capodaglio, G., K. H. Coale and K. W. Bruland. "Lead Speciation in Surface Waters of the Eastern North Pacific," *Mar. Chem.*, In press.

CHEMICAL EQUILIBRIUM ANALYSIS OF LEAD AND BERYLLIUM SPECIATION IN HAZARDOUS WASTE INCINERATORS

Alexander P. Mathews
Kansas State University, Manhattan, Kansas

1. INTRODUCTION

Solid and hazardous waste incinerators may contain toxic organic chemicals, solvents, pesticides, and small quantities of toxic metals such as arsenic, antimony, beryllium, cadmium, lead and mercury. Historically, landfilling has been the least expensive and widely used method for the disposal of these wastes. However, the recent discovery of several landfill sites that have contaminated groundwater resources, and the realization that these sites pose long-term threat to the environment have prompted governmental initiatives on reducing risks from such hazardous waste disposal practices. The Hazardous and Solid Waste Act Amendments (HSWA) to the Resource Conservation and Recovery Act (RCRA) were enacted by the U.S. Congress in 1984 in response to this perceived need. Under the provisions of this act, all RCRA listed wastes must be passed through an analysis to assess the feasibility of banning these wastes from further land disposal.

Incineration is one of the major alternatives for the treatment of these wastes, and can effectively destroy the hazardous organic constituents. However, the increased use of incineration as a method of disposal is a cause for concern, since the fate of trace metals in incineration systems is not fully understood. Toxic metals such as beryllium and lead are not destroyed during combustion, but they can be transformed into compounds that can partition into the gas and solid phases in the incinerator. Metals may form volatile species in the combustion zone and subsequently condense onto flyash particulates, or they may be present in gas phase as vapors. The release of these particulates and vapors into the atmosphere can have significant adverse human health and environmental effects.

The speciation of trace metals in the combustion zone into the gas phase and condensed phases, and the subsequent condensation mechanisms that are operative as the hot gases are quenched, will determine the nature of emissions from hazardous waste systems. The type of metal species that are formed will depend on a number of factors including the waste and fuel characteristics and the incinerator operating conditions. Speciation will not only affect the amount of metal that is volatilized, but it may also affect air pollution control system efficiencies because of the different condensation temperatures and mechanism for the different species formed.

At present, there is little information on metal emissions from hazardous waste incinerators. The data available from trial burn studies [1], research facilities, [2] and field installations [3, 4] are quite limited and varied in nature, due to the variation in waste feed and incinerator operating conditions at different sites. A methodology for predicting the fate of trace metals during incineration would be useful to operators in developing effective control strategies, and to governmental agencies in developing regulations and criteria for granting permits.

This paper examines the major mechanisms involved in the fate and transformation of toxic metals incineration systems, and employs thermodynamic analysis to determine the type of inorganic species formed under various waste feed conditions and combustion temperatures. Research from this analysis can be used to evaluate potential emissions from hazardous waste incineration systems, to develop control strategies, and to assess the suitability of incineration for the disposal of RCRA listed wastes.

2. METAL TRANSFORMATIONS IN INCINERATION SYSTEMS

Several different intrinsic phenomena are involved in controlling the fate of trace metals in incineration systems. Figure 1 illustrates the major mechanisms that control the behavior in the combustion zone followed by cooling or quench zones. These mechanisms are examined to some extent further in the following discussion. Operating variables such as temperature, residence time, turbulence and factors such as waste and fuel composition, will determine which of the above mechanisms will predominate under specific operating environments.

The inorganic species in the waste particle or droplet will undergo different transformations in the combustion zone depending on whether they are present in the waste in free form or in a less readily accessible matrix. Toxic metals may be present in general as their oxides. They may also be in the waste combined with organic compounds or with elements such as chlorine and sulfur. The waste particle or droplet may be subject locally to both

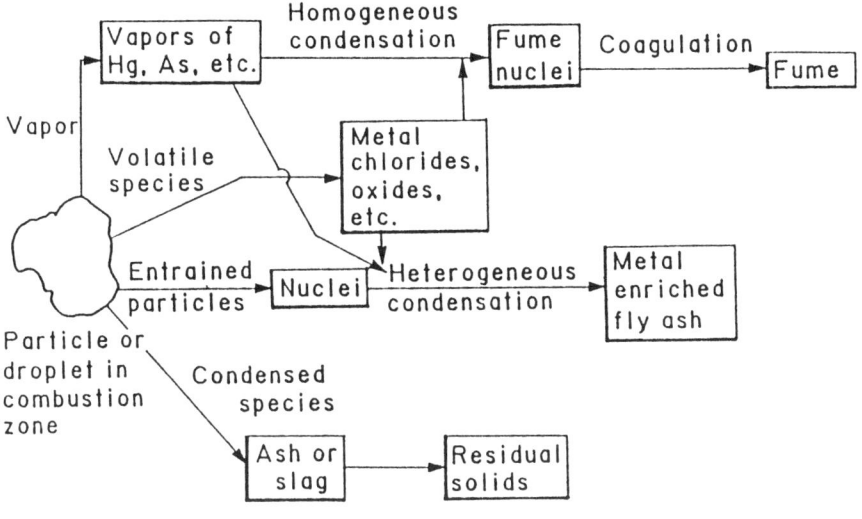

FIGURE 1. Metal Speciation and Condensation Mechanisms

reducing and oxidizing environments. Under reducing conditions, the oxidized metal species may be reduced by carbon monoxide to volatile metal or metal oxide species [5]. The presence of trace elements such as chlorine, bromine, sulfur, etc., will also result in the production of volatile species that are removed from the combustion zone in the gas phase. In addition, some metals such as As, Hg, Pb, etc. will vaporize at the incinerator operating temperature.

Condensed species in the form of ash or slag will be formed under oxidizing conditions. In the case of liquid hazardous wastes fired in liquid injection furnaces, all of the ash will be present in the gas phase as fine particulates. However, with both liquid and solid hazardous wastes, the formation of cenospheres and entrainment of fine particulates into the gas phase will occur.

The temperature of the combustion gases will drop as the hot gases move through the quench zone to the air pollution control device. As the temperature drops, condensation of the volatilized species will occur via homogeneous or heterogeneous condensation mechanisms. Homogeneous nucleation of fine particles of inorganic vapor species may occur under rapid cooling conditions as when the gas is passed through a quench zone [6]. Fumes may also be produced when a nonvolatile metal oxide is subjected to a reducing atmosphere to form volatile species, followed by an oxidizing atmosphere to form the nonvolatile species again [5]. Particles in the submicron range are generally formed by hemogeneous condensation.

Heterogeneous condensation is the deposition of volatilized species and vapors onto entrained fly ash particles and fumes. The number and size distribution of existing particles in the gas phases will determine the rate of heterogeneous condensation and the enrichment of metals on these particles. Smaller particles will generally have higher concentration of metals than larger particles. Moreover, the concentration of metals near the particle surface will be much higher than the average particle concentration [7, 8]. Heterogeneous condensation will generally produce particles that are larger in size than 1 μm and these particles can be more effectively removed by the air pollution control equipment.

3. METAL EMISSIONS FROM INCINERATORS

There is very limited information on the emission of metals in the form of vapors or as particulates from hazardous waste incinerators. Data for stack emissions for lead from trial burn studies [1] indicate a range from 7% to 48% of the lead in the waste feed. The concentration of lead in fly ash particulates ranged from 5,300 μg/gm to 98,000 μg/gm. The wide variation in the emissions data underscores the importance of attempting to elucidate the major mechanisms involved in the formation, condensation and capture of volatile metal species in hazardous waste incineration systems.

Coal-fired power plants and municipal solid waste incinerators are also significant contributors to the toxic metal content of aerosols in urban areas [9-13]. Many toxic elements such as Pb, Cd, and Cr are found at high concentrations in the micron and submicron particles [8]. Particles of diameter less than 1 μm deposit predominantly in the alveolar regions of the lungs, from where the toxic metals are rapidly absorbed into the bloodstream with an efficiency ranging from 50% to 80%. The effective toxicity of these ambient aerosol particles will depend on the type of volatile metal species deposited on them, and the metal enrichment factors.

Beryllium, lead and their compounds rank high in terms of their toxic and carcinogenic effects. Inhalation of beryllium laden air causes berylliosis, a disease that damages the lungs. The threshold limit value (TLV) for beryllium is 2 μg/m^3 [14]. Ingestion of lead at toxic levels can produce carcinogenic and mutagenic effects, and can also affect organs such as the kidneys and the heart [15]. TLV for lead is 150 $\mu g/m^3$.

4. MODELING APPROACH

The type of analysis required to determine the amount of toxic metals that may be potentially discharged in the gas. liquid and solid phases include heat balance calculations, reaction or equilibrium analysis to determine species formation, evaluation of condensation mechanisms, and capture efficiency of air pollution control equipment. This paper will address the formation of volatile species under various incinerator operating temperatures and excess air ratios.

Three interrelated parameters affect the destruction and removal efficiency (DRE) for organic constituents, and the inorganic species transformations in the incinerators. These are the combustion temperature, residence time in the incinerator and turbulence. As the incineration temperature is increased, the reaction rates for various transformations will rapidly increase, to approach equilibrium conditions. The residence time in the incinerator will affect the extent of conversion for the various reactions occurring in the combustion zone. In typical hazardous waste incinerators, residence time may not be a limiting factor affecting equilibrium conversion.

The diffusion of oxygen and other elements to the reacting species, and the diffusion of combustion products such as carbon dioxide, carbon monoxide and volatilized species away from the combustion zone, may affect transformation rates in some instances. This may not be a major factor with liquid injection furnaces burning liquid hazardous wastes with low ash content. However with solid or semi-solid wastes and liquid hazardous wastes with high ash content, mass transfer limitations may limit approach to equilibrium. The analysis in this paper is based on the assumption that the residence time and the turbulence in the incinerator are sufficient so that mass transfer or reaction kinetics do not limit the rapid approach to equilibrium.

4.1 Thermodynamic Analysis

Thermodynamic analysis for the reacting species is based on the premise that at equilibrium the free energy of the system must be at a minimum. For a series of reactions involving the various

78 CHEMISTRY OF TOXIC METALS

elements in the combustion zone, the criterion for equilibrium can be represented by the following S equations.

$$\sum_{i=1}^{N} v_{ij} \mu_i = 0, \quad j = 1,2...S \qquad (1)$$

where, v_{ij} are the stoichiometric coefficients for each species in the reaction equation, μ_i is the chemical potential of each species at the incineration temperature, and N is the number of species considered. The S reactions include reactions between elements to form gases and condensed phases.

The conservation of mass equations for the M elements in the system are:

$$n_i = n_i^\circ + \sum_{j=1}^{S} v_{ij} \xi_j, \quad i = 1...N \qquad (2)$$

where, n_i° and n_i are the initial number of moles and equilibrium moles of species i, and ξ_j is the reaction extent.

The chemical potential μ_i is related to the standard chemical potential μ_i° by equation (3).

$$\mu_i = \mu_i^\circ + RT \ln \left(\frac{n_i}{n_T} \right), \quad i = 1...N \qquad (3)$$

where, n_T is the total number of moles, R is the universal gas constant and T is the absolute temperature. The (N+S) nonlinear equations represented by the equilibrium and stoichiometric formulations (1) to (3) can be solved to obtain the equilibrium composition at any given temperature and elemental composition.
The main input data to the program include estimation of the species that are likely to be formed along with standard chemical potential data, and elemental composition of input streams to the incinerator. The standard chemical potential μ_i° can be obtained from sources such as the JANAF Thermochemical Tables (16) at any temperature using the following equation.

$$\mu_i^\circ (T) = \Delta H^\circ_{f298} + T\left(\frac{G^\circ - H^\circ_{298}}{T} \right) \qquad (4)$$

where, ΔH_f° is the standard enthalpy of formation, and data for the thermodynamic function $(G^\circ - H^\circ_{298})/T$ are available in the Tables as a function of temperature. Thermochemical data published by Barin et.al. [17] also provide the above information.

The elemental composition of fuel and waste feed entering the combustion zone is obtained using a heat balance model. This model uses enthalpies of the input streams, enthalpies of combustion products, heats of combustion, and an estimated heat loss of five percent from the incinerator. Waste feed data such as heating value, moisture and ash contents, and C,H,O,N,S and Cl contents for the waste and the fuel, are included in the analysis. The output from this model is the elemental composition of waste and fuel required to operate the incinerator at a specified temperature and excess air rate.

5. MODEL SIMULATIONS AND ANALYSIS

Computer simulations were made for an organic waste with a chlorine content of 7.5% and low ash (0.5%) and moisture contents (8.4%). The carbon, hydrogen and oxygen content for the waste are 64%, 8.4% and 11.2% respectively. Both the waste and fuel had negligible quantities of sulfur. The heating value of the waste is 20,130 kJ/kg. The concentration of beryllium and lead in the waste were assumed to the 2000 ppm and 500 ppm respectively.

The operating temperatures considered in the simulation ranged from 1100°K to 2000°K. A range of air supply values in excess of the stoichiometric combustion requirements were used. For each case, the fuel requirement to maintain a specified operating temperature and excess air ratio was obtained by utilizing a search routine coupled with the heat balance model. The output from this model provided quantities of the seven elements C,H,O,N,Cl, Be, and Pb for each simulation run. A total of 30 chemical species that included the combustion products, inorganic metal species, excess oxygen, and nitrogen, were included in the analysis. The JANAF Thermachemical Tables were the major source for determining which chemical species are likely to be formed.

5.1 Effect of Incinerator Operating Conditions On Be Speciation

The effect of incinerator operating temperature on beryllium speciations at an excess ratio of 25% is shown in Figure 2. The species formed are solid beryllium oxide and beryllium dihydroxide, along with minor quantities of beryllium chloride and beryllium hydroxide. Beryllium oxide is the major species at temperatures below 1600°K, and it's concentration decreases rapidly at higher temperatures. Gaseous beryllium dihydroxide is produced increasingly at temperatures greater than 1400°K. Thus, the emission of beryllium in the stack gases can be minimized by using two stage combustion systems, such as the rotary kiln operated at a lower temperature, followed by an afterburner at high temperature.

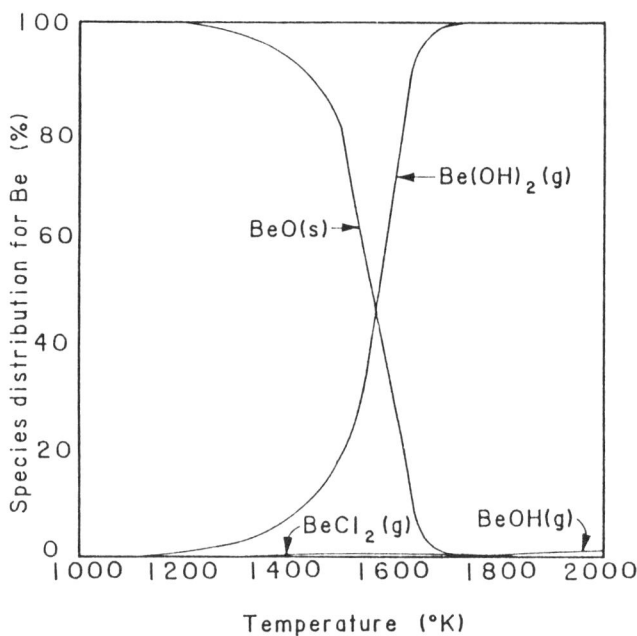

FIGURE 2. Beryllium Species Distribution at 25% Excess Air Ratio.

The effect of excess air rate on beryllium speciations is shown in Figure 3 for two incinerator temperatures. The dashed lines are the simulation results for a temperature of 1600°K and the solid lines for 1500°K. As the excess air ratio is increased, the less desirable beryllium dihydroxide species is formed at increasing concentration levels. The optimum excess air ratio to be used is a function of temperature. Rotary kilns are typically operated at excess air ratios greater than 100%, and this may result in the emission of increased levels of beryllium. The presence of chlorine or other halogens is not likely to have any major impact on beryllium speciation.

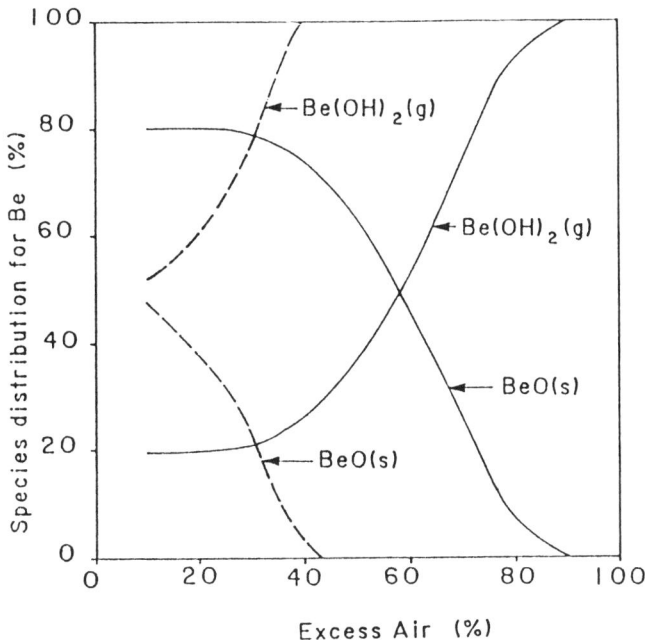

FIGURE 3. Beryllium Species Distribution at 1500°K (-) and 1600°K (--).

5.2 Effect of Incinerator Operating Conditions on Pb Speciation

The major lead species formed at various temperatures is shown in Figure 4 for an excess air ratio of 25%. Lead speciates predominantly into the chloride species lead tetrachloride ($PbCl_4$) and lead chloride ($PbCl_2$) at lower temperatures, and into lead oxide (PbO) and lead vapor at the higher temperatures. $PbCl_4$ is the major species at temperatures below 1300°K. $PbCl_2$ concentration increases gradually from 1200°K and attains a maximum value at approximately 1500°K. The location of the peak and the value of the maximum $PbCl_2$ concentration are functions of the excess air rate used. At temperatures greater than 1400°K lead oxide species becomes important, and it becomes the major species at temperatures higher than 1700°K.

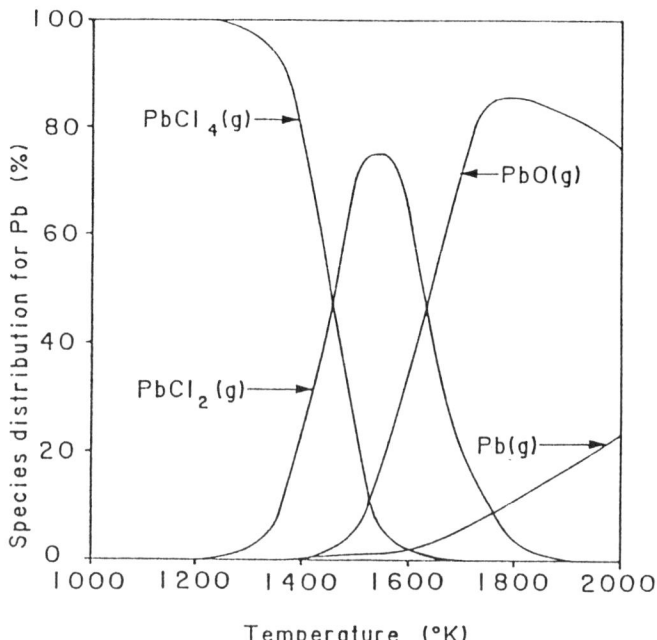

FIGURE 4.   Lead Species Distribution at 25% Excess Air Ratio.

The presence of halogens in the waste will limit the formation of any solid lead species. Thus, it would be desirable to limit the incineration of halogenated hazardous wastes containing any significant amounts of lead. The condensation mechanisms for various species and particle formation and coagulation rates are not known at present. Further studies must be undertaken to determine which of these species are likely to condense to particles that are readily removed by the air pollution control equipment. Since homogeneous condensation will, in general, produce submicron particles, incineration of liquid hazardous waste with low ash content can be expected to generate lead fume particles that are less readily captured.

6.   CONCLUSIONS

The increased use of incineration as a method of hazardous waste disposal can result in increased levels of toxic metals in the atmosphere. The quantity and nature of metal emissions will depend on the type of metal species formed in the combustion zone, and the efficiency with which the volatilized species are removed by the air pollution control equipment.

Chemical equilibrium analysis was used to predict the formation and concentration of beryllium and lead species under different operating conditions. Beryllium speciates into solid beryllium oxide at the lower temperature. At high temperatures and high excess air ratios beryllium is volatilized as beryllium dihydroxide. The presence of chlorine in the waste will not significantly affect beryllium speciation. Lead is volatilized readily as the halogenated species at the lower combustion temperatures, and as lead oxide and lead vapor at higher temperatures. Further studies must be undertaken to determine the condensation mechanisms for the various species, and the type and size distribution of particles formed, to facilitate the efficient capture and removal of these metals from the gas phase.

REFERENCES

1. Trenholm, A., P. Gorman, and G. Jungclaus. "Performance Evaluation of Full-Scale Hazardous Waste Incinerators," Final report, USEPA Contract No. 68-02-3177, (1984).

2. Lee, J.W., R.W. Ross, R.H. Vocque, J.W.Lewis, and L.R. Waterland. "Distribution of Trace Element Emissions from the Liquid Injection Incinerator Combustion Research Facility," PB87-224689, National Technical Information Service, U.S. Department of Commerce, (1987).

3. Fuhr, H. "Hazardous Waste Incineration at Bayer, AG," *Hazardous Waste and Hazardous Materials*, 2(1):1-5 (1985).

4. Kristensen, A. "Operating the Rotary Kiln Incinerators at Kommunekemi," *Hazardous Waste and Hazardous Materials*, 2(1):2-21 (1985).

5. Quann, R.J. and A.F. Sarofim. "Vaporization of Refractory Oxides During Pulverized Coal Combustion," Nineteenth Symposium (international) on Combustion/The Combustion Institute (1982). pp. 1429-1440.

6. MeNallan, M.J., G.J. Yurek, and J.F. Elliott. "The Formation of Inorganic Particulates by Homogeneous Nucleation in Gases Produced by the Combustion of Coal," *Combustion and Flame*, 42:45-60 (1981).

7. Davison, R.L., D. Natusch, J.R. Wallace, and C.A. Evans. "Trace Elements in Fly Ash, Dependence of Concentration of Particle Size," *Environ. Sci. Tech.*, 13(8):1107-1113 (1974).

8. Natusch, J.F.S., J.R. Wallace, and C.A. Evans. "Toxic Trace Elements: Preferential Concentration in Respirable Particles," *Science*, 183:202-204 (1974).

9. Law, S.L. and G.E. Gordon. "Sources of Metals in Municipal Incinerator Emissions," *Environ. Sci. Tech.* 13 (4):432-438 (1979).

10. Greenberg, R.R., W.H. Zoller, and G.E. Gordon. "Composition and Size Distribution of Particles Released in Refuse Incineration," *Environ. Sci. Tech.*, 12(5):566-573 (1978).

11. Sabbioni, E., L. Goetz, and G.Bignolia. "Health and Environmental Implications of Trace Metals Released from Coal-Fired Power Plants: An Assessment Study of the Situation in the European Community," *The Sci. of the Total Env.*, 40:141-154 (1984).

12. Coles, D.G., R.C. Ragaini, J.M. Ondov, G.L. Fisher, D. Silberman, and B.A. Prentice. "Chemical Studies of Stack Fly Ash from a Coal-Fired Power Plant," *Environ. Sci. Tech.* 13(4):455-459 (1979).

13. Markowski, G.R. and R. Filby. "Trace Element Concentration as a Function of Particle Size in Fly Ash from a Pulverized Coal Utility Boiler," *Environ. sci. Technol.*, 19(9):796--804 (1985).

14. Wilber, C.G., *Beryllium- A Potential Environmental Contaminant* (C.C. Thomas, Publishers, Springfield, IL , 1980).

15. National Academy of Sciences, *Lead in the Human Environment*, (National Academy Press, Washington, D.C. 1980).

16. Stull, D.R. and H. Prophet, Eds. *JANAF Thermochemical Tables* (U.S. Government Printing Office, Washington D.C. 1971).

17. Barin, I., O. Knacke and O. Kubashewski, *Thermochemical Properties of Inorganic Substances*, (Springer-Verlag Publishers, New York 1977).

DISCUSSION OF:
CHEMICAL EQUILIBRIUM ANALYSIS OF LEAD AND BERYLLIUM SPECIATION IN HAZARDOUS WASTE INCINERATORS

Sabino A. Bufo
University of Bari, Bari, Italy

INTRODUCTION

It is well known that incineration of plastic, wood products, fossil fuels, and organic wastes may result in atmospheric emission of toxic metals [1]. Låg [2] has emphasized the possibility of significant pollution of soil by certain municipal incinerators within a distance of about 10km and urges that the location and design of such facilities be given careful consideration.

Prediction of the impact of toxic metal pollution on the environment can be approached by first identifying the different chemical species which contribute to total concentration of each metal, and then, to distinguish the non labile from labile forms, the latter being able to react under certain conditions with material found in the environment.

Several models have been proposed to assess metal speciation in aqueous solutions, soil, and sewage sludge [3-9]. Moreover, composition of incinerator emissions and total concentration of metals from several particulate sources have been determined [10-12], but specific information on the speciation of metals in the combustion zone of incinerators is not found in the literature.

The two metals of concern here are extremely dangerous to human health. Almost all the presently known Beryllium compounds are acknowledged to be toxic in both the soluble and insoluble forms and produce a disease affecting the entire body [13]. Evidence is also present that Beryllium in the soil is toxic to plant life [14]. The major source of Beryllium in the atmosphere is industrial combustion processes [15], and it can exist as dust, mist or fumes. Control devices for Beryllium, such as venturi scrubbers, paked towers, cyclones, electrostatic precipitators, and bag collectors can be used. The choice of control equipment depends on the nature of the emission and the size of the operation. In regard to the options for disposal of Beryllium and its compounds, recycling is the most desirable [13]. Liquid or solid wastes containing

Beryllium particulates which are too dilute to recycle are buried on private property or in public landfills. Often, waste is first burned to produce insoluble, chemically inert oxides, a process which is easily and safely performed, providing that the exhaust gases are scrubbed to remove any particulates. These procedures have been verified [16] and at present seem adequate. In other words, waste should be converted into chemically inert oxides using incineration and particulate collection techniques. These oxides may then be landfilled [16].

Lead is widely used in industry and therefore problems are not uncommon among industrial workers and ordinary citizens. Inorganic lead compounds are very insoluble, a property which influences the grade of toxicity and possibility of contamination [17-19]. Organic lead, such as tetra-alkyl lead compounds, is a special problem in itself. Sources of lead emitted into the atmosphere as a consequence of the activities of man have been reviewed by Snyder et al. [20]. Emission of the metal from waste incineration and open burning has also been observed, ranging from 320 to 5,800 tons of Pb [21-22].

Because metallic lead is insoluble in water and not attacked by most inorganic acids or bases, lead in small quantities is often disposed of in landfills [16]. Though this disposal method is satisfactory under most conditions, it is not recommended because lead is lost to future use. In addition, under certain conditions, lead is solubilized as either a chloride or sulfate, which are both slightly soluble in water. Since lead oxides are insoluble in water, and are also stable under ordinary temperatures, disposal of lead oxide wastes by means of landfill is acceptable. Disposal of soluble lead compounds such as the acetate, nitrate and nitrite by landfill technique is however, inadmissible. Even the slightly soluble lead chloride should not be disposed of by landfill.

DISCUSSION

The previous discussion serves as a general introduction to the paper presented by Dr. Mathews, and in particular, to the model proposed for the speciation of metals emitted by incinerators.

The mobilization of metals in the atmosphere by means of enrichment of particulates and gas combustion emitted by chimneys, is controlled by phenomena of volatilization and condensation, and indirectly, by two parameters which influence these processes: the halide content of wastes which influences the formation of particularly volatile metallic halides, and the combustion temperature which determines the amount of volatilization.

Therefore, we can state that the analysis which is necessary to determine the quantity of toxic metals which could be emitted, must be based on the exact composition of these wastes, on the possible formation reactions of the species, on the mechanisms of condensation, and finally, on the efficiency of the control and collection systems of the emitted substances.

The model proposed by Mathews seems to formulated well enough if one accepts the concept that the chemical reactions and the phase transformations within the combustion temperature range in the incinerator are equilibrium processes, and that this equilibrium is reached fairly quickly, and also that the transfer of mass does not influence these processes. These limitations are acceptable if we consider that this is practically the first attempt to present a model for chemical processes which occur in incinerator systems.

Having accepted the previous statements, also the thermodynamic analysis proposed seems to be exact, complete and well generalized, so that the successive applications to Pb and Be are to be considered only as examples. In fact, an extension to other metals more commonly found in both urban and industrial wastes would be useful and interesting; furthermore, organic substances, even if not considered here, constitute a serious problem in incinerator emissions.

Perhaps in equation (2) the significance of the reaction extent should be better explained.

In addition, further explanation of the model used for the estimation of the elemental composition for the optimum functioning of the incinerator at a given temperature and excess air rate might be necessary. However, these problems are beyond the theme of this convention.

On the other hand, we observe with alarm that the simulation trial confirms the formation of Beryllium dihydroxide at both high temperatures and major air excess; owing to its volatility, this species is more dangerous than Beryllium oxide which is formed at lower temperatures. Also for Lead, the formation of volatile species is confirmed, among which are found halides, which are extremely dangerous for reasons stated previously.

CONCLUSIONS

In conclusion, the model, which in input simply needs the estimation of the species which could be formed together with standard chemical potentials, and the elemental composition of fuel and waste feed, can furnish an evaluation of the percentage of formation of chemical species in equilibrium at various temperatures and excess air rate, thereby assisting operators to develop strategies for more adequate control techniques.

REFERENCES

1. Tiller, K.G. "Heavy Metals in Soils and Their Environmental Significance," in "Heavy Metal Pollution of Soil," *Advances in Soil Science*, 9:114-142 (1989).

2. Låg, J. "Soil Pollution by Cadmium from Incineration Plants," *Ambio*, 14:356-373 (1985).

3. Sinzburg, G. "Calculation of All Equilibrium Concentrations in a System of Competing Complexation," *Talanta*, 23:149-152 (1976).

4. Jenne, E.A. "Chemical Modeling in Aqueous Systems," *ACS Symposium Series No. 93*. Washington, DC, American Chemical Society (1979).

5. Stumm, W. and J.J. Morgan. *Aquatic Chemistry. An Introduction Emphasizing Chemical Equilibria,* (John Wiley & Sons, London, 1981).

6. Legret, M., D. Demare, and P. Marchandise. "Speciation of Heavy Metals in Sewages Sludges," in *Proc. of the 4th Int. Conf. Heavy Metals in the Environment*, 350-353. Edinburgh, CEP (1983).

7. Lake, K.L., P.W. Kirk and J.N. Lester. "Fractionation, Characterization, and Speciation of Heavy Metals in Sewage Sludge," *J. Environ. Qual.*, 13:175-183 (1984).

8. Forstner, U. *Chemical Forms and Reactivities of Metals in Sediments, in Chemical Methods for Assessing Bioavailable Metals in Sludges and Soil,* ed. by R. Leschber., R.D. Davis and P. L'Hermite, 1-31, (Elsevier Applied Science Publishers, London, 1985).

9. Kirk, P.W.W., D.L. Lake, J.N. Lester, T. Rudd, and R.M. Sterrit. *Metal Speciation in Sewage, Sewage Sludge and Sludge-Amended Soil and Seawater. A Review,* ed. by J.A. Campbell. WRC Environment TR226, Medmenham, WRC (1985).

10. Greenberg, R.R., W.H. Zoller, and G.E. Gordon, "Composition and Size Distributions of Particles Released in Refuse Incineration," *Environ. Sci. Tech.*, 12:566-577 (1978).

11. Coles, D.G., R.C. Ragaini, J.M. Ondov, G.L. Fischer, D. Silberman, and B.A. Prentice. "Chemical Studies of Stack Fly Ash from a Coal-Fired Power Plant," *Environ. Sci. Tech.*, 13:455-459 (1979).

12. Markowski, G.R. and R. Filby. "Trace Element Concentration as a Function of Particle Size in Fly Ash from a Pulverized Coal Utility Boiler," *Environ. Sci. Tech.*, 19:796-804 (1985).

13. Durocher, N.L. "Air Pollution Aspects of Beryllium and Its Compounds," Report PB 188,078, Springfield, Va., Nat. Tech. Information Service (1969).

14. Luckey, T.D., B. Venugopal and D. Hutcheson. *Heavy Metal Toxicity, Safety and Hormonology,* (Academic press, New York, 1975).

15. Booz-Allen Applied Research, Inc. "A Study of Hazardous Waste Materials, Hazardous Effects & Disposal Methods," Report PB-221,466, Springfield, Va., Nat. Tech. Information Service (1973).

16. Powers, P.W., "How to Dispose of Toxic Substances and Industrial Wastes," Park Ridge, N.J., Noyes Data Corp (1976).

17. Kehoe, R.A. "Industrial Lead Poisoning," in: *Hygiene and Toxicology,* F.A. Patty, ed., 2nd ed., (New York, John Wiley, 1962).

18. Chisholm, J.J., Jr. "Lead Poisoning," *Scientific American,* 224:15-23 (1971).

19. Kinnison, R.R., "Lead-In Search of the Facts," *Env. Sci. Tech.*, 10:644-649 (1976).

20. Snyder, R.B., D.J. Wuebbler, J.E. Pearson, and B.J. Ewing. " A Study of Environmental Pollution by Lead," Document No. 71-7, Chicago, Illinois, Illinois Inst. of Environmental Quality (1971).

21. Goldberg, A.J. "A Survey of Emissions and Controls for Hazardous and Other Pollutants," Wash, D.C., Environmental Protection Agency (1973).

22. Davis, W.E. and Assoc. "Emission Study of Industrial Sources of Lead Air Pollutants," 1970, Leawood, Kansas (1973).

NUCLEATION AND CRYSTAL GROWTH STUDIES ON PRECIPITATION OF CADMIUM HYDROXIDE FROM AQUEOUS SOLUTIONS

J. W. Patterson and B. Luo
Illinois Institute of Technology, Chicago, Illinois

D. Marani, and R. Passino
Water Research Institute, Rome, Italy

1. ABSTRACT

This paper presents preliminary results of a study designed to improve the knowledge of the fundamental principles governing cadmium hydroxide precipitation as well as the subsequent solid/liquid separation processes in wastewater treatment. In particular, the present work was addressed to the study of the kinetics of $Cd(OH)_2$ precipitation in terms of nucleation and crystal growth rates. Physical characteristics of the precipitate particles were studied in terms of surface charge and particle size distribution. $Cd(OH)_2$ precipitation has been investigated both in the absence and in the presence of citric acid as a model inhibiting agent.

The preliminary results show that $Cd(OH)_2$ precipitation is very fast in the absence of an inhibiting agent. In this instance, CSTR tests seem to be more appropriate than batch tests for kinetic studies. At low base/metal molar ratios, the precipitate is stabilized in a colloidal form by a positive surface charge.

In agreement with classical nucleation theory, nucleation rate was found to be highly dependent on the supersaturation.

2. INTRODUCTION

Heavy metals such as cadmium, copper, chromium, and lead, are used in many industries including metal plating and finishing, petroleum refining and organic chemical production [Fassett, 1980]. The widespread use of these metals has led to concern over detrimental effects on human health and the environment caused by the discharge of these toxic elements into receiving waters.

The promulgation of direct discharge standards, as well as pretreatment standards for the discharge of industrial waste into

Publicly Owned Treatment Works (POTW's), has been aimed at controlling these metal concentrations. With this regulatory incentive, much interest has been placed upon the development and improvement of removal technologies, including chemical precipitation, electrodeposition, adsorption, ion exchange, solvent extraction, and others [Patterson, 1985].

The most widely applied technology for metals removal is hydroxide precipitation [Patterson, 1987]. This process has been favored for its low cost and ease of implementation, as well as its ability to remove a broad spectrum of metals concurrently.

While attempts have been made to define selective precipitation schemes for metals from thermodynamic data [Jenke and Diebold, 1983], these have been largely unsuccessful due to incomplete understanding of thermodynamics with regards to coprecipitation. Kinetic limitations are also cited as reasons for lack of conformity of results with predictions [Karra et al., 1983].

The degree of aggregation of the suspension produced in the precipitation process is primarily influenced by the early stages of the process, which include nucleation and crystal growth [Walton, 1967].

Knowledge of hydrolysis and precipitation kinetics may help to design selective precipitation processes based on kinetic rather than thermodynamic control. A better understanding of nucleation and crystal growth kinetics can help to promote favorable precipitate characteristics for better sludge handling. Promotion of differences in floc densities for different metal precipitates could provide a means for separation based on differential settleability [Patterson, 1987].

The nucleation and crystal growth of metal hydroxide/oxide are of importance in view of the large application of precipitation processes. Alkaline precipitation processes, because of their complex nature, are usually applied on an empirical basis rather than being designed on sound scientific principles. While metal insolubilization is quite easily controlled by means of pH, the subsequent step of precipitate separation from the soluble phase is often quite difficult, due to the colloidal nature of the precipitate.

The degree of aggregation of the suspension produced in the precipitation process is primarily influenced by the early stages of the process, which include nucleation and crystal growth [Walton, 1967]. Crystal growth and, overall, nucleation rates are highly affected by the supersaturation of the solution and by its chemical composition. Unlike in the crystallization process, where supersaturation is held at low values, the precipitation processes, as applied in industrial practice, usually occur at extremely large relative supersaturation [Nyvlt, 1985]. Under these conditions nucleation is enhanced at the expense of crystal growth; a large number of primary particles are generated, which aggregate in amorphous secondary particles. In addition, at high supersaturation values a metastable solid phase is precipitated first, which later undergoes aging and transformation processes toward a more stable phase [Walton, 1967].

Citric acid and its salts are often used to clean the surface of metals because they cause less damage to the substrate metals than more corrosive and hazardous strong acids [Bradley, 1975; Blume, 1977]. It is also reported that citrates could assist in the plating of many metals as buffers and complexing agents [Peters and Ku, 1987]. One of the problems in the use of complexing agents is the difficulty in precipitating complexed metals out of waste water during treatment [EPA, 1980].

This paper presents preliminary results of a study designed to improving the knowledge of the fundamental principles governing metal hydroxide precipitation as well as the subsequent solid/liquid separation processes in wastewater treatment. In particular, the present work addressed the study of the kinetics of $Cd(OH)_2$ precipitation in terms of nucleation and crystal growth rates. Cadmium has been chosen as a model metal, because of its widespread presence in industrial wastewaters and the environmental concern posed by its high toxicity. Physical characteristics of the precipitate particles were studied in terms of surface charge and particle size distribution. $Cd(OH)_2$ precipitation has been investigated both in the absence and in the presence of citric acid as a model inhibiting agent.

3. BACKGROUND

3.1 Solubility Phenomena

When a sufficient amount of hydroxide ion is added to a supersaturated metal salt solution, metal hydroxide precipitate is formed, with a disordered lattice [Feitknecht and Schindler, 1963]. Metal hydroxide can occur in amorphous and several crystalline modifications. The tendency to form amorphous precipitate increases with the valency of the metal ion.

Often, when hydroxide ions are added to a metal salt solution, hydroxide salts are first formed in an active state. If less than an equivalent amount of alkali is added, the active forms pass into the inactive ones upon aging. In the presence of anions forming hydroxide salts of considerable stability, precipitation occurs at a lower pH than with anions showing little tendency to form hydroxide salts.

For cadmium hydroxide solubility in aqueous solution, the basic reaction can be written in the form of

$$Cd(OH)_{2(s)} = Cd^{2+} + 2OH^- \tag{1}$$

The conventional solubility product for equation (1) is given by:

$$K_{so} = [Cd^{2+}][OH^-]^2 \tag{2}$$

The solubility of metal hydroxides in aqueous solutions varies over a wide range. It is dependent on factors such as pH, ionic strength, the presence of anions, temperature and complexing agents in the solution. The reported K_{so} of cadmium hydroxide range from $10^{-12.9}$ to $10^{-14.4}$ [Smith and Martell, 1976, Sillen and Martell, 1971, Feitknecht and Schindler, 1963]. Based on the experimental con-ditions (i.e., temperature, ionic strength, and the form of precipitate) and theoretical conditions, a literature value of $K_{so} = 3.98 \times 10^{-15}$ was chosen for $Cd(OH)_2$ (Sillen and Martell, 1971) in this study.

Although precipitate formation is predicted as thermodynamically favorable at or above saturation values (i.e., when the ionic activity product is greater than or equal to the K_{so} for the solid phase), precipitation will generally not occur until a significant degree of supersaturation is reached [Nancollas and Purdie, 1964]. The supersaturation ratio may be defined as:

$$\text{Supersaturation Ratio (S)} = \frac{[Cd^{++}][OH^-]^2}{K_{so}} \tag{3}$$

Three crystalline forms of $Cd(OH)_2$ are known [Feitknect and Schindler, 1963]. These are β-$Cd(OH)_2$, which can be precipitated out by excess base, and is the stable form at 25°C; and α-$Cd(OH)_2$ and γ-$Cd(OH)_2$ which can be formed under special conditions but are eventually converted in water to β-$Cd(OH)_2$. Several studies have been made of the variation of the solubility of β-$Cd(OH)_2$ with pH, but the results show considerable scatter even within a single study [Baes and Mesmer, 1976].

The fresh and aged β form of cadmium hydroxide precipitates do not show any significant morphological differences [Feitknect and Schindler, 1963].

3.2 Nucleation and Crystal Growth Theories

The mechanism of formation of precipitate has been reported to occur in three stages: nucleation, crystal growth, and aging.

The number, size, structure and morphology of precipitated crystals are primarily controlled by nucleation. However the nucleation process is not well understood, owing to the extremely small size of nuclei (5 - 20 A) and difficulty of making physical observations [Walton, 1967].

Nucleation has been traditionally classified into two theoretical types [Walton, 1967]. The precipitation of a solid from a homo- geneous solution (in the absence of preexisting solid substrate) is called homogeneous nucleation. Homogeneous nucleation may be conceptualized as a stepwise buildup of ionic or molecular monomers to form clusters [Walton, 1965]. These associations are due to random statistical fluctuation in supersaturation. Above the critical supersaturation point, stable nuclei will form which then serve as centers for crystal growth.

Heterogeneous nucleation occurs in the presence of a preexisting surface. Because virtually all aqueous solutions contain free particles of various types, most nucleation is heterogeneous.

In heterogeneous nucleation, the nuclei form on the surface of seed crystals and the nucleation occurs at much lower supersaturations than homogeneous nucleation due to the catalytic action of the extraneous surfaces. In sediments, because of a very high density of potential seeds, it is likely that nucleation is almost always heterogeneous (Baes, 1986). The same should be true, although to a lesser extent, in wastewater treatment.

Nucleation kinetics have been studied in terms of Gibbs free energy. This model has been advanced by Volmer (1926), Becker and Doering (1935), Frenkel (1946), and Turnball and Fischer (1949). In order for nucleation to proceed, an activation energy barrier must be crossed. For a system at supersaturation conditions, the formation of precipitate will lead to a lowering of the system free energy until equilibrium is reached. Counter balancing this effect is an increase in interfacial free energy associated with the formation of a surface. The activation energy can be derived as [Stumm and Morgan, 1981]:

$$\Delta G^* = \frac{16\gamma^3 V^2}{3[kT\ln(s)]^2} \quad (4)$$

where: γ is the interfacial energy,
V is the "molecular" volume, which is the volume of a

molecule or an ion,
k is Boltzmann's constant,
T is the absolute temperature,
s is the supersaturation

and the nucleation rate can be expressed as:

$$J = C\exp\left(\frac{-16\gamma^3 V^2}{3kT[kT\ln(s)]^2}\right) \quad (5)$$

where C is a factor related to the efficiency of collision of ions or molecules.

As it is very difficult to obtain quantitative information about the ΔG^* in equation (4), it has become standard practice in industrial crystallization to use an empirical power law function of supersaturation [Nielsen, 1964] at a given temperature. A proposed equation of this form is expressed as [Randolph and Larson, 1971]:

$$B_o = ks^i \quad (6)$$

where s is the supersaturation,
B_o is the nucleation rate,
k is a kinetic constant.

Once nucleation has occurred, spontaneous crystal growth may take place. The growth may be divided into two distinct phases [Snoeyink and Jenkins, 1980]: (1) the movement of solute from bulk solution to the crystal/liquid interface by means of diffusion or convection; (2) adsorption of solute onto the solid surface and incorporation into the crystal lattice. The size of precipitate particle is affected by the mechanisms of new particle formation and particle growth. The form of the particle size distribution is dependent on the growth kinetics. Many theories on the mechanism of crystal growth have been proposed and these are described in detail by Buckley (1957) and by Verma (1953).

In order to determine nucleation and crystal growth rates, an analysis of the crystal population density could be performed according to Randolph and Larson (1971) on the basis of assuming that the McCabe ΔL-law holds (growth rate independent of crystal size). The population balance of a CSTR reactor leads to the equation:

$$n = n^o \exp(-L/G\theta) \quad (7)$$

in which L = size of the crystal,
         n = crystal population density related to size L,
         n° = nuclei population density,
         G = crystal growth rate,
         θ = residence time.
By plotting log n versus L, n° and G values can be determined. In addition, the nucleation rate B_o can be calculated from the equation:

$$B_o = n°G \qquad (8)$$

Complexing agents and other foreign ions may affect the metal precipitation process both from thermodynamic and kinetic points of view. Sulfate ion has been demonstrated to affect aluminum hydrolysis-precipitation, inducing the precipitation at pH values lower than in the absence of sulfate [Hek et al., 1978]. Helz and Horzempa (1983) have shown that EDTA, a chelating agent, may have a kinetically inhibiting effect on the precipitation process as well. Peters and Ku (1986) have investigated the effect of citric acid, a weak chelating agent, on the precipitation of metal sulfides and hydroxides. While equilibrium solubilities were not significantly affected by the presence of citrate, particle sizes were found to be decreased. It was reported [Packter and Jakubowski, 1985] that the precipitation of most sparingly soluble metal salts may be modified by addition of inorganic and organic anions and polyanions that complex the metal cation. The rates of nucleation and the kinetics of crystal growth (after the induction periods) may be reduced while the morphologies and sizes of the final precipitate crystals may be changed appreciably. Hek, et al. (1978) reported that the presence of sulfate influences the composition, texture and structure of the aluminum precipitate, apparently in some mode associated with sulfate ion adsorption onto the actively forming aluminum precipitate solid phase.

4. EXPERIMENTAL

Preparation of the solutions: 1M NaOH, 0.1M $Cd(NO_3)_2$ and 0.1M citric acid stock solutions were prepared by dissolving the appropriate amounts of analytical grade reagents in twice-distilled water. Solutions for batch and CSTR tests were prepared by dilution of stock solutions. For the precipitation test in the presence of citric acid, the cadmium nitrate solution and citric acid solution were premixed before each experiment. The solutions were tested for the actual concentration of NaOH, cadmium, and citrate with alkalimetric titrations, atomic absorption spectroscopy (AAS), and combustion/infrared laboratory total organic carbon analyzer, respectively.

Completely Stirred Tank Reactor (CSTR) tests: Figure 1 shows the experimental apparatus used in the CSTR tests. The cylindrical plexiglass reactor had an effective volume of 325 ml. In the lid of the reactor were inserted: nitrogen gas inlet tube, impeller shaft, pH electrode, Cd and reference electrodes, NaOH and cadmium solutions inlet tubes, PenKem System 3000 and Hiac/Royco particle counter sampling tubes. Sampling of the suspension for the analysis of the total and soluble (filterable) residual metal was performed through the lid as well.

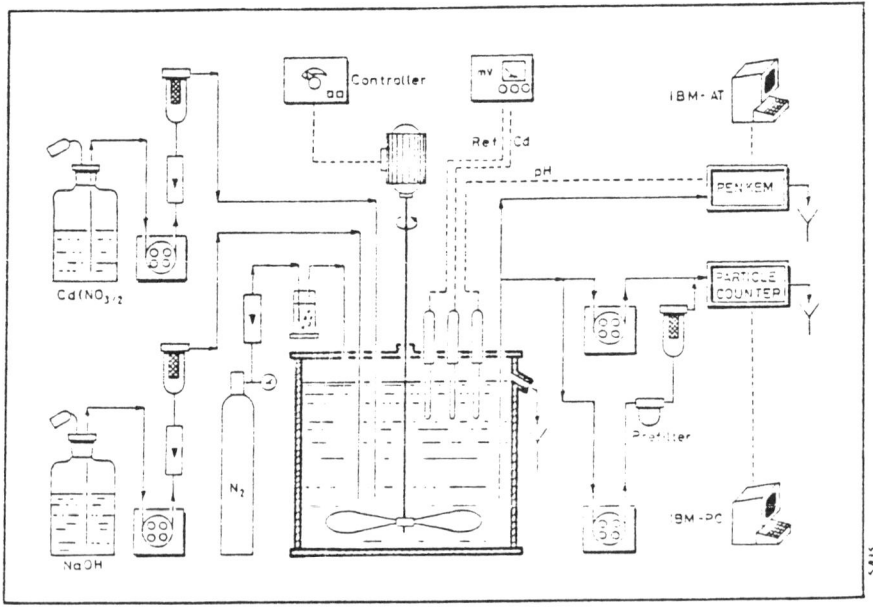

FIGURE 1. Experimental Apparatus for CSTR Tests.

Influent NaOH and cadmium solutions were fed with peristaltic pumps at the same constant flowrate. Before entering the reactor the solutions passed through 0.2 micron filters. The reactor was stirred with an impeller at 250 rpm and was kept under a nitrogen blanket. The following analytical parameters were monitored during each run: pH, conductance, turbidity, total, filterable (0.45 um) and free cadmium concentrations, particle concentration and size distribution, and electrophoretic mobility of the precipitate.

A series of step-down tracer studies were performed, using potassium chloride (KCl) as tracer, in order to check the degree of mixing of the reactor in the test conditions described above. The results showed that the reactor follows very closely the ideal behavior of a CSTR reactor [Levenspiel, 1972]. With the impeller rate at 250 rpm, thorough mixing was assured in the reactor, even at residence times as low as 15 seconds.

Batch tests: A 2500 ml plexiglass cylinder with four internal baffles was used for the batch tests. At time zero, 1 liter of NaOH solution and 1 liter of metal solution, both prefiltered through a 0.2 micron capsule filter, were flash mixed in the reactor. The reactor was maintained under a nitrogen blanket and the reaction was monitored with the following measurements: pH, conductance, turbidity, total and residual soluble (filterable) metal, particle concentration, particle size distribution and electrophoretic mobility of the precipitate solids.

Analytical measurements: pH, conductance, turbidity and electrophoretic mobility were measured using a PenKem System 3000, which is equipped with a laser beam for turbidity and electrophoretic mobility measurements. Free cadmium concentrations ($[Cd^{++}]$) were measured potentiometrically, using a Cd electrode Orion model 94-48 and a reference electrode Orion model 90-01, connected to a potentiometer Orion model 901.

Total and soluble (filterable) residual metal concentrations ($[Cd]_T$ and $[Cd]_{sol}$, respectively) were determined via atomic absorption spectroscopy. 10 ml samples for the total metal concentration analysis were withdrawn from the reactor using a syringe. After acidification, the sample was transferred into a sampling vial for the subsequent AAS analysis. The same general procedure was used to take samples for soluble (filterable) metal concentration measurements. In this case a disposable 0.2 or 0.45 um membrane filter was inserted between the sampling tube and the syringe to remove the precipitated solids.

Particle concentration and size distribution were measured using a Hiac/Royco Model 4300 particle counter equipped with the Model 246 laser probe (lower detection limit equal to 0.5 um). The instrument employs a flow through cell so that the particles may be measured in real time. In order to have particle concentrations within acceptable limits, a dilution system was used (as shown in Figure 1). The diluting solution had the same composition as in the reactor so that the dilution process would not alter the characteristics of the solid phase. This solution, after having been sequentially filtered through a gross filter, 0.45 micron filter and 0.2 micron capsule filter, was mixed with the sample stream immediately before entering the laser probe.

5. RESULTS AND DISCUSSION

5.1 Metal Precipitation in the Absence of an Inhibiting Agent

Before the CSTR experiments were initiated, a series of batch tests were performed at preset values of total metal concentration and base/metal molar ratios. The results showed that both pH and residual metal concentration reach essentially constant values soon after flash mixing of the reagents. These constant values were quite close to the theoretical equilibrium values calculated with a speciation computer program [Westall et al., 1976]. As far as the filterable residual metal is concerned, no difference was noted between 0.45 um and 0.2 um membrane filters. Such results suggest that in the absence of interfering agents, the precipitation of cadmium hydroxide is very fast. As a consequence, batch tests, as described above, are not suitable to study the kinetics of the reaction, in that the response time for the measurements of most of the variables of interest can be of the same order or longer than the reaction time.

A preliminary series of CSTR tests was performed in order to identify experimental conditions (in terms of total metal concentration and base/metal molar ratio) which are suitable for kinetic studies. Figure 2 reports the results of one set of CSTR tests performed with the same influent solutions (hence at the same total metal concentration and base/metal molar ratio), but varying the CSTR residence time. This Figure shows that, with increasing residence time, both soluble residual metal concentration and pH are in a decreasing trend until reaching a constant value at large residence time. The effect of residence time (θ) on residual metal concentration and on pH strongly suggests that below 100 seconds, the system is far from equilibrium conditions. In order to avoid flocculation and to be in a dynamic condition such that reliable data on the kinetics of the reaction can be obtained, it was found in this research that the optimum operative metal concentrations for cadmium hydroxide precipitation are from 1E-4 M to 1E-3 M.

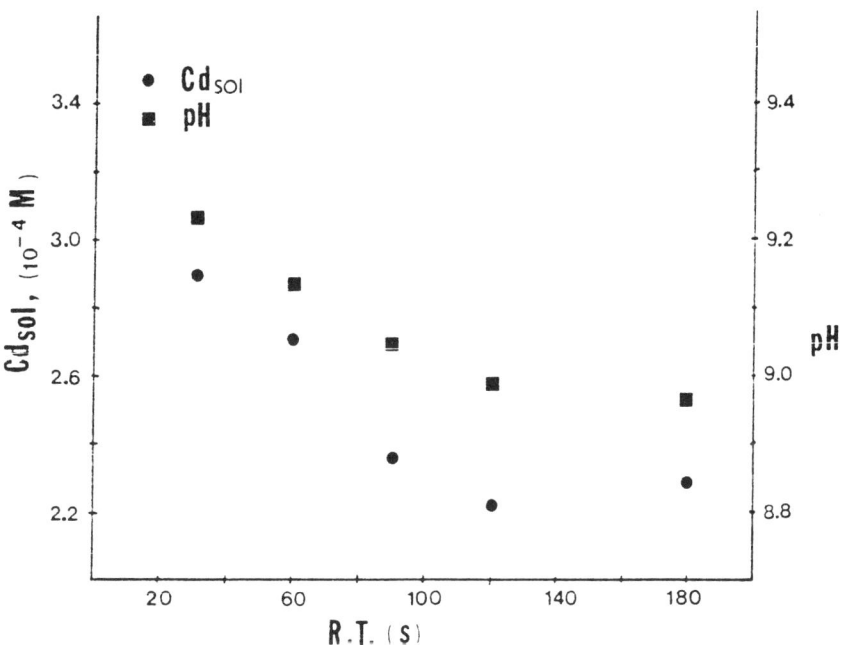

FIGURE 2. Effect of Residence Time in CSTR Precipitation Tests at Steady State Conditions.

Figure 3 shows the characteristics of the precipitate, in terms of electrophoretic mobility, versus base/metal molar ratio (OH/Cd) and pH. The precipitate has a positive surface charge that decreases as the OH/Cd ratio increases. However, even in the presence of a large excess of base (OH/Cd = 5.3) the surface charge is still positive, or, in other words, pH values are still lower than the isoelectric point of the cadmium hydroxide, which is reported to be near 12 [Parks,1965]. In addition to the surface charge, the aggregation state of the precipitate could be observed through the microscopic system of the PenKem. For CSTR tests, these observations showed a finely dispersed colloidal precipitate. Significant aggregation of the particles was observed only at high base/metal molar ratio or long residence times. Hence, the surface charge stabilizes the precipitate particles in a colloidal form, at least in the experimental conditions of a CSTR test with short residence time.

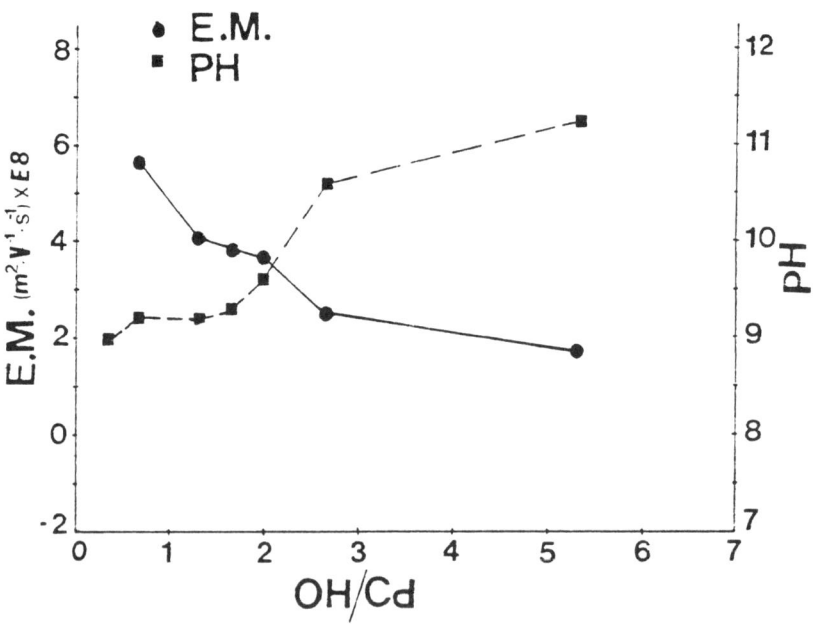

FIGURE 3. Electrophoretic Mobility (E.M.) and pH Behavior in CSTR Tests on $Cd(OH)_2$ Precipitation as a Function of OH/Cd Molar Ratio.

Table 1 summarizes the results of CSTR tests on $Cd(OH)_2$ precipitation performed under dynamic conditions in the absence of an inhibiting agent. Column 8 of Table 1 reports the supersaturation ratio for each run. This value has been calculated from the steady state values of free metal ion activity and pH, according to equation (3).

TABLE 1. *Results of CSTR Tests for $Cd(OH)_2$ Precipitation.*

(1) RUN #	(2) θ s	(3) pH	(4) $[Cd^{++}]$ M	(5) $[Cd]_{sol}$ M	(6) B_o #/ml-min	(7) G um/min	(8) S	(9) Prected Counted %
72	30	9.24	2.00E-4	2.90E-4	3.83E8	0.258	15.2	10.9
73	60	9.14	1.91E-4	2.73E-4	3.36E7	0.253	9.1	16.1
74	90	9.05	1.89E-4	2.34E-4	1.02E7	0.223	6.0	21.9
75	120	8.99	1.91E-4	2.17E-4	3.48E6	0.236	4.6	36.5
76	180	8.96	1.88E-4	2.31E-4	2.06E6	0.193	3.9	54.8
77	30	9.26	2.38E-4	3.26E-4	7.98E8	0.191	19.8	6.7
78	60	9.24	1.99E-4	3.18E-4	2.81E8	0.116	15.1	10.3
79	90	9.22	1.73E-4	2.98E-4	3.42E7	0.117	12.0	15.5
80	120	9.12	1.73E-4	3.05E-4	1.05E7	0.160	7.5	28.8
81	180	9.06	1.68E-4	3.04E-4	-	-	5.6	4.9
82	30	9.54	3.90E-5	9.40E-5	-	-	11.8	0.7
83	60	9.53	3.93E-5	9.80E-5	-	-	11.3	1.9
84	90	9.51	4.31E-5	1.05E-4	-	-	11.3	2.9
85	120	9.47	4.90E-5	1.09E-4	-	-	10.7	2.7
86	180	9.44	5.52E-5	1.09E-4	-	-	10.5	2.8
87	60	9.35	1.02E-4	2.19E-4	3.29E7	0.165	12.8	7.9
88	90	9.32	9.53E-5	2.16E-4	-	-	10.4	4.3
89	180	9.30	9.45E-5	2.12E-4	-	-	9.4	1.5

*For run #72-76, Cd_T = 4.40E-4 M, Base/Cd = 1.01
run #77-81, Cd_T = 5.16E-4 M, Base/Cd = 0.78
run #82-86, Cd_T = 2.06E-4 M, Base/Cd = 1.01
run #87-89, Cd_T = 4.88E-4 M, Base/Cd = 0.86

Columns 6 and 7 of Table 1 report the nucleation rates and crystal growth rates determined from the population balance via a particle counter with 0.5 um detection limit. Figure 4 shows a typical population balance for a CSTR test on $Cd(OH)_2$ precipitation, which is in satisfactory accordance with Randolph and Larson's model (equation (6)). It is important to recall that the population balance shown in Figure 4 is related only to the upper portion of the continuous size distribution of the particles suspended in the reactor. Assuming spherical particles with a density equal to the value for the crystalline $Cd(OH)_2$, a conservative estimate of the mass of the particles counted can be made. When compared with the mass of precipitate determined via filtration with 0.45 um membrane filter, the amount of precipitate counted appears to be usually a small percentage of the precipitate retained by the filter (column 9 in Table 1). Taking into account the conservative assumption for the precipitate density and that both analytical techniques have similar detection limit, one would expect to have the response from the particle counter (in terms of amount of precipitate) higher or, at least, equal to that from the filtration technique. The unexpected small percentage reported in column 9 could be due to the filtration procedure which is likely to build up a cake of precipitate over the membrane filter, so that the effective cut-off of the filter is much lower than the nominal value.

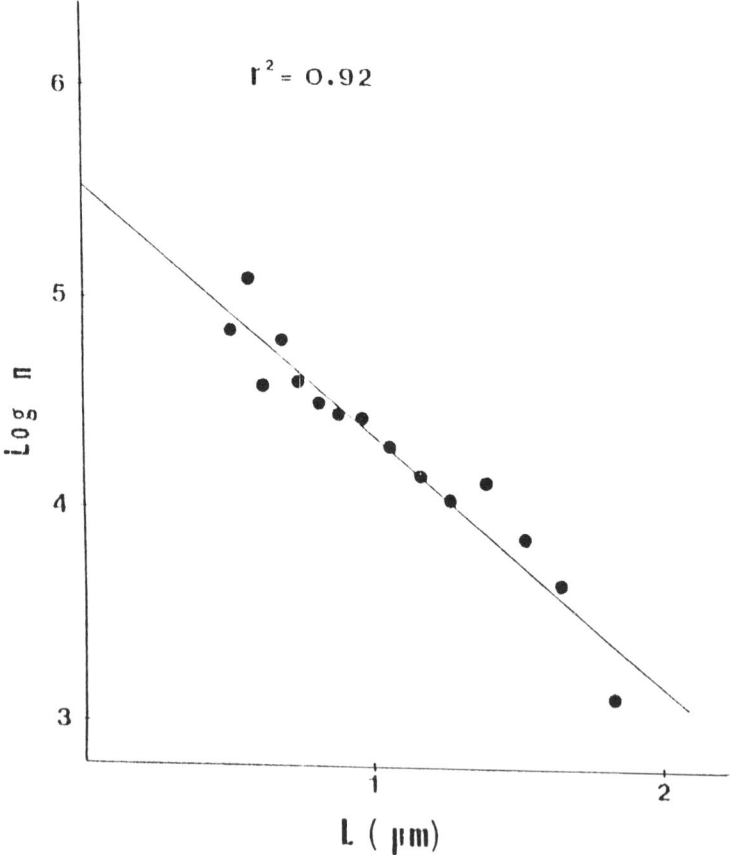

FIGURE 4. Population Balance for a CSTR Test on $Cd(OH)_2$ Precipitation.

When the percentage of the precipitate counted is very small, the population balance determined through the particle counter no longer can be taken as representative of the entire population of particles in the reactor. Taking 5% as a reasonable threshold, only the data related to higher percentages of precipitate counted were calculated for columns 6 and 7 of Table 1.

The nucleation data, selected according to the aforementioned criterion, are represented in Figure 5, which suggests a linear model for nucleation rate versus supersaturation. The least squares method gives the following model:

$$\text{Log}(B_o) = 4.15 + 3.48 \text{ Log}(S) \tag{9}$$

which is statistically highly significant with an F-ratio = 43.6 versus a critical $F_{0.99}$ = 11.3. This model can explain over 90% of the total variability of the nucleation rate data. In agreement with nucleation theory [Walton, 1967; Nyvlt, 1985] the slope of the line of Figure 5 shows that the nucleation rate is dramatically affected by the supersaturation. However, from the physical point of view, it must

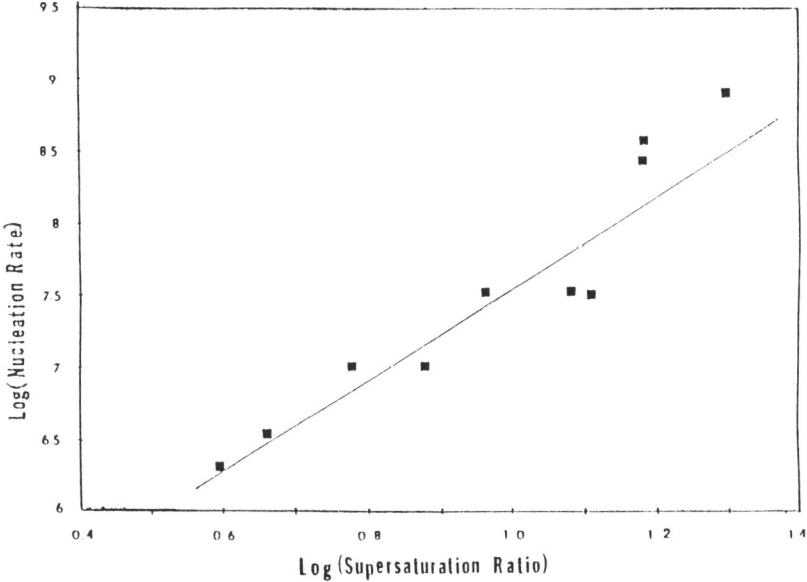

FIGURE 5. Nucleation Rate vs Supersaturation.

be noted that the model allows significant nucleation to occur even in undersaturated solutions (supersaturation ratio $\leq$ 1). This inconsistency can probably be explained by the fact that the power law kinetic expression for nucleation cannot be extrapolated to very large range of supersaturation [Nielsen, 1964].

Unlike nucleation rate, crystal growth does not show a strong correlation with the supersaturation condition, at least for the preliminary set of data presented in Table 1. Seemingly, additional data are needed in order to separate the effect of supersaturation on crystal growth rate from that of other independent variables.

5.2 Metal Precipitation in the Presence of Citric Acid

Typical trends of soluble and free metal concentrations during a batch precipitation test in the presence of citric acid are illustrated in Figures 6 and 7. The experimental conditions for Figure 6 and 7 were: total cadmium concentration = 5.74E-4 M, base/Cd molar ratio = 2.76, citrate/Cd molar ratio = 0.27. In comparison with tests in the absence of citric acid, where quasi-equilibrium conditions are reached within a few minutes, the precipitation reaction is dramatically slowed by the presence of citric acid. Figure 6 shows that, for the experimental conditions described above, the precipitation is still underway after one hour of reaction. The slow increase in turbidity is consistent with the slow decrease of residual filterable metal concentration. It is interesting to compare the trends of pH and free metal ion activity in Figure 7. Because pH remains substantially constant during the test, and assuming that citrate concentration in solution is constant too, the slow decrease in free metal activity cannot be attributable to a change in speciation of the soluble metal, but must be due to the change in total soluble metal concentration shown in Figure 6.

In order to study the effect of citric acid concentration on the precipitation rate, a series of batch tests was performed at the same conditions for all the experimental independent variables, but varying citric acid concentration. Table 2 reports the results of these tests both in terms of final soluble (filterable) metal concentration and in terms of half-reaction times. The latter values are the time length needed to reach 50% of the theoretical precipitation yield expected at the experimental conditions used. They vary from values as low as a few minutes to more than one hour at moderate citrate concentrations (citrate/cadmium molar ratio 0.3 - 0.5).

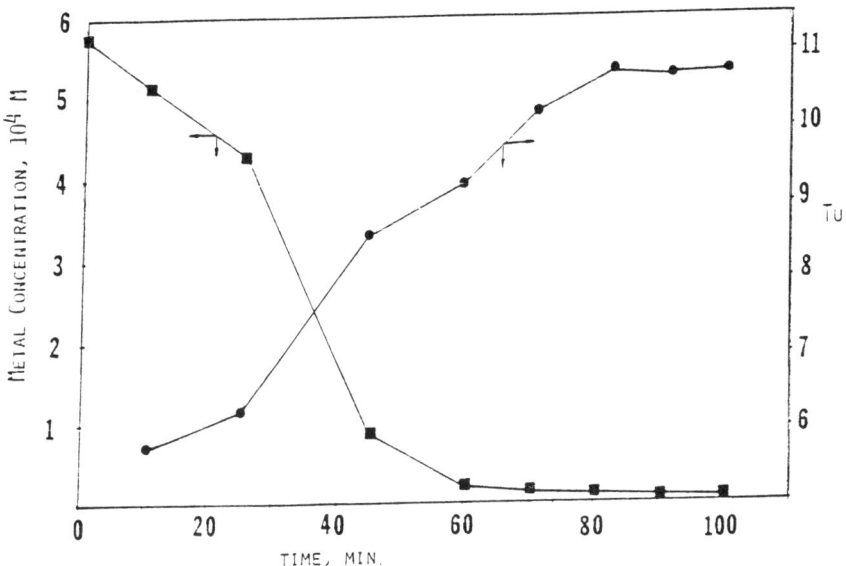

FIGURE 6. Soluble Metal Concentration and Turbidity Trend in a $Cd(OH)_2$ Precipitation Batch Test in the Presence of Citric Acid.

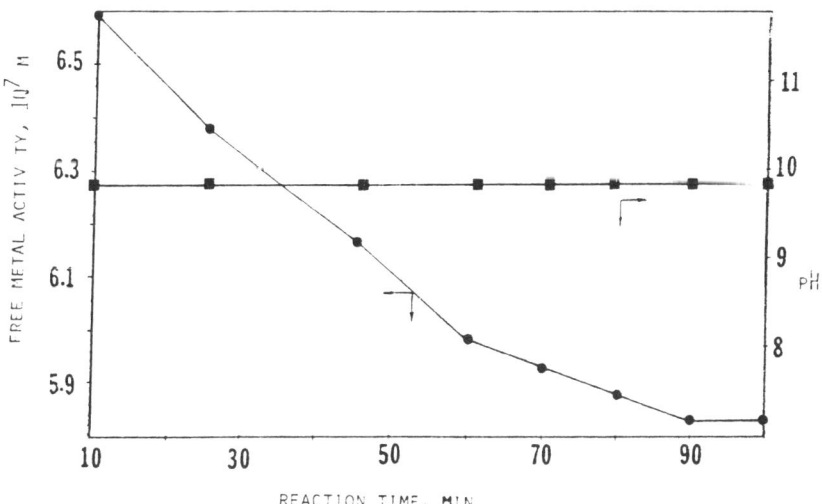

FIGURE 7. Free Metal Activity and pH Trend in a Batch $Cd(OH)_2$ Precipitation Test in the Presence of Citric Acid.

TABLE 2. *Results of Batch Precipitation Tests in the Presence of Citric Acid.*

Run #	Cd_T M	OH/Cd	Cit./Cd	Cd_{sol}	$t_{1/2}$ min
1	8.68E-04	1.15	0.14	4.60E-04	2.0
2	9.44E-04	1.06	0.28	6.63E-04	8.8
3	5.74E-04	2.00	0.27	5.00E-05	12.5
4	5.74E-04	2.76	0.27	6.00E-06	32.5
5	5.74E-04	3.29	0.27	3.90E-04	---
6	5.00E-04	1.82	0.20	9.00E-05	2.5
7	5.00E-04	1.82	0.30	1.76E-04	3.8
8	5.00E-04	1.82	0.40	3.39E-04	4.4
9	5.00E-04	1.82	0.50	4.97E-04	>120

Figure 6 seems to suggest that citric acid inhibits the precipitation of $Cd(OH)_2$. Two hypotheses for the inhibiting effect of citric acid were evaluated:
- slow release of free Cd cation from a Cd-citrate complex;
- particle growth inhibition.

For an explanation of the mechanism of inhibition, the complexation reaction was, first of all, taken into account. Using the computer program MINEQL (Westall,1975) and stability constants from Sillen and Martell (1971), the theoretical speciation of Cd(II) in an influent solution, containing $Cd(NO_3)_2$ at 5.74E-4 M and citric acid/Cd ratio equal 0.27, gives only 14.7% of metal complexed with citric acid. After mixing this solution with an equal volume of NaOH at OH/Cd ratio 2.76, more than 85% of Cd is theoretically hydrolyzed. Even assuming that the Cd-citrate complexation reaction is kinetically preferred above the hydrolysis reaction, only 27% of the metal could be complexed with citrate anion after flash mixing. Hence the hypothesis of slow release of free Cd from Cd-citrate complexes can not be invoked to explain the inhibiting effect of citric acid under these experimental conditions. An alternative hypothesis, which is worth checking with additional batch and CSTR tests, seems to be the adsorption of citrate anion at the particle/solution interface.

On the basis of such considerations, the inhibition of particle growth rate was deemed the most plausible explanation of the low precipitation rates experienced in the presence of citric acid. Crystal growth models suggest that the overall precipitation reaction rate can be either diffusion controlled or surface reaction controlled

[Doremus, 1970]. Nielsen (1964) proposed a so-called chronomal method, in order to ascertain whether the growth is diffusion controlled. This method assumes that the particles produced during a batch precipitation test are spherical, monodisperse, and that particle concentration remains constant. The diffusion chronomal, I_D, is given by:

$$I_D = \alpha^{-1/3}(1-\alpha)^{-1} d\alpha \tag{10}$$

where: α = degree of reaction

(i.e. $(Cd_T - Cdsol(t))/(Cd_T - Cdsol(eq))$

For a diffusion-controlled process, a plot of I_D vs time should give a straight line. Figure 8 shows the chronomal vs time plot obtained from the data of Figure 6. The nonlinearity of this plot suggests a growth mechanism in which surface reaction plays a major role. In this conditions, the inhibiting effect of citric acid could be most likely due to the adsorption of citrate anion on the precipitate surface. This hypothesis will be thoroughly explored with further batch and CSTR experiments performed in the presence of citric acid.

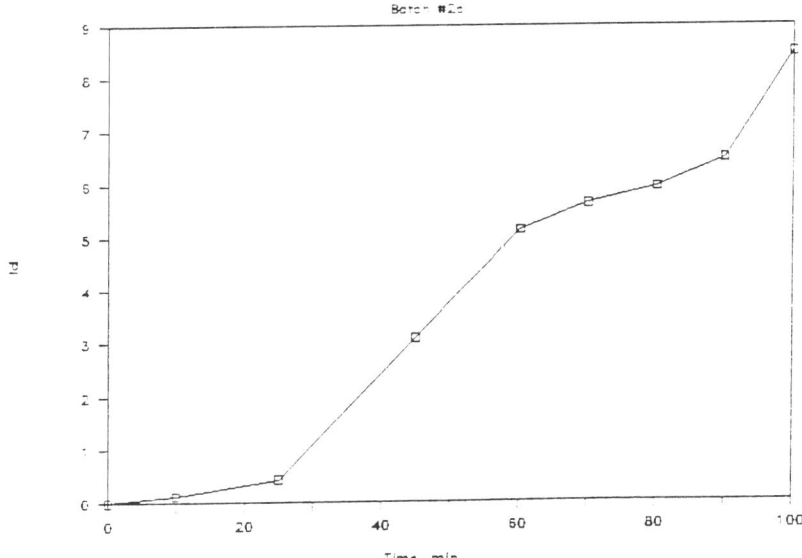

FIGURE 8. Diffusion Chronomal vs Time for a Batch $Cd(OH)_2$ Precipitation in the Presence of Citric Acid.

6. CONCLUSIONS

From the preliminary results obtained in the experimental study of $Cd(OH)_2$ precipitation, the following conclusions can be drawn:

-$Cd(OH)_2$ precipitation is very fast in the absence of inhibiting agent. In this case CSTR tests seem to be more appropriate than batch tests for kinetic studies.

-At low base/metal molar ratios, the precipitate is stabilized in a colloidal form, by a positive surface charge.

-In the absence of particle aggregation (low base/metal molar ratios and short residence times), the particles population balance in the CSTR reactor can be used to determine nucleation and crystal growth rates.

-In agreement with classical nucleation theory, nucleation rate appears to be highly dependent on the supersaturation conditions in the reactor, which can be determined through the potentiometric measurements of pH and free metal ion activity.

-$Cd(OH)_2$ precipitation is slowed dramatically in the presence of citric acid. Theoretical considerations tend to rule out Cd-citrate complexation as a plausible explanation of the inhibiting effect.

-The analysis of batch precipitation data suggests that the mechanism of inhibition may be surface adsorption which depresses crystal growth rate.

Additional batch and CSTR experiments are underway in order to establish a satisfactory nucleation and crystal growth model for $Cd(OH)_2$ precipitation both in the presence and in the absence of an inhibiting agent.

ACKNOWLEDGMENTS

The authors wish to acknowledge the financial support of the U.S. Environmental Protection Agency through the Industrial Waste Elimination Research Center of Illinois Institute of Technology.

REFERENCES

1. Baes, C.F. and R.E. Mesmer. *The Hydrolysis of Cations*, (John Wiley and Sons, New York, N.Y., 1976).

2. Bales, R. C. "Surface Chemistry in Water Treatment: Reactions at the Solid - Liquid Interface," *J. AWWA*, 59, (1986).

3. Bradley, G.W.,et al., "Investigation of Ammonium Citrate Cleaning Solvents," in *Proceedings of the 36th International Water Conference*, November 4-6, Pittsburgh, PA. (1975).

4. Blume, W.J. "Citric Acid Based Chemical Cleaning Process," *Meter. Perform.*, 16(3):15-19 (1977).

5. Buckley, H.E. *Crystal Growth*, (Wiley, N.Y.,1957).

6. Becker, R. and W. Doering. "The Kinetic Treatment of Nuclear Formation in Supersaturated Vapors," *Annales der Physik*. 24:719-752 (1935).

7. Doremus, R.H. "Precipitation and Crystal Growth from Solution," *Croatica Chemica Acta*, 42:293-297 (1970).

8. EPA. "Development Document for Proposed Existing Source Pretreatment Standards For The Electroplating Point Source Category," EPA 440/1-78/085 (1978).

9. Fassett, D.W. "Cadmium," in *Metals in the Environment*, ed. H. A. Waldron, (London: Academic Press, 1980) pp. 61-110.

10. Feitknecht, W. and P. Schindler, "Solubility Constants of Metal Oxides, Metal Hydroxides and Metal Hydroxide Salts in Aqueous Solution," *Pure Appl. Chemistry*, 6:130-199 (1963).

11. Frenkel, J. "Kinetic Theory of Ligands," *Oxford University Press*, Chapter VII, New York, N. Y. (1946).

12. de Hck, H., R.J. Stol and P.L. de Bruyn, *Journ. Coll. and Interfacial Science*, 64:72-89 (1978).

13. de Hek, H. et al. "Hydrolysis-Precipitation Studies of Aluminum (III) Solutions. 3. The Role of Sulfate Ion," *J. Am. Water Works Assoc.* 651:562-568 (1978).

14. Helz, G.R. and L.M. Horzempa, *Water Research*, 17:167-172 (1983).

15. Jenke, D.R. and F.E. Diebold. "Recovery of Valuable Metals from Acid Mine Drainage by Selective Titration," *Water Research*, 17:1585-1590 (1983).

16. Karra, S.B., C.N. Haas, V. Tare and H.E. Allen, "Kinetic Limitations on the Selective Precipitation Treatment of Electronic Wastes," *Water, Air and Soil Pollution*, 24:253-265 (1983).

17. Levenspiel, O. *Chemical Reaction Engineering*, 2nd edition, (John Wiley and Sons, New York, N.Y. 1972).

18. Nielsen, A. E. *Kinetics of Precipitation*, (Macmillan, N.Y. 1964).

19. Nyvlt, J. et al. *The Kinetics of Industrial Crystallization*, (Elsevier, Amsterdam, 1985).

20. Packter, A. and J. Jakubowski. "The Precipitation of Sparingly-soluble Metal Chromates from Aqueous Solution in the Presence of Chelating Organic Anions: Induction Periods and Nucleation Rates," *Cryst. Res. Technol.*, 20:1063-1072 (1985).

21. Parks, G.A. *Chemical Reviews*, 65:178-198 (1965).

22. Patterson, J.W. *Industrial Wastewater Treatment Technology*, 2nd edition, (Butterworth Publishers, Stoneham, MA., 1985).

23. Patterson, J.W. "Metals Speciation and Recovery," in *Metals Speciation, Separation and Recovery*, Patterson and Passino, Eds., (Lewis Publishers Inc., Chelsea, Michigan, 1987).

24. Peters R.W. and Y. Ku. in *Metal Speciation, Separation, and Recovery*, Patterson J.W. and Passino R.," Eds., (Lewis Publishers Inc., Chelsea, MI., 1987).

25. Randolph, A.D. and M.A. Larson "Theory of Particulate Processes," (Academic Press, New York, N.Y. 1971).

26. Sillen, L.G. and A.E. Martell. "Stability Constant of Metal Ions Complexes," (Chemical Society, London. 1971).

27. Smith, R.M. and A.E. Martell. "Critical Stability Constants," *Inorganic Complexes*, Volume IV. (Plenum, N.Y., 1976).

28. Snoeyink, V.L. and D. Jenkins, *Water Chemistry*, (John Wiley and Sons, New York, N.Y. 1980).

29. Stumm, M. and J.J. Morgan, *Aquatic Chemistry*, 2nd ed. (John Wiley and Sons, New York, N.Y. 1981).

30. Turnbull, D. and J.C. Fischer. "Rate of Nucleation in Condensed Systems," *J. Chem. Phys.* 17:71-73 (1949).

31. Verma, A. R, *Crystal Growth and Dislocations*, (Butterworth, London, 1953).

32. Volmer, M. *Kinetik der Phasenbildung*, (Edwards Bros., Ann Arbor, MI., 1926).

33. Walton, A.G. "Nucleation of Crystals from Solution," *Science*, 148:601-607 (1965).

34. Walton, A.G. *The Formation and Properties of Precipitates*, (John Wiley and Sons, New York, N.Y. 1967).

35. Westall, J.C., J.L. Zachary and F.M. Morel, "MINEQL, a Computer Program for the Calculation of Chemical Equilibrium Speciation of Aqueous Systems," (Technical Note Number 18), Ralph M. Parsons Laboratory, Massachusetts Institute of Technology, (Cambridge, Mass. 1976).

DISCUSSION OF:
NUCLEATION AND CRYSTAL GROWTH STUDIES ON PRECIPITATION OF CADMIUM HYDROXIDE FROM AQUEOUS SOLUTIONS

Roger A. Minear
University of Illinois, Urbana, Illinois

INTRODUCTION

What the paper is. The paper by Patterson, Luo, Marani and Passino is an interesting study of the cadmium hydroxide precipitation process. The authors are to be complimented on their work. As the authors state, the results are preliminary but may, as the authors state further, improve the knowledge of fundamental principles regarding metal hydroxide precipitation. I cannot disagree. However, I view the role of formal discussor to be that of a provocatuer in the spirit of challenging scientific investigation and, thus, stimulating constructive discussion. I intend for my discussion to be probing toward that end, not antagonistic.

What the paper is not. The paper is not particularly related to solid/liquid separation process, although the authors state this as one of the study designs. For this reason, I would like to provoke response and discussion both from the authors and the audience as to how these results will relate to improved separation processes.

NUCLEATION AND CRYSTAL GROWTH

I agree with the authors' observation that any practical consideration of nucleation must focus on heterogeneous nucleation. However, the question I did not see addressed in this research is the matter of initial heteronuclei or microparticle concentration and the effect of variation in this concentration on the observed behavior, if any. While Stumm and Morgan (1981) emphasize that the influence of such particles on heterogeneous nucleation probably relates more to lattice matching and atomic distances rather than chemical similarity, the question remains as to how this variable can be controlled. Obviously, the immediate concern is with the experimental systems under study and reproducability of experimental results, but, extrapolation to "real" systems where microparticle variation is likely to be much greater could be very difficult. While the authors express a desire to go beyond empiricism in precipitation treatment research, this factor and other solution variables represent formidable barriers to transfer of laboratory results to practical treatment application.

In the research, use of double distilled water may aid in reduction of background particles, but 0.2 µm filtration will not control particles. For example, my research group has experienced accumulation of calcium carbonate particles down stream of a 0.2 µm filter in ultrafiltration concentrates of a hard water stream when cation exchange did not precede the ultrafiltration step. Since we were using a 1000 dalton ultrafiltration membrane, we know it was not a problem of increasing ion concentration with subsequent precipitation.

Unfortunately, with a lower detection limit of 0.5 µm particle size with the research instrumentation, no direct information on the quantity of particles in the original experimental system is available. The question to be raised is what effect could this have on the k and i values in equation 6 of the paper? ($B_Q = ks^i$) Are the initial heteronuclei concentrations a factor in defining the slope and hence intercept of figures 4 and 5 of the paper? Uncertainty in N_g is large. Is this a problem in translation of the work to other laboratories? To real treatment applications? So, my principal question is how might the importance of background solution microparticles be evaluated? Where does the research go from here to deal with variations in initial hetero-nuclei concentrations. In the real world, these concentrations will likely be quite variable. As an aside, I might suggest Light Scattering measurements on the starting solutions to obtain background values for the hetero-nuclei concentrations.

COMPLEXATION EFFECTS

With respect to the complexation effects, I have to agree with the authors that particle growth inhibition is the only logical mechanism since the citrate was always present in considerably less than stoichiometric amounts. The question that has not been addressed is where the citrate forms complexes; on the cadmium hydroxide surface or the original heteronuclei or, more likely, both. The attempt to use the "chromal method" to differentiate between diffusion and surface reaction control of crystal growth is inconclusive. I say this primarily because the assumptions of the method are likely not met, namely that spherical, monodisperse particles are present which remain at a constant number concentration.

Two alternate approaches are suggested. One is to evaluate crystal growth rate constants at varying temperatures in the presence of citrate and evaluate the activation energy, i.e., plot of ln k versus 1/T. This information should provide some insight as to whether chemical reaction or diffusion is controlling. High activation energy is commonly accepted as evidence of chemical reaction control and vice versa.

Alternatively (or in addition) maintain control of the diffusion layer thickness by specific experimental design, i.e., mixing variations. This approach could also aid in differentiating the two processes.

HYDROXIDE VERSUS CARBONATE

Another question that has not been addressed is whether the hydroxide is in fact the desirable precipitate or even that which is formed under practical conditions. The theoretical solubility curve of Figure 1 for $Cd(OH)_2(S)$ varies slightly with choice for K_{SO} and the ionic strength. However, in the presence of carbonate, the thermodynamically stable solid form is the carbonate, $Cd\ CO_3(S)$.

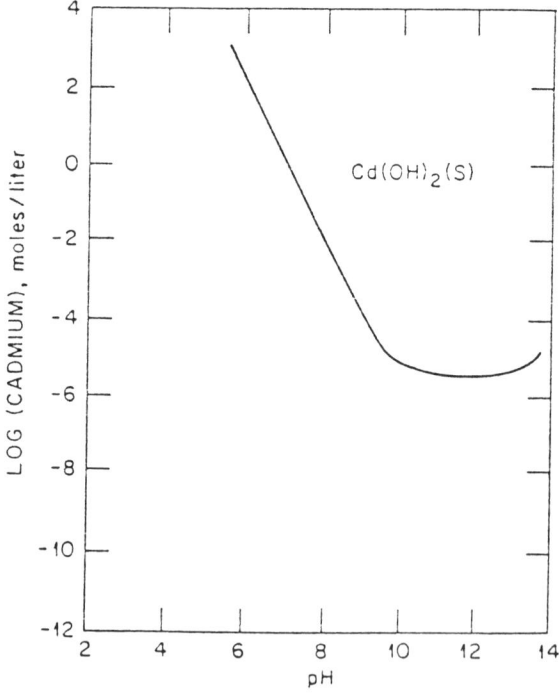

FIGURE 1.   Theoretical Cadmium Hydroxide Solubility Curve.

Figure 2 demonstrates the solid phase dominance regions for a very high carbonate concentration of 0.063M or an alkalinity of 6300. Variation of solubility with additional increases in ionic strength is also represented.

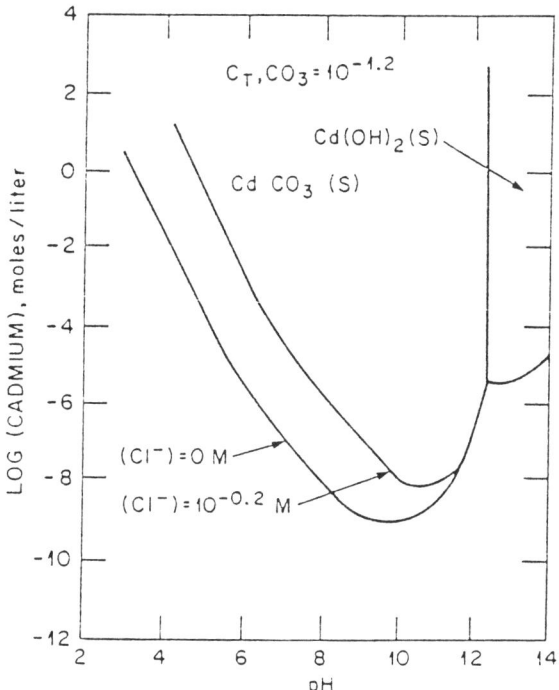

FIGURE 2. Theoretical Cadmium Carbonate/Cadmium Hydroxide Phase Diagram.

To illustrate the importance of alternative solid phase considerations, I want to turn to some earlier work done in my laboratory.

Butcher's (1985) work examined different precipitation environments in what was principally a study of cadmium cyanide waste treatment by alkaline chlorination. As a part of the work, he studied alkaline precipitation in the absence of cyanide at varying carbonate concentrations and with both sodium and calcium hydroxides. Figure 3 presents the experimental design. Unfiltered samples were intended to be a crude simulation of natural clarification by gravity setting.

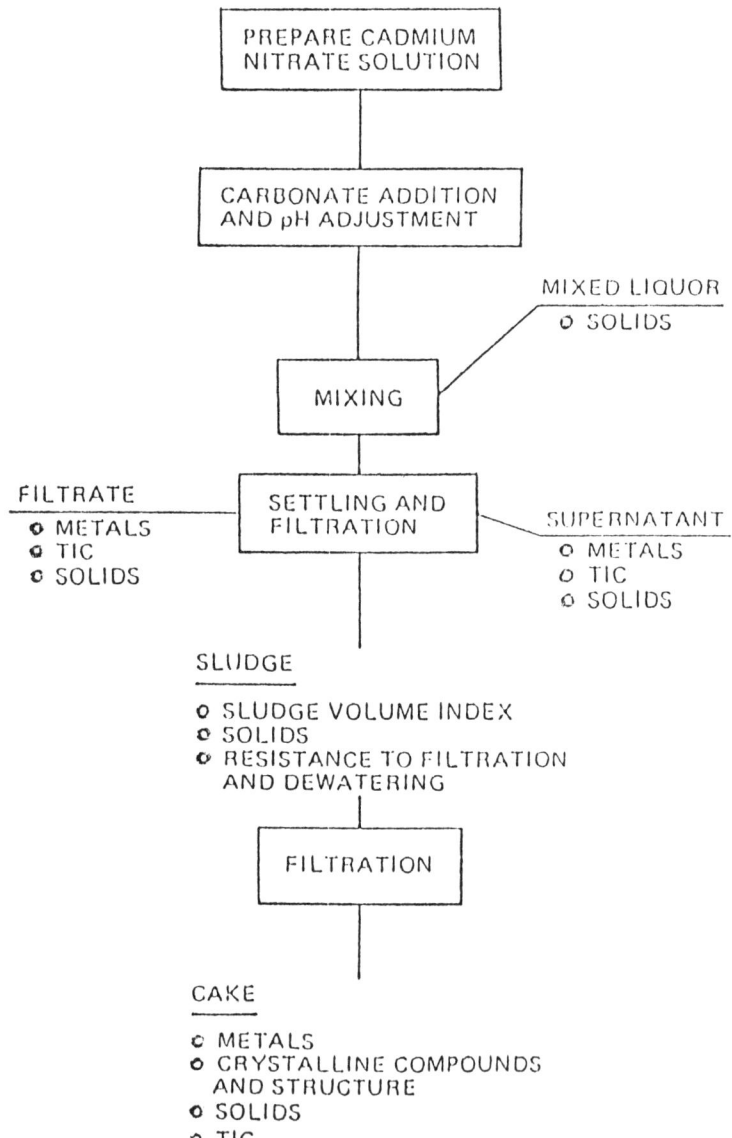

FIGURE 3. Experimental Design, Non-Cyanide Phase.

Figures 4 and 5 are representations of the theoretical solubilities at varying carbonate concentrations for $Cd(OH)_2$ and $CdCO_3$ (lines) and the theoretical values of actual samples based upon the measured pH and carbonate concentrations. Ionic strength of the solutions was included in the calculations. Measured cadmium concentrations (determined by atomic absorption spectroscopy) are superimposed on the diagram. Figure 4 has no calcium in the system. Figure 5 contains calcium. These additions were to simulate use of NaOCl and $Ca(OCl)_2$ respectively.

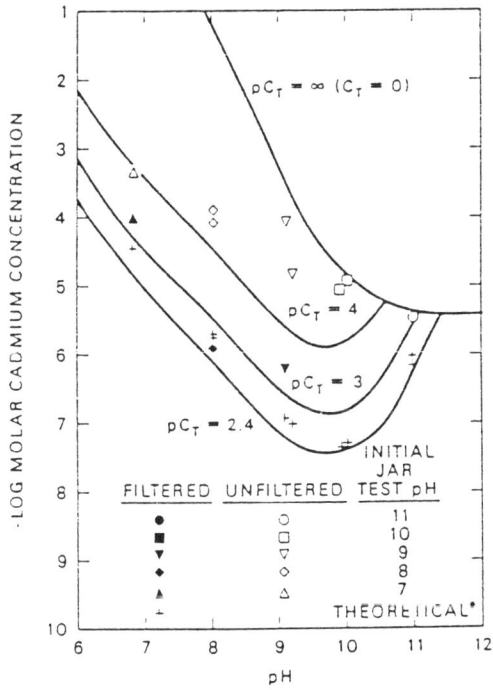

FIGURE 4. *Comparison of Theoretical Solubility Values with Experimental Cadmium Date Obtained in Dilute Cadmium Jar Tests with Sodium Chloride Additions. (*Theoretical Cadmium Concentration Representing Individual Jar Test C_T and pH values.)*

What is important to note is the evidence that carbonate solids appear to be formed, especially at lower pH but even at high pH from the solubility relationships. X-ray diffraction results on the solid phases shown in Table 1, strongly support this conclusion since Otavite was in evidence in all samples. The dominance of Calcite for the $CaCl_2$ systems was expected.

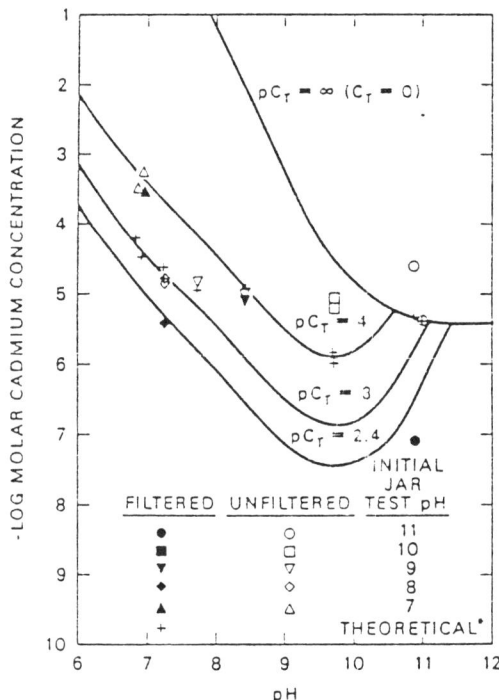

FIGURE 5. *Comparison of Theoretical Solubility Values with Experimental Cadmium Data Obtained in Dilute Cadmium Jar Tests with Calcium Chloride Additions. (*Theoretical Cadmium Concentration Representing Individual Jar Test C_T and pH Values.)*

Reaction and holding times in these systems are no greater than the citrate studies of Patterson, Luo, Marani and Passino, so there is some basis for comparison.

Interestingly enough, when actual $Cd(CN)_2$ solutions were treated with NaOCl (Figure 6) and $Ca(OCl)_2$ (Figure 7), filtered samples implied cadmium carbonate formation at all pH values. However, x-ray diffraction results presented in Table 2, which were conducted on the settled solids, do show the presence of cadmium hydroxide forms at pH 11.

The implication is that in real systems, some carbonate is always present and it would seem pertinent to direct future effort to this area.

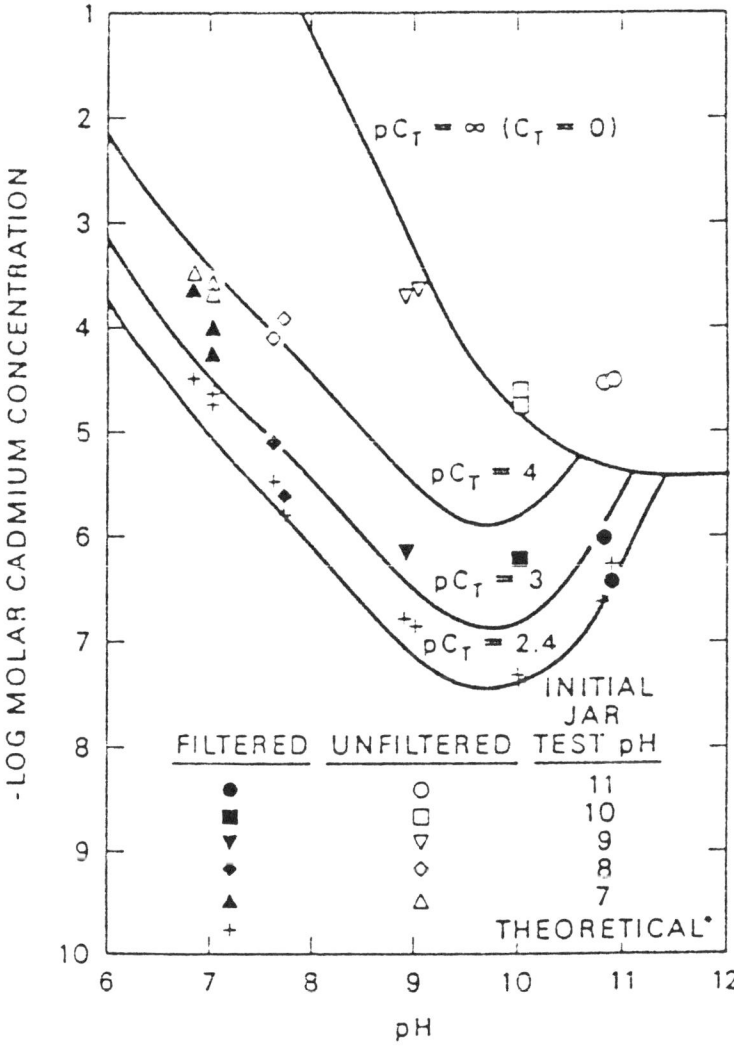

FIGURE 6. Comparison of Theoretical Solubility Values with Experimental Cadmium Data Obtained in Cadmium Cyanide Jar Tests with Sodium Hypochlorite Additions. (*Theoretical Cadmium Concentrations Representing Individual Jar Test C_T and pH Values.)

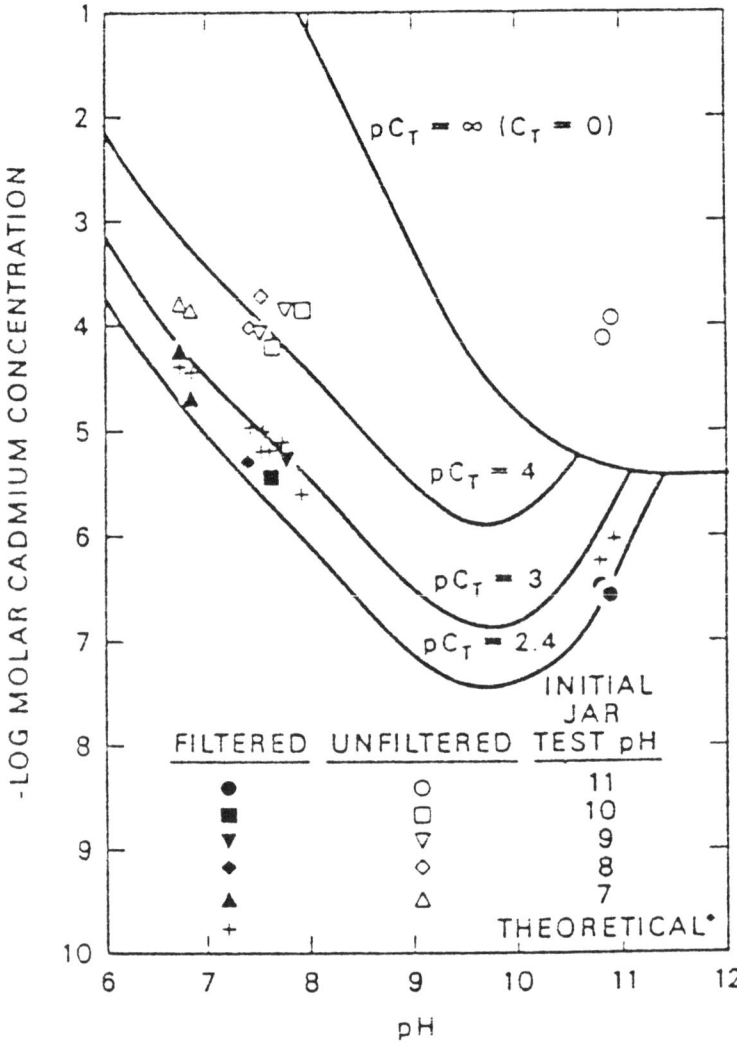

FIGURE 7. Comparison of Theoretical Solubility Values with Experimental Cadmium Data Obtained in Cadmium Cyanide Jar Tests with Calcium Hypochlorite Additions. (*Theoretical Cadmium Concentration Representing Individual Jar Test C_T and pH Values.)

TABLE 1. Results of Analyses on Precipitates from Dilute Cadmium Jar Tests.

Jar Test pH	%[1] Cadmium	% Calcium	% Sodium	X-ray Diffraction Patterns		
				1st	2nd	3rd
NaCl						
11				Otavite[2]		
11	61.0	19.5[3]	0.06	Otavite		
10	58.5	0	0.06	Otavite		
10				Otavite		
CaCl$_2$						
11	20.6	27.9	0.04	Calcite[4]	Otavite	(5)
11	21.2	28.5	0.05	Calcite	Otavite	β-Cd(OH)$_2$
10	21.8	26.2	0.10	Calcite	Otavite	
10	21.9	26.3	0.09	Calcite	Otavite	
9	23.9	23.9	0.03	Calcite	Otavite	
8	40.7	14.0	0.02	Calcite	Otavite	

[1] Percent by weight in dried precipitate (typical).
[2] Otavite is the crystalline form of cadmium carbonate.
[3] Possible contamination of sample from preparing XRD samples.
[4] Calcite is one crystalline form of calcium carbonate.
[5] This secondary pattern could not be identified. Possibly a form of cadmium hydroxide.

TABLE 2. Results of Analyses on Precipitates from Cadmium Cyanide Jar Tests.

Jar Test pH	% Cadmium	% Calcium	% Sodium	X-ray Diffraction Patterns		
				1st	2nd	3rd
NaOCl						
11	73.0	0	0.072	γ-Cd(OH)$_2$	β-Cd(OH)$_2$	Otavite
11	71.0	0	0.060	γCd(OH)$_2$	Otavite	β-Cd(OH)$_2$
10	64.9	0.017	0.345	Otavite		
10	65.6	0.0054	0.135	(1)		
9	65.0	0	0.113	Otavite		
9	65.0	9.0	0.088	(1)		
8	62.4	0	0.086	(1)		
7	62.1	0	0.118	(1)		

REFERENCES

1. Butcher, B.T. "Destruction of Cadmium Cyamide Waste By Alkaline Chlorination Treatment," MS Thesis, University of Tennessee, Knoxville, TN (1985).

2. Stumm, W. and J.J. Morgan. *Aquatic Chemistry*, (New York: John Wiley, 1981). 780 pp.

EFFECT OF SPECIATION ON THE RATES OF OXIDATION OF METALS

Frank J. Millero,
Rosenstiel School of Marine and Atmospheric Science, University of Miami, Miami, Florida

1. INTRODUCTION

Trace metals can participate in a wide range of chemical and biological reactions in industrial and natural waters. The speciation of metals in these waters is important because the different forms of many metals have different chemistries. For example, the biological toxicity and availability is strongly affected by the redox state and chemical form of a given metal. Copper, for example, is toxic to many marine organisms when it is in the free or uncomplexed state. Iron and manganese are needed by phytoplankton for growth. The oxidized forms of these metals are insoluble and are quickly scavenged from surface waters. Although thermodynamic speciation calculations can yield the most probable redox and complexed form of a given metal, the redox form is normally controlled by the rates of oxidation and reduction. Recent kinetic measurements [1-9] have been made on the oxidation of Cu(I) and FE(II) with O_2 and H_2O_2 in natural waters. These studies have demonstrated that the formation of ion-pairs can either accelerate or decrease the rates of oxidation. The formation of $FeOH^+$ and $Fe(OH)_2^{\circ}$ accelerates the rate of oxidation of Fe(II); while the formation of $CuCl^{\circ}$, $CuCl_2^-$ and $CuCl_3^{2-}$ causes the rates of oxidation of Cu(I) to decrease. In the present paper we will demonstrate how the speciation of metals affects the rates of oxidation of metals and nonmetals and how these effects can be incorporated into the rate equations.

2. BACKGROUND

A knowledge of ionic interactions is necessary to obtain an understanding of the chemical processes that occur in natural and industrial waters. Short range ion-water and ion-ion interactions have been shown to affect equilibrium processes such as acid-base, solubility and redox reactions. The physical-chemical properties of natural waters are also affected by ionic interactions. Since ionic interactions affect the state or speciation of metal ions, many have suggested that biological activity if affected by ionic interactions. In recent years we have shown that ionic interactions can affect the rates of chemical processes in natural waters.

The effect of ionic interactions on the rates of ionic reactions of metals and nonmetals can be affected by the major and minor ionic components of natural waters. The rates of oxidation of metals can be affected by the anions (Cl^-, OH^-, SO_4^{2-}, HCO_3^-) in aqueous solutions. For example, the formation of the ion pairs.

$$Cu+ \; + \; Cl^- \; \longrightarrow \; CuCl^° \quad (1)$$

$$Fe^{2+} \; + \; SO_4^{2+} \; \longrightarrow \; FeSO_4^° \quad (2)$$

causes the rates of oxidation of Cu(I) and Fe(II) with O_2 and H_2O_2 to decrease. The formation of the ion pairs

$$Cu^+ \; + \; HCO_3^- \; \longrightarrow \; CuHCO_3^° \quad (3)$$

$$Fe^{2+} \; + \; OH^- \; \longrightarrow \; FeOH^+ \quad (4)$$

causes the rates of oxidation of Cu(I) and Fe(II) to increase. The rates of oxidation of non metals can be affected by the cations (H^+, Mg^{2+}, Ca^{2+}, etc.). For example, the formation of the ion complexes

$$HS^- \; + \; H^+ \; \longrightarrow \; H_2S \quad (5)$$

$$O_2^- \; + \; Mg^{2+} \; \longrightarrow \; MgO_2^+ \quad (6)$$

causes the rates of oxidation of H_2S to decrease and the disproportionation of HO_2 to H_2O_2 to decrease. In the next section we will give some examples of how these ionic interactions affect the rates and how one can account for these effects.

3. EFFECT OF pH ON THE RATES OF OXIDATION

The pH of a solution can affect the state of metals and non-metals in aqueous solutions. The increase of the pH can affect metals by providing ligands such as OH^- and CO_3^{2-} that can form ion pairs. The increase of the pH can affect the form of an acid by causing it to ionize. The effect of pH on the oxidation of H_2S

$$H_2S + O_2 \dashrightarrow \text{Products} \quad (7)$$

$$d[H_2S]_T/dt = -k[H_2S]_T[O_2] \quad (8)$$

is shown in Figure 1 [10]. At a pH below 4, the rate of oxidation

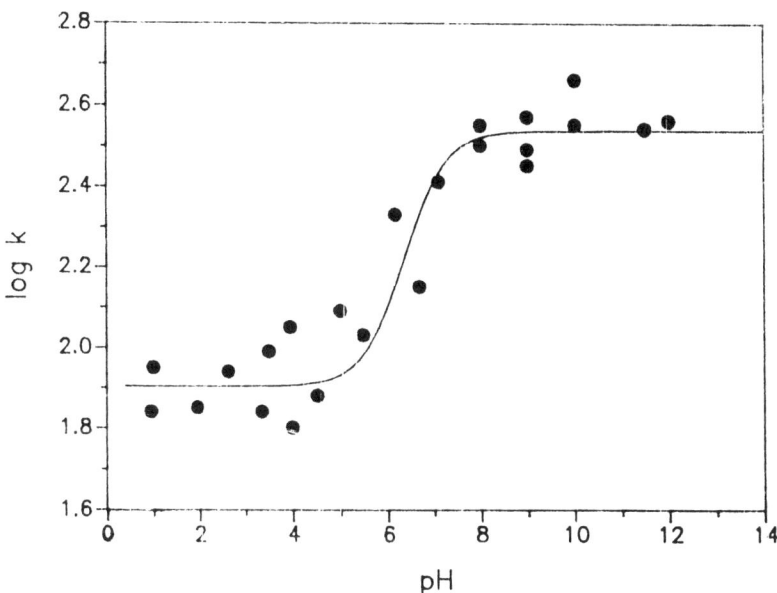

FIGURE 1. Effect of pH on the Oxidation of H_2S with O_2.

is not affected by pH; however, the rate increases dramatically between a pH of 6 to 8. Above a pH of 8, the rate is unaffected by an increase in the pH. This change in the rate as a function of pH is related to the ionization of H_2S.

$$H_2S \dashrightarrow H^+ + HS^- \quad (9)$$

The increase in the rate is due to the formation of HS- which is more reactive than H_2S. The overall rate constant (Equation 8) is given by

$$K = K_0 \alpha_{H_2S} + K_1 \alpha_{HS} \quad (10)$$

where α_i represents the fraction of species i, and K_0 and K_1 are the rate constants for the oxidation of H_2S and HS^-, respectively. Combining the dissociation constant for the ionization of H_2S

$$K_1 = [H^+][HS^-]/[H_2S] \tag{11}$$

with Equation 10 gives

$$k/\alpha_{H_2S} = k_0 + k_1 K_1/[H^+] \tag{12}$$

A plot of k/α_{H_2S} versus $K_1/[H^+]$ can be used to determine k_0 and k_1, the rate constant for the oxidation of H_2S and HS^-, respectively. The straight line fit gives $K_0 = 80$ h M^{-1} and $K_1 = 344$ h M^{-1}. The smooth curve in Figure 1 gives the values of K calculated from Equation 10.

The oxidation of H_2S with hydrogen peroxide

$$H_2S + H_2O_2 \longrightarrow \text{Products} \tag{13}$$

is given by the rate equation

$$d[H_2S]_T/dt = -K[H_2S]_T[H_2O_2] \tag{14}$$

The values of log k as a function of pH for the oxidation of H_2S with H_2O_2 are shown in Figure 2 [11].

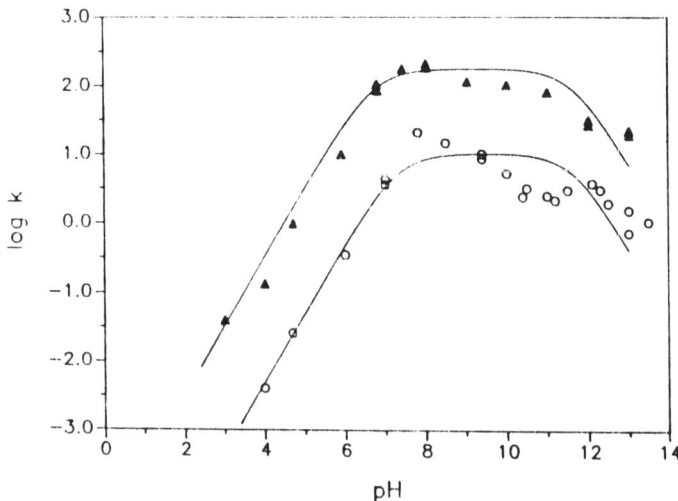

FIGURE 2. *Effect of pH on the Oxidation of H_2S with H_2O_2 at 45°C (Δ) and 5°C (O).*

From a pH = 2 to 8, the rate increases in a near linear manner with an increase in pH. As with the oxidation with oxygen, this increase is due to the ionization of H_2S (Equation 9). The values of log k determined in this study do not appear to level off at a low pH as found for the oxidation with O_2 (see Figure 1). A plot of K/α_{H_2S} versus $K_1/[H^+]$ for the oxidation of H_2S with H_2O_2 gives $K_O = 0$ min^{-1} M^{-1} and $K_1 = 36.2$ min^{-1} M^{-1}. The values of log k at a pH above 8 decrease, unlike the values for the oxidation with O_2. This decrease at a pH greater then 10 can be attributed to the ionization of hydrogen peroxide

$$H_2O_2 \longrightarrow H^+ + HO_2^- \tag{15}$$

which has a pK = 11.6 for the ionization. If the reaction of HS$^-$ with HO_2^- is small, the decrease in k can be attributed to

$$k = k_1 \alpha_{H_2S} \alpha_{H_2O_2} \tag{16}$$

where

$$\alpha_{H_2O_2} = [H^+]/([H^+] + K_{H_2O_2}) \tag{17}$$

The addition of this correction term improves the fit above a pH = 10, but does not explain the nearly linear decrease above pH = 8. Other factors such as the formation of polysulfide ions (HS_n^-) and S^{2-} may be important.

The effect of pH on the oxidation of Fe(II) with O_2 and H_2O_2 has also been studied [2,3,5,12,13]. The pseudo first order rate constant for the oxidation of Fe(II) with O_2

$$d[Fe(II)]/dt = -k_1[Fe(II)] \tag{18}$$

has been shown by Stumm and Lee [14] to be second order with respect to pH in water. This second order dependence of the oxidation of Fe(II) has also been shown to be obeyed in seawater, NaCl and NaClO$_4$ as a function of temperature and ionic strength [3,5,12]. The NaCl and NaClO$_4$ results are shown in Figure 3. The second order dependence of the rate of oxidation of Fe(II) with O_2 can be attributed to the rate determining step being

$$Fe(OH)_2^o + O_2 \longrightarrow Fe(OH)_2^+ + O_2^- \tag{19}$$

Since the concentration of Fe(OH)$_2$° is proportional to the $[H^+]^2$ or $[OH^-]^2$, the overall rate constant is also proportional to $[H^+]^2$.

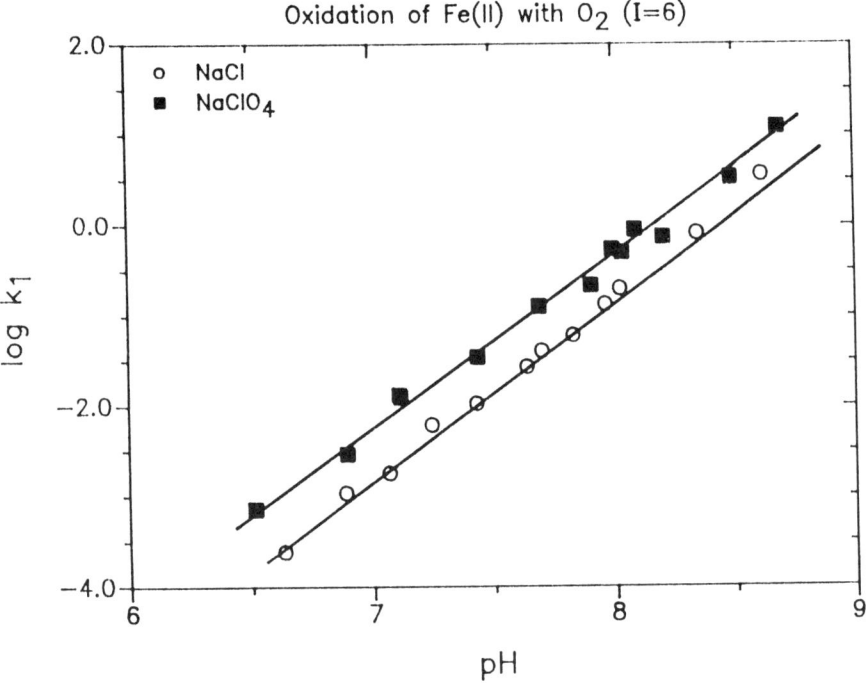

FIGURE 3. *Effect of pH on the Oxidation of Fe(II) with O_2.*

The pseudo first order rate constant for the oxidation of Fe(II) with H_2O_2 shows a more complicated pH dependence (see Figure 4). In the low pH range the rate is independent of pH and gradually become directly proportional to the $[H^+]$. These results indicate that the rate of oxidation of Fe(II) with H_2O_2 is due to the oxidation of

$$Fe^{2+} + H_2O_2 \xrightarrow{k_0} \text{Products} \tag{20}$$

$$Fe(OH)^+ + H_2O_2 \xrightarrow{k_1} \text{Products} \tag{21}$$

The overall rate constant is given by

$$k = k_0 \, \alpha_{Fe} + k_1 \, \alpha_{Fe(OH)} \tag{22}$$

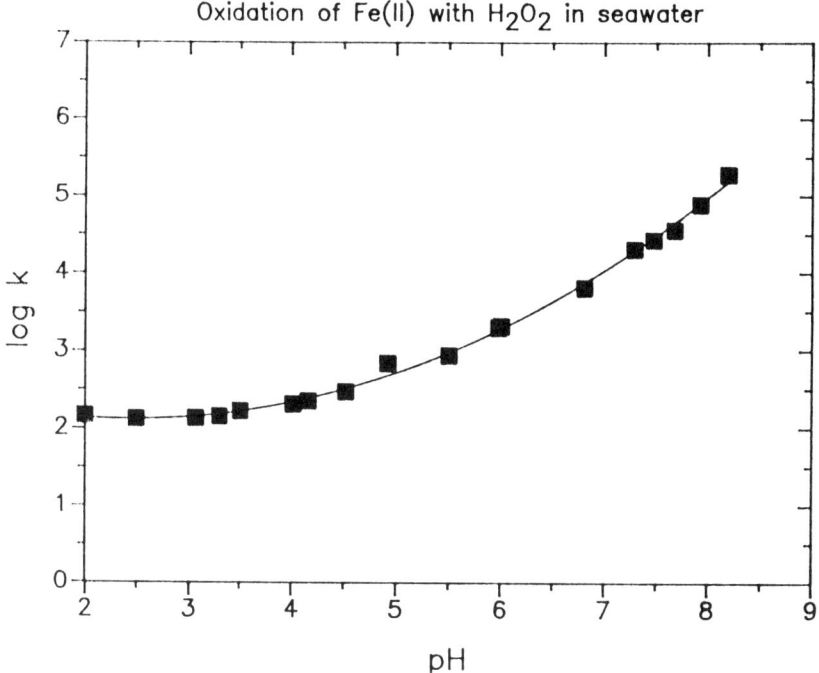

FIGURE 4. *Effect of pH on the Oxidation of Fe(II) with H_2O_2.*

where k_0 and k_1 are the rate constants for the oxidation of Fe^{2+} and $Fe(OH)^+$ and α_{Fe} and $\alpha_{Fe(OH)}$ are the molar fraction of these species. The linearized form of Equation 22 has been used to determine $k_0 = 10^{2.17}$ s^{-1} M^{-1} and $k_1 = 10^{6.77}$ s^{-1} M^{-1}. The smooth curve in Figure 4 has been calculated from Equation 22 using these values of k_0 and k_1. Since the effect of pH on the oxidation of Cu(I) with O_2 and H_2O_2 are very small, hydrolysis species of Cu(I) are not involved in the oxidation [7,17,18].

4. EFFECT OF THE MAJOR IONIC COMPONENTS OF NATURAL WATERS ON THE RATES OF OXIDATION

The major ionic components of natural waters can also effect the rates of oxidation of Fe(II) and Cu(I) with O_2 and H_2O_2 [2,5,7,8,9,12,13]. The effect ionic strength on the oxidation of

Fe(II) in NaCl and seawater solutions is shown in Figure 5. The values in seawater are lower than the results in NaCl at the same ionic strength. This decrease is related to the formation of Fe(II) ion pairs

$$Fe^{2+} + Cl^- \longrightarrow FeCl^+ \qquad (23)$$

$$Fe^{2+} + SO_4^{2+} \longrightarrow FeSO_4^\circ \qquad (24)$$

that are less reactive to oxidation. To investigate this decrease, we have made some measurements as a function of composition at a constant pH and ionic strength (0.7m). The results are shown in Figure 6.

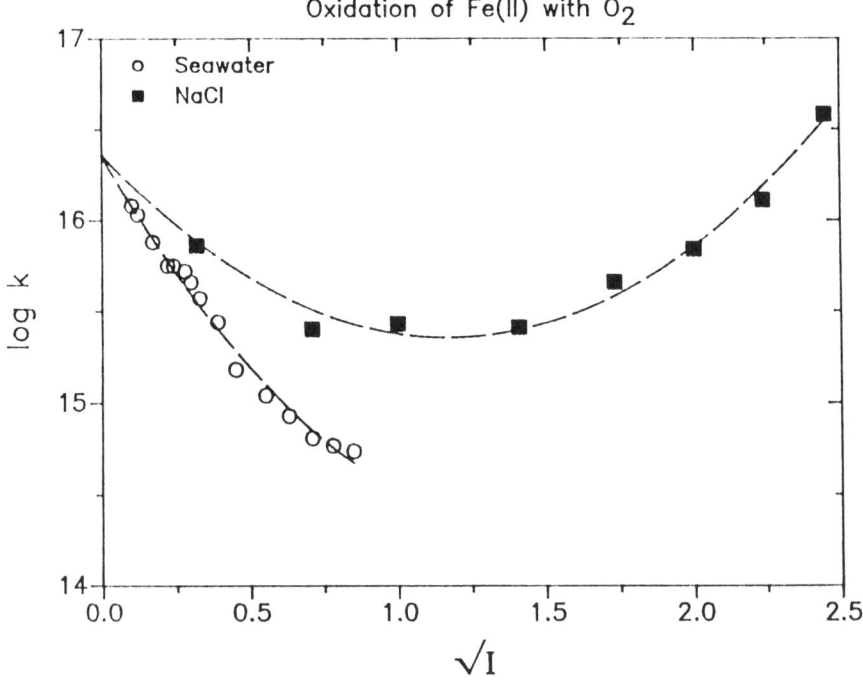

FIGURE 5. The Effect of Ionic Strength on the Oxidation of Fe(II) with O_2 in seawater and NaCl.

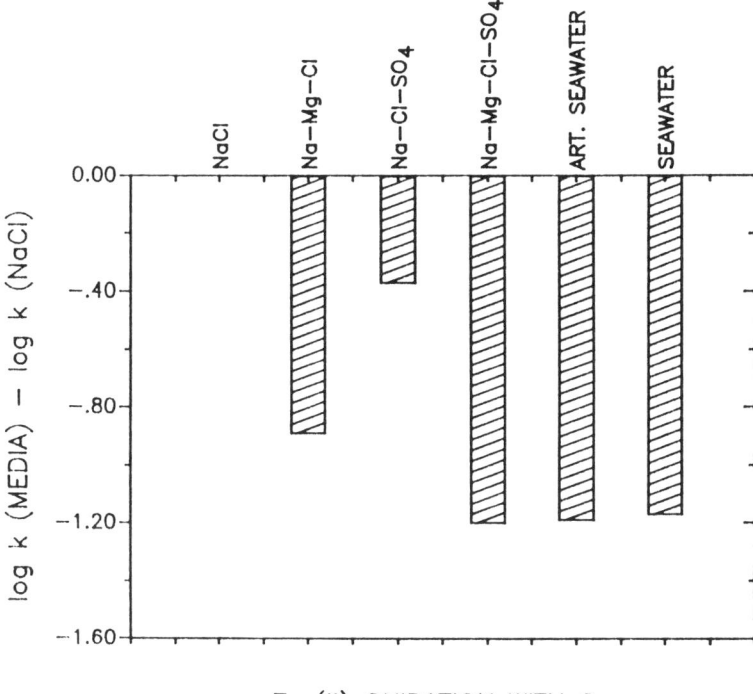

Fe (II) OXIDATION WITH O_2

FIGURE 6. The Oxidation of Fe(II) with O_2 in various Ionic Media.

The replacement of Na^+ with Mg^{2+} (0.0547m) causes a decrease in the rate. This effect was not expected. This decrease in the rate with added Mg^{2+} may be related to the formation of $MgCO_3^°$ which will decrease in the concentration of $FeHCO_3^+$ and $FeCO_3^°$ species which, as discussed later, cause the rates to increase.

The replacement of Cl^- by SO_4^{2+} (0.0293m) causes the rate of oxidation to decrease. This can be attributed to the formation of the $FeSO_4^°$ ion pair. This decrease can be used to estimate the stability constant for the formation of $FeSO_4^°$

$$k_{NaCl}/k_{NaClSO_4} = 1 + \beta_{FeSO_4}[So_4^{2+}] \tag{25}$$

Our results give β_{FeSO_4} = 500 which is in reasonable agreement with the infinite dilution value (β = 132), but larger than expected at I =0.7m of β = 10 [16].

A solution made up of Na^+, Mg^{2+}, Cl^-, and So_4^{2+} at their concentrations in seawater gives the same results as artificial and real seawater.

The effect of various anions on the oxidation of Fe(II) has been determined by making measurements in NaCl- NaX mixtures at a constant ionic strength. The results [12] are shown as function of the concentration of the added anions in Figure 7.

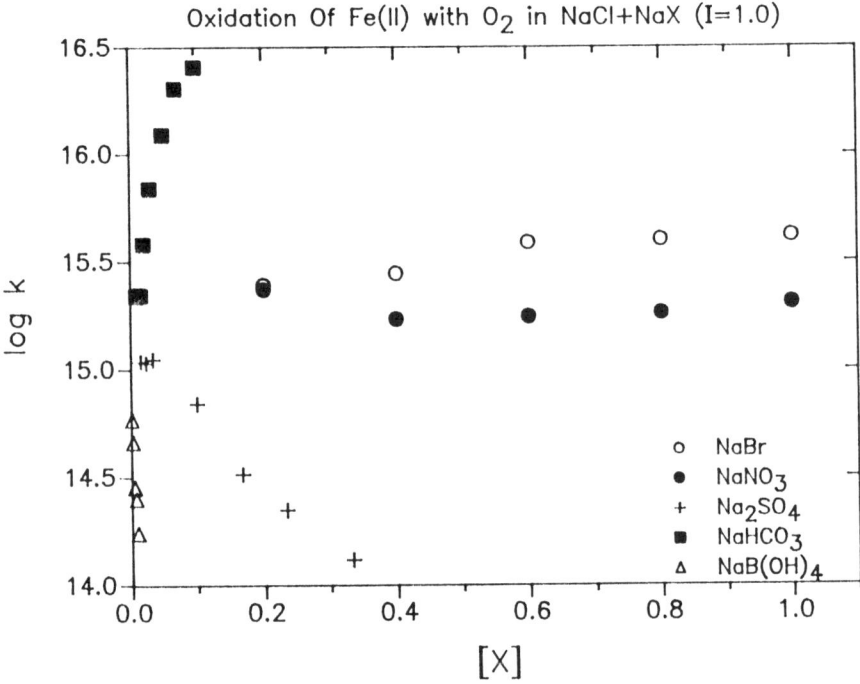

FIGURE 7. *The Effect of Various Anions on the Oxidation of Fe(II) with O_2 in NaCl-NaX Mixtures.*

The addition of HCO_3^- causes the rate to increase, while the addition of SO_4^{2+} and $B(OH)_4^-$ causes the rate to decrease.

The large increase due to the addition of HCO_3^- can be attributed to the formation of $FeCO_3^°$, which has a faster rate of

$$Fe^{2+} + CO_3^{2-} \dashrightarrow FeCO_3^° \tag{26}$$

oxidation than Fe(OH)$_2^\circ$. Since a reliable β_{FeCO_3} is not available at the present time, it is not possible to make a reliable estimate of the rate of oxidation of FeCO$_3$. The slight increases due to the addition of the anions Br$^-$, NO$_3^-$, and ClO$_4^-$ is due to the weaker interaction of these anions with Fe^{2+} compared to Cl$^-$.

The strong decrease in the rate due to the addition of SO$_4^{2-}$ and B(OH)$_4^-$ can be attributed to the formation of the FeSO$_4^\circ$ and FeB(OH)$_4^+$ ion pairs. If one assumes that the ion pairs cannot be oxidized readily, the decrease in the rates can be given by Equation 25. The experimental values of the rate constants give log β = 1.8 for FeSO$_4^\circ$ and log β = 3.2 for FeB(OH)$_4^+$ at I = 1.0. The value for FeSO$_4^\circ$ is in good agreement with the infinite dilution value of log β = 2.1 [16]. To the best of my knowledge literature values are not available for the FeB(OH)$_4^+$ ion pair.

The rates of oxidation of Cu(I) with O$_2$ and H$_2$O$_2$ have been made in seawater, NaCl and sea salts [7,8,9,17]. The measurement were made as a function of pH, ionic strength, and temperature. a comparison of the results for seawater and NaCl solution are shown in Figure 8. The large ionic strength dependence in both seawater

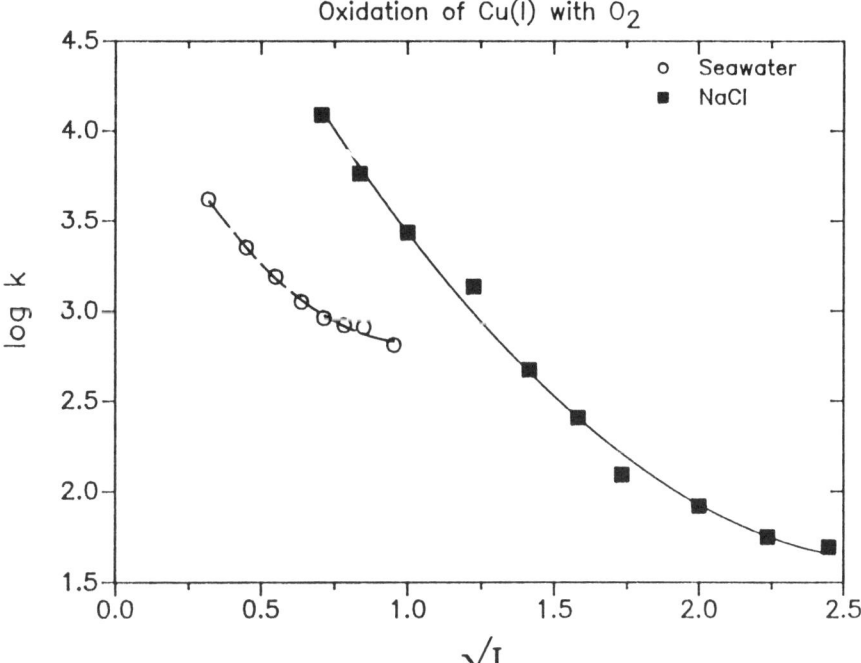

FIGURE 8. *The Effect of Ionic Strength on the oxidation of Cu(I) with O$_2$ in Seawater and NaCl.*

NaCl is related to the changes in the concentration of Cl⁻. This chloride dependence can be analyzed by assuming that the various Cu(I) chloro complexes have different rates of oxidation [2]. This gives

$$Cu^+ + O_2 \xrightarrow{k_0} \text{Products} \tag{27}$$

$$CuCl^\circ + O_2 \xrightarrow{k_0} \text{Products} \tag{28}$$

$$CuCl_2^- + O_2 \xrightarrow{k_0} \text{Products} \tag{29}$$

$$CuCl_3^{2-} + O_2 \xrightarrow{k_0} \text{Products} \tag{30}$$

The observed rate constant is given by

$$k = k_0 \alpha_{Cu} + k_1 \alpha_{CuCl} + k_2 \alpha_{CuCl_2} + k_3 \alpha_{CuCl_3} \tag{31}$$

where k_i is the rate constant and α_i is the molar fraction of species i. The substitution of the stepwise stability constants, β_i, for the formation of the various ion pairs gives

$$k/\alpha_{Cu} = k_0 + k_1 \beta_1 [Cl^-] + k_2 \beta_2 [Cl^-]^2 + \ldots \tag{32}$$

Values of k/α_{Cu} for measurements in NaCl-NaClO₄ mixtures at a constant ionic strength are shown in Figure 9. The linear behavior indicates that Cu^+ and $CuCl^\circ$ are the reactive species. Similar results were found for the oxidation of Cu(I) with H_2O_2 [17]. Measurements made over a wider range of ionic strengths indicate that the $CuCl_2^-$ species becomes important at high ionic strengths. Measurements in NaBr and NaI solutions indicate the ion pairs $CuBr^\circ$ and CuI° have similar rates of oxidation to $CuCl^\circ$ [9]. The difference of the overall rates of oxidation of Cu(I) in Cl⁻, Br⁻, and I⁻ solutions ($k_{Cl} > k_{Br} > k_I$) is related to the differences in the stability constants ($\beta_{CuCl} > \beta_{CuBr} > \beta_{CuI}$).

The results for the oxidation of Cu(I) in seawater at the same Cl⁻ concentration are lower than the values in NaCl (see Figure 8). To elucidate these differences measurements have been made in various sea salt mixtures. The results are shown relative to the measurements in NaCl in Figure 10.

The addition of SO_4^{2-} shows no effect, while the addition of Mg^{2+} and Ca^{2+} causes the rate to decrease. The addition of HCO_3^- causes the rate to increase. The solution containing Na^+, Mg^{2+}, Cl^-

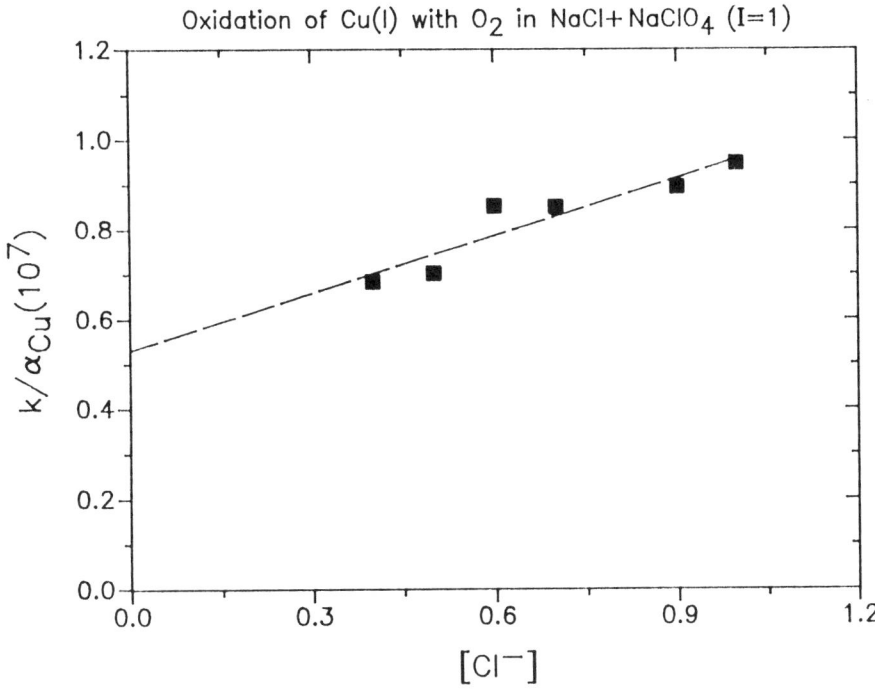

FIGURE 9. Values of k/α_{Cu} as a Function of the Concentration of Cl^- for the oxidation of $Cu(I)$ with O_2.

and HCO_3^- gives rates in agreement with the seawater results. Measurements of the effects of added Mg^{2+} and HCO_3^- have been made over a wide range of ionic strengths [8,17]. These results indicate that the effects are diminished at higher ionic strengths.

The decrease in the rate of oxidation caused by the addition of Mg^{2+} and Ca^{2+} can be attributed to the slow exchange of MgL complexes with Cu^{2+}

$$MgL + Cu^{2+} \longrightarrow CuL + Mg^{2+} \tag{33}$$

This slow exchange may cause the overall oxidation rates of Cu(I) to be slower due to the back reaction of Cu(II) with H_2O_2. Other ligands give similar rates at concentrations sufficient to complex the Cu(II) formed. The increase in the rate of oxidation due to the addition of HCO_3^- and CO_3^{2-} may be related to the formation of

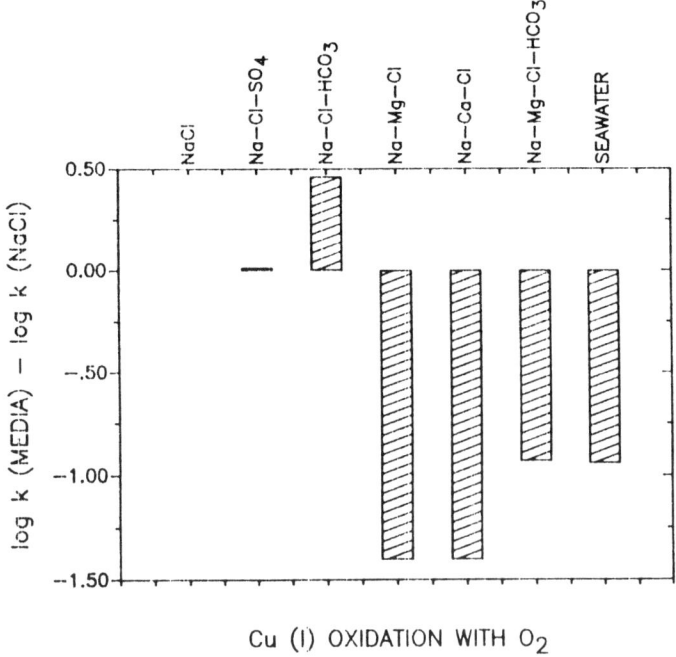

FIGURE 10. *The Oxidation of Cu(I) with O_2 in Various Ionic Media.*

$CuCO_3^\circ$ which slows down the back reaction. Another possibility is that the increase is due to the formation of Cu(I) carbonate ion pairs

$$Cu^+ + HCO_3^- \dashrightarrow CuHCO_3^\circ \qquad (34)$$

$$Cu^+ + CO_3^{2-} \dashrightarrow CuCO_3^- \qquad (35)$$

These ion pairs may be more reactive than $CuCl^\circ$. Differences in the back reactions of CuEDTA and $CuCO_3^\circ$

$$CuEDTA + O_2^- \dashrightarrow Cu^+ + O_2 + EDTA \qquad (36)$$

$$CuCO_3^\circ + O_2^- \dashrightarrow Cu^+ + O_2 + CO_3^{2-} \qquad (37)$$

with O_2^- or H_2O_2 could also have different rates. If Equation 36 is faster than Equation 37, the overall rate of oxidation of Cu(I) will be faster in CO_3^{2-} solutions. Further kinetic and thermodynamic data are needed to elucidate the carbonate effects. Even though the cause of how Mg^{2+}, Ca^{2+} and HCO_3^- affect the rates of oxidation of Cu(I) is uncertain, the correction of the seawater results for these effects yields rate constant s that agree [18] with the measured values in NaCl solutions (see Figure 11).

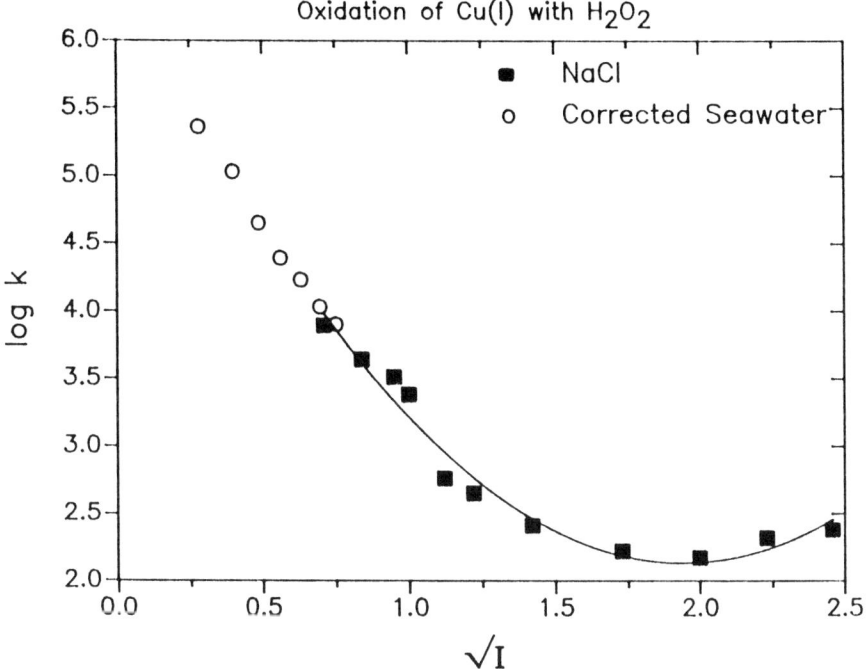

FIGURE 11. *Comparisons of the Oxidation of Cu (I) with H_2O_2 in NaCl and Seawater after Corrections are made for the Effect of Mg^{2+} and HCO_3^-.*

In conclusion, our recent studies of the rates of oxidation of Fe(II) and Cu(I) clearly demonstrate how the formation of ion complexes can affect the rates. Future rate studies for other metals are needed to examine how changes in the composition of natural waters can affect the rates of other redox processes.

ACKNOWLEDGMENTS

The author wishes to acknowledge the support of the Office of Naval Research (N00014-87-0116) and the Oceanographic section of the National Science Foundation (OCE86-00284) for this study.

REFERENCES

1. Moffett, J.W. and R.G. Zika. "Oxidation Kinetics of Cu(I) in Seawater: implications for the Existence in the Marine Environment," *Mar. Chem.* 13:239-251(1983).

2. Millero, F.J. "The Effect of Ionic Interactions on the Oxidation of Metals in Natural Waters," *Geochim. Cosmochim.* Acta 49:547-553 (1985).

3. Millero, F.J., S. Sotolongo, and M. Izaguirre. "The Oxidation Kinetics of Fe(II) in Seawater," *Geochim. Cosmochim.* Acta 51:793-801 (1987).

4. Moffett, J.W. and R.G. Zika. "Reaction Kinetics of Hydrogen Peroxide with Copper and Iron in Seawater," *Environ. Sci. Technol.,* 21:804-810 (1987).

5. Millero, F.J., M. Izaguirre, and V.K. Sharma. "The Effect of Ionic Interactions on the Rates of Oxidation in Natural Waters," *Mar. Chem.* 22:179-191 (1987).

6. Millero, F.J. "Estimate of the Life Time of Superoxide in Seawater," *Geochim. Cosmochim. Acta.*, 51:351-353 (1987).

7. Sharma, VK. and F.J. Millero. "The Oxidation of Cu(I) in Seawater," *Environ. sci. Technol.* 22:768-771 (1988).

8. Sharma, V.K. and F.J. Millero. "Effect of Ionic Interactions on the Rates of Oxidation of Cu(I) with O_2 in Natural Waters," *Mar. Chem.* 25:141-161 (1988).

9. Sharma, V.K. and F.J. Millero. "Determining the Stability Constant of Copper(I) Halide Complexes from Kinetic Measurements," *Inorg. Chem.* 27:3256-3259 (1988).

10. Millero, F.J., S. Hubinger, M. Fernandez and S. Garnett. "Oxidation of H_2S in Seawater as a Function of Temperature, pH, and Ionic Strength," *Environ. Sci. Technol.* 21:439-443 (1987).

11. Millero, F.J., A.L. Laferriere, M. Fernandez, S.Hubinger, and J.P. Hershey. "Oxidation of Hydrogen Sulfide with Hydrogen Peroxide," *Environ. Sci. Technol.* 23:209-213 (1989).

12. Millero, F.J. and M. Izaguirre. "The Effect of Ionic Strength and Ionic Interactions on the Oxidation of Fe(II)," *J. Solution Chem.* 18:000-000 (1989).

13. Millero, F.J., and S. Sotolongo. "The Oxidation of Fe(II) with H_2O_2 in Natural Waters," *Geochim. Cosmochim. Acta.*, in press.

14. Stumm, W. and F.F. Lee. "Oxygenation of Ferrous Iron," *Ind. and Eng. Chem.* 53:143-146 (1961).

15. Sung, W. and J.J. Morgan. "Kinetics and Products of Ferrous Oxygenation in Aqueous Solutions," *Environ. Sci. Technol.* 14:561-568 (1980).

16. Kester, D.R., R.H. Byrne, and Y.J. Liang. "Redox Reactions and solution Complexes Of Iron in Marine Systems," *Marine Chemistry in the Coastal Environment* (ed. T. Church). A.C.S. Pub. Washington K.C., 56-79 (1975).

17. Sharma, V.K. and F.J. Millero. "The Oxidation of Cu(I) in Electrolyte Solutions," *J. Solution Chem.*, 17:581-599.

18. Sharma, V.K.and F.J. Millero. "The Oxidation of Cu(I) with H_2O_2 in Natural Waters," *Geochim. Cosmochim. Acta.*, in press

DISCUSSION OF:
EFFECT OF SPECIATION ON THE RATES OF OXIDATION OF METALS

Giuseppe Macchi
Water Research Institute, CNR, Rome, Italy

INTRODUCTION

The physical and chemical forms of metals control their mobility and circulation in the environment, including their interactions with biota and, consequently, their bioaccumulation and toxicity. Speciation information would be also useful in order to forecast and interpret the behavior of pollutants in treatment processes: precipitation, ion-exchange, dialysis, etc.

Many difficulties can emerge in the obtainment of reliable speciation data when the matrix is very complicated and at experimental level. However a wide literature exists for natural waters.

Recently Betti and Papoff reviewed such a topic and conclude that, by the light of the present knowledge, the attempts to develop analytical procedures to distinguish different chemical species of metals have not yet produced satisfactory results, at least from a general point of view. At that purpose I would like to report their statement: "Chemical speciation of individual species is possible whenever the separation does not involve element exchanges between molecular forms during the separation time. This is possible for organometallic compounds and for molecular species containing the element in different valence state" [1].

These difficulties led to the development of speciation programs based on equilibrium thermodynamic models. However, the suitability of such theoretical models to describe metals species distribution in natural water ecosystems needs further considerations.

The comparison between some analytical results and thermodynamic forecasts have shown the importance of the influence of biological processes [2] and kinetics of redox-systems [3] as major factors determining the presence of some reduced metal species in natural waters.

This is the case of the distribution ratios between oxidized and reduced species of arsenic, chromium, selenium in sea water, which show concentrations of reduced species too high with respect to the thermodynamic expected values [4].

The paper of Professor Millero deals with problems of general interest by evidencing the strong relationship between speciation and kinetics.

Even if the paper is chiefly applied to seawater, it deals with problems of general scope and provides an example of how to handle thermodynamic and kinetic data in order to obtain useful information.

GENERAL REMARKS

I would like to note that in the paper there is no part devoted to experimental contents nor, in some cases, is the value of conditional equilibrium constants used for calculations reported. I feel that such omissions have to be ascribed to the fact that the paper consists more of a general review of various results concerning different redox systems than a work devoted to a specific subject. Probably most of this information can be found in the cited literature. However, a short note on this aspect of the work would improve the reading for people who are not familiar with this topic and with the corresponding literature.

IONIC INTERACTION AND KINETICS

Professor Millero reports in his paper some notions concerning the influence of pH, ionic strength and major constituents on the oxidation kinetics of H_2S, Cu(I) and Fe(II) by O_2 and H_2O_2. These notions are focused on natural water systems.

The reduced forms of copper and iron in oxygenated natural waters may be related to the metabolism of biological organisms such as algae and bacteria which, after they die, decompose and release these elements at the low valence state. The distribution between oxidized and reduced species of the two metals may be influenced not only by the dissolved O_2 concentration, but also by the concentration of dissolved H_2O_2, whose presence in surface water is due to photochemical processes [5]. For this purpose Zika and coworkers measured concentration values of H_2O_2 in the order of $10^{-7} - 10^{-8}$ M. To this regard, it would be interesting to know whether in surface ocean waters the distribution between oxidized and reduced forms of Cu and Fe agree with thermodynamic equilibria or is it controlled by kinetic factors.

I would like to draw Professor Millero's attention to Figure 7 of his paper. I feel it is useful to spend some words on this subject, because, in a previous paper, it was assumed that the species ferrous carbonate has a slower rate of oxidation than the dissolved ferrous hydroxide in apparent contradiction with results shown on Figure 7 [6].

The importance of considering the oxidation of hydrogen sulphide by H_2O_2 in open sea is less evident, because this compound originates mainly from sediments, and is oxidized from dissolved oxygen before it comes in contact with H_2O_2. Even in the Black Sea, and some other anoxic basins, this is the case.

However the oxidation with H_2O_2 can be very important in shallow waters, lagoons, marshes, rivers, fjords and lakes especially during the summer stratification of waters.

Furthermore, other very important sources of H_2S are those concerning certain types of industrial wastes, for example, acid mine drainage, petroleum wastes, leather tanning wastes, and certain textile wastes.

All these implications indicate the economic and ecologic importance of the presence of H_2S in an aqueous environment.

The kinetics and mechanism oxidation of H_2S by H_2O_2 in acid solution was studied by Hoffman, which proposed this process to eliminate the "rotten egg" odor due to H_2S generation in municipal sewage treatment, and to avoid corrosion in concrete sewer lines [7]. O'Brien and Birkner [8] studied the oxidation of H_2S with O_2.

Another aspect stressed in the paper is the influence of major elements, anions and cations. This aspect is very important not only for natural waters but also in the development of technological processes. For example, the separation and recovery of some metals after an acid extraction through a redox process can be more or less easy according to the acid that has been chosen because it may give origin to species which can have a faster or slower kinetics.

REFERENCES

1. Betti, M. and P. Papoff. "Trace Elements: Data and Information in the Characterization of an Aqueous Ecosystem," *CRC. Critical Review in Analytical Chemistry*, Vol. 19, 4:271-322, (1988)

2. Andreae, M.O. "Arsenic Speciation in Seawater and Interstitial Waters; the Influence of Biologic-Chemical Interactions on the Chemistry of a Trace Element," *Limnol. Oceanog.*, Vol. 24, 3:440-452, (1979).

3. Stumm, W. "What is the pE of the Sea?" *Thalassia Jugoslavica*, Vol. 14, 197-208, (1978).

4. Pettine, M., T. La Noce, and A. Liberatori. "La Speciazione di As, Cr e Se in Mare: Analisi dei Fattori che la Controllano," *Proceedings of 3rd SITE Congres.*, (Oct. 21-24, 1987 Siena, Italy) in press.

5. Cooper, W.J., R.G. Zika, R.G. Petasne, and J.M.C. Plane. "Photochemical Formation of H_2O_2 in Natural Waters Exposed to Sunlight," *Environ. Sci. Technol.*, Vol. 22, 1156-1160, (1988).

6. Millero, F.J., M. Izaguirre, and V.K. Sharma. "The Effect of Ionic Interaction on the Rates of Oxidation in Natural Waters," *Mar. Chem.*, Vol. 22, 179-191, (1987).

7. Hoffman, M.R. "Kinetics and Mechanism of Oxidation of Hydrogen Sulphide by Hydrogen Peroxide in Acid Solution," *Environ. Sci. and Technol.*, Vol. 11, 1:61-66, (1977).

8. Dennis, J.O. and F.B. Birkner. "Kinetics of Oxygenation of Reduced Sulphur Species in Aqueous Solution," *Environ. Sci. and Technol.*, Vol. 11, 12:1114-1120, (1977).

PART II: PRECIPITATION PHENOMENA

EXPERIMENTS ON THE SIMULTANEOUS OXIDIZING EXTRACTION PROCESS OF Fe(II)

Díaz J. M. and A. E. Fernández
Department of Chemical Engineering, University of Oviedo, Oviedo, Spain

Aguayo A. T. and J. Viguri
Department of Chemical Engineering, University of Pais Vasco, Bilbao, Spain

1. SUMMARY

Analysis of oxidizing extraction process must be made based on the kinetics of the constituting oxidation of the reactives and extraction of the products. Both processes for Fe(II) oxidation with actylacetone have been studied, kinetic equations being proposed. The rates of the different processes involved, including oxidizing extraction, were compared.

2. INTRODUCTION

Iron elimination from aqueous solution is commonly carried out both in the treatment of waste water and in the post-leaching step of hydrometallurgical processes. The presence of iron dissolved in natural waters is generally attributed to the dissolution of rocks and minerals. The amount of iron in these waters usually increases because the streams from several industrial, mining or domestic waste waters which flow into them are loaded with dissolved iron. Otherwise, the leaching step in hydrometallurgical processes yields aqueous solution of the valuable metals, and from those solutions, iron must be separated as soon as possible because most of the extractants for valuable metals are not selective with respect to iron.

Most of the methods proposed for iron removal require a previous oxidation to ensure that all the metal present is in the ferric form, more easily separable than the ferrous one. In natural water, it is common to remove the iron solved by precipitation of a ferric salt and subsequently a filtration of it. Organic matter present in the medium can stabilize the colloidal solution formed or increase the solubility of the salt making the iron elimination very difficult.

Classical industrial methods for iron elimination, named after the compounds of precipitates formed (i.e. Jarosite, Goethite and Haematite), imply the formation of a solid phase too. These are effective methods but require high work temperatures and the solid byproduct can adsorb other valuable metals dissolved not being interesting then for hydrometallurgical processes. In addition, these solids are difficult to handle.

In this work, iron elimination from aqueous solutions by oxidation, avoiding the formation of precipitate, and including the simultaneous oxidation/extraction has been studied. The oxidizing extractive method has already been treated by different authors in the literature for several specific currents [1]. In such a method, instead of a solid byproduct, we get, after stripping, a concentrated solution of Fe(III) from which we could try to achieve the recovery of iron. The analysis of the method requires a separate study of oxidation of Fe(II) by an air flow and subsequently a study of the extraction kinetics of Fe(III).

3. EXPERIMENTAL METHODS

Acetylacetone HA, a β-diketone, has been chosen as complexant and extractant dissolved in toluene. This kind of organic complexing agent presents an acid and a basic group in its molecule, and in the literature this agent has been mentioned as a quantitative extractant for iron, capable of separating ferric and ferrous iron [2]. Oxygen, being transferred from air, was used as oxidant.

When a simultaneous process of oxidation and extraction takes place, the reaction produced can be expressed as:

$$4\ Fe(II) + 4H^+ + O_2 \rightarrow 4\ Fe(III) + 2\ H_2O \tag{1}$$

$$Fe(III) + 3\ (HA)_o \rightleftharpoons (FeA_3)_o + 3\ H^+ \tag{2}$$

A considerable advantage of the process is the autoregulation of the pH. No neutralizing reagent is required to shift the extraction equilibrium to the right because H+ are consumed in the oxidation reaction. On the other hand, the extraction process gives the aqueous phase enough protons to avoid the precipitation of the Fe(III) as a ferric salt.

The experimental set-up consists of a four baffled stirred tank of 0.085 m in diameter. The mechanical impeller was six-blade turbine type and air feed was introduced through a ten-hole distributor in the bottom of the vessel. The dimensions of the equipment are in accordance with the standard ones proposed by Rushton [3]. The reactor is placed inside a thermostatic bath to control the temperature of the reaction system ±0.2°C. The gas circuit is such that it allows work either with air or with nitrogen being previously humidified to avoid water dragging from the reactor. The gas flow rate could be measured in the range of 0-5.4 x 10^{-5} m^3/s being 2.4 x 10^{-5} m^3/s with 11.6 rps stirring speed and 20°C in standard experiments.

The ferric working solutions were prepared dissolving $FeCl_3 \cdot 6H_2O$ with distilled and deionized water adjusting the pH values with solutions of NaOH and $HClO_4$. The ferrous ones were prepared disssolving Mohr salt, $(NH_4)_2Fe(SO_4)_2$, in standard solutions of $HClO_4$. The distilled deionized water must be previously boiled to avoid the oxidation of Fe(II). Nitrogen was maintained flowing threw all the solutions previously to the experiments and analysis. In standard experimental conditions initial iron concentration in aqueous phase was 10^{-3} $Kmol/m^3$ and 8 X 10^{-3} $Kmol/m^3$ of HA.

Analysis of Fe(II) and Fe(III), both of which forming colored complexes with acetylacetone, was made by v-uv spectrophotometry. The required condition $[ML_n]/[M] = \beta_n [L]^n \geq 10^2$, could be reached by varying the pH and extractant concentration values, with β being the conditional formation constant. In the extraction experiments the best values of the wave length and pH found for measures of Fe(III) in aqueous phase were λ-442 nm and pH = 5, fulfilling the Beer Lambert law in the range of 0-30 ppm. For the organic phase, the optimum λ found was 357 nm, and the extraction of the ferric iron was carried out with toluene, at pH = 5 and [AH] = 0.1 $Kmol/m^3$.

In the oxidation experiments, both Fe(II) and Fe(III) are present and complex in the medium, and spectrophotometric method was adjusted consisting of the search of an optimum λ that maximized the difference of absorbance between FeA_3 and FeA_2. This value was 450 nm, and with this value, a calibration curve was constructed for a total amount of iron of 10^{-3} $Kmol/m^3$. To study the simultaneous oxidizing extraction process, we have established a calibration method based on the total amount of iron versus the absorbance at 450 nm for different proportions of Fe(III)/Fe(II) in the aqueous phase. This figure, together with the calibration curve

152 PRECIPITATION PHENOMENA

of the organic phase would give the iron concentration of either ferrous or ferric species in the aqueous phase.

4. EXPERIMENTAL RESULTS AND DISCUSSION

Different reactions can be involved in these processes of oxidation/extraction. First of all, the simple homogeneous oxidation, expressed by reaction (1) as was studied by Stumm and Lee [4], and whose kinetics, they found follows $r_1 = k(T) \cdot [Fe(II)] \cdot [O_2] \cdot [OH^-]^2$. In this work, we study the oxidation of Fe(II) with HA present in the aqueous phase:

$$Fe^{2+}(HA) + O_2 + OH^- \rightarrow Fe^{3+}(HA) \qquad (3)$$

Moreover, if no complexing agent is present and the pH is high, the oxidation of Fe(II) gives raise to a third phase formed due to the precipitation of $Fe(OH)_3$ represented by:

$$Fe^{2+} + OH^- + O_2 + [Fe(OH)_3] \rightarrow Fe(OH)3 \qquad (4)$$

This reaction has a kinetics that could be composed by two contributions, one homogeneous like that of Stumm and Lee, and other one heterogeneous which depends on the amount of solid precipitated. Finally the simple extraction of Fe(III) with acetylacetone, HA, dissolved in toluene is expressed as

$$FeA_3(aq) \rightleftharpoons FeA_3(org) \qquad (5)$$

If changes in the concentration of Fe(II), HA, pH and temperature are introduced, different combinations of the presented reactions could appear.

4.1 Iron Oxidation

Regimes Analysis

The oxidation of iron by oxygen from Fe(II) to Fe(III) is a typical gas-liquid reaction. To obtain an overall equation for the oxidation rate two effects must be considered at the same time, mass transfer and chemical reaction. If one of these effects is faster than the other, the oxidation rate can be approximated to the slowest one. So, the type of the reaction regime must be studied and the dimensionless Hatta number, Ha, is frequently used to characterize it [5]. For this particular case, $Ha = (k_1 D_A C_{BO})^{1/2}/k_L$.

A previous revision of the oxidation rate data for different values of ferrous iron concentration against pH value compared with the maximum diffusional rate, Figure 1, gives us an intuitive idea about the different controlling regimes. Calderbank correlation [6] has been used to obtain k_L and then the limiting value of the mass transfer rate.

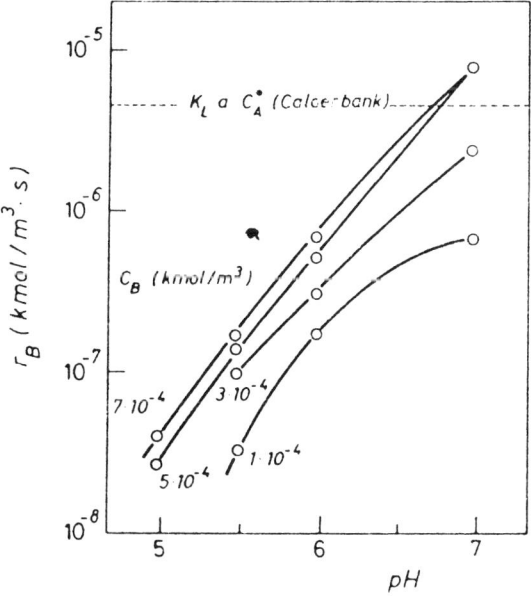

FIGURE 1. Observed Oxidation Rate of Fe(II) in G - L System Compared with the Maximum Mass Transfer Rate of Oxygen.

154 PRECIPITATION PHENOMENA

From Figure 1, it is deduced that to study the reaction avoiding a possible control by mass transfer, the pH value must not be high. This can be found in the range pH = 5- 5.5 - 6, where the reaction rate is controlled by the chemical reaction as the value of Hatta number showed. For pH = 7, the condition of negligible reaction on the film for diffusional regime is fulfilled, but the other necessary condition, that is the bulk oxygen concentration being equal to zero, is not strictly carried out. So the observed kinetic equation $-r_B = k'_L.a.C_A^*$ will be related as:

$$\frac{1}{k'_L.a} = \frac{1}{k_L.a} + \frac{1}{k_1.C_B} \quad (6)$$

Representing $1/k'_L.a = C_A^*/r_B$ vs $1/(k_1.C_B)$, Figure 2 allows us to see in a qualitative way the dominant regime depending on Fe(II) concentration and in the range of diffusional control a value of $k_L.a = 0.029$ s^{-1}, very close to that of Calderbank, is obtained.

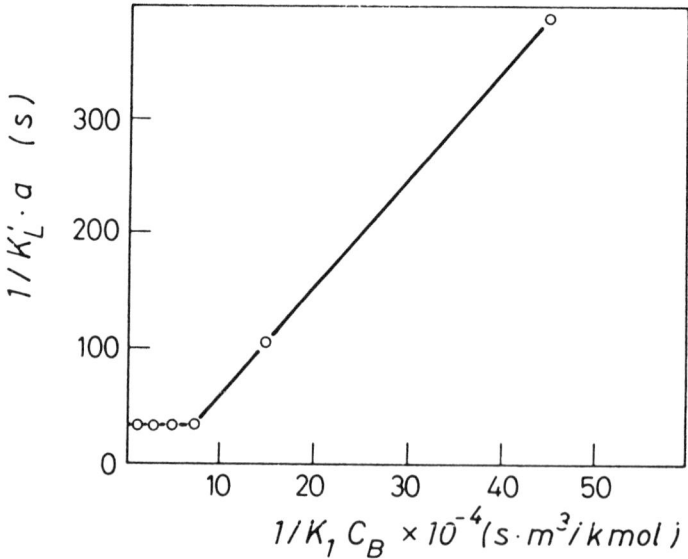

FIGURE 2. *Change of Reaction Regime with Fe(II) Concentration in the Oxidation Process at pH = 7.*

For the same value of pH = 7, an experiment was carried out in a cell reactor where a, interfacial area, is a known value 0.125 cm^{-1}. Results showed that at the highest experimental concentrations, the reaction could begin to be controlled by the diffusion.

Kinetic Study

First of all, the effect of HA aqueous concentration was tested. For a quantity of HA higher than that required for iron complexing, 8 x 10^{-3} or 4 x 10^{-2} Kmol/m^3, no substantial effects on the oxidation were found, and then, all the experiments were carried out with 8 x 10^{-3} Kmol/m^3 of HA.

In the kinetic regime, at pH values equal to or lower than 6, the results could be fitted to a first order kinetic equation with respect to Fe(II) concentration for each pH value. The straight lines obtained in the representation of $\ln C_{B_o}/C_B$ vs time, Figure 3, shows the good match of the equation for different values of the pH.

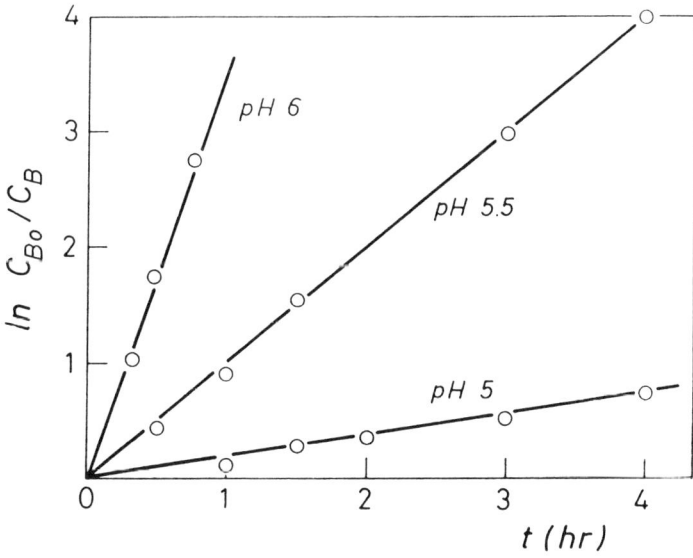

FIGURE 3. Fitting of Results in de Fe(II) (HA) Oxidation to a First Order Kinetic Equation.

The k_1 calculated values were higher than those given in the classical Stumm and Lee work [4]. This higher reaction rate is due to the presence of HA corresponding to its complexing capacity both on Fe(II) and Fe(III) giving coupled processes with the oxidation. The effect of the pH on the oxidation rate was studied taking logarithms in the equation $k_1 = k_{1n} \cdot [OH^-]^n$ and adjusting the experimental values of $\log k_1$ and pOH to a straight line obtaining an approximate slope value of 1.3.

The way k_1 at pH = 6 is modified by temperature in the range 20 to 40°C could be fitted to an Arrhenius type expression $k_1 = A.\exp(-E_a/R.T)$. The calculated values A = 54.6 min^{-1} and Ea = 3970 cal/mol, show a low activation energy resulting from the coupling of the processes and the activity of species. At higher temperatures than the ones tested here, diffusion could also begin to play a role in the reaction control. It has also been shown that Fe(III) has some catalytic effect on the reaction as has demonstrated the fact that when 10^{-3} $Kmol/m^3$ concentration of Fe(III) is introduced initially in the system, a similar kinetic equation is obtained, but k_1 being 60% higher.

Precipitation Limits

Precipitation of Fe(III) at high pH value was avoided due to the HA complexing effect, surpassing the $Fe(OH)_3$ formation. This one depends strongly on the pH value. Previous experiments [7] were made for oxidation in precipitating hydroxide presence without complexant. The study of the kinetic regime showed that the chemical reaction control was shifted to more acidic values. The reaction rate can be defined as a sum of a homogeneous rate assimilated to the Stumm and Lee oxidation rate and a heterogeneous one: $r = r_{ho} + r_{he}$. Figure 4 gives a comparison between the three oxidation reactions that have been considered, homogeneous, homogeneous in the presence of HA and heterogeneous.

In these results, when the proportion of precipitate increases (percentages in Figure 4), the kinetic constant also increases for a given pH. This is in accordance with the heterogeneous rate which depends on the amount of solid phase present and consequently on the quantity of Fe(III) adsorbed. Working at pH = 6 and at a stirring speed of 11.6 rpm, a decrease of the oxidation rate was obtained with the increase of the holes diameter and the decrease of the gas flow rates, that produce a change of the interfacial area of the bubbles and of the solid adsorbent phase.

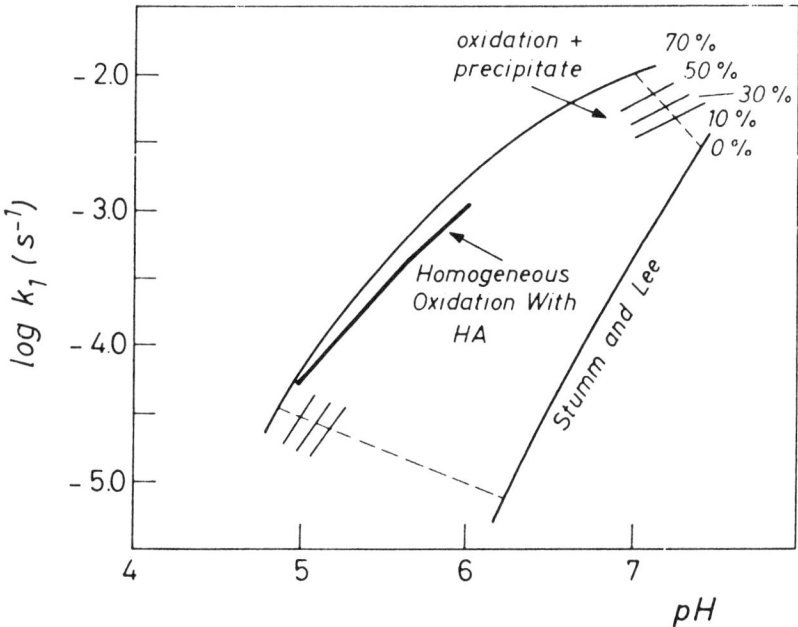

FIGURE 4. First Order Kinetic Constant for Fe(II) Oxidation, with HA Complexing Agent and in its Absence, with and without $Fe(OH)_3$ ppdo.

Two additional experiments have been made with a quantity of HA lower than the necessary amount to complex the Fe(III) formed by the oxidation. This makes the process a combination of the reactions (1), (3) and (4). In reference to the standard experiments, in the first experiment, the initial Fe(II) concentration was five times higher while in the second experiment HA concentration was reduced ten times. In Figure 5, the appearance of precipitate can be followed by the change in the slope of $\ln C_{Bo}/C_B$ versus time (k_1 value).

4.2 Liquid Extraction

Extraction Equilibrium

The prime interest in the study of Fe(III) extraction by HA in toluene is the equilibrium data. Temperature was maintained at 20

± 0.2°C and concentrations were measured after enough time to ensure that the equilibrium had been reached. Previously, from experiments at pH = 5 it was shown that 1% of HA in the organic phase is enough to get the maximum possible extraction with the standard initial concentration of Fe(III) of 10^{-3} Kmol/m^3, and this concentration of HA in organic phase was used in the extraction experiments.

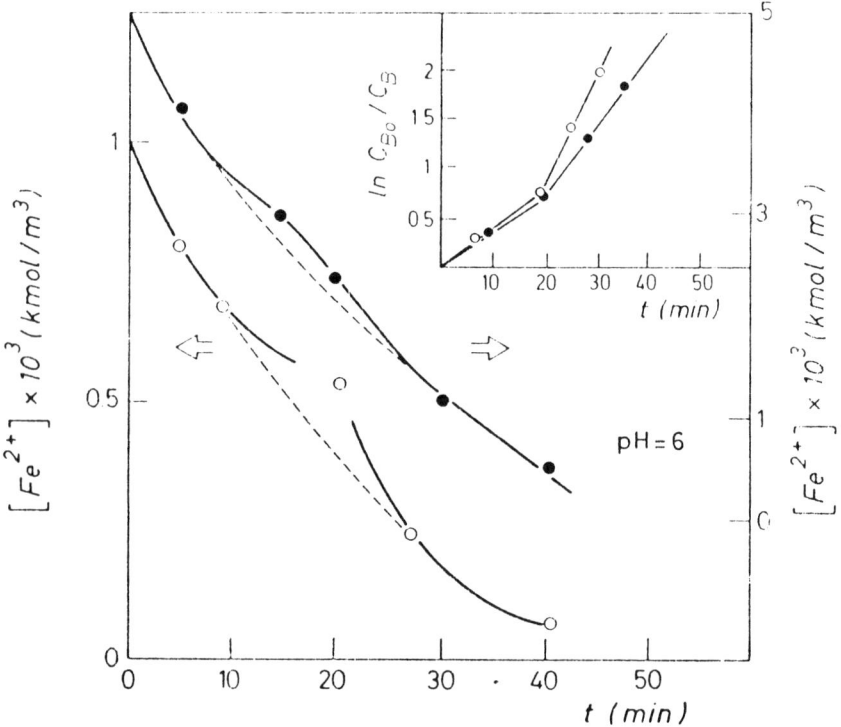

FIGURE 5. Oxidation Kinetics of Fe(II) with Deficit of HA for Fe(III) Complexation. ○ $[HA] = 8 \times 10^{-4}$ Kmol/m^3; ● $[Fe(II)] = 5 \times 10^{-3}$ Kmol/m^3.

Equilibrium results have been obtained measuring the pH value after the extraction in the aqueous phase and its representation as distribution coefficient (log) vs the pH value, is shown in Figure 6. For the extraction of the complex the equilibrium can be studied from the expression:

logD = logKe + 3 log[HA]$_o$ + 3 pH.

In Figure 6, three well-defined zones, described in the literature [8] are shown. The horizontal range, 5.5 < pH <7, indicates the FeA_3 is present both in the organic and aqueous phase. the transition zone between upward curve and horizontal curve does not show a sharp edge as would be expected because intermediate complexes have been formed in the aqueous phase. The downward curve is due to anionic complexes as MA_{n+1} present in the aqueous phase. Figure 6 also shows that in the range 4.3 < pH <7.7, the percentage of extraction E is higher than 99%.

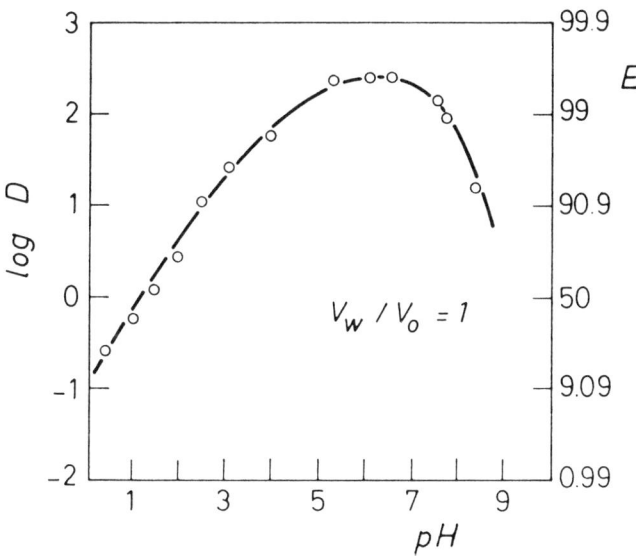

FIGURE 6. Distribution Coefficient and Percentage of Extraction for Fe(III) Extraction with HA in Toluene.

Extraction Kinetics

Kinetic experiments for the extraction process were achieved at three different pH values, 1-3-5. In addition to a stirring rate of 11.6 rps, as in oxidation and oxidizing extraction experiments, a gas flow rate of 2.4×10^{-5} m^3/s was introduced into the vessel bottom. In all the experiments, the equilibrium was reached within 5 minutes.

This high extraction rate is due to the mechanism involving the formation of the aqueous complex of Fe(III). The results were adjusted to a first order kinetics for the concentration of ferric iron finding a good fitting for the three pH values. The mechanism proposed [9] for extraction with β-diketone was:

$$(HA)_o \rightleftharpoons HA \tag{7}$$

$$Fe(OH)^{2+} + HA_{enol} \rightleftharpoons FeA^{+2} + H_2O \tag{8}$$

$$FeA^{2+} + 2\ HA_{enol} \rightleftharpoons FeA_3 + 2H^+ \tag{9}$$

$$FeA_3 \rightleftharpoons (FeA_3)_o \tag{10}$$

And for this mechanism, with hydrolyzed iron the kinetic equation (10) can be proposed. If hydrolysing constant $k_H = 10^{-3.05}$ is taken, values of K in a large range are obtained [10].

$$-d[Fe^{3+}]/dt = k.[Fe^{3+}].[(HA)_O].(1+K_H.[H^+]^{-1}) \tag{11}$$

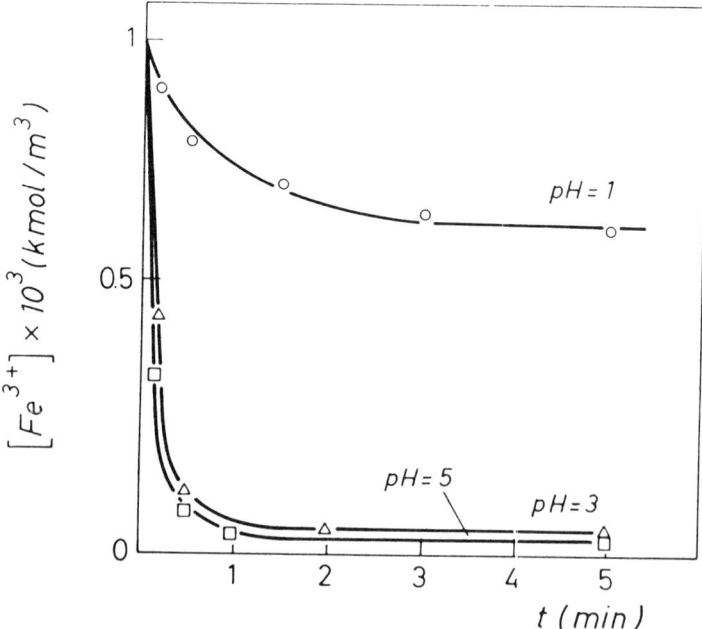

FIGURE 7. *Kinetics of Fe(III) Extraction in a Mechanically Stirred Tank and Additional Air Bubbling.*

Extractive Oxidation

Three experiments at 20°C using three different values of pH were carried out at pH = 5, 6, 7 as is shown in Figure 8. A special spectrophotometric calibration graph was previously obtained. Adjusting these results to a first order kinetic equation with respect to the ferrous iron concentration, did not give a good fit as can be observed looking at the results at pH = 5 in Figure 8.

The process is probably controlled by the chemical reaction because the extraction kinetics are nearly instantaneous in the pH range studied. The increase of the oxidation-extraction rate compared with the simple oxidation could lead to an influence of diffusional control. The presence of a new phase, the organic one, can also modify the gas-liquid interfacial area a, and the reaction rate.

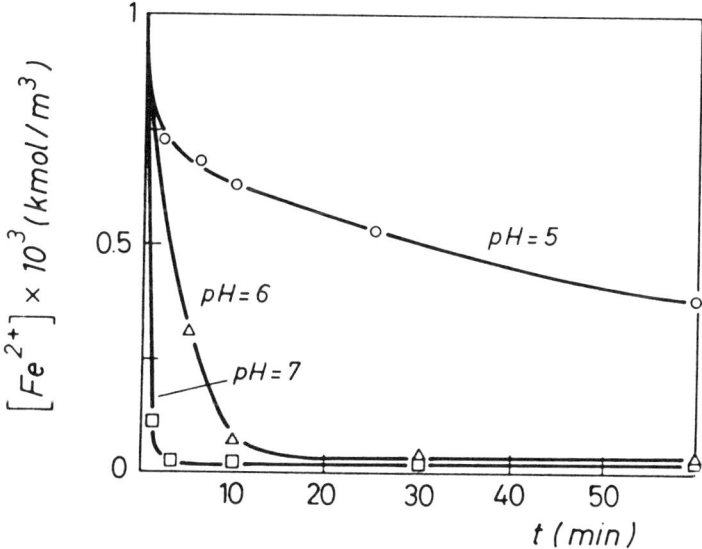

FIGURE 8. Kinetics of the Simultaneous Oxidation/Extraction Process in Standard Conditions.

To test if the extractive kinetics is able to extract all the Fe(III) being formed, the oxidation rate at some time for a given Fe(II) concentration value and the extraction rate for Fe(III) that could be formed until that time, should be compared. Figure 9 shows the fast extraction rate increasing with the amount of dissolved Fe(III) and also shows the oxidation-extraction rate faster than the homogeneous oxidation one.

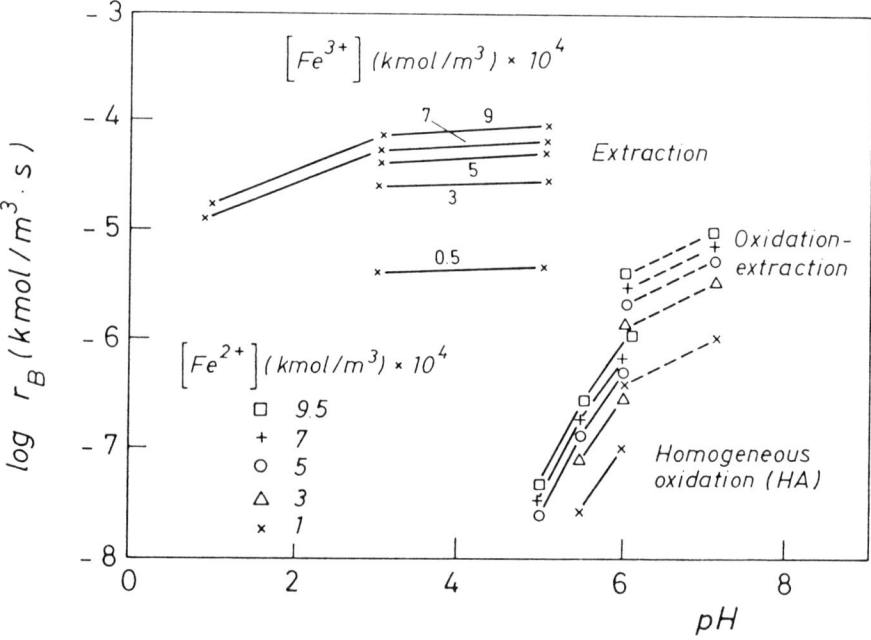

FIGURE 9. *Comparison Between the Rates of Fe(II) Oxidation, Fe(III) Extraction and the Combined Oxidation/Extraction Process.*

5. SYMBOLS

a	Gas-liquid interfacial area.
A_*	Frequency factor.
C_A	Concentration of the dissolved gas, oxygen, at the interphase.
C_{B0}	Initial concentration of Fe(II) in solution.
D	Distribution coefficient.
D_A	Diffusivity of the dissolved gas, oxygen.
E_a	Activation energy.
Ha	Dimensionless Hatta number.

HA   Acetylacetone.
k   Rate constant for extraction kinetics.
Ke   Extraction equilibrium constant.
K_H Constant of the hydrolysis equilibrium.
k_L Liquid-film mass transfer coefficient.
k_L' Effective liquid-film mass transfer coefficient.
k_1 Rate constant for first order reaction.
r_B Rate of reaction per unit volume.
r_{ho} Rate of the homogeneous oxidation per unit volume.
r_{he} Rate of the heterogeneous oxidation per unit volume.
V_O Volume of the organic phase.
V_w Volume of the aqueous phase.
β'_n Conditional formation constant.
λ Wavelength.

REFERENCES

1. A. Gaunand, G. Bouboukas, and H. Renon. "Iron Elimination in Cupric Chloride Hydrometallurgical Processes by Oxidizing Extraction," *Chem. Eng. Sci.*, 42(5):1221-1227 (1987).

2. Habashi, F., *Principles of Extractive Metallurgy.* (New York: Gordon and Breach, 1970).

3. Nagata S., *Mixing. Principles and Applications.* (Tokyo: Halsted Press, 1975).

4. Stumm, W. and G.F. Lee. "Oxygenation of Ferrous Iron," *Ind. Eng. Chem.*, 52(2):143-146 (1961).

5. Danckwerst, P.V., *Gas Liquid Reactions.* (New York: McGraw-Hill, 1970).

6. Calderbank, P.H. and M.B. Moo-Young. "The Continuous Phase Heat and Mass Transfer Properties of Dispersions," *Chem. Eng. Sci.*, 16:39-54 (1961).

7. Viguri J., A.T. Aguayo, and J.M. Diaz. "Oxidación del Hierro en Fase Acuosa Con Formación Simultánea de Precpitado," *Technol. Agua,* 21:49-53 (1985).

8. Zolotov, Y.A., *Extraction of Chelate Compounds.* (Ann Arbor, London, 1970).

9. Morrison and Freiser, *Solvent Extraction in Analitical Chemistry.* (John Wiley, New York, 1957).

10. Laddha G.S. and T.E. Degaleesan, *Transport Phenomena in Liquid Extraction.* (McGraw-Hill, New York, 1978).

DISCUSSION OF:
EXPERIMENTS ON THE SIMULTANEOUS OXIDIZING EXTRACTION PROCESS OF Fe(II)

A. Lopez and D. Petruzzelli
Water Research Institute, Rome, Italy

INTRODUCTION

Selective removal of iron from liquid mixtures of multivalent ions has been and still is a challenge for the scientific community. Most efforts in the field are based on the possibility of sequestering ferric ions by a selective chelating agent which usually is fixed on a polymeric matrix (chelating resins) or dissolved into a water insoluble solvent. As for chelating resins applications, so far, large scale plants have not yet been developed even though, efficient resins are currently used in analytical chemistry for Fe(II)-Fe(III) separation [1, 2] and, in laboratory scale plant, for Fe(III) selective removal and recovery from acid eluates of tannery sludges containing also Al(III) and Cr(III) ions [3].

On the other hand, also liquid-liquid extraction processes, mainly developed for the selective recovery of valuable metal ions (i.e., noble metals) [4], have been proposed for ferric ions selective removal. The authors of the paper under discussion, in fact, present an experimental study on the kinetic characterization of a process named "simultaneous oxidizing extraction of Fe(II)." In such a process, in order to avoid the interference of irons during the recovery of valuable metals present in liquid streams coming from post-leaching step of hydrometallurgical processes, the oxidation of Fe(II) to Fe(III), the complexation of ferric ion by a proper complexing agent and the liquid-liquid extraction of the formed complex are simultaneously accomplished.

DISCUSSION

Two main observations arise after reading the paper. The first one concerns the process, and so it is not directly addressed to the authors. We wonder if, for large scale plants, the process is economically convenient and technologically appropriate. In other words, we feel that the adoption of the liquid-liquid extraction technology, if can be justifiable for the recovery of precious metals (i.e., gold, palladium, platinum), seems to be rather "sophisticated" for iron. On the basis of recent results, ion-exchange technology seems to be a valid alternative [3]. A detailed comparative costs-analysis should be done especially taking into account the possibility of recovering extractant and/or regenerant solutions. This step is an important requirement for the development of cyclic separation procedures and could affect the whole economics.

The second observation is more specific and concerns the rate constant for the homogeneous oxidation of ferrous ions. In pure water, for such reaction the proposed general rate law [5] is:

$$-d[Fe(II)]/dt = k[OH^-]^2 P_{O_2}[Fe(II)]$$

at constant pOH and P_{O_2} we shall obtain a first-order equation:

$$-d[Fe(II)]/dt = k_1[Fe(II)]$$

where $\quad k_1 = k[OH^-]^2 P_{O_2}$

Other than by temperature, pH, P_{O_2} and [Fe(II)], the oxidation rate has been demonstrated to be affected by ionic strength and by specific ionic interactions [6-8]. An increase of ionic strength, with $NaClO_4$, from 9 to 110 mM increases the half time for oxidation from 18 to 38 min. as reported by Sung et al. [6].

Furthermore, Tamura et al. [8] have found that several ions have a significant retarding influence on the oxygenation rate. The oxygenation half times are in the sequence:

$$SO_4^= > I^- > Br^- > Cl^- > NO_3^- > ClO_4^-$$

This sequence has been ascribed to the different oxidation rate of the various ion-pairs that the various anions form with Fe(II).

The authors who have studied the oxidation of Fe(II) in an aqueous phase containing a complexant agent selective towards Fe(III), seem to have disregarded the above reported influences.

From our point of view, instead, we believe that also for the investigated system ionic strength and specific ionic interactions should be considered.

In other words, the "matrix effect" relative to ionic speciation of ferrous ions and ionic strength of the solution, could play a relevant role in determining the kinetic behavior of the system.

Accordingly, this latter aspect should be taken in a better account and a deeper insight into its influence on the overall oxidation rate should be substantiate.

CONCLUSIONS

The authors report a well designed and correctly carried out experimental kinetic characterization of a simultaneous oxidizing-extraction process of ferrous ions.

In order to make such study complete, "matrix effect" and the possibility of recovering the extractant solution should be more properly taken into account.

This is much more evident if we consider that the matrix of real streams submitted to the process shall never be a pure aqueous phase.

REFERENCES

1. Hodgkin. J.H. and R. Eibl. "Ferric-ion Chelation by Aminophenol Resins," *Reactive Polymers*, 4:285-291 (1986).

2. Chanda, M., K.F. O'Driscoli, and G.L. Rempel. "Polybenzimidazole Resin-based New Chelating Agents. Ferric Ion Selectivity of Resins with Immobilized Oligoamines," *Reactive Polymers*, 9:277-284 (1988).

3. Petruzzelli, D., G. Macchi., M. Pagano., R. Passino., M. Santori, and G. Tiravanti. "Cromium, Aluminum, Iron Separation from Tannery Sludge by Selective Ion-exchange," Poster presented at Metals Speciation, Separation and Recovery Int. Conf., Rome, Italy, (May 15-20, 1989).

4. M. Grote., U. Huppe, and A. Kettrup. "Solvent Extraction of Noble Metals by Formazans," *Hydrometallurgy*, 19:51-68 (1987).

5. Reference [4] in the paper under discussion.

6. Sung, W. and J.J. Morgan. "Kinetics and Product of Ferrous Iron Oxygenation in Aqueous Systems," *Env. Sci. and Tech.* 14(5):561-568 (1980).

7. F.J. Millero. "The effect of Ionic Interactions on the Oxidation of Metals in Natural Waters," *Geochim. Cosmochim. Acta.* 49:547-553 (1985).

8. Tamura, H., K. Goto, and M. Nagayama, *J. Inorg. Nucl. Chem.* 38:113-117 (1976).

HYDROLYSIS, PRECIPITATION AND AGING OF COPPER(II) IN THE PRESENCE OF NITRATE

J.W. Patterson, R.E. Boice and C. Petropoulou
Illinois Institute of Technology, Env. Eng. Dept., Chicago, Illinois.

D. Marani, G. Macchi
Water Research Institute, Rome, Italy.

1. INTRODUCTION

The study of the hydrolysis-precipitation of copper is of importance in view of the wide application of the alkaline precipitation process for copper removal from industrial wastewaters. Copper treatment is predominantly accomplished by precipitation [Patterson, 1985].

Despite its wide application, the precipitation process is still applied on an empirical basis. Intriguing deviations in the performance of precipitation technology are observed in time, in space and specific application, and in comparison with chemical theory [Patterson, 1987]. A number of explanations for such anomalies can be postulated, among which are: 1) unanticipated interactions among co-existing chemical species [Bowers et al., 1981], and 2) kinetically controlled behavior, within the real-time spans of minutes or hours represented by the residence times of the processing units [Karra et al., 1983; Jenista, 1984]. While metal insolubilization is quite easily controlled by means of pH, the subsequent step of precipitate separation from the soluble phase is often quite difficult, due to the colloidal nature of the precipitate, which also affects the volume of sludges to be disposed of. The need to optimize the precipitation process on sound scientific principles warrants a deeper investigation of the several concurrent or sequential phenomena involved.

The general sequence of a precipitation process involves induction of supersaturation, nucleation, crystal growth, and several aging processes such as Ostwald ripening, recrystallization and agglomeration [Salutsky, 1959]. Nucleation and growth rates, colloidal stability, and aging processes are the main factors which affect the morphology and the state of aggregation of the precipitates [Walton, 1967].

The presence of such anions as nitrate, sulfate and chloride, which are commonly found in industrial wastewaters, may affect Cu(II) hydrolysis-precipitation both from thermodynamic and from kinetic standpoints. In analytical separations it is well established that foreign ions take part in several phenomena which affect the composition and purity of the precipitates: adsorption, postprecipitation, occlusion, and coprecipitation [Kenner and O'Brien, 1971]. In addition, several anions may induce precipitation of basic cupric salts, with a clearly defined stoichiometry and solubility product [Feitknecht and Schindler, 1963]

This paper discusses the results obtained in a research project aimed at studying chemical composition and physico-chemical properties of precipitates obtained from alkaline precipitation of copper from cupric nitrate solutions. Changes in precipitate properties have been studied as a function of reagents mixture composition and mixing protocol, as well as reaction and aging time.

2. BACKGROUND

2.1 Phenomenology of Cu(II) Hydrolysis-Precipitation

Upon addition of dilute alkali to a cupric salt solution, a blue gelatinous mass is obtained, which is a precipitate containing up to 20 moles of water to one mole of copper [Weiser, 1935]. The fresh gel is unstable, and if allowed to age in contact with the mother liquor it loses water and changes in color from blue to green, brown, and, finally, black. Neville and Oswald (1931) showed that the blue and the black precipitates give entirely different x-ray diffraction patterns, which correspond to $Cu(OH)_2(s)$ (blue) and to $CuO(s)$ (black), respectively. Hence, upon aging at room temperature, the highly hydrated $Cu(OH)_2(s)$ is slowly converted to a only slightly hydrous $CuO(s)$.

The conversion of $Cu(OH)_2(s)$ to $CuO(s)$ was investigated by Schindler et al. (1965) from a thermodynamic standpoint. Taking into account the influence of molar surface S (m^2/mole) upon solubility, at 25 °C and 0.2 M ionic strength they determined:

$Cu(OH)2(s)$: $\log K_{s0} = 8.92 + 4.8 \times 10^{-5} S$ $\quad \gamma = 410$ ergs/cm^2 $\quad$ (1)
$CuO(s)$ $\quad$: $\log K_{s0} = 7.89 + 8.0 \times 10^{-5} S$ $\quad \gamma = 690$ ergs/cm^2 $\quad$ (2)

According to Schindler et al. (1965), at sufficiently large subdivision of the solids (diameter ≤ 4 nm) $Cu(OH)_2$(s) may become more stable than CuO(s), due to the larger interfacial energy (γ) of the oxide as compared to the hydroxide. This can explain why $Cu(OH)_2$(s) crystals nuclei are formed in the initial phase of precipitation. Upon subsequent growth of nuclei into larger particles, CuO(s) becomes more stable and hydroxide is converted into oxide according to the reaction :

$$Cu(OH)_2(s) = CuO(s) + H_2O(l) \tag{3}$$

2.2 Solubility of Cupric Hydroxide-Oxide Precipitates

Table 1 shows the large variability of equilibrium data which can be found in the literature for some soluble and insoluble Cu(II) hydrolysis products. An explanation of the large variability of K_{s0} values reported for $Cu(OH)_2$(s) is given by Gulens et al. (1984), who, studying Cu(II) hydrolysis by copper ion-selective electrode, found a quite significant effect of aging time on the value of K_{s0}. They concluded that a significant number of the K_{s0} values reported in the literature can not be considered as "thermodynamic" constants, because they are referenced to a metastable solid system.

In this paper experimental results are discussed in relation to the theoretical equilibrium composition of the system under study. The theoretical evaluations were performed with the help of the MINEQL computer program [Westall et al., 1976]. Unless otherwise stated, the equilibrium constants reported in Table 2 were used for theoretical equilibrium computations.

2.3 Basic Cupric Salts

Studies of copper hydrolysis are complicated by the fact that a number of basic salts of Cu(II) are quite insoluble [Baes and Mesmer, 1976]. The electrometric and phase rule studies of Britton (1925) on the precipitation of copper hydroxide showed the formation of basic salts of copper(II), when NaOH was slowly added to a solution of cupric sulfate, chloride, bromide, or nitrate. Feitknecht and Schindler (1963) reported the stoichiometric composition and the solubility constant of basic salts of copper(II) with nitrate, sulfate, chloride, bromide and perchlorate.

With carbonate anion, Cu(II) forms two uniquely stable carbonate complexes and two basic salts (malachite $Cu_2(OH)_2CO_3$(s) or azurite $Cu_3(OH)_2(CO_3)_2$(s)) [Schindler, 1967; Smith and Martell,

1976]. According to the reported equilibrium constants, solubility of Cu(II) in air-saturated water (P_{CO_2} = -3.52) is governed by the solid $CuO_{(s)}$ over the whole pH range [Baes and Mesmer, 1976]. However, at higher carbonate concentrations either malachite or azurite can be the predominant solid phase in the low pH region [Stumm and Morgan, 1981].

TABLE 1. *Reported Equilibrium Constants for K_{s0} ($Cu(OH)_2(s)$), K_{s0} ($CuO(s)$), B_1 ($CuOH^+$), and B_2 ($Cu(OH)_2$), at 25°C and Zero Ionic Strength. (Reactions are reported in Table 2).*

Reference	$\log K_{s0}$ $Cu(OH)_2(s)$	$\log K_{s0}$ $CuO(s)$	$\log B_1$	$\log B_2$
Schindler et al (1965)		7.65		-13.96
			-8.06	
Sillen and Martell (1977)	8.66			
			-7.7	
Smith and Martell (1976)	8.68			
Baes and Mesmer (1976)	8.64	7.62	<-8	<-17.3
Vuceta and Morgan (1977)	10.7			-13.7
Heijne and Lindend (1978)	9.6			
Stella et al. (1979)	10.3			-16.0
Paulson and Kestner (1980)			-7.96	-16.24
Stumm and Morgan (1981)		7.65	-8.0	
			-7.4	
Gulens et al. (1984)	10.3-(*)			-16.0

TABLE 2. Reactions of Concern and Selected Equilibrium Constants Used in Theoretical Equilibrium Computations.

Reaction	Symbol	logK	logK$_f$(*)	Reference
$Cu(OH)_2(s) + 2H^+ = Cu^{+2} + 2H_2O$	K_{s0}	8.7	-8.7	Smith and Martell (1976)
$CuO(s) + 2H^+ = Cu^{+2} + H_2O$	$'K_{s0}$	7.6	-7.6	Baes and Mesmer (1976)
$Cu_2(OH)_3(OH_3)_s + 3H^+ = 2Cu^{+2} + NO_3^- + 3H_2O$	$"K_{s0}$	9.2	-9.2	Feitknecht and Schindler (1963)
$Cu^{+2} + H_2O = CuOH^+ + H^+$	B_1	-8.0	-8.0	Baes and Mesmer (1976)
$Cu^{+2} + 2H_2O = Cu(OH)^{20} + 2H^+$	B_2	-16.0	-16.0	Gulens et al. (1984)
$2Cu^{+2} + 2H_2O = Cu_2(OH)_2 +2 + 2H^+$	B_{22}	-10.4	-10.4	Baes and Mesmer (1976)
$Cu^{+2} + 3H_2O = Cu(OH)^-_3 + 3H^+$	B_3	-26.3	-26.3	Stumm and Morgan (1981)
$Cu^{+2} + 4H_2O = Cu(OH)_4^{-2}$	B_4	-39.6	-39.6	Baes and Mesmer (1976)
$Cu^{+2} + CO_3{-2} = CuCO_3^0$	$'B_1$	6.7	6.7	Smith and Martell (1976)
$Cu^{+2} + 2CO^{-2}{}_3 = Cu(CO_3)_2^{-2}$	$'B_2$	9.9	9.9	Smith and Martell (1976)
$CO_3^{-2} + H^+ = HCO_3^-$	K_1	10.3	10.3	Smith and Martell (1976)
$HCO_3^- + H^+ = H_2CO_3^*$	K_2	6.3	16.6	Smith and Martell (1976)
$H_2O = H^+ + OH^-$	K_w	-14.0	-14.0	Smith and Martell (1976)

* K_f: equilibrium constant for the reaction written in terms of formation of the soluble or solid species, for computation with the MINEQL computer program (example: $Cu^{+2} + H_2O - 2H^+ = CuO(s)$).

The method of mixing of the reagents was found to have a significant effect both on the physical characteristics and on the chemical composition of the precipitate. Britton (1943) noted that slow addition of alkali to copper chloride and bromide solutions forms a pale blue, heavy, finely divided precipitate at 1.5 moles of alkali per mole of copper. However, if the alkali is added quickly without adequate mixing, a gelatinous precipitate is formed containing 1.8 moles of alkali per mole of copper. Similarly, in studies of copper sulfate titration by sodium hydroxide, it was found that rapid addition necessitated the use of a larger alkali quantity for complete copper precipitation and consequently the gelatinous precipitate obtained was more basic than the blue, non-gelatinous, precipitate produced on slow addition of the alkali [Britton, 1925].

3. EXPERIMENTAL

3.1 Batch Tests

Batch precipitation tests were performed in a plexiglass cylindrical reactor into which equal volumes of NaOH and $Cu(NO_3)_2$ solutions were mixed. Three mixing procedures were used:

A - fast simultaneous addition of both solutions under intense stirring;

B - addition of NaOH solution to $Cu(NO_3)_2$ solution at a constant rate of 0.001 moles/liter-hour;

C - addition of $Cu(NO_3)_2$ solution to NaOH solution at a constant rate of 0.0005 moles/liter-hour.

Procedure A was used only in a preliminary series of experiments, but was later discarded because of uncontrollable solution dispersion conditions.

For mixing procedures B and C, NaOH and $Cu(NO_3)_2$ solutions were added with peristaltic pumps, while the rate of addition was monitored through flowmeters. Before entering the reactor, the solutions were filtered through 0.2 micron capsule filters to retain particulate impurities. During and after mixing of the reagents, the reactor was stirred with an impeller at 250 rpm and was kept under a nitrogen blanket.

Two series of batch tests were performed, in which the final concentration of $Cu(NO_3)_2$ was about 0.0005 M and 0.005 M, respectively. The predefined final base/copper molar ratios ranged from 0.81 to 3.4.

The batch reactions were closely monitored for several hours (up to 19 hours). Afterwards, the suspension was put into a stoppered plastic bottle for further analyses after aging.

3.2 Continuous Stirred Tank Reactor (CSTR) Tests

CSTR precipitation tests were performed in the experimental apparatus shown in Figure 1. The cylindrical plexiglass reactor had an effective volume of 526 ml. In the lid of the reactor were inserted: nitrogen gas inlet tube, impeller shaft, pH electrode, Cu, nitrate and reference electrodes, NaOH and Cu solution inlet tubes, PenKem System 3000 and Hiac/Royco particle counter sampling tubes. Sampling of the suspension for the analysis of the total and soluble (filterable) residual metal was performed through the lid as well. Influent NaOH and $Cu(NO_3)_2$ solutions were fed with a peristaltic pump which was equipped with two identical pump heads in order to provide the same constant flow rate for both solutions. Before entering the reactor the solutions were filtered through 0.2 micron capsule filters. The reactor was stirred with an impeller at 250 rpm and was kept under a nitrogen blanket.

3.3 Analytical Measurements

The following analytical parameters were monitored during each run: pH, conductivity, turbidity, total and soluble metal concentration, free metal ion activity, and electrophoretic mobility of the precipitate. In addition, in CSTR experiments free nitrate ion activity and particle concentration and size distribution were measured. Concentration and size distribution of particles were analyzed for parallel studies on nucleation and crystal growth kinetics, whose preliminary results are not reported in this paper.

pH, conductivity, turbidity, and electrophoretic mobility were measured using a PenKem System 3000 [Swenson, 1987]. The pH electrode was inserted directly into the reactor, while at preset time intervals a sample was automatically loaded from the reactor into the PenKem chamber to measure conductivity, turbidity, and electrophoretic mobility. Free nitrate ion activity was measured using a nitrate specific electrode Orion model 93-07 and a double junction reference electrode Orion model 90-02, connected to an Orion potentiometer model 901.

Free copper ion activity was measured using a copper specific electrode Orion model 94-29 and a double junction reference electrode Orion model 90-02. The manufacturers' suggested procedure for calibration of such selective electrodes (measurements of sequentially diluted, but unbuffered standard Cu^{++} solutions) leads inevitably to a $> 10^{-7}M$ limit of Nernstian response [Orion, 1979]. This is due to trace impurities from various sources, including the electrode itself. However, Avdeef et al. (1983) have shown that, in buffered standard solutions, selective copper electrodes have a consistent linear Nernstian response in the range pCu 3-19. In this work, the electrode was calibrated according to the manufacturers' procedure in standard $Cu(NO_3)_2$ solutions $> 10^{-7}M$. On the basis of Avdeef et al. (1983) findings, we could with confidence extrapolate this calibration to determine Cu^{++} activity values much lower than $10^{-7}M$. In fact, because our samples contained total (soluble and insoluble) metal concentrations larger than $10^{-5}M$, they could be considered as buffered metal solutions.

Total and soluble (filterable) copper concentrations were measured via Atomic Absorption Spectroscopy (AAS). 10 ml samples for the total metal concentration analysis were withdrawn from the reactor using a syringe. 0.1 ml of 8 M HNO_3 were added into the syringe and, after shaking the syringe, the sample was transferred into a sampling vial for the subsequent AAS analysis.

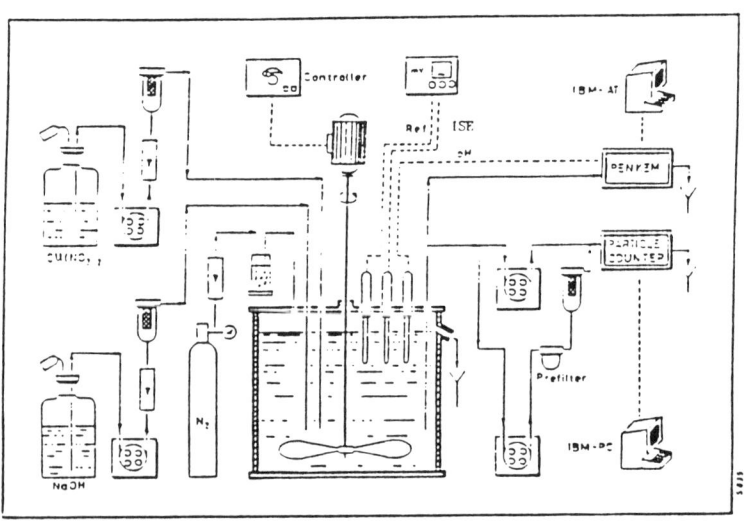

FIGURE 1.   *Experimental Apparatus for CSTR Tests.*

The same general procedure was used to take samples for soluble (filterable) metal concentration measurements. In this case a disposable 0.2 or 0.45 μm membrane filter was inserted between the sampling tube and the syringe to remove the precipitated solids.

In order to determine the composition of the precipitate, an aliquot of suspension was sampled from the reactor and filtered on a 0.45 μm membrane filter. The precipitate on the filter was washed repeatedly with double distilled water, and then redissolved with an excess of standard HCl solution. The acid solution was subsequently analyzed for the copper and nitrate content, whereas the OH content of the precipitate was determined through the backtitration of the excess HCl with NaOH.

Three analytical methods were compared for the analysis of the nitrate in the redissolved precipitates: a) the spectrophotometric method based on the direct measure of the absorbance of UV radiation at 220 nm [Standard Methods, 1985, Method 418A], b) the colorimetric method based on the preliminary reduction of nitrate to nitrite [Standard Methods, 1985, Method 418C], and c) the ion chromatographic method which uses an anion exchange resin to sequentially separate the several anions present in the sample [Standard Methods, 1985, Method 429]. The spectrophotometric UV method was discarded because it showed a significant interference by the soluble Cu(II) [Boice, 1988]. The correction for such inorganic interferences, as suggested by Armstrong (1963) and Navone (1964), involves a time consuming procedure. A comparison between the results of methods b) and c) provided evidence of the substantial equivalence of the two analytical procedures. Hence, the more simple and automated method c) was selected for analysis of nitrate content in the precipitate

In order to check for carbonate contamination, inorganic carbon (IC) analyses were performed in the precipitate suspensions. IC was measured by using a Dohrman, Xertex Model DC-90 Total Organic Carbon Analyzer. In addition, carbonate content in the initial NaOH solutions was determined via titration with HCl to pH 8.3 and pH 4.4 end points.

4. RESULTS AND DISCUSSION

4.1 Cupric Hydroxide Precipitation and Aging

Figure 2 shows the typical trends of the parameters measured during and after the addition of a solution of NaOH 0.0012 M to an equal volume of $Cu(NO_3)_2$ 0.001 M. The comparative analysis of the trends of the variables measured suggests that the overall precipitation process is a result of several reaction steps. Figure 2 can be divided into three sections characterized by particular trends in pH, specific conductance (SC) and turbidity measurements, which suggest three sequential phases of aging. Phase 1 is characterized by a sharp decrease of pH, whereas in phase 2 pH remains essentially constant.

The rapid consumption of OH^- in phase 1 can be explained by postulating the initial formation of a fresh basic cupric nitrate, which then is converted into hydroxide by replacing nitrate ion with OH^-. In support of this hypothesis, Figure 3 reports the theoretical behavior of the system under consideration during the addition of 1 liter of NaOH 1.2×10^{-3} M to 1 liter of $Cu(NO_3)_2$ 1×10^{-3} M (base/copper final molar ratio = 1.2). The abscissa α indicates the fraction of NaOH added with respect to the final amount. For speculative purposes, the theoretical composition of the system at each stage of addition has been calculated assuming that the equilibration of the system (to a metastable state) is much faster than the rate of base addition. The potential precipitation of basic cupric nitrate and $Cu(OH)_2(s)$ ($logK_{s0}$ = 9.1 is used for the active, small size, hydroxide as suggested by this work) is taken into consideration, whereas the stable, aged CuO(s) is not allowed to form for kinetic reasons. During the slow addition of NaOH solution to an equal volume of $Cu(NO_3)_2$ 1×10^{-3} M solution, nitrate concentration is diluted from the initial 2×10^{-3} M concentration to the final 1×10^{-3} M concentration. Figure 3 shows that the high concentration of nitrate anion is expected to induce precipitation of cupric hydroxy nitrate, initially. With further NaOH addition the metastable $Cu(OH)_2(s)$ becomes the predominant solid.

In reality, during the second part of NaOH addition, two simultaneous phenomena can be hypothesized: precipitation of cupric hydroxide and slow conversion of the initially formed cupric hydroxy nitrate into cupric hydroxide. In the experimental conditions of Figure 2, such conversion seems to be completed during phase 1, as suggested by the rapid decrease in pH. This hypothesis is supported by the results of analyses of nitrate content in fresh and aged precipitates obtained in the first series of experiments (final Cu(II) concentration = 5×10^{-4}M). In these precipitates, nitrate was not appreciably detected (nitrate/copper molar ratio < 0.03), except for a small residual (nitrate/copper about 0.06) in samples less than an hour old, obtained upon addition of less than stoichiometric amount of base. With further aging of the suspension, no nitrate was detected in the latter precipitate samples either.

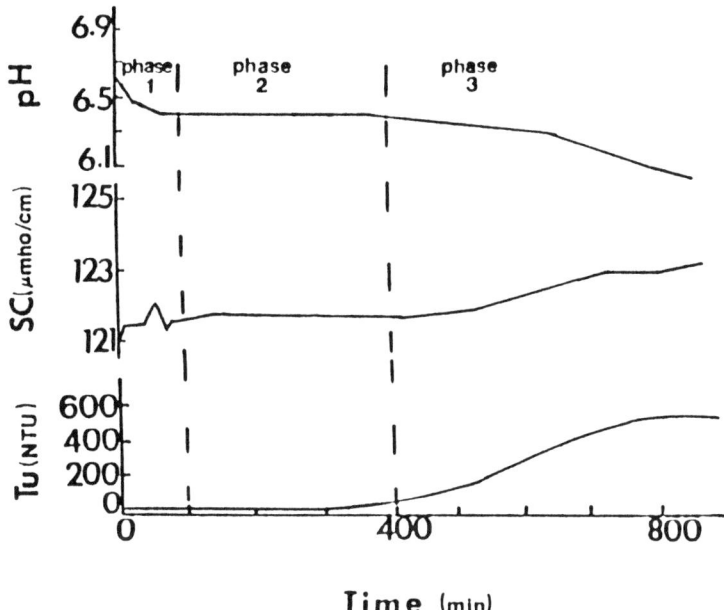

FIGURE 2. *Trends of pH, Conductivity (SC), and Turbidity (Tu) after Completion of Reagents Mixing in a Batch Test with Slow Addition of NaOH Solution to $Cu(NO_3)_2$ Solution. Base/copper Molar Ratio = 1.2.*

The blue precipitate present during phase 1 seems to be representative of the gelatinous and highly hydrated precipitate identified as $Cu(OH)_2(s)$ by the studies of Neville and Oswald (1931). $Cu(OH)_2(s)$ at very small particle size is more stable than CuO(s) [Schindler et al., 1965]. The particles are thermodynamically active and, after achieving a critical size, begin converting into the more stable CuO(s). The conversion-dehydration reaction (3) can, in fact, be hypothesized as occurring in phase 2, during which pH and conductance remain substantially constant. This hypothesis is in agreement with the observation that a gradual color change in the solution occurred during this period, from blue to brown.

In parallel with the color change, a shift in turbidity trend is observed within phase 2. These turbidity measurements reflect the 90° scattering of a laser beam of $\lambda = 6328$ Å. According to the Rayleigh and Mie theories, the intensity of light scattering is a complex function of number and size of suspended particles, and of the difference in refraction index between suspending and suspended phases. In the system under consideration, because the total amount of suspended material remains supposedly constant, the sharp turbidity increase most probably reflects a change in refraction index of the precipitate, which gradually passes from a highly hydrous cupric hydroxide to a very dense, slightly hydrated, cupric oxide (density of cupric hydroxide = 3.37 g/cm^3, index of refraction unknown, density of cupric oxide = 6.4 g/cm^3, index of refraction = 2.63 [Weast, 1972]).

Another interesting characteristic of phase 2 was the constant value of the Ion Activity Product (IAP) expressed in terms of IAP = $\{Cu^{++}\}\{H^+\}^{-2}$. In addition, IAP values related to phase 2 resulted independent of the particular test and in the range of logIAP = 9.1 ± 0.2. Converting this value to a $logK_{s0}$ value at 0.2 M ionic strength and substituting in equation (1) of Schindler et al. (1965), we can calculate a molar surface of 9,600 ± 4,200 m^2 and, assuming monodispersed spherical particles of density 3.37 g/cm^3, we obtain a size in the order of 23 ± 10 nm. These values seem to be quite close to the size of conversion calculated by Schindler et al. (1965), which is about 4 nm.

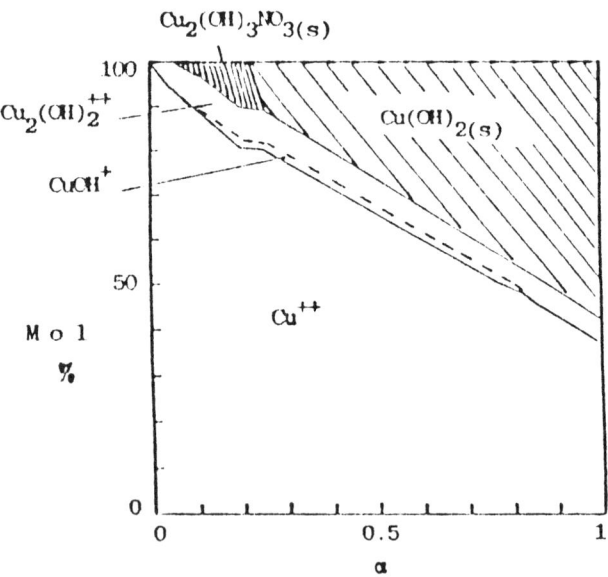

FIGURE 3. *Theoretical Distribution Diagram of Soluble and Insoluble Cu(II) Species as a Function of Extent of Addition of 1 Liter of NaOH 1.2E-3M to 1 Liter of 1.0E-3M Cu(NO$_3$)$_2$ (α = base added/base final). The rate of equilibration of the system is assumed much faster than the rate of NaOH addition. Log K_{s0} = 9.1 for metastable Cu(OH)$_2$(s) (this work).*

Residual filterable metal concentrations began declining during phase 2, suggesting that some CuO(s) particles had reached a size which can be retained on the filter. During the initial increase of the turbidity a marked difference was noted in the residual metal concentrations from 0.2 and 0.45 micrometer membrane filters, indicating that a non negligible fraction of the particles was in the size range between the cut-off thresholds of the two filters. As expected, upon further growth of the particles, both residual concentrations tended asymptotically toward the same equilibrium value.

IAP values decreased asymptotically in phase 3. In this phase the growth of particles is likely to occur mainly via Ostwald ripening, in which larger particles grow at the expense of smaller ones, which redissolve. As a consequence of Ostwald ripening, a reduction in the specific surface area of the particles is expected. According to equation (2), the decrease of IAP in phase 3 can be attributed to the decrease of molar surface area of CuO(s) particles.

The trends described above are reproducible for the precipitation tests performed by adding a less than stoichiometric amount of NaOH to the copper solutions. However with increasing base/copper molar ratio in the reaction mixture, the sequential reaction steps, evidenced in Figure 2, appears to be shortened and the overall precipitation/aging process seems to be accelerated. On the basis of observations on color changes, it seems that the rate of hydroxide/oxide conversion is faster at base/copper molar ratio close to the stoichiometric ratio. When the NaOH added is larger than the stoichiometric value, flocculation is clearly visible during the slow addition of NaOH solution soon after the stoichiometric ratio is passed.

Figure 4 shows pH and conductivity trends in a solution of cupric nitrate, soon after the addition of an excess of NaOH. From a comparison with Figure 2, an apparent contrast in the trends of conductivity is observed. In phase 3 of Figure 2, conductivity increases in connection with a decrease of pH, whereas in Figure 4 both pH and conductivity decrease. By taking into account the equivalent conductivity of the major ionic species, the trends in conductivity can be related to the changes in ionic composition of the suspension. The apparent contradiction is explained by the data of Table 3, which shows a general satisfactory agreement between the measured trends in conductivity and those predicted by the concentration changes of the major ionic species. With a base/copper ratio < 2, Cu^{++} is in excess and the conductivity trend primarily reflects changes in the concentration of free copper. With excess base, changes in conductivity result from changes in the concentration of OH^-.

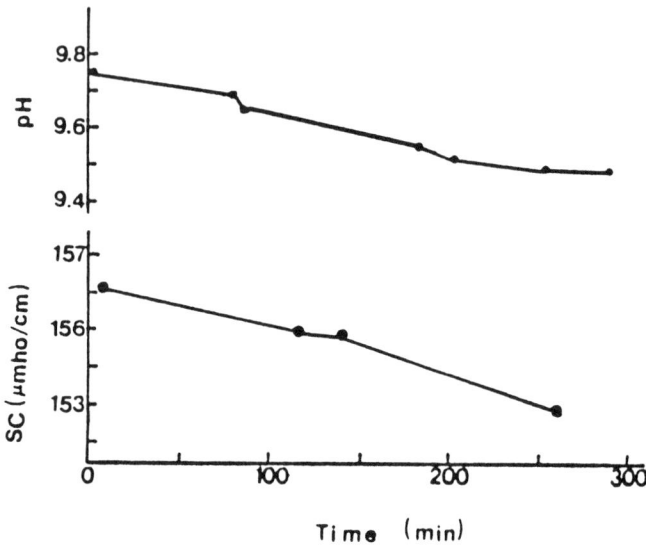

FIGURE 4. *Trends of pH and Conductivity (SC) after Completion of Reagents Mixing in a Batch Test with Slow Addition of NaOH Solution to $Cu(NO_3)_2$ Solution. Base/Copper Molar Ratio = 2.34.4*

TABLE 3. *Comparison of Measured and Predicted Conductivity Changes in Batch Tests. Predicted Changes are Calculated from Concentration Changes of the Major Ionic Species.*

Base Copper	Method of Addition	Phase	Major Ion	Measured ΔSC (mS/m)	Calculated ΔSC (mS/M)
1.2	B	I	H^+	0.05	0.01
		I-III	Cu^{++}	0.20	0.24
1.2	C	I	Cu^{++}	-0.25	0.19
		III		0.10	0.44
1.53	B	I	Cu^{++}	nc	0.07
		III		0.20	0.37
1.94	B	I	Cu^{++}	nc	0.06
2.34	B		OH^-	-0.35	0.46
2.34	C		OH^-	nc	0.04
2.5	B		OH^-	-0.21	0.11
3.4	B		OH^-	-0.27	0.80
3.4	C		OH^-	nc	0.35

nc = no change
Order of addition: B = base-to-copper; C = copper-to-base

4.2 Effect of Reagents Mixing Procedure

In order to demonstrate the effect of the mixing protocol, Figure 5 can be compared with Figure 2. The precipitation test of Figure 5 was performed by adding the cupric nitrate solution to the NaOH solution. All other experimental conditions were the same as those of Figure 2. During the addition of excess of metal ion to a NaOH solution, the system passes through the isoelectric point of cupric hydroxide-oxide, and visible flocs are formed during the addition. The visible flocs, which rapidly settle in the Pen Kem chamber, do not contribute to the turbidity measurements, which are very low soon after the addition. However, the presence of such particulate metal is evidenced by the low measurements of residual soluble metal concentration. Later on, the visible flocs seemingly dissolve or disperse, and the system seems to go through the same sequence of processes described in phases 2 and 3 of Figure 2.

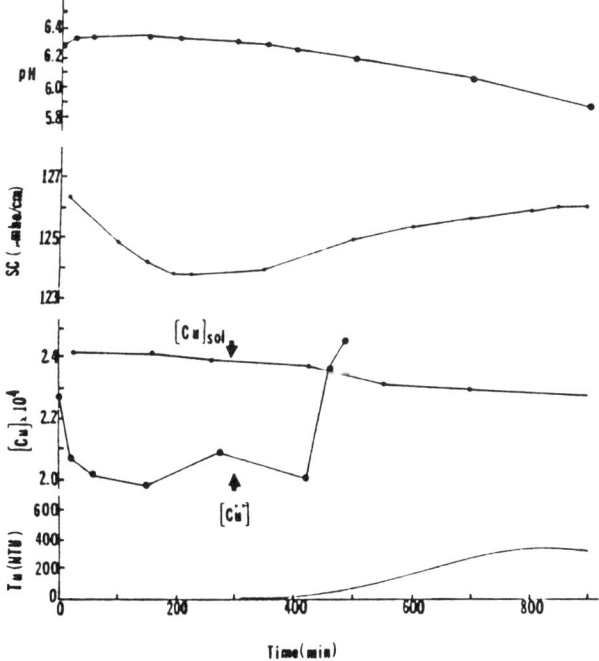

FIGURE 5. Trends in pH, Conductivity (SC), Free and Soluble Metal Concentration, and Turbidity (Tu) after Completion of Reagents Mixing, in a Batch Test with Slow Addition of Metal Solution to Base Solution. Base/Copper Molar Ratio = 1.2.

In Figure 5 a sharp initial decrease of free metal ion activity is observed after the addition of metal solution to the alkaline solution. This may reflect excess metal ion adsorption onto the surface of the precipitate previously formed. After a phase of essentially constant concentration, free cupric ion shows a sudden concentration increase in parallel with the increase of turbidity measurements. The increase in free metal ion activity in parallel with the increase in turbidity, which was noted in other runs as well, can be the result of either a real increase of metal ion concentration and of an analytical error. During the phase of turbidity increase, the system is still slowly approaching equilibrium conditions, with decrease of specific surface area of the precipitate. As a consequence, desorption of excess metal ion from the surface of the precipitate could produce a real increase in free metal ion activity in solution. However, in order to explain the sharp increase shown in Figure 5 we may also invoke an analytical error due to the light-sensitive membrane of cupric electrode [Orion, 1979]. It was found that a decrease in illumination conditions can cause a significant positive error in the response of the electrode [Boice, 988]. Hence, the increase in free metal ion activity concomitant with the increase in turbidity may be due also to the decrease in illumination around the electrode with increasing sample turbidity. This positive error may explain why the last measurements of free metal ion concentration appear to be unreasonably higher than total soluble metal concentrations.

Figure 5 shows also that in this case the conductivity trend consists of a first phase of sharp decrease, a second short phase of constant value and a third phase of slow but steady increase. Table 3 shows that in this run conductivity changes mainly reflect changes in concentration of free metal ion in solution. However, while the agreement between the measured change and the predicted change is quite satisfactory in phase 1, in phase 3, apart from agreement on the positive sign of change, a large difference in the absolute value of ΔSC is observed. This provides further evidence that in the measurements of the final sharp increase in free metal ion concentration an analytical error is superimposed on a real concentration increase.

Figure 6 provides an additional comparison between two batch tests performed with two addition orders and otherwise identical conditions. In this case, the base is in excess of the total metal in solution (base/metal molar ratio = 2.34). In either order of addition, a constant decrease of pH is evidenced with reaction/aging time. The decrease in pH is paralleled by a decrease in conductivity, because, at high pH values, OH^- becomes the major contributor to the conductivity (see Table 3). Figure 6 demonstrates a significant effect of the order of reagents addition both on pH and on conductivity measurements. In general, the metal-to-base addition order seems to result in a system initially closer to equilibrium, in comparison with the base-to-metal addition order.

The comparison of Figure 6 seems to suggest that in the slow addition of NaOH to $Cu(NO_3)_2$ the precipitate undergoes a more complex sequence of aging than in the reverse order of addition. In the former case, Figure 3 demonstrates that even precipitation of the cupric hydroxy nitrate is feasible at the beginning of the addition process. Hence, this hypothesized that the aging of the precipitate obtained with slow addition of NaOH implies the conversion of cupric hydroxy nitrate into hydroxide, followed by hydroxide to oxide conversion.

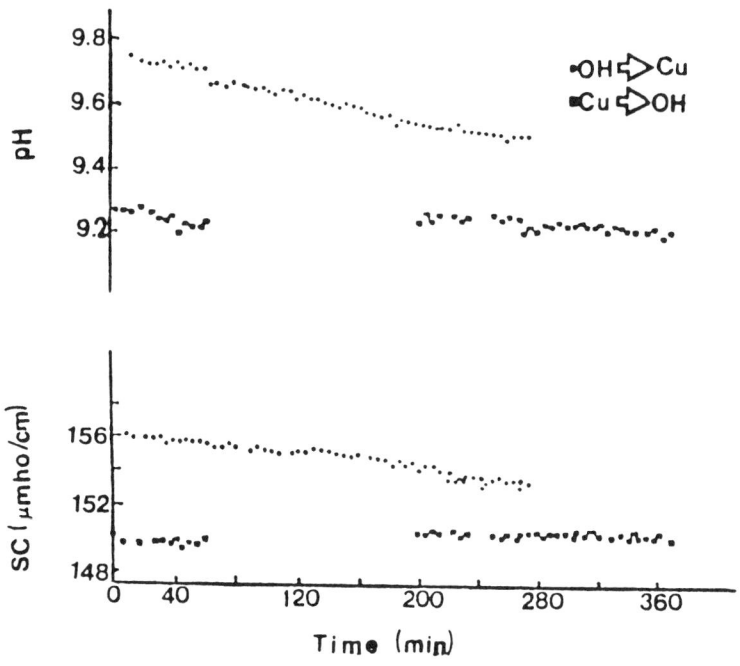

FIGURE 6. *Comparison of Trends of pH and Conductivity (SC) Following Reagents Mixing with Two Different Orders of Addition. Base/Copper Molar Ratio = 2.34.*

4.3 Thermodynamic Considerations

Figure 7 reports log{Cu^{++}} vs pH results at the end of the mixing process, and in aged suspensions (for several months). Results of the preliminary experiments with simultaneous addition of both reagents are also included in this plot. The fresh samples give values quite scattered and far from the theoretical line for CuO(s) (logK_{s0} = 7.6). This confirms the previous observation that the freshly obtained precipitate is a metastable, quite active solid, whose "non-thermodynamic" K_{s0} depends heavily upon mixing procedure and reaction time.

In contrast, the aged samples are all very close to the theoretical line for the equilibrium of CuO(s) (log{Cu^{++}} = logK_{s0} - 2pH). A least squares regression of log{Cu^{++}} vs pH yields the following 95% confidence intervals for aged samples:

slope = -1.98 ± 0.02 which encompasses the expected stoichiometric coefficient for H^+ (-2);

intercept = 7.8 ± 0.3 which encompasses the literature value of 7.6 for logK_{s0} of CuO(s).

The results in Figure 7 demonstrate two major points:
1. The potentiometric measurements in the aged samples give values of stoichiometric coefficient and K_{s0} quite close to those reported for CuO(s) formation. The slight difference with literature values is within the 95% confidence interval of our values. This means that the experimental error due to the lower illumination conditions in turbid samples can be neglected (small compared with overall random error of potentiometric measurements), at least in the aged samples where the precipitate was mainly in the aggregate state.
2. The free copper results for the aged solutions show close agreement to the theoretical CuO(s) equilibrium throughout the concentration range measured (0.36 × 10^{-13} - 3.0 × 10^{-4}M), even though the calibration was conducted in the range of 10^{-6} - 10^{-4}M. This supports the assumption that the electrode calibration can be extrapolated to measure very low Cu^{++} concentrations, in systems buffered with high total (soluble and insoluble) Cu(II) concentration.

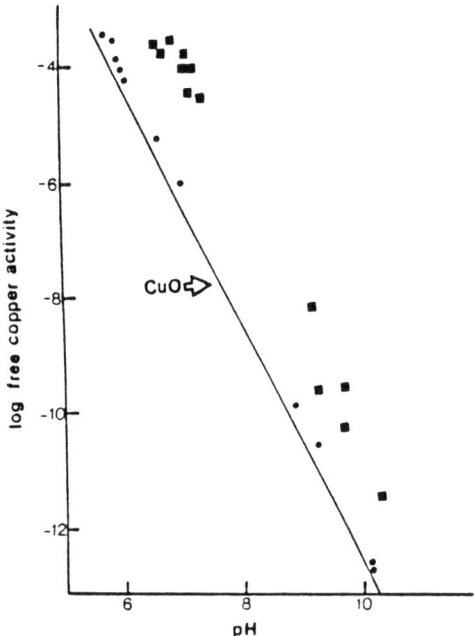

FIGURE 7. *Free Cupric Ion Activity vs pH in Fresh (■) and Aged (●) Precipitate Suspensions. Solid line Represents the Predicted Equilibrium Values for CuO(s).*

In Figure 8 the experimental pH values in aged samples are compared with the theoretical titration of a 5.0×10^{-4}M $Cu(NO_3)_2$ solution with NaOH. In the samples with less than stoichiometric additions of base, pH values are in close agreement with the theoretical values related to the formation of CuO(s). In contrast, in samples with excess of base the actual pH values are lower than expected. As we have already seen above, Figure 7 supports the presence of CuO(s) over the whole base/copper range investigated. Hence, low pH values must be attributed to some side reactions involving H^+. Despite the precautions taken to avoid uptake of atmospheric CO_2, the presence of significant concentrations of carbonate species (up to 1.8×10^{-4}M in the samples at high base/copper molar ratio) was determined in the aged suspensions through the analysis of inorganic carbon [Boice, 1988]. The presence of carbonate can be attributed mainly to impurities in the initial solutions (in NaOH solution, most likely) and to CO_2 uptake from the atmosphere during sample handling in the long aging period. The dashed line of Figure 8 is related to the CuO(s) equilibrium assuming that 10% of the alkalinity added with NaOH is in the form of carbonate. In the latter case, a closer agreement of the experimental values with the theoretical predictions seems to validate the hypothesis that the experimental/theoretical discrepancy in the high pH region is due to the presence of carbonate species in equilibrium with CuO(s).

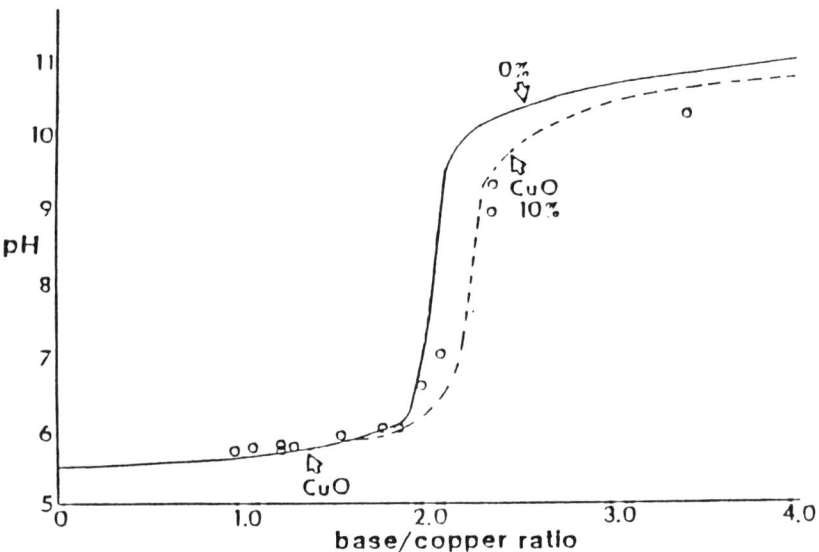

FIGURE 8. pH of Aged Suspensions vs Base/Copper Molar Ratio in the Reagent Mixture.
____*theoretical prediction for CuO(s) precipitation.*
--- *theoretical prediction assuming 10% carbonate alkalinity in NaOH solution.*

As for the solubility of Cu(II) hydrolysis species, Figure 9 compares the results of residual filterable copper with two theoretical models related to the system $CuO(s)-H_2O$. For batches with excess base, residual soluble metal concentration results were below the detection limit (2×10^{-7} M) of the analytical method used (AAS). In this region, theoretical computations of the solubility of CuO(s) give expected residual values either below or above the experimental detection limit, depending upon the value chosen for $\log B_2$. The results of this investigation seem to be in agreement with the low values of B_2 proposed by Baes and Mesmer (1976), Stella et al. (1979) and by Gulens et al. (1984) (see Table 1).

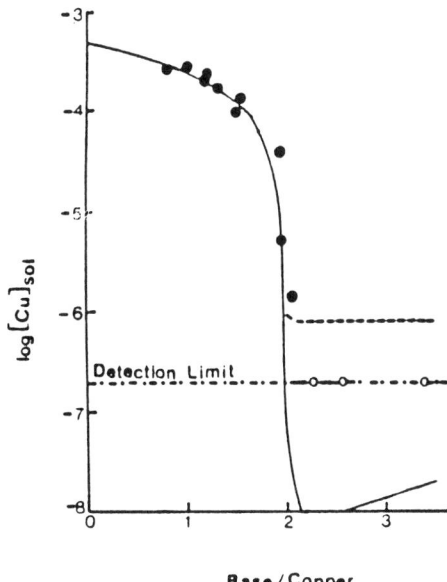

FIGURE 9. Residual Filterable Copper Concentration vs Base/Copper Molar Ratio in the Reagents Mixture. O = Below Detection Limit.
___ *theoretical prediction with* $\log B_2 = -16.0$. *[Stella et al., 1979]*
---- *theoretical prediction with* $\log B_2 = -13.7$. *[Vuceta and Morgan, 1977]*

4.4 Basic Salt Precipitation

A second series of batch tests was performed in order to check the potential formation of basic cupric salts at higher $Cu(NO_3)_2$ concentrations. Both base-to-metal and metal-to-base orders of addition were used. In this series of tests, total copper concentration was of the order of 5×10^{-3}M at the end of the reagent mixing step. Table 4 summarizes the composition of the precipitate obtained at several aging times.

Table 4 once again highlights the significant effect of mixing conditions and aging time upon precipitate composition. Comparing the two runs at base/copper = 1 but with reversed sequence of reagent addition, we can see a cupric hydroxy nitrate is readily obtained upon addition of base to the metal solution. In contrast, the addition of metal solution to the base gives a metastable hydroxide, which is very slowly converted into the basic nitrate. An opposite trend is observed at base/copper = 2.12 with slow addition of base to the metal solution. At such high base/metal ratio, the initially formed cupric hydroxy nitrate is unstable and, upon aging, is slowly converted into hydroxide form.

As for the aged precipitates, the preliminary results reported in Table 4 seem to indicate a stable cupric hydroxy nitrate component at low base/metal molar ratios in the reagents mixture. At higher ratios the hydroxide-oxide system seems to be more stable. In order to compare these results with theoretical predictions, a theoretical speciation was performed for a solution containing 5×10^{-3} M $Cu(NO_3)_2$ and predefined amounts of base. When the potential formation of basic nitrate or oxide are taken into account, $CuO(s)$ ($\log K_{s0} = 7.6$) is always indicated to be the most stable solid phase, independently of the base/copper molar ratio. However, if $Cu(OH)_2(s)$ ($\log K_{s0} = 8.7$) is allowed to form instead of $CuO(s)$, we obtain the distribution diagram reported in Figure 10. In this case a stable cupric hydroxy nitrate precipitates at low base/copper ratios, whereas cupric hydroxide becomes the predominant solid phase at higher base/copper ratios. The results of Table 4 show a noticeable inconsistency with the theoretical predictions of Figure 10, in that a stable cupric hydroxy nitrate may precipitate even when, from the thermodynamic standpoint, it is indicated to be a less stable solid phase than $CuO(s)$.

TABLE 4. Chemical Composition of the Precipitate Obtained in a Series of Batch Tests at Total Cu(II) Concentration = $(5 \pm 0.3) \times 10^{-3}$ M.

Total Cu(II) M	Base Copper	Order of Addition	Precipitate Composition after Several Reaction times (Nitrate/Copper Molar Ratio)					
			(a)	1d	2d	4-5d	6-7d	2wks	4wks
5.00E-3	1.00	OH ⇒ Cu	0.48		0.46	0.49			
5.26E-3	1.80	OH ⇒ Cu	0.52			0.52			
5.21E-3	1.84	OH ⇒ Cu	0.51				0.44		
5.00E-3	2.00	OH ⇒ Cu	0.19	0.11					
4.85E-3	2.12	OH ⇒ Cu	0.45		0.27	0.11			
5.00E-3	3.00	OH ⇒ Cu	0.00					0.00	
5.00E-3	1.00	Cu ⇒ Cu	0.03			0.35		0.41	0.41
5.13E-3	0.90	Cu ⇒ Cu	0.02						
5.00E-3	2.00	Cu ⇒ Cu	0.00						
4.86E-3	2.12	Cu ⇒ Cu	0.01						

(a) soon after reagents mixing.

The unexpected formation of a stable cupric hydroxy nitrate was also confirmed in a series of CSTR tests performed at $Cu(NO_3)_2$ concentrations in the range of 3.39 - 9.0 × 10^{-3}M and at base/copper molar ratios less than 2. This basic salt resulted even after long aging time (4 - 6 months). In addition to pH and free cupric ion activity, in these tests the potentiometric measurement of free nitrate ion activity was performed as well.

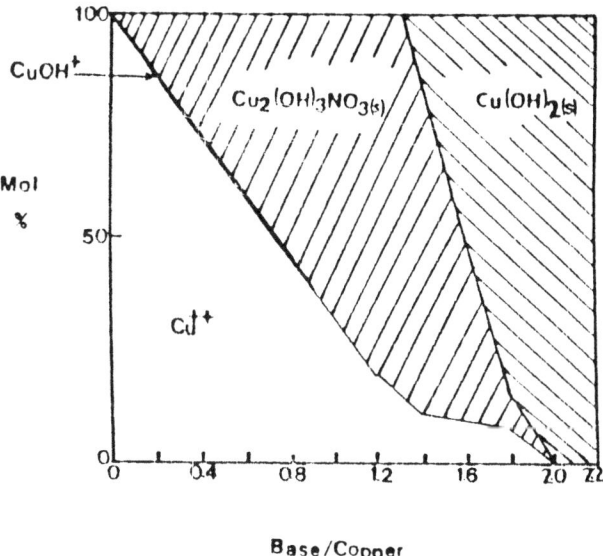

FIGURE 10. Theoretical Distribution Diagram of Soluble and Insoluble Cu(II) Species in a 5 × 10^{-3}M $Cu(NO_3)_2$ Solution, as a Function of Base Added (Base/Copper = Molar Ratio in Reagents Mixture).

This allowed determination of the Ion Activity Product (IAP) of the suspensions both with respect to cupric hydroxide-oxide (IAP_{OX}) and with respect to the basic salt cupric hydroxy nitrate (IAP_{BS}). LogIAP_{BS} (IAP_{BS} = $\{Cu^{++}\}^2\{H^+\}^{-3}\{NO_3^-\}$)) and Log$IAP_{OX}$ (IAP_{OX} = $\{Cu^{++}\}\{H^+\}^{-2}$) are reported in columns 4 and 5, and in column 6 of Table 5, respectively. LogIAP_{BS} results are reported both for fresh (column 4) and for aged (column 5) precipitates. It is interesting to note the marked change in IAP values due to the ageing process. In addition, Table 5 shows that IAP values of fresh precipitates are quite scattered, depending upon experimental conditions such as base/copper ratio and residence time. In contrast, final IAP values in the aged samples do not seem to depend upon the initial experimental conditions. This suggests that such values are related to a quite stable, inactive solid phase.

TABLE 5. *Ion Activity Product (IAP) of the Precipitate Obtained in CSTR Tests at Steady State Conditions.*

$[Cu]_T$ reaction	Base/Copper in	Res. Time	log(LAP_{BS}) (LAP = $\{Cu^{++}\}^2\{H^+\}^{-3}\{NO_s^-\}$)		log (LAP_{OX}) (LAP = $\{Cu^{++}\}\{H^+\}^{-2}$)
M	Mixture	min	Fresh	Aged	Aged
3.37E-3	0.84	8.0	10.60	8.35	8.26
9.98E-3	0.87	1.0	10.20	8.48	8.14
4.60E-3	1.84	1.0	10.88	8.36	8.31
9.00E-3	0.96	1.0	10.34	8.37	8.39
4.43E-3	0.80	1.0	10.02	8.54	8.40
4.79E-3	0.75	2.0	9.32	8.34	8.18

For the aged precipitates, the $logIAP_{BS}$ values seem to be lower than the reported $logK_{s0}$ for the cupric hydroxy nitrate ($logK_{s0}$ = 9.2 from Feitknecht and Schindler, 1963). On the other hand, $logIAP_{OX}$ values seem to be lower than the solubility of cupric hydroxide ($logK_{s0}$ = 8.7) but higher than the solubility of cupric oxide ($logK_{s0}$ = 7.6). This confirms the inconsistency with the theoretical predictions, which was already noted from the results of Table 4.

The limited preliminary data reported in Tables 4 and 5 do not allow us to hypothesize any plausible explanation for the inconsistency of our results with respect to the thermodynamic predictions. However, this inconsistency seems to be worth further investigation. Such investigation should provide a statistical evaluation of IAP values related to a large number of aged samples, as well as additional informations on the aged solid phases (chemical composition, crystallographic data).

5. SUMMARY AND CONCLUSIONS

The widely applied alkaline precipitation process for Cu(II) removal from wastewaters involves a quite complex set of concurrent and sequential phenomena. The aim of this investigation was a better understanding of such phenomena, in order to optimize the precipitation process on the basis of sound scientific principles. Two series of batch Cu(II) precipitation tests have been performed through mixing of base (NaOH) with metal solutions, at $Cu(NO_3)_2$ concentration levels of 5×10^{-4}M and 5×10^{-3}M, respectively.

In the presence of nitrate, the potential formation of three different Cu(II) precipitates must be taken into account: cupric hydroxide, cupric oxide, and cupric hydroxy nitrate. On the basis of theoretical solubility considerations, cupric oxide is expected to be the most stable solid phase at both levels of $Cu(NO_3)_2$ concentration experimented. Results, however, show that the chemical composition and solubility of fresh precipitates may significantly deviate from thermodynamic predictions. The properties of fresh precipitates heavily depend upon initial composition of the reagents mixture (base/copper molar ratio) and upon the experimental protocol for reagents mixing (whether base is slowly added to cupric nitrate or vice versa).

Results obtained in 5×10^{-4}M $Cu(NO_3)_2$ solutions suggest that cupric hydroxy nitrate may preferentially precipitate during the first part of slow addition of base to metal solution. Further addition of base induces conditions favorable for precipitation of hydroxide, while the previously formed basic salt is converted into hydroxide as well. Soon after completion of reagents mixing, a quite slow aging process is evidenced in the stirred suspension by changes in pH, conductivity, turbidity, free and soluble copper concentration. The concurrent measurement of these parameters during aging has resulted very meaningful in identifying a likely sequence of phenomena involved. Seemingly, cupric hydroxide nuclei grow up to a critical size, at which they become less stable than cupric oxide (small particles of cupric oxide are unstable due to high surface tension). This induces the conditions for conversion of the blueish, highly hydrated hydroxide into the black, less hydrated cupric oxide. Eventually, after months of aging, the properties of the suspensions resulted in close agreement with those theoretically predicted for cupric oxide systems.

Potentiometric measurements on aged precipitate suspensions are consistent, within the experimental error, with reported solubility products for cupric oxide ($\log K_{s0} = 7.6$). Literature values for the stability constant of the hydrolysis species $Cu(OH)_2^o$ range from $10^{-13.7}$ to $10^{-16.2}$. Our results on residual soluble copper concentration suggest that a value in the lower range may be more appropriate.

Results obtained in 5×10^{-3}M $Cu(NO_3)_2$ solutions show that, at low base/metal molar ratios, stable cupric hydroxy nitrate is produced upon aging. This seems to be in contrast with theoretical predictions, which indicate that, even at such high nitrate concentrations (10^{-2}M), cupric oxide is the most stable solid phase. Potentiometric measurements on such aged suspensions indicate systems undersaturated with respect to the basic salt, but oversaturated with respect to cupric oxide. Further investigation on hydrolysis-precipitation of Cu(II) in the presence of high nitrate concentrations is needed for a plausible explanation of the apparent inconsistency between our results and thermodynamic predictions.

In conclusion, this investigation points out the importance of ageing phenomena on chemical composition and physicochemical characteristics of the solid phase obtained in the alkaline precipitation of Cu(II). Chemical composition and solubility of metastable fresh precipitates may be quite far from theoretical predictions based on thermodynamic standpoint. In industrial practice, in which relatively short residence times are allowed for precipitation to occur, such intermediate forms can be even more important than the thermodynamically stable precipitates.

ACKNOWLEDGMENTS

The authors wish to acknowledge the financial support of the U.S. Environmental Protection Agency through the Industrial Waste Elimination Research Center.

REFERENCES

1. Armstrong, F.A.J. "Determination of Nitrate in Water by Ultraviolet Spectrophotometry", *Anal. Chem.*, 35:1292-1294 (1963).

2. Avdeef, A., J. Zabronsky, and H.H. Stuting. "Calibration of Copper Ion Selective Electrode Response to pCu 19," *Anal. Chem.*, 55:298-304 (1983).

3. Baes, C.F. and R.E. Mesmer. *The Hydrolysis of Cations*, Chapter 12, (John Wiley & Sons, New York, 1976).

4. Bowers, A.R., G. Chin, and C.P. Huang. "Predicting the Performance of a Lime-Neutralization/Precipitation Process for the Treatment of Some Heavy Metal-Laden Industrial Wastewaters," *Proceedings of the 5th Mid-Atlantic Industrial Waste Conference*, Ann Arbor Science, Michigan, (1981).

5. Britton, H.T.S. "An Electrometric and a Phase Rule Study of Some Basic Salts of Copper," *J. Chemical Society*, 12:2796-2807, (1925).

6. Britton, H.T.S. "Application of Electrometric Methods," *Chemical Society Annual Reports*, 40:43-44 (1943).

7. Boice, R.E. "Solution and Precipitate Characteristics for Hydrolysis-Precipitation of Cu(II) in the Presence of Nitrate", M.S. Thesis, Illinois Institute of Technology, Chicago, Il. (1988).

8. Feitknecht, W. and P.Schindler, *Pure Appl. Chemistry*, 6:130-199 (1963).

9. Gulens, J., P.K. Leeson, and L. Seguin. "Kinetic Influences on Studies of Copper(II) Hydrolysis by Copper Ion-Selective Electrode," *Analytica Chimica Acta*, 156:19-31 (1984).

10. Heijine, G.J.M. and W.E. van der Linden. "The Formation of Copper Sulfide-Silver Sulfide Membranes for Copper(II)-Selective Electrodes. Part III. The Electrode Response in the Presence of Complexing Agents," *Analytica Chimica Acta*, 96:13-22 (1978).

11. Karra, S.B., C.N. Haas, V. Tare and H.E. Allen. "Kinetic Limitations on the Selective Precipitation Treatment of Electronics Wastes", *Water, Air and Soil Pollution*, 24:253-265 (1983).

12. Kenner, C.T. and R.E. O'Brien, *Analytical Separation and Determinations*, Chapter 7, (Macmillan, New York 1971).

13. Jenista, J.L., "The Effect of Carbonate on the Solubility of Zinc and Copper," M.S. Thesis, Illinois Institute of Technology, Chicago, Il. (1984).

14. Navone, R. "Proposed Method for Nitrate in Potable Waters," *J.A.W.W.A.*, 56:781-783 (1964).

15. Neville and Oswald, *J. Phys. Chem.*, 35, 60 (1931).

16. Orion Research, Inc., *Instruction Manual for Cupric Ion Electrode*, (Cambridge, MA 1979).

17. Patterson, J.W., *Industrial Wastewater Treatment Technology*, 2nd edition, (Butterworth Publishers, Stoneham, MA, 1985).

18. Patterson, J.W. "Metals Separation and Recovery" in *Metals Speciation, Separation, and Recovery*, Patterson and Passino Eds., (Lewis Publishers Inc., Chelsea, Michigan, 1987).

19. Paulson A.J. and D.R. Kestner, "Copper(II) Ion Hydrolysis in Aqueous Solution," *Journal of Solution Chemistry*, 9 (4) 269-277 (1980).

20. Salutsky M.L., "Treatise on Analytical Chemistry," the *Interscience Encyclopedia*, Part 1, Volume 1, Chapter 18, (New York, 1959).

21. Schindler, P., H. Althaus, and W. Minder, *Helv. Chim. Acta*, 48,1204 (1965).

22. Schindler, P. "Heterogeneous Equilibria Involving Oxides, Hydroxides, Carbonates and Hydroxide Carbonates," in *Equilibrium Concepts in Natural Water Systems, Advances in Chemistry Series*, No. 67, American Chemical Society, Washington, D.C. (1967).

23. Sillen, L.G. and A.E. Martell. "Stability Constants of Metal Ion Complexes," *Special Publication*, No. 25, Chemical Society, London (1971).

24. Smith, R.M. and A.E. Martell, "Critical Stability Constants," Vol.4, *Inorganic Complexes*, (Plenum Press, New York, N.Y. 1976).

25. *Standard Methods For the Examination of Water and Wastewater 5th Edition*, APHA-AWWA-WPCF, (1985).

26. Stella, R. and M.T. Ganzerli-Valentini, "Copper Ion Selective Electrode for the Determination of Inorganic Copper Species in Fresh Waters", *Analytical Chemistry*, 51 (13) 2148-2151 (1979).

27. Stumm, W. and J.J. Morgan, *Aquatic Chemistry*, 2nd Ed., (John Wiley and Sons, New York, N. Y. 1981).

28. Swenson, P., "Nucleation and Particle Growth Studies on Cadmium Hydroxide Precipitation," M.S. Thesis, Illinois Institute of Technology, Chicago, Il. (1987).

29. Vuceta, J. and J.J. Morgan, "Hydrolysis of Cu(II)", *Limnology and Oceanography*, 22:742-746 (1977).

30. Walton, A.G., *The Formation and Properties of Precipitates*, (John Wiley and Sons, New York, N. Y. 1967).

31. Weast, R.C., *Handbook of Chemistry and Physics*, 53rd Ed., (CRC Press, Cleveland, Ohio, 1972).

32. Weiser, H.B., "The Hydrous Oxides and Hydroxides", *Inorganic Colloid Chemistry*, Vol. II Chapter V, (John Wiley & Sons, New York, 1935).

33. Westall, J.C., J.L. Zachary, and F.M. Morel, *MINEQL, a Computer Program for the Calculation of Chemical Equilibrium Speciation of Aqueous Systems*, Technical Note Number 18, Ralph M. Parsons Laboratory, Massachussetts Institute of Technology, Cambridge, Mass. (1976).

DISCUSSION OF:
HYDROLYSIS, PRECIPITATION AND AGING
OF COPPER(II) IN THE PRESENCE OF NITRATE

L. Campanella
Department of Chemistry, University of Rome, "La Sapienza", Italy

All the nitrates are soluble with the exception of some basic nitrates which are soluble only in strong acid solutions. The values of the stability constants of the complexes of nitrate are generally very low, in the range between 1.5 and -0.5. Also the solubility constants of the known basic nitrates are rather low (2.5 for bismuth, 2.2 for barium).

The precipitation equilibrium from $Cu(NO_3)_2$ in basic media can be represented as:

$$Cu(NO_3)_2 + OH^- \rightarrow Cu(OH)(NO_3) + NO_3$$

$$Cu(OH)(NO_3) + OH^- \rightarrow Cu(OH)_2 + NO_3$$

as NO_3 is the anion of a strong acid solubility of $Cu(NO_3)_2$ is not depending on pH, but as OH^- is the anion of a very weak acid, the water, the solubility of the basic nitrates is a function of pH, so that their formation may be investigated by the study of the influence of pH on the solubility:

$$Cu(OH)(NO_3) + H^+ \rightarrow Cu^{++} + NO_3 + H_2O$$

The addition of H^+ results in solubilization and dissociation of the basic nitrate.

The solubility is a higher solubility product as H^+ concentration, and as lower as the acid dissociation constant.

The mixed compounds can be imagined to be formed according to:

$Cu^{++} + H_2O \rightarrow Cu(OH)^+ + H^+$	pK = 7.3
$mCu(OH)^+ + pNO_3 \rightarrow Cu_m(OH)_m(NO_3)_p$	m = p
$2Cu^{++} + H_2O \rightarrow Cu_2O^{2+} + 2H^+$	pK = 10.5
$mCu_2O^{2+} + pNO_3 \rightarrow mCu_2O(NO_3)_2$	m = p/2

The interpretation furnished by the paper is substantially in terms of precipitation of mixed compounds (cupric hydroxy nitrate).

Adsorption Hypothesis

According to the Paneth-Fajans-Hahn rule a crystal lattice is able to adsorb other salts having with it a common ion. This phenomenon is as higher as less soluble is the compound between the common ion and the adsorbed ion of opposite sign. So silver iodide adsorbs better silver acetate than silver nitrate. Adsorption of a compound is further ruled by the electrolytic dissociation of it and has the ability to deform the adsorbed ion and also adsorption processes with the decrease of the former and with the increase of the latter. All these considerations applied to our system seem to advise us to conclude that adsorption is scarcely probable but it must be recalled that no other competitive adsorption can occur and so if conditions are favorable, adsorption can be observed as not being an easily adsorbed species. What happens is the precipitate of mixed composition is warmed when suspended or is formed in the presence of concentrated supporting electrolyte ions, and are able to compete in the adsorption processes.

Another point of interest concerns the exchanging power of the precipitates. In the considered system, the two involved anions are both univalent, so that there is no prevalence of one over the other according to the selectivity scale, but yet this phenomenon can attain meaning values if compared with pure coprecipitation phenomena. Probably the choice of sulphate in place of nitrate could aid in the comprehension of the reaction.

The influence of temperature also needs to be considered. Adsorption is decreased by increasing temperature. If a stoichiometric compound is formed, its composition must not be dependent on the experimental temperature, so experiments at different temperatures will have useful results.

Occlusion Hypothesis

By substitution of a cation or of an anion of the crystal lattice with a cation or an anion of the same charge (in this case OH^- possibly replaced by NO_3) and similar, dimensions-mixed crystals are formed. These can be described as solid solutions. Probably the comparison between favorable situations to this phenomenon, anions of similar dimension and unfavorable can clarify this aspect. The formation of the mixed crystals is inhibited by an excess of the primary ion and this can be related to the influence of OH^- concentration on the chemical composition of the precipitate. On the other hand, in the presence of other cations such as in a real matrix, these are surely occluded proportionally to the concentration of the reticular anions.

Some Observations

It should be very interesting to perform a systematic study to compare the composition of the solid phases when their precipitation occurring from homogeneous solutions with the one carried out in normal analytical conditions, by slowly adding the precipitating agent to a diluted hot-stirred solution of the ion to be precipitated (copper in this case).

The behavior of the system $Cu^{++}/OH^-/SO_4$, is substantially represented by the formation of two mixed compounds, copper sulphate tribasic and copper sulphate bibasic respectively, of formula $Cu_4H_6O_{10}S$ $[Cu_4(OH)_6(SO_4)]$ and $Cu_3H_4O_8S$ $[CU_3(OH)_4SO_4)]$ both occurring in nature as minerals; brochanite being the first and antlerite the second.

For some aspects, the results obtained for nitrate in the paper agree as tribasic copper nitrate is evidenced, and little doubt about the nonevidence for bibasic $CU_3(OH)_4(NO_3)_2$ (molar ratio = 0.66) from the adopted methodology is justified.

$CU(OH)_3(NO_3)$ is one of the salts for which x-ray powder data were standardized by ASTM. It could be useful to insulate the solid and to analyze it by an x-ray method.

Conclusive Remarks

Finally, the paper on one hand puts into evidence the importance of the aging phenomena but on the other hand points out the role of the intermediate compounds formed before the final one. The connection between the results and the possible applications for removal or recovery processes is obvious, but how is it possible to transfer the results obtained in synthetic matrices to real ones, especially when solubility equilibria, so easily influenced by the medium are considered? Conditional constants (but how to know them?) seem more likely to be applied than thermodynamic ones.

REFERENCES

1. Ginehn's Copper, 8th Ed., 60BB, 164 (1958).

2. Denk, Z. Leschorn, *Anorg. Chem.* 336, 58 (1965).

3. Glemser, Sauer, in *Handbook of Preparative Inorganic Chemistry*, Vol. 2, G. Brauer Ed., (Academic Press, New York, 1965), p. 1012.

4. Addison, Hathaway *J., Chem. Soc.*, 3099 (1958).

5. Frear, *Chemistry of Pesticides*, (VanNostrand, New York, 1951).

6. ASTM, X-Ray Powder Date File, Set 1-5 (1960).

7. Shaver, *Anal. Chem.* 28, 2015 (1956).

8. Schlezter, Sillen, *Acta Chem. Scand.* 4, 1322 (1950).

9. Hahn, *Applied Radiochemistry*, (Cornell University Press, Ithaca, 1939).

10. Walton, Walden, *J. Amer. Chem. Soc.* 68, 1742 (1946).

ROLE OF COPPER COMPLEXATION ON TREATMENT EFFICIENCY AND DESIGN OF TWO INDUSTRIAL WASTESTREAMS

Kostas Saranteas
Sr. Chemical Engineer, Polaroid Corporation

Irvine W. Wei
Assoc. Professor, Civil Engineering Dept. Northeastern University

1. ABSTRACT:

An organo-metallic dye manufacturing process train generates two major copper containing wastestreams. Preliminary bench scale tests indicate that chemical precipitation by sulfide or hydroxide techniques is only effective for one of the streams while activated carbon treatment is only effective for the other stream. Potentiometric and treatment results show that the reason for such treatment selectivity is that copper exists in its free ion/inorganic complex form for the precipitation favorable waste while it exists in its organo-metallic form for the carbon favorable waste. A segregated stream treatment is proposed for pilot studies, and is expected to yield copper removal at greater than 99% levels for both streams.

2. INTRODUCTION

In recent years, a number of specialized treatment processes have been developed and reported in the literature on heavy metals removal from industrial waste streams [4,5,7,8,9,10,12,13,14,17]. EPA's most recent report on effluent limitation guidelines for heavy metals [6] indicates that chemical precipitation by hydroxide and/or sulfide techniques is considered the best available technology. Such conclusion was based, however, largely on work published from the Metal Finishing Industries, where most heavy metals appear in their free ion or relatively unstable inorganic complex form. Industries such as organo-metallic dye manufacturers use starting materials that are "by definition" highly stable metal forms, and therefore, possibly highly resistant to chemical precipitation. EPA agreed in the same report that such industries may need to develop "unique" technologies for heavy metals treatment, and also effluent standards may have to be established on a case by case basis.

The present paper describes how two copper containing waste streams originating from such a dye manufacturing process respond to chemical precipitation and activated carbon treatment. The first waste, due to its low initial pH (0.45) is referred to, here, as the "Acidic" waste and the second waste, due to its high initial pH (13.5) is referred to as the "Basic" waste. The starting material for this multi-step process is copper pthalocyanine, an organic pigment specifically identified by EPA as a reagent generating complex metal bearing wastes.

The scope of this work included: Characterize the nature of copper in the two wastestreams (free ion, inorganic complex, organic complex); Test the effectiveness of chemical precipitation by the hydroxide and soluble sulfide techniques; Test the effectiveness of activated carbon treatment and the effect of various operating parameters (carbon type, temperature, pH) on carbon adsorption capacity; Examine the effectiveness of a combined stream treatment; Recommend a treatment system for pilot plant studies.

3. SOME THEORETICAL CONSIDERATIONS

3.1 Cupric Ion Electrode

The cupric ion electrode consists of a sensing element containing sulfides of copper and other metals bonded into an epoxy body [11]. When the electrode is in contact with a cupric solution, an electrode potential develops across the sensing element. This potential, which depends on the cupric ion concentration in solution, is measured against a constant reference potential with a digital MV meter. The measured potential corresponding to the level of cupric ion in solution is described by the Nernst equation:

$$E = E_o + S \log (A) \qquad (1)$$

Where E is the measured electrode potential (MV). E_o is the reference potential (a constant), A is the cupric ion activity and S is the electrode slope (25-30 MV per decade for Cu^{++}). The cupric ion activity is related to its concentration by the activity coefficient

$$A = \gamma C \qquad (2)$$

The activity coefficient is a variable largely dependent on ionic strength of the solution. Ionic strength is related to ion concentration Ci and ion charge Zi by:

$$\mu = \tfrac{1}{2} \sum_i (C_i\, Z_i^2) \qquad (3)$$

If the background ionic strength is high and constant relative to the sensing ion electrode, the activity coefficient is constant and activity is directly proportional to concentration.

3.2 Chemical Precipitation

Copper, just like most heavy metals, can be removed from a solution by a precipitating agent such as a sulfide or hydroxide ion according to a reversible reaction of the general form:

[Dissolved Copper] + [Precipitating Agent] ⇌ [Metal Salt]

In addition to the above reaction, however, dissolved copper acts as a Lewis acid (electron pair acceptor) and interacts with any dissolved species that could be considered Lewis bases (electron pair donor) and forms various dissolved metal complex ionic or non-ionic substances. Such Lewis bases can be either organic or inorganic in nature and can dramatically influence the effectiveness of the chemical precipitation treatment: If both the precipitating and complexing agents are present in a metal solution, then the form that will prevail is the one with the highest equilibrium constant.

3.3 Activated Carbon Treatment

It is well known that when activated carbon particles are placed in a solution containing an organic solute and the slurry is agitated or mixed to give adequate contact, the adsorption of the solute occurs. The solute concentration will decrease from an initial concentration C to an equilibrium value Ce, if the contact time is sufficient to ensure that equilibrium has been reached.

A number of empirical formulas have been developed to describe the adsorption equilibria [2,3]. The most commonly used formula is the so called Freundlich Isotherm, and one of its forms is:

$$\log \frac{x}{m} = \log K + \frac{1}{n} \log C_e \qquad (4)$$

Where X is the solute mass adsorbed, M the adsorbent mass used, and K,n empirical constants. This is the equation of a straight line of which the slope is $\frac{1}{n}$ and intercept is log K at Ce = 1. Although ideally straight line isotherm plots are obtained, there may be occasional departures from linearity [2]. A concave down, non-linear isotherm may be obtained if a non-adsorbable impurity is present. A sudden change in slope indicates, usually, more than one component present, each adsorbing at a different rate.

3.4 Experimental Procedure

Both chemical precipitation and activated carbon tests were performed in a batch-wise fashion as illustrated in Figure 1. A waste sample of 100 ml size was transferred to a 250 ml glass beaker. The waste was stirred using a magnetic stirrer apparatus for a predetermined period of time. Treatment chemicals were transferred, by pipette, from fresh stock solutions for the chemical precipitation tests, or by simple addition of a pre-weighed sample in the case of the activated carbon experiments. The whole test took place inside a constant temperature room at some predetermined temperature level. At the end of each test the treated sample was filtered via a 0.45 μm Buchner funnel setup and the filtrate was taken for subsequent copper and organics analysis. It is important to note here that the filtration particle retention capacity was limited to 0.45 μm, therefore, what is referred to as "residual" copper in the results section represents both dissolved and colloidal copper together. An Orion model 407A pH/ion analyzer was used to monitor the pH and also to serve as the potentiometer for the cupric ion electrode.

The cupric ion electrode used was an Orion Model 94-29. The reference electrode for the same measurement was an Orion electrode Model 90-02. Dissolved copper analysis was done using a Perkin-Elmer 560 atomic absorption spectrophotometer. Finally an Ionics Model 555 organic carbon analyzer was used for TOC analysis.

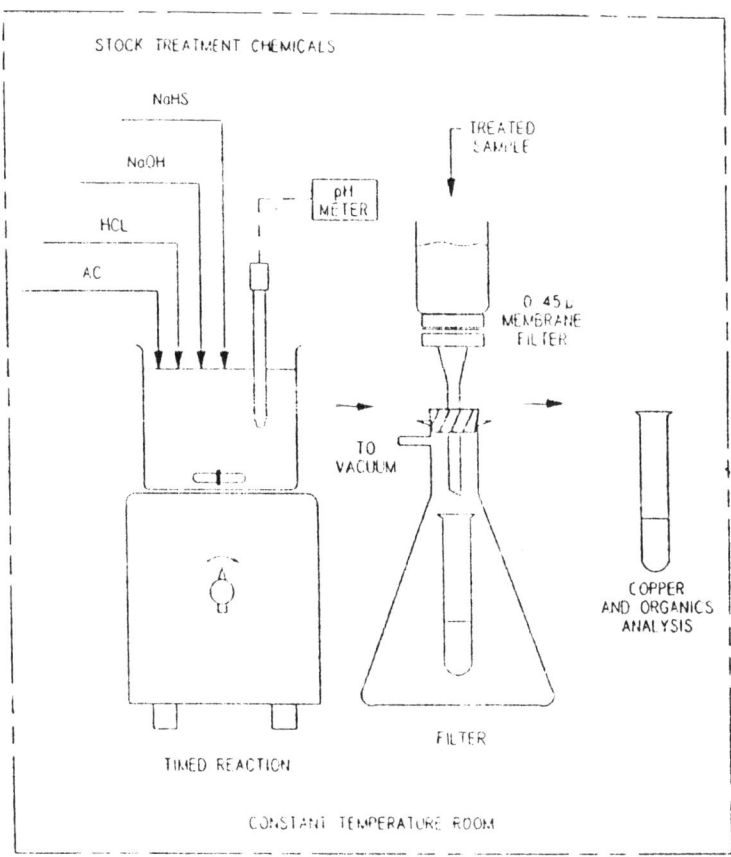

FIGURE 1. *Experimental Procedure for Batch-Reaction Studies.*

4. RESULTS AND DISCUSSION

4.1 Free Copper Analysis

Following standard procedures for the calibration of the cupric ion electrode [11], various Cu++ dilutions were prepared using a standard 1000 ppm solution. The electrode potential difference was recorded for each solution and Figure 2 shows the resulting cupric ion electrode calibration line. Based on this calibration line, the cupric ion concentration of the waste samples was extrapolated using the same electrode technique. Table 1 summarizes these results and it also includes a combined waste measurement.

TABLE 1.   Free Copper Concentration of Waste Samples.

Waste Type	Dilution*	pH(Final)	MV Reading	Cu^{++}(ppm)	Cu_T(ppm)	$\%Cu/Cu^{++}$	TAS
"Acidic"	1:307	2.1	137	63.0±16.1	139.5±16.1	45.2	
"Basic"	1:103	3.2**	25	0.0±0.04	156.9±12.5	0.0	
"50/50 Mixture"	1:103	2.3	143	36.0±4.6	148 ± 13.7	24.3	

*   Following manufacturer's recommendations, dilutions were used to minimize chloride ion interference effects and maintain pH within recommended range.

** pH was adjusted by using H_2SO_4.

The potentiometric results show quite clearly that the "Basic" waste contains no free copper at all, while approximately 50% of the "Acidic" waste copper is in its free ion form. The fact that only 24.3% of copper is free in the combined waste sample is consistent with the individual waste analysis.

4.2 Chemical Precipitation

Hydroxide Precipitation

Screening tests on the "Basic" waste hydroxide precipitation for pH values between 9 and 12 yielded no measurable copper removal even after 48 hours reaction time. The "Acidic" waste, however, did respond to hydroxide precipitation. Table 2, Figure 3 summarize the results of residual copper vs. adjusted pH after 48 hours reaction time, while Table 3, Figure 4 show the kinetic profile of the reaction around the minimum solubility pH. While the minimum solubility pH of the waste copper is well within the theoretical minimum solubility pH, the minimum residual copper concentration is not. The asymptotic behavior of residual copper around 2 ppm

TABLE 2. pH Effect on Residual Copper Concentration (Hydroxide Precipitation).

("Acidic" Waste, RXN Time: 48 Hrs.)

pH (Final)	Res. Cu (ppm)
8.5	8.67 ± 0.138
9.0	4.79 ± 0.052
9.5	3.16 ± 0.055
10.0	1.25 ± 0.094
11.0	2.36 ± 0.126
11.5	2.60 ± 0.157
12.0	3.15 ± 0.060
12.5	6.29 ± 0.658
13.0	10.61 ± 0.828

TABLE 3. Copper Hydroxide Precipitation Kinetics of "Acidic" Waste Around Minimum Solubility pH ($^-10$).

Reaction Time (Hrs)	Residual Copper (ppm)		
	PH ADJ. =9.75	PH ADJ. = 10.00	PH ADJ.=10.25
1	14.88 ± 0.126	14.10 ± 0.155	13.60 ± 0.156
2	10.37 ± 0.184	8.97 ± 0.215	7.64 ± 0.137
4	5.53 ± 0.159	8.16 ± 0.208	5.49 ± 0.206
6	3.35 ± 0.100	5.55 ± 0.098	5.47 ± 0.147
24	3.26 ± 0.071	3.92 ± 0.064	2.72 ± 0.049
72	2.75 ± 0.068	2.65 ± 0.060	2.18 ± 0.034

is probably due to the presence of some colloidal copper hydroxide that passed the filtration setup, and therefore, was accounted as "dissolved" copper.

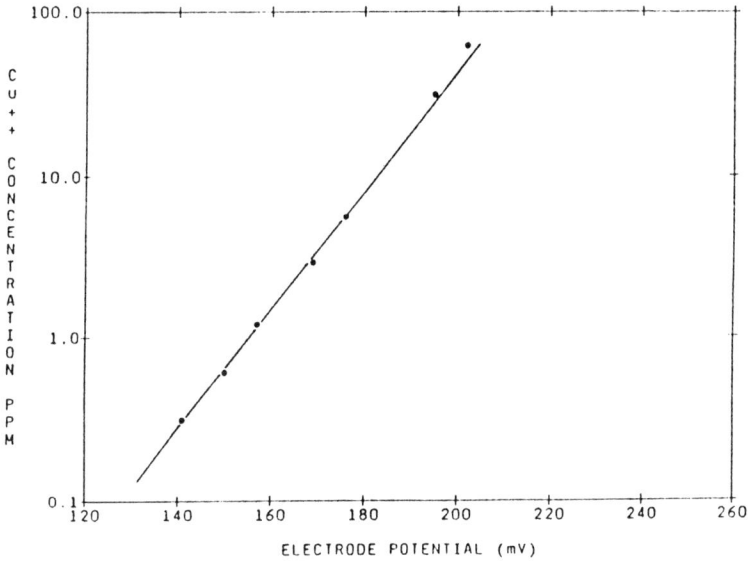

FIGURE 2. Cupric Electrode Calibration Line.

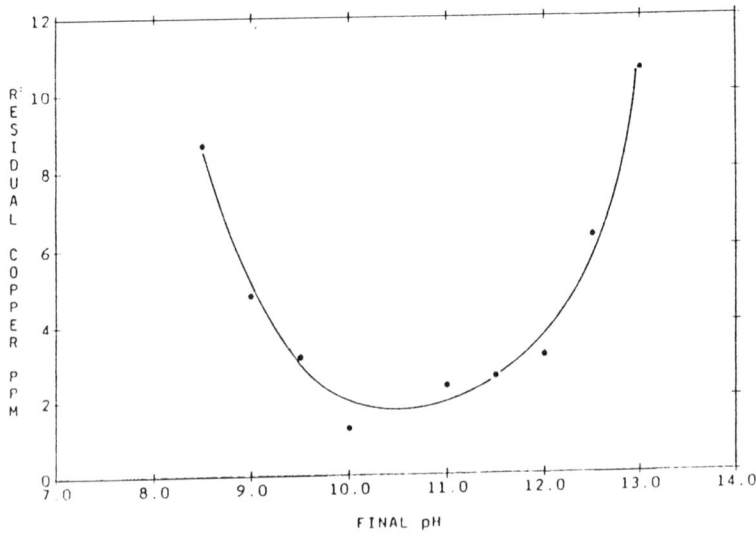

FIGURE 3. pH Effect on Copper Concentration (Hydroxide Precipitation) ("Acidic Waste").

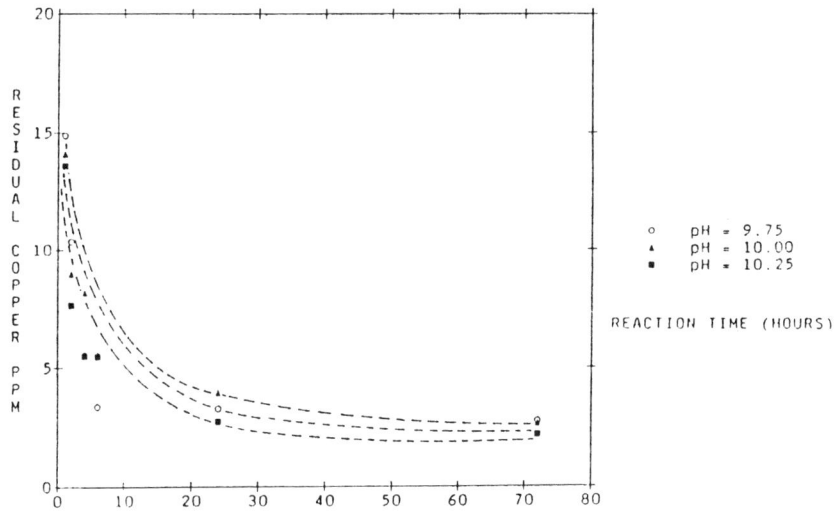

FIGURE 4. *Copper Hydroxide Precipitation Kinetics Around Minimum Solubility pH ("acidic" Waste).*

Sulfide Precipitation

Screening tests, using a NaHS solution as the sulfide source for treating the "Basic" waste showed no measurable dissolved copper concentration reduction. The "Acidic" waste, however, responded dramatically to sulfide treatment showing instantaneous precipitation as quickly as the sulfide reactant was added to the waste sample. Table 4, 5 and Figure 5 show the effect of reactant dose and adjusted pH on residual copper and organics concentration.

During the sulfide treatment tests, it was observed that the precipitant appearance and sample filterability were strongly affected by the reactant dose, sample pH and reaction time. Figures 6,7,8 illustrate some of these effects in a semi-quantitative manner. Figure 6 shows the strong dependence of precipitant size on reactant dose, with such dependence less dramatic (but with larger particles) as the sample pH is adjusted around 8.0.

Figure 8 shows the effect of reaction time on filterability of the "Acidic" waste and the residual copper concentration. Filterability here was expressed as a cycle time required to filter a fixed amount of treated sample through a fixed filtration apparatus under a fixed level of vacuum as the driving force. The fact that filterability changes, over approximately the first 20 minutes of reaction time, while the dissolved copper concentration remains constant, probably suggests some kind of particle sized redistribution in the solid phase, The simplifying assumption here is that filterability is directly related to particle size with increasing particle size leading to decreasing specific surface area of the solids, and therefore, decreasing frictional losses and filtration resistance. Similar behavior of a phase transformation while the reaction system is in chemical equilibrium has been reported by Peters et al. [14].

4.3 Activated Carbon

Both wastes were treated with various doses of pulverized PWA activated carbon (obtained from Calgon Corp.) in the constant temperature room for 24 hours. After carbon filtration, treated samples analysis for copper and organics yielded some new trends as illustrated in Figure 9: Although both "Acidic" and "Basic" waste organic carbon were effectively removed by the carbon, it was only the "basic" copper (opposite to chemical precipitation results) that responded favorably to the carbon treatment.

TABLE 4. Effect of Reactant (s^{--}) Dose on Residual Copper and Organics Concentration for the "Acidic" Waste. (pH = 0.45, T = 30°C, RXN Time = 1 Hr., Sample Size = 100 ml).

Sample	Reactant Dose (ml)*	Stoich. Demand**	TOC (ppm) (Measured)	TOC (ppm) (Corrected)***	RES. Cu (ppm)
1	0	0	795	795	139.5 ± 16.1
2	1	0.3x	646	652	124.8 ± 3.6
3	2	0.6x	687	700	119.3 ± 1.2
4	3	1 x	662	682	110.24 ± 1.2
5	4	1.3x	658	684	102.9 ± 1.2
6	5	1.6x	615	646	98.3 ± 0.7
7	7	2.3x	628	672	84.2 ± 0.8
8	10	3.2x	687	756	47.8 ± 0.8
9	20	6.5x	602	722	1.37 ± 0.02
10	40	12.9x	453	634	0.45 ± 0.01
11	100	32.3x	417	834	0.23 ± 0.01

* Reactant Solution: 4 gr NaHS per liter D.I. H_2O.
** 3.1 ml. reactant is the stoichiometric demand per 100 ml Sample per 140 ppm Cu^{++} concentration for the elementary Reaction:
Cu^{++} S^{--} = Cu S↓.
*** Corrected for dilution effects due to reactant addition.

TABLE 5. Effect of Reactant (S^{--}) Dose on Residual Copper After pH Adjustment for the "Acidic" Waste. (pH = 8.0, T = 30°C, RXN Time = 1 Hr.).

Sample	Reactant Dose		Residual Copper (ppm)
	(ml)	Stoich. Demand	
1	0	0	139.5
2	3	1.0x	3.0
3	5	1.6x	2.06
4	10	3.2x	1.73
5	15	4.8x	1.22
6	20	6.5x	1.12
7	40	12.9x	0.82

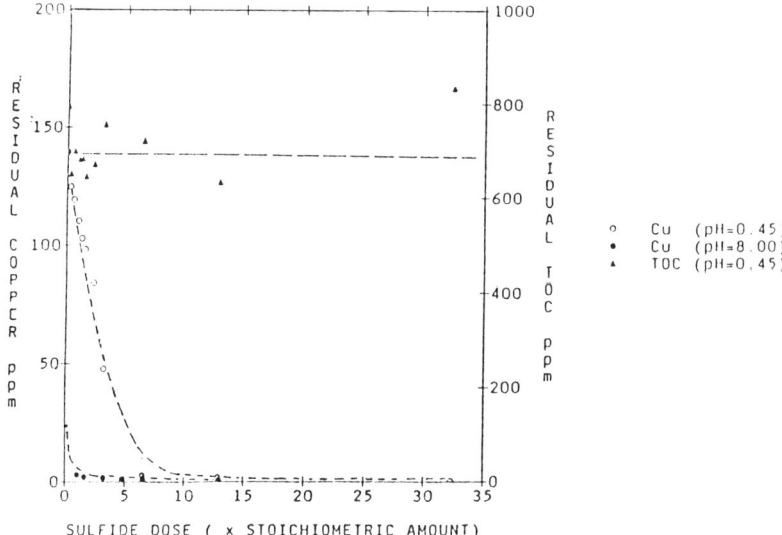

FIGURE 5. Effect of Sulfide Dose on Effluent Copper and Organics ("Acidic Waste").

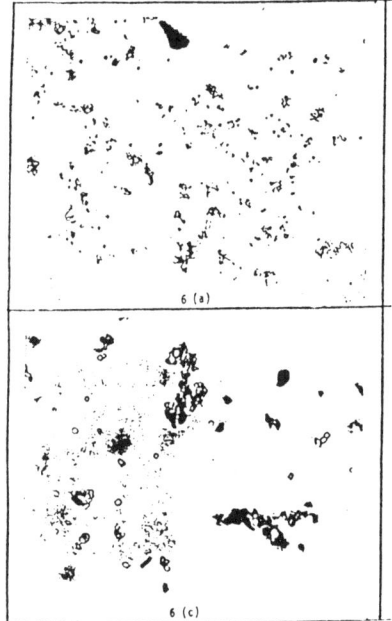
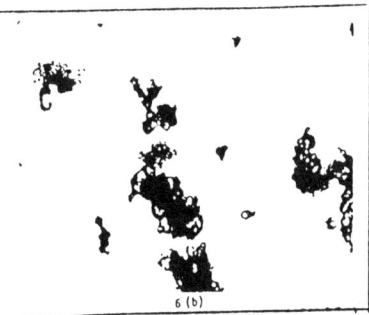

FIGURE 6. *Micrographs Showing Effect of Sulfide Dose on Particle (CuS) Size: (pH = 0.45 Unadjusted)*

Micrograph	Dose	Appr. Size (μm)
6(a)	2.8x	3
6(b)	6.5x	30
6(c)	12.9x	13

FIGURE 7. *Micrographs Showing Effect of Sulfide Base on Particle Size (pH Adjusted: 8.0)*

Micrograph	Base	Appr. Size (μm)
7(a)	6.5x	60
7(b)	12.9x	35
7(c)	17.0x	50

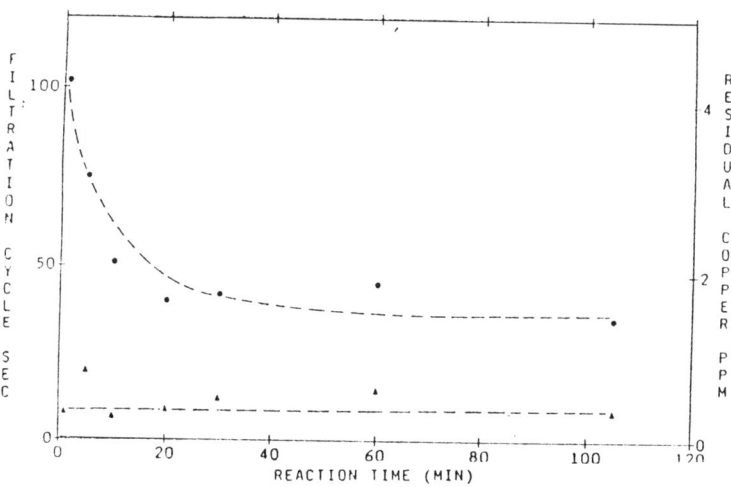

FIGURE 8. *Effect of Reaction Time on Filtration Cycle and Residual Copper ("Acidic Waste").*

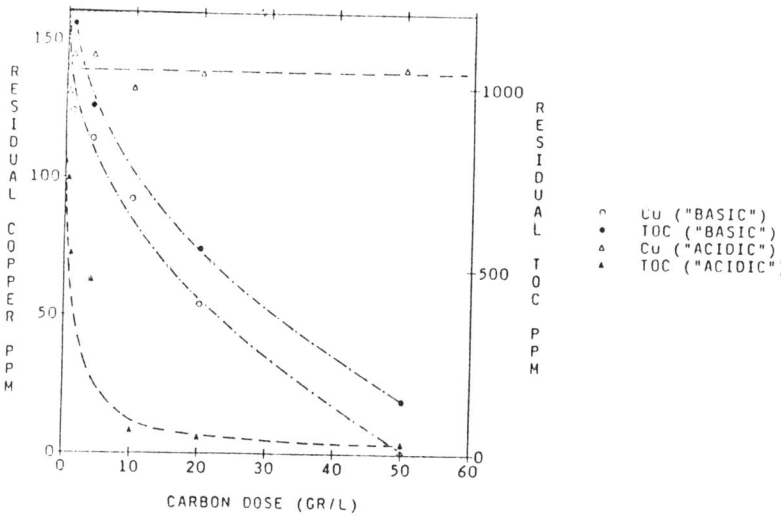

FIGURE 9. *Effect of Carbon Dose on Copper and TOC Removal for Both Wastes (T = 10°C. Unadjusted pH's).*

Not only the "basic" waste dissolved copper was effectively treated by carbon but a strong correlation between residual copper and residual organics with carbon dose was clearly observed. Such correlation was not evident at all in the "acidic" waste results suggesting that probably the "basic" copper is organometallic in nature while the "acidic" copper is not. Further screening tests in the same batch fashion were performed to optimize some of the parameters influencing adsorption capacity. Figure 10 illustrates the adsorption kinetics of the carbon treatment.

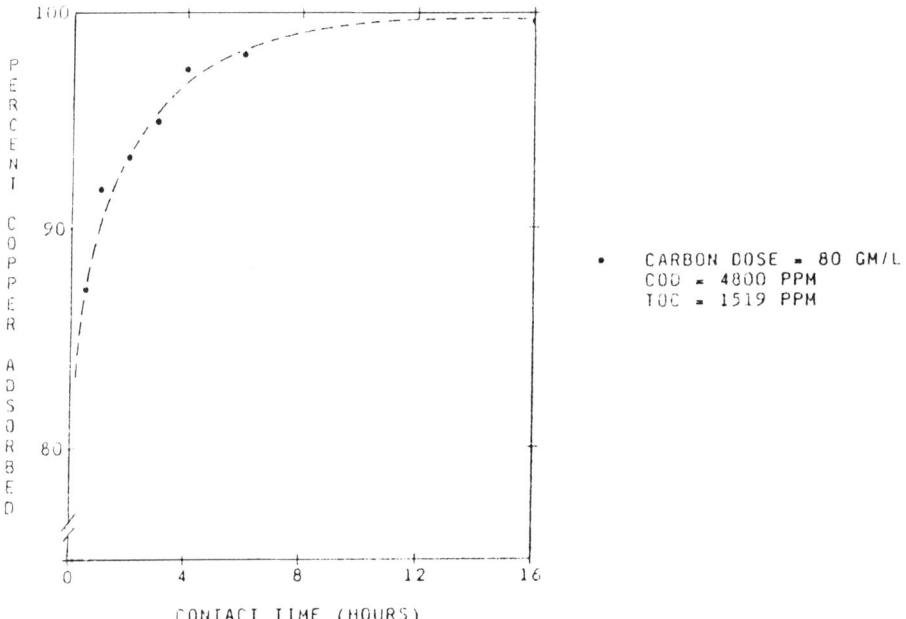

FIGURE 10. Kinetics of Dissolved Copper Adsorption with Activated Carbon ("Basic Waste").

Figure 11 shows the effect of carbon type on adsorption capacity and Figures 12, 13 show the temperature and pH adjustment effects on treatment efficiency.

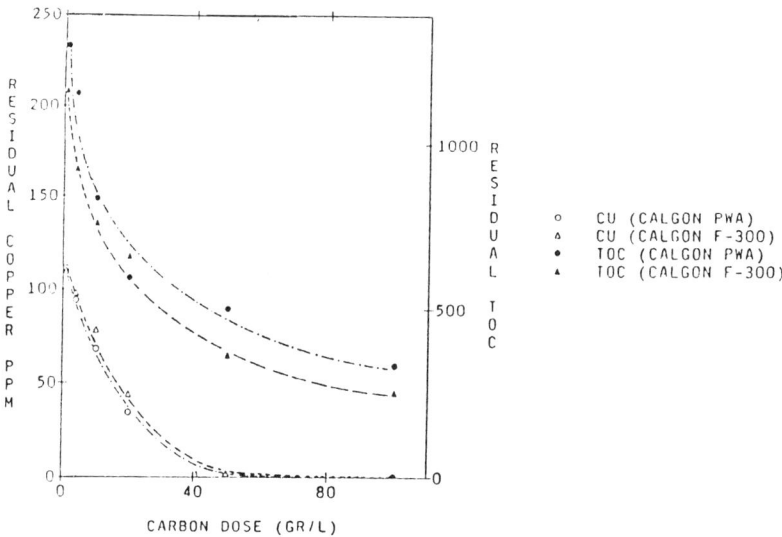

FIGURE 11. Effect of Carbon Type on Copper and TOC Removal (pH = 13.4; "Basic" Wastes; T = 30°C).

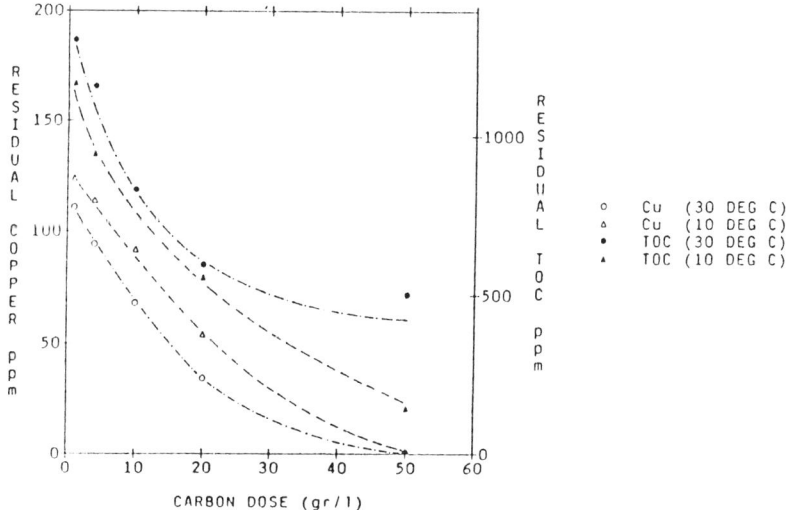

FIGURE 12. Temperature Effect on Carbon Adsorption Capacity for Copper and TOC Removal (pH = 13.4; "Basic" Waste; PWA Carbon).

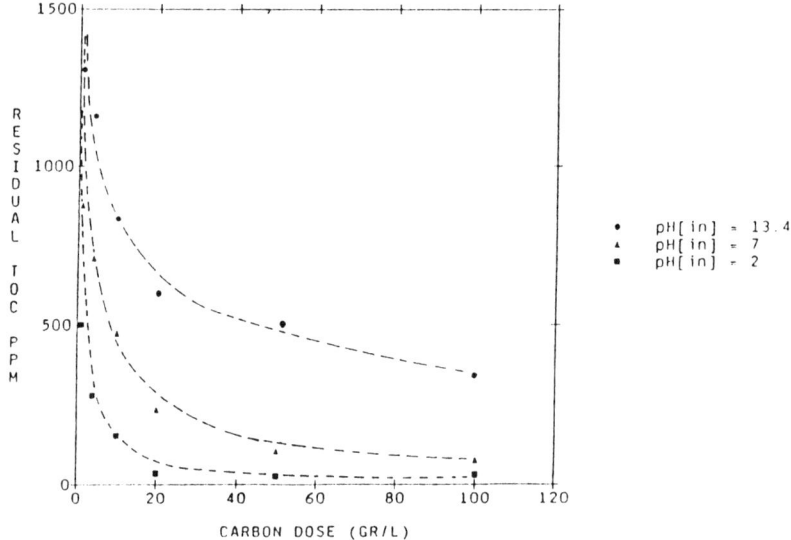

FIGURE 13 (a): *Effect of pH on Carbon Adsorption Capacity for TOC Removal ("Basic" Waste).*

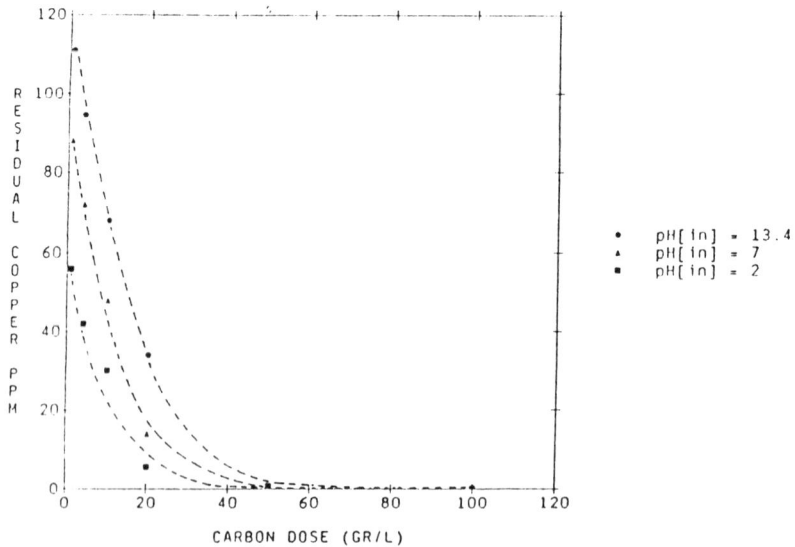

FIGURE 13 (b): *Effect of pH on Carbon Adsorption Capacity for Copper Removal ("Basic" Waste).*

Based on these results, a new series of experiments were performed under the "best" possible conditions from a practical point of view (lower pH, higher temperature) to get the highest adsorption capacity. Figures 14, 15 show the resulting Freundlich isotherms from one of these tests. A linear regression analysis of the resulting isotherms gives an isotherm slope of 1.570 with a correlation coefficient of 0.959 for TOC treatment and an isotherm slope of 1.499 with a correlation coefficient of 0.983 for the copper treatment.

A modified non-linear correlation between residual TOC and copper isotherms can be also concluded if it is observed that the TOC isotherm shows at least two distinct breaks suggesting three different organic species present and adsorbed at different rates and capacities, while the copper isotherm shows one break suggesting at least two copper containing organometallic species present. Further analytical work into identifying the "non-copper containing"/"poorly adsorbing" organic species indicated that it was methylene chloride which is the standard solvent used in the process train, and has indeed a poor standard adsorption capacity of only 1.6 mgr of solvent per gram of carbon addition.

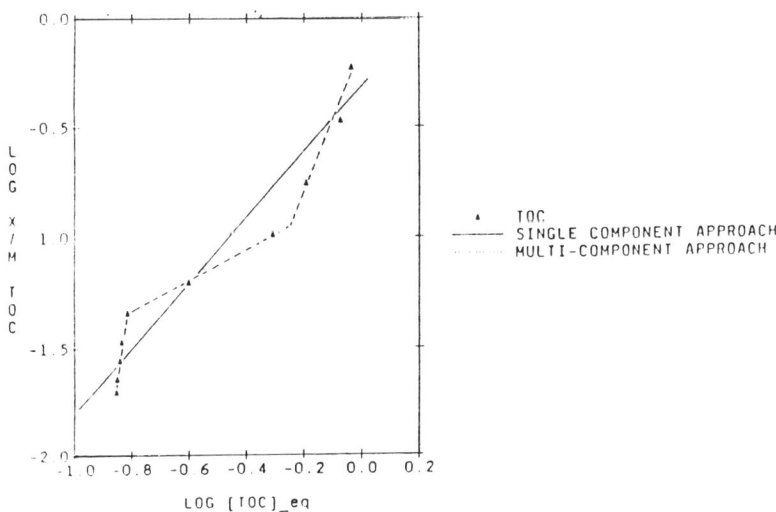

FIGURE 14. Freundlich Isotherm of "Basic" Waste TOC Treatment with PWA Carbon (pH = 6.0; T = 30°C).

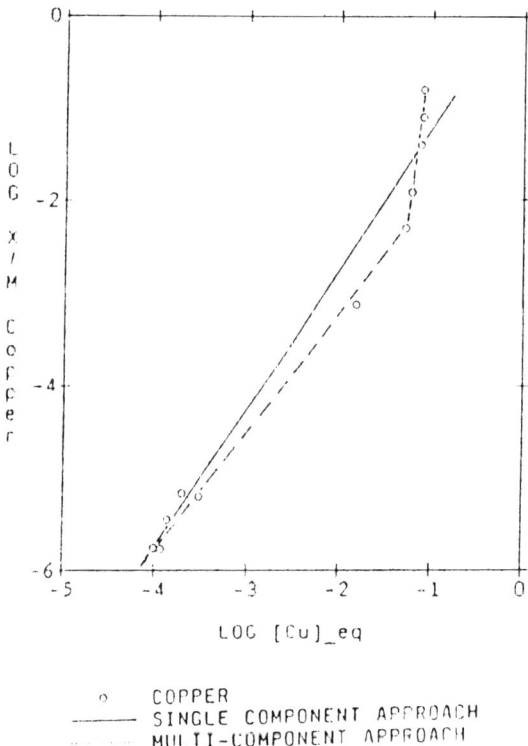

FIGURE 15. Freundlich Isotherm of "Basic" Waste Copper Treatment with PWA Carbon (pH = 6.0; T = 30°C).

4.4 Combined Waste Treatment

A new waste sample was prepared by combining 50% by volume of each one of the two wastes. Such waste was treated with both sulfide precipitation treatment and the activated carbon treatment. Figures 16, 17 show the results of the combined waste tests. These results, following the expected trends from the segregated treatments, show a copper removal of only about 50% suggesting the impracticality of combining the two streams since they only increase the waste volume, but not the treatment efficiency (They don't exclude a sequential treatment, however).

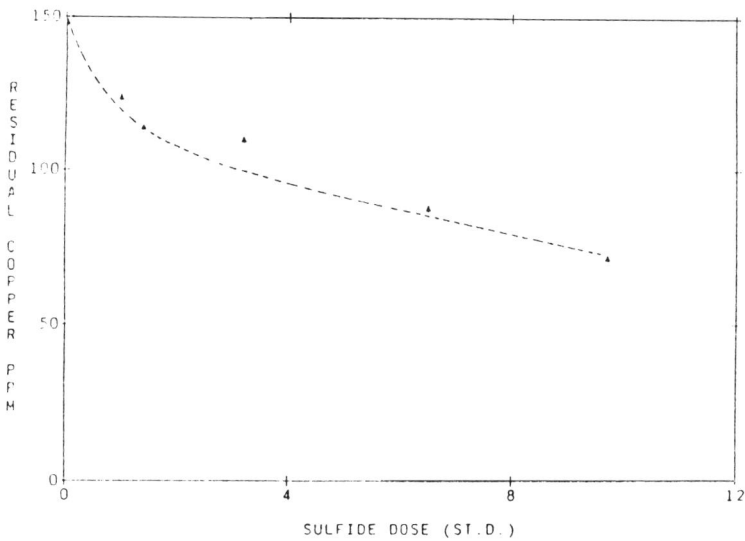

FIGURE 16. Effect of Sulfide Treatment on Residual Copper for a 50% "ACIDIC" / 50% "BASIC" Combined Waste (pH=0.8; T=30°C).

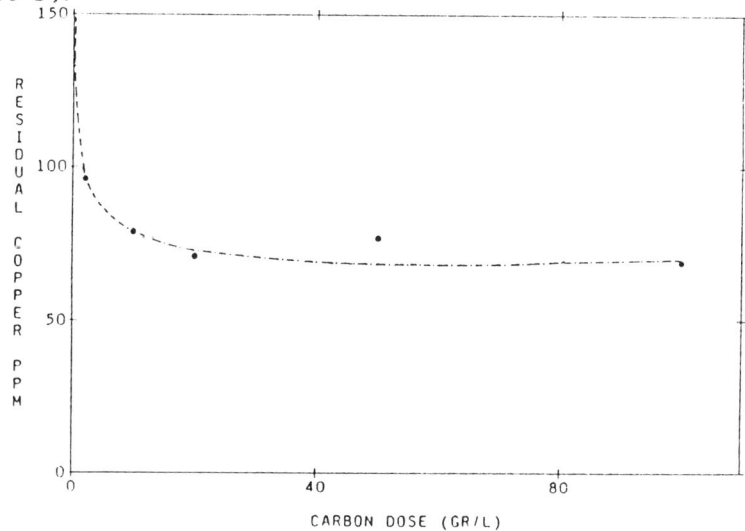

FIGURE 17. Effect of Carbon Treatment on Residual Copper for a 50% "ACIDIC" / 50% "BASIC" Combined Waste (pH = 0.8; T = 30°C).

5. CONCLUSIONS

The results of both chemical precipitation and activated carbon treatments for copper removal from the two wastestreams are summarized in Table 6.

TABLE 6.   Copper Treatment Comparison Table.

Waste Type	Hydroxide Precipitation		Sulfide Precipitation		Activated Carbon	
	$Cu_{T,OUT}$ (ppm)	% Removal	$Cu_{T,OUT}$ (ppm)	% Removal	$Cu_{T,OUT}$ (ppm)	% Removal
"Acidic" (Cu_{IN} = 139.5 ppm)	1.25	99.1	0.23	99.8	135.5	0.0
"Basic" (Cu_{IN} = 156.9 ppm)	156.9	0.0	156.9	0.0	0.11	99.9
50-50 Mixture (Cu_{IN} = 148.0 ppm)	-	-	72.0	48.6	69.0	46.6

In conclusion, none of the three treatments tested is effective for both wastes. The reasons for such ineffectiveness are independently explained from the analytical and treatment results. The "Acidic" waste contains copper in its free ion and relatively unstable inorganic complex form, and that's the reason chemical precipitation is an effective treatment. The "Basic" waste contains copper exclusively in its highly stable organometallic form, and that's the reason it resists chemical precipitation, but shows strong affinity for carbon.

Figure 18 shows a proposed treatment system for pilot scale studies based on the results from this work. It involves a segregated stream treatment in order to minimize waste volume handling, followed by a stream combination to facilitate final pH adjustment.

The pilot studies will focus on a number of unanswered questions and scale-up issues. Some of the major ones are:
a) What is the effect of process variability on treatment efficiency; b) Solid/liquid separation performance of the precipitation reaction and need for flocculating agent use; c) How real is the H S from the excess sulfide use, and the need for an oxidizing agent use; d) Carbon column performance optimization.

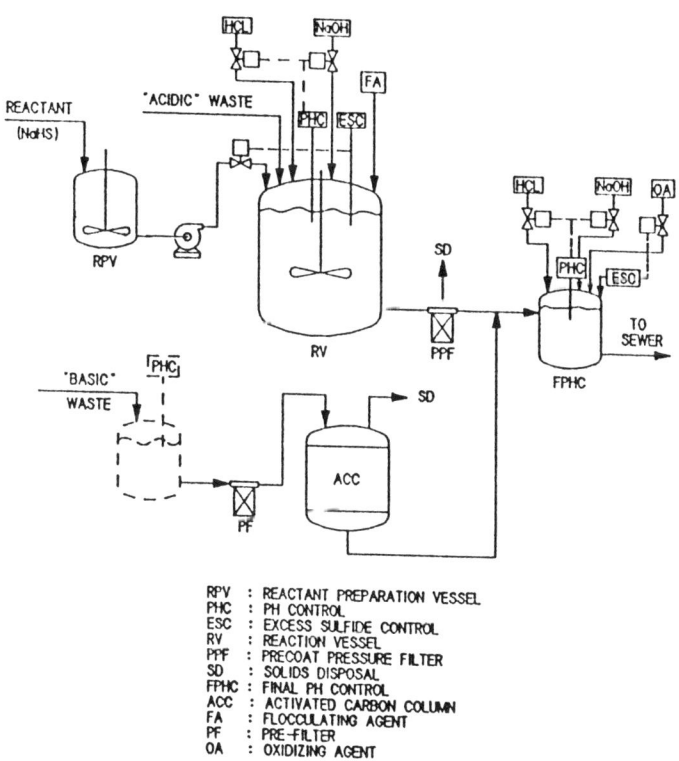

FIGURE 18. *"Semi-Combined"* Copper Wastewater Treatment Preliminary Flow Diagram.

REFERENCES

1. Benefield, L.D., J.F. Judfins and B.L. Weand, *Process Chemistry for Water and Wastewater Treatment*, (Prentice Hall Publ., 1982).

2. Calgon Corp., "The Laboratory Evaluation of Granular Activated Carbons for Liquid Phase Applications," Technical Bulletin Series. Bulletin 23-60b, (1986).

3. Cheremisinoff, P. and F. Ellerbush. "Carbon Adsorption Handbook," *Ann Arbor Science.* (1978).

4. Dean J.G., F.L. Bosqui, and K.H. Laouette. "Removing Heavy Metals from Wastewater," *Env. Sci. Techn.*, 7:692-694, (1972).

5. EPA Publications. "Summary Report: Control and Treatment Technology for the Metal Finishing Industry; Sulfide Precipitation," EPA 625/8-80-003, (1980).

6. EPA Rules and Regulations. "40 CFR parts 414 and 416-OCPSF Category Effluent Limitations Guidelines, Pretreatment Standard and New Source Performance Standards; Final Rule," Vol. 52, No. 214, (Nov. 1987).

7. Hoffman, M. "Thermodynamic, Kinetic, and Extrathermodynamic Considerations in the Development of Equilibrium Models for Aquatic System".

8. Inoue, Y.and M. Munemori. "Coprecipitation of Mercury (II) with Iron (III) Hydroxide," *Env. Sci. Techn.*, 13:443-445, (1979).

9. Kim, B.M.and P.A. Amodeo. "Calcium Sulfide Process for Treatment of Metal Containing Wastes," *Env. Progress.* 2(3):175-180 (1983).

10. McDonald, C.W. and R.S. Bajwa. "Removal of Toxic Metal Ions from Metal Finishing Wastewater by Solvent Extraction," 12(4):435-445, (1977).

11. Orion Research, *Cupric Electrode, Model 92-29 Instruction Manual.* Technical Bulletin Series.

12. Patterson, J. et al. "Carbonate Precipitation for Heavy Metal Pollutants," *J. WPCF*, (Dec. 1977).

13. Peters R., Young Ku and D. Bhattacharaya. "Evaluation of Recent Treatment Techniques for Removal of Heavy Metals from Industrial Wastewaters," *AIChE Symposium Series*, No.243, Vol. 81, (1985).

14. Peters, R., Young Ku, and Tsun-Kuo Chang. "Heavy Metal Crystallization Kinetics in an MSMPR Crystallizer Employing Sulfide Precipitation," *AIChE Symposium Series*, No 240, Vol. 80, (1984).

15. Reynolds, T. *Unit Operations and Processes in Environmental Engineering*, (PWS Publishers, 1982).

16. Shoeyink, V.L. and D. Jenkins. *Water Chemistry*, (John Wiley and Sons Publ., 1982).

17. Whang, J.S., D. Young, and M. Pressman. "Soluble Sulfide Precipitation for Heavy Metals Removal from Wastewaters," *Env. Progress*, 1(2):110-113, (1982).

DISCUSSION OF:
ROLE OF COPPER COMPLEXATION ON TREATMENT EFFICIENCY AND DESIGN OF TWO INDUSTRIAL WASTESTREAMS

Alan R. Bowers
Department of Civil and Environmental Engineering, Vanderbilt University

Soon H. Cho
Department of Environmental Engineering, Ajou University, Suwon, Korea

INTRODUCTION

The authors have presented an interesting case study for the treatment of waste containing organometallic compounds, specifically copper phthalocyanine, $Cu(C_8H_4N_2)_4$. Organometallic wastes represent unique treatment difficulties and little is known about the aqueous chemistry of most of these compounds.

Two wastes were compared, one containing 45.2% "free" copper ("Acidic" waste) and the other containing no "free" copper ("Basic" waste). A 50:50 mixture was also prepared and treated. No detailed analysis of the wastes was reported, inhibiting a comprehensive evaluation of the author's data. The wastes have come from an industrial site, i.e. real, not synthetic, and it is not uncommon nor unreasonable for industries to prohibit the publication of exact details. However, copper phthalocyanine is commercially available in reasonably pure form (Aldrich Chemical, Inc., Milwaukee, Wisconsin). Some experimental data for synthetic wastewaters would have been much more helpful from a general standpoint.

Free Copper Analysis

The authors have reported data on "free" copper (as Cu^{+2}) and total copper based on specific ion electrode analysis and atomic absorption (see Table 1). Although this has some usefulness, especially for the "Basic" waste where none appeared in the "free" form, the results for the "Acidic" waste is questionable. For the "Acidic" waste, only 45.2% was reported in the "free" form. However, this waste was diluted by a 307:1 ratio and the pH was fixed at 2.1. According to the authors, dilution serves to eliminate interference from chloride ions, but it must be realized that dilution will also reduce the concentrations of metals and ligands and perturb the equilibrium that exists in solution. This equilibrium is also highly pH dependent and the fraction of "free" copper is certain to change significantly over the range of treatability (9.5 < pH < 11.5: from Figure 3).

It is obvious from the treatability data presented that the character of the "Acidic" waste is entirely different than the "Basic" waste. For the "Basic" waste, no copper was removable by precipitation, either as hydroxides or sulfides, while the copper was almost completely removed (see Table 2, 1.25 ppm at pH 10.0) from the "Acidic" waste by hydroxide precipitation. This indicates that the 54.8% of the copper which was not "free" (Table 1) exists as a weak complex (probably organic) which was not strong enough to prevent precipitation of copper hydroxide. On the other hand, the "Basic" waste must have contained the copper phthalocyanine, which is an irreversibly associated organometallic compound. The reviewers recommend that a comparison of "free" copper as a function of pH using the specific ion electrode would have yielded two distinct behaviors for these wastes. For example, this has been done by Elliott and Huang for a copper-NTA system [18,19].

Chemical Precipitation

The authors compared hydroxide and sulfide precipitation for the two wastes, and a mixed waste.

In the case of "Basic" waste, neither form of precipitation worked due to the characteristics of the organometallic compound, which has the copper irreversibly tied up. This should not have been unexpected.

For the "Acidic" waste, the copper was removed by hydroxide precipitation. In these experiments, the lowest obtainable copper concentration was 1.25 ppm (Table 2), which is well above the theoretical minimum solubility of copper hydroxides, i.e. less than 0.1 ppm (19). The authors attribute this to the formation of small colloids (< 0.45 microns), which passed through their filters. However, these reviewers have found that even 0.01 micron particles are retained by the 0.45 micron filters, since they clog rather quickly. Instead, we would propose that some weak complex-forming agents are present (as indicated by the authors in Table 1) which are able to compete with hydroxides at pH 10.0, leaving a soluble residual copper complex in solution. These complexes may many times be disrupted by the addition of ferric iron as a coprecipitant species which can out-compete copper for the complex-forming agents, and/or provide a secondary adsorbing colloid for the copper complexes [21,22,23]. It should also be pointed out that further results during kinetic studies indicated that the minimum solubility for copper hydroxides was greater than 2.18 ppm (See Table 3). This further emphasizes the magnitude of the complex formation problem and demonstrates some variability in the waste characteristics and behavior.

The sulfide was somewhat more successful for precipitation of the acidic waste. However, the authors chose to use only two pH values, 0.45 and 8.0, to evaluate sulfide. The lower pH value, although achieving low residual copper concentrations 1.37 to 0.23 ppm (Table 4), required excessive sulfide dosages (6.5 to 32.3 x stoichiometric, respectively). At pH 8.0, comparable results were obtained using smaller dosages (3.2 x 12.9 x stoichiometric). On a theoretical basis, the precipitation of sulfides is an extremely pH dependent phenomenon, or based upon the equilibrium equations for sulfide precipitation with Cu^{+2} and S^{-2} as the active species:

$$[Cu][S] = K_{sp}/(\alpha_2 \gamma_o) \tag{1}$$

where

Cu = soluble copper hydroxide species, Cu^{+2}, $CuOH^+$, $Cu(OH)_2$, etc.

S = soluble sulfide acid species, H_2S, HS^-, S^{-2}.

α_2 = fraction of soluble sulfide acid species as S^{-2}

γ_o = fraction of soluble copper hydroxide species as Cu^{+2}

or:

$$\alpha_2 = K_{a1}K_{a2}/(K_{a1}K_{a2} + K_{a1}[H^+] + [H^+]^2)$$

and, ignoring polynuclear species:

$$\gamma_o = 1/(1 + K_1[OH^-] + \beta_2[OH^-]^2 + \beta_3[OH^-]^3 + \beta_4[OH^-]^4)$$

All of these constants are available for copper sulfide from standard reference sources [24]:

$$\log K_{sp} = -36.1^* \qquad \log K_1 = 6.3^*$$

$$\log \beta_2 = 12.8$$

$$pK_{a1} = 6.61 \qquad \log \beta_3 = 14.5$$

$$pK_{a2} = 13.8 \qquad \log \beta_4 = 15.6$$

*all values at $I = 1.0$ M @ $25°C$ except for K_{sp} and K_1, which are reported at $I = 0$ M.

The conditional solubility product is then defined as:

$$K_{sp}\text{-conditional} = K_{sp}/(\alpha_2 \gamma_o)$$

The calculated values, based upon the constants reported for CuS, H_2S and Cu^{+2} species are shown in Figure 19. The conditional solubility product reaches a minimum value at pH 7.7, indicating the minimum solubility of CuS around this point. The pH of 0.45 used for one set of experimental data (Table 4) has a much higher conditional solubility product, log K_{sp}-conditional = -16.5, indicating a much higher potential solubility due to competition from other ligand species having lower acidity constants (< 6.61) than bisulfide (HS^-). A rigorous evaluation of sulfide precipitation should have included more experimental data around the minimum solubility point. This may have resulted in reduced chemical dosages of sulfide to obtain acceptable results. In addition, poor attention was paid to the physical properties, and thus the separability of the copper sulfide particulates that formed. Figure 8 shows the filtration time for a "fixed amount of sample. . . under a fixed level of vacuum" as a function reaction time. Great care must be taken in interpreting and using these results for the design of a treatment system. First, the physical experimental conditions are ill-defined. The reactor configuration (vortex formation?) and degree of mixing and shear (velocity gradient) are unspecified. These will greatly influence the rate of particle flocculation as well as the final size and shape [25,26]. Second, it must be pointed out that sampling and handling procedures prior to the particle size micrographs (Figure 7) are extremely important. It has been shown that even pipetting can disturb the particle configuration and size [27].

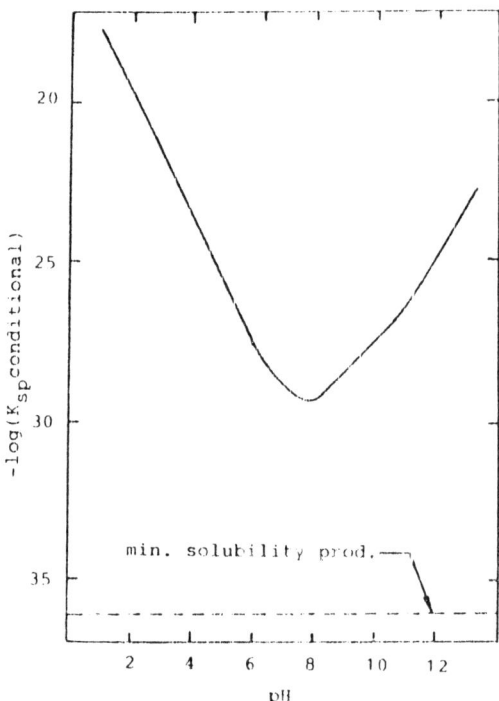

FIGURE 19. *Conditional Solubility Product as a Function of the pH.*

Activated Carbon

Activated carbon was evaluated for removal of the organometallic compounds, since precipitation was not successful for the "Basic" waste. The carbon forms used were each successful in removing both copper and most of the total organic carbon, with pH playing little role in the removal of copper (Figure 13b). However, treating the "Basic" waste with activated carbon might not be an appropriate technique for several reasons:

1. Even though the dissolved copper waste was treated to some degree by activated carbon, further treatment or a much larger amount of activated carbon will be required to meet the effluent standard (Figure 11).

2. According to Figures 13(a) and 13(b), at least 60 g/L of activated carbon is required for the efficient removal of both TOC and copper from the "Basic" waste. This carbon dosage is excessive, indicating that adsorption is not favorable.

3. Regeneration of the carbon has not been discussed by the authors. However, it does not appear that pH adjustment may be used to effectively strip off the copper organometallic compounds (See Figures 12 and 13). High temperature regeneration probably will not be effective, since the copper will be left behind in an inorganic form to poison the carbon surface. Solvent regeneration and recovery of the organometallic may be possible but needs investigation.

TABLE 7. Summarizes the Cost of Activated Carbon for the Treatment of "Basic" Waste Based on the Experimental Results with no Regeneration of the Carbon.

Type of Activated Carbon	Cost^a(\$/1,000 gal)
PWA	1,040
F-300	515

a: Based on a dosage of 60 g/L using a cost of \$2.08/lb for PWA and \$1.03/lb for Filtrasorb 300. Costs obtained from Calgon, Inc., Pittsburgh, PA, April 11, 1989.

By comparison, total treatment costs of \$40/1,000 gal (powdered activated carbon) and \$16/1,000 gal (granular activated carbon) for control of volatile organic hydrocarbons, have been reported [28]. Considering other reagents, capital costs and carbon disposal costs (hazardous or non-hazardous), the total treatment cost of the process proposed is unacceptable.

SUMMARY

The authors have investigated waste treatment schemes for the removal of copper in the organometallic form, "Basic" waste, and in the "free" form, "Acidic" waste. For the "Acidic" waste, copper was removed by hydroxide or sulfide precipitation. However, the "Basic" waste could not be treated by precipitation, due to the organometallic form of copper. Subsequently, activated carbon was shown to be an effective adsorbent at high dosages (40 to 60 g/L).

Although effective, regeneration was not discussed and activated carbon adsorption appears to be economically unfavorable. Other treatment alternatives should have been considered. Ion exchange is not possible, since copper phthalocyanine is nonionic. Other alternatives should include chemical oxidation for example, using catalyzed hydrogen peroxide, ozone, permanganate, or a wet air oxidation process. These could oxidize the organic portion, leaving copper in a "free" or weakly complexed state that could be precipitated. In addition, solvent extraction could be evaluated as a treatment alternative, or for regeneration of the spent activated carbon.

Although this review has been somewhat critical of the authors' treatability choices, it must be recognized that little is known about the treatment of organometallic wastewaters or the behavior of these compounds, either in the environment or during conventional treatment processes. The authors are to be commended for adding substantially to the literature regarding the behavior or organometallics.

REFERENCES

18. Elliott, H.A. and C.P. Huang. "Adsorption of Some Copper(II)-Amino Acid Complexes at the Solid-Solution Interface. Effect of Ligand and Surface Hydrophobicity," *Environ. Sci. Technol.*, 14, 87, (1980).

19. Elliott, H.A. "The Adsorption of Cu(II) at the Solid- Solution Interface: Effect of Complex Formation," Ph.D. Dissertation, (University of Delaware, Newark, Delaware, 1979).

20. Bowers, A.R., G. Chin, and C.P. Huang. "Predicting the Performance of a Lime-Neutralization/Precipitation Process for the Treatment of Some Heavy Metal-Laden Industrial Wastewaters." *Proceedings of the 5th Mid-Atlantic Industrial Waste Conference,* C.P. Huang, Editor, Ann Arbor Science, Ann Arbor, Michigan, 51, (1981).

21. Kolthoff, I.M. "Theory of Coprecipitation," *J. Phys. Chem.*, 36, 860, (1932).

22. Bowers, A.R. and C.P. Huang. "Role of Fe(II) in Metal Complex Adsorption by Hydrous Solids," *Water Research*, 21, 757, (1987).

23. Ortiz, C.A. "Removal of Soluble Copper and Chromium by Coprecipitation with Iron(III)," Ph.D. Dissertation, (Vanderbilt University, Nashville, Tennessee, 1987).

24. Smith, R.M. and A.E. Martell. *Critical Stability Constants*, (Plenum Press, Inc., New York, New York, 1974).

25. Ives, K.J., Editor, "The Scientific Basis of Flocculation," *Proceedings of the NATO Advanced Studies Institute*, (Sijthoff and Noordhoff, Inc., Alphen aan den Rijn, Netherlands, 1978).

26. Pandya, J.D. and L.A. Spielman. "Floc Breakage in Agitated Suspensions: Effect of Agitation Rate," *Chem. Eng. Sci.*, 38, (1983).

27. Gibbs, R.J. and L.N. Konwar. "Effect of Pipetting on Mineral Flocs," *Environ. Sci. Technol.*, 16, 119, (1982).

28. Patterson, J.W. *Industrial Wastewater Treatment Technology*, 2nd Ed., (Butterworth Publishers, Boston, Massachusetts, 1985).

PART III: METAL SPECIATION AND COMPLEXATION IN NATURAL SYSTEMS

SPECIATION OF ALUMINUM IN GEOTHERMAL BRINES: COMPARISON OF DIFFERENT METHODOLOGIES

M. Achilli, G. Ciceri, and R. Ferraroli
CISE S.p.A. - Tecnologie Innovative, Segrate (MI), Italy

G. Culivicchi and S. Pieri
ENEL/UNG (Italian Electricity Board/National Geothermic Unit), Pisa Italy

1. SUMMARY

A comparison of different methods for aluminum speciation in geothermal brines is reported and discussed with the view of the optimization of an accurate and easy in field operable procedure to separate monomeric from polymeric forms of the element.

The tested methods involve complexation with pyrocatechol violet (PCV) (3,3', 4' -trihydroxyfuchsone sulphonic acid), retention on strong cationic resin or inorganic exchangers beds, chelation with supported 8-hydroxyquinoline (8-HQ), filtration or ultrafiltration, solvent extraction. Some of the selected methods have been studied as a function of several parameters (Al concentration, flow rate, pH, temperature, brine composition, concentration of competitive binding agents).

The results show the selectivity of PCV in complexing monomeric Al and the usefulness of supported 8-HQ to preconcentrate total Al from geothermal brines. The polymeric fraction can be quantitatively retained on a cationic resin if the monomeric forms are previously chelated with PCV. Neither filtration (0.01 μm) nor ultrafiltration (10,000 dalton nominal weight cut-off) were able to separate monomeric from polymeric forms of Al.

Emphasis is given to the preparation of Al standard solutions with defined monomer/polymer ratios. The stability of these solutions was studied as a function of aging, temperature and CO_2 content.

2. INTRODUCTION

Aluminum speciation in geothermal brines is important to quantify the geochemical reactions affecting rocks dissolution in a given well. In the field of electricity production the knowledge of the relationship between chemical composition of the liquid phase and the geological properties of the circulation environment, makes easier the understanding of the hydrogeothermal peculiarities affecting the characteristics of a power plant and its management.

Gibbsite, Kaolinite and various forms of Al $(OH)_3$ are considered to be the major source of Al in geothermal brines [1,2], where it may exist as monomer of polymeric hydroxo-complexes, depending on both its age in the dissolved phase and the characteristics of the considered well (temperature, pressure, brine composition, etc.). Fluoride ions can compete with hydroxyde as complexing ligands, while phosphate complex products are likely to be excluded because of their usual low concentration with respect to total Al concentration in solution [3].

^{27}Al Nuclear Magnetic Resonance was used [4,5] to study the distribution of Al forms in solution showing the Al speciation to be strictly dependent on Al concentration and on the $[OH^-]/[Al]$ ratio (r). $[Al_{13}(OH)_{24}(H_2O)_2]^{7+}$ ("Al$_{13}$") and the dimmer $[Al_2(OH)_2(H_2O)_8]^{4+}$ are the most important species in concentrated solutions, while at lower Al concentrations, the "Al$_{13}$" form was observed only when r is higher than 0.25. The nature of polymeric Al was also investigated by Buffle et al.[6] and Akitt and Farthing [7,8] concluding that at least 80% of total polymer content was "Al$_{13}$".

Several methods have been proposed to separate monomeric from polymeric Al, such as solvent extraction of oxinate in methylisobutylketone (MIBK) [9,10] to separate fast from slow reacting species in solution. In other studies [11] monomeric Al concentration was determined by colorimetry using "ferron" chelating agent, or PCV without [12] or with [13] a cation exchange step to separate non-labile (organic) and labile (inorganic) monomeric Al.

In this work some methods are evaluated and compared for the optimization of an accurate and in field operable procedure for Al speciation in geothermal brines. Since, often, total Al concentration in geothermal brines may be less than 0.1 mg/l, a preconcentration step is needed before instrumental analysis by ICP-AES.

The proposed procedure is based on the retention of polymeric Al on a cationic resin after complexation of the monomeric fraction with pyrocatechol violet. Monomeric Al is then preconcentrated by retention on supported 8-HQ, eluted with 1M HCl/0.1 M HNO_3 and analyzed by ICP-AES. Total Al is determined on a different aliquot of sample after preconcentration on 8-HQ.

Finally, a simple analytical scheme for aluminum speciation in geothermal brines, at different total concentrations of the element, is illustrated.

3. EXPERIMENTAL

3.1 Apparatus

Plexiglass columns (5 mm i.d.; 10 cm length, Figure 1) filled with strong cationic exchanger where used to retain polymeric Al. Brine was fluxed through the column by means of a peristaltic pump (Carl Roth KG-Karlsruhe, 60 rpm).
Spectrophotometric measurements were performed using a Pye-Unicam PU8800 spectrophotometer.
Al was determined by ICP-AES, using a Plasma-Therm mod. 5,000 sequential ICP.
Acid cleaned (HCl 1M) polypropylene glassware was used.
Millipore ultrafilter systems Ultrafree PF, with polisulphone membrane, were used for ultrafiltration experiments.

3.2 Reagents

1M concentrated Al solution was prepared dissolving 24.12 g of $AlCl_3 \cdot 6H_2O$ (Carlo Erba RPE Analytical grade) in 50 ml of deionized and vacuum degassed water, bringing to volume in a 100 ml polyethylene calibrated flask.

TABLE 1. *Composition of the Synthetic Brine.*

Parameter	Concentration ($\mu g/ml$)
Ca	400
Mg	70
Na	3,000
K	300
Fe	10
NH_4^+	60
Cl^-	4,800
SO_4^{2-}	1,000
H_3BO_3	1,600

The 1 M Al solution was used to prepare dilute (3.34×10^{-2} M) solutions with $[OH^-]/[Al]$ ratio (r) of 0, 1 and 2, as reported by Bretsch [4]. The composition of these solutions was checked with the PCV colorimetric method (see below) before every set of laboratory runs.

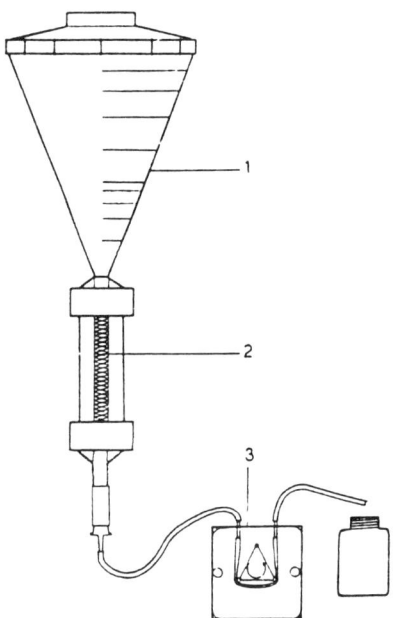

*FIGURE 1. Apparatus Used for Column Experiments.
1: Reservoir; 2: Column; 3: Peristaltic Pump.*

These solutions were used for the experimental tests after dilution with water or geothermal brine.

Real geothermal brines were collected in the zone of Larderello geothermal fields, while synthetic brine had the composition reported in Table 1.

Strong cationic exchanger resin was BDH Dowex 50W-x12, 50-100 mesh, H^+ form, converted to Na^+ form before use by treatment with saturated NaCl solution until the effluent became neutral.

Silica gel and XAD-7 supported 8-HQ were prepared following Hill [14] and Janak [15,16] respectively.

Al_2O_3 was Carlo Erba RPE Analytical Grade Reagent.

Pyrocatechol violet solution (0.038% w/V) was prepared dissolving 76 mg of pyrocatechol violet (Carlo Erba RPE Analytical Grade) in about 30 ml of water, transferring the solution in a 200 ml calibrated flask and filling up with deionized water. This solution was stored in the dark.

Hexamethylenetetramine buffer solution (30% w/V) was obtained dissolving 75 g of examine (Carlo Erba RPE Analytical Grade) in about 150 ml of water, filtering the undissolved material if present. The solution was transferred in a 250 ml calibrated polyethylene flask, 4 ml of concentrated ammonia solution (30% w/w) and then concentrated HCl until pH 6-6.2 were added then filling up with deionized water.

Acetic acid-sodium acetate solution was obtained mixing 223 ml of 10 M NH_3 and 115 ml of glacial acetic and, adjusting to pH 8.3 with NH_3 or acetic acid, and then diluting to 1 l with deionized water.

All other reagents were Carlo Erba Analytical grade.

3.3 Preliminary Studies

The influence of temperature on the composition of the hydroxy-Al solutions was evaluated at 20°C and 70°C determining the monomeric Al content by the PCV colorimetric method. The influence of temperature on the stability of aged solutions ($r = 1$) was checked in the range 20-70°C recording the variations in monomeric Al concentration determined using the PCV colorimetric method.

The effect of aging and CO_2 content on the composition and stability of Al solutions was also evaluated.

The empirical formula, the formation constant and the thermal stability of the complex Al/PCV were studied.

The molar ratio Al/PCV in the complex was determined spectrophotometrically in solutions at constant Al concentration (0.2 μg/ml) increasing PCV/Al ratio.

The formation constant was estimated on the basis of the calculated Al/PCV ratio.

Thermal stability of the complex was evaluated spectrophotometrically for temperature up to 60°C and time up to 2 hours.

The influence of fluoride concentration (in the range 0.4-36 μg/ml) on Al-PCV complex formation was studied at increasing Al (0.2-0.6 μg/ml) and PCV concentration in geothermal brine solutions.

The capacity of both silica gel and XAD-7 supported 8-HQ was determined by batch treatment with Al or Cu solutions, measuring the free element after equilibration and filtration.

3.4 Pyrocathechol Violet (PCV) Colorimetric Method

This method was previously described by Seip et al. [1]. It is based on the different complexation kinetics of monomeric vs polymeric Al with pyrocatechol violet at pH 6-6.2.

Experimentally, 5 ml of solution, containing up to 1.8 µg/ml of Al, were added to 1 ml of PCV solution and 2 ml of examine buffer, respectively. The absorbance, at 585 nm) of the resulting product was measured, after dilution with deionized water or geothermal brine to 25 ml, and compared with that of standard Al solutions and blanks. The measurements were performed after about five minutes for a complete color development: a longer time must be avoided since polymeric Al could react too.

3.5 Adsorption on Alumina

Tests were performed both in deionized water and in synthetic brine. Al solutions, with different r values, were passed at 2 and 3.5 ml/min., by means of a peristaltic pump, through polypropylene columns (10 mm i.d.; 40 mm length) filled with alumina. The monomeric Al content was determined in the initial and in the final solutions using the PCV method, while total Al was determined by ICP-AES. No Al was released by fluxing the column with 1M $HCl/0.1M$ HNO_3.

3.6 Filtration and Ultrafiltration

Tests were performed both in deionized water and in synthetic brine. Cellulose acetate (250,00 dalton molecular weight cut-off) and polysulphone membranes (10,000 dalton molecular weight cut-off) were used. In the latter case the Al solution (1.8 µg/ml) was forced through the filter by manual pressure applied on the surface with a syringe. Monomer content was determined by the PCV colorimetric method before and after filtration.

3.7 8-HQ Complexation-MIBK Extraction Method

This method is described by Barnes [9]. It consists of the complexation of monomeric Al with 8-HQ and extraction of the resulting product with MIBK. Experimentally, 400 ml of filtered (0.10 µm) geothermal brine were added with 5 ml of 8-HQ solution 5%, concentrated ammonia until pH 8.3 is reached, 5 ml of acetic acid-sodium acetate solution and 20 ml of MIBK. After separation, Al was determined by ICP-AES directly in the organic phase or after acid back-extraction with 6M HCl. Iron interferes at concentration less than 0.4 µg/ml, but can be previously chelated with hydroxylamine-ortophenantroline at pH 4.

Tests were performed using Al solutions with different (r) values in deionized water.

3.8 Ion Exchange - PCV Complexation Method

Tests were performed passing Al solutions (in synthetic or in real geothermal brine), previously treated with PCV at pH 6-6.2, through plexiglass columns (6 mm i.d.; 10 cm length) filled with Dowex 50W-x12 strong cationic exchanger, Na^+ form. Al retention yield was calculated by determining the element in the initial and the final solutions by ICP-AES and by the PCV colorimetric method. Retention yield was measured at different flow rates as well as the speciation efficiency (0.4-17 ml/min.).

Fluoride competition was studied treating a solution (Al = 0.6 μg/ml) in geothermal brine of increasing fluoride concentrations (up to 15 μg/ml) and determining Al colorimetrically before and after the column treatment.

An estimate of the practical capacity of the used cationic resin was obtained passing a 0.87 μg/ml Al solution through the column and determining the final concentration at established times: plotting Al concentration vs volume of the solution passed through the column, it was possible to obtain the capacity and the maximum sample volume the column can tolerate without regeneration. Both Al elution and resin regeneration were performed using 1M HCl/0.1M HNO_3 acid mixture.

3.9 Chelation with Supported 8-HQ

The influence of sulphides on Al retention efficiency on XAD-7 supported 8-HQ was evaluated by equilibrating for 2 hours 0.5g of resin with a solution containing about 3 mg/ml of sulphide. After filtration the exchanger was further treated with a 0.6 mg/ml Al solution and the capacity was calculated as above reported. The influence of sulphide on the Al retention kinetics was studied measuring residual Al concentration as a function of time of contact with the exchanger in solution containing also 80 μg/ml of sulphides.

Silica gel 8-HQ was used for speciation tests in geothermal brines.

Before use the exchanger (4g) was washed, in sequence, with 1M HCl/0.1M HNO_3, 5M HCl or concentrated HCl and then with water until neutral reaction. The influence of the different washing solution on the subsequent Al retention as well as the retention yield as a function of pH (range 5.5-7.5) and the effects of the temperatures (4 to 75 °C), time of contact and amount of exchanger on Al retention were also evaluated.

Absorbed Al recovery was studied as a function of the eluent volume and its time of contact with the exchanger.

4. RESULTS AND DISCUSSION

4.1 Hydroxy-Al Solutions of Different [OH⁻]/[Al] Ratio

The content of monomeric Al in the solutions of different $[OH^-]/[Al]$ ratio (determined using the PCV colorimetric method) is reported in Table 2, as well as literature data [5].

In order to verify the achieving of equilibrium condition, the pH variation (pH decreases when polymer content increases) of the solutions was recorded for 24 hours after preparation. The results showed a gradual variation of pH in the solutions with r = 1 (from 3.8 to 3.2) and r = 2 (from 4.1 to 3.7), that took about 15 hours to complete. No significant variation was observed when r = 0. Notwithstanding this, we followed the literature suggestions [5], that established five days as the time requested to obtain solutions of stable composition.

[OH⁻]/[Al]	This work			Literature[4]	
	mon. (% ± σ)	pol. (% ± σ)	n	mon.(%)	pol.(%)
0	99.7 ± 2.3	0.3 ± 0.007	10	100	0
1	60.3 ± 8.2	39.7 ± 5.4	11	54 - 64	36 - 46
2	22.8 ± 4.6	77.2 ± 15.6	10	16 - 17	83 - 84

n: sample size.

TABLE 2. Composition of Al Aqueous Solutions Prepared as Described in the Paper. Monomer Content Was Determined with the PCV Colorimetric Method.

Time (hours)	[OH⁻]/[Al]					
	0		1		2	
	Monomer (%)		Monomer (%)		Monomer (%)	
	20°C	70°C	20°C	70°C	20°C	70°C
0	100	100	78	67	37	29
1	--	--	--	67	--	--
2	--	--	--	65	--	--
15	--	--	--	--	33	28
120	100	98	64	63	27	29
192	97	100	66	63	26	26

TABLE 3.   Monomer Content in Al Solutions at Different "r" Prepared at Different Temperatures as a Function of Time.

Table 3 reports results regarding monomeric Al determination in solution prepared at different temperatures, as a function of time. Solutions with r = 1 or 2, prepared at 70°C, reached the equilibrium just after preparation, while when prepared at room temperature (20°C) they required less than 5 days. No temperature effect was observed for solutions with r = 0: equilibrium was reached just after preparation. Monomer final concentration was not significantly affected by the initial solution temperature. It is possible to conclude that temperature affects only the kinetics of the reactions involved but not its final composition.

In Table 4 results regarding the effect of the variation of temperature on Al speciation in a 5-days aged solution (r = 1) are reported. No variations were observed in the monomer content after about 4 hours of heating at 70°C and also after cooling at room temperature; thus the solutions are thermally stable, at least in the actual experimental conditions.

Heating			Cooling		
Time (hours)	°C	Monomer (%)	Time (hours)	°C	Monomer (%)
0	20	58	0	70	57
1	70	60	1	50	58
2.5	70	59	1.5	25	59
4.5	70	57	2	22	58

TABLE 4.   Effect of Variation of Temperature on the Composition of Aged Solution in Synthetic Brine (r = 1).

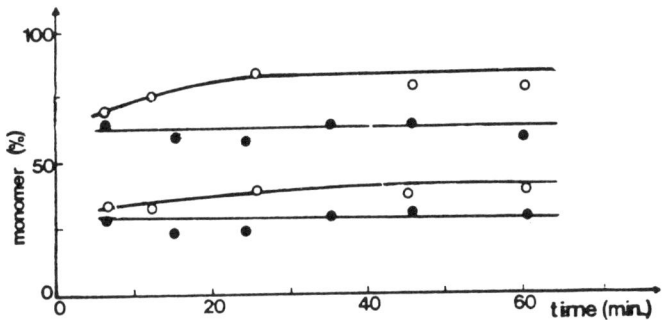

FIGURE 2. CO_2 Effect on Speciation of the Al Solution. ○
- Dilution with Deionized Water; • - *Dilution with Deionized, CO_2
Free Water.*

Figure 2 shows the results regarding the influence of CO_2 content on the composition of Al solutions with r = 1 and 2. Dissolved CO_2 produced an alteration in the monomer content (resulting in an increasing of the monomer percentage of about 30%) that reached the maximum 30 minutes after dilution.

4.2 Pyrocatechol Violet (PCV) Colorimetric Method

Preliminary tests on Al-PCV complex

Preliminary studies showed the molar ratio Al/PCV in the complex to be 1/3, at pH 6-6.2, according to literature [17]. The latter can be explained considering the molecular structure of pyrocatechol violet (Figure 3). The sulphonic group does not participate in forming the complex [17]. Al is coordinated by the oxygen of two or to OH groups [18], while the remaining OH groups are not likely to be available because of steric reasons. Being Al ion esacoordinated (octahedral structure) three moles of PCV will complex 1 mole of Al. The complex appeared non electrically charged passing through anionic exchanger bed, in agreement with the structure derived on the basis of the acid dissociation constant reported in literature [19].

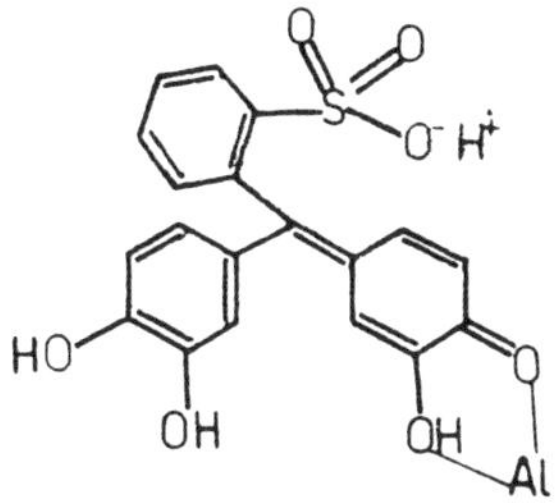

FIGURE 3. Pyrocatechol Violet Structure.

Taking into account the calculated Al/PCV ratio as 1/3 and considering a molar absorptivity of $6*10^4$ $l*mol^{-1}*cm^{-1}$, in good agreement with literature values [20], a concentration formation constant of $10^{20.3}$ was computed, in good agreement with literature data [19].

The Al complex is thermal stable at least up 60°C; in fact, no variations in spectrophotometric characteristics was observed after 2 hours of heating at this temperature.

A competition between PCV and fluoride in binding Al is to be expected; in fact, the formation constant of AlF_6^{3-} (the strongest Al fluorocomplex in aqueous solution) is 10^{20}, that is close to that of Al-PCV previously reported.

The extent of fluoride ions competition on the Al complexation is reported in Figure 4, that shows a large decrease in the absorbance of a 0.6 µg/ml Al solution for fluoride concentration less than 5 µg/ml. At Al concentration of 0.2 µg/ml the decrease becomes significant only for higher fluoride concentrations. In this view, the quantity of PCV added plays a very important role as verified measuring the absorbance of solutions having an increasing concentration ratio PCV/Al (till 7.5) for an initial Al concentration of 0.6 µg/ml, after fluoride addition; up to 15 µg/ml of fluoride did not interfere.

Tests on standard Al solutions show that in the range 0.09-0.35 µg/ml the instrumental response is linear. The relative standard deviation (σ) to within batches (five determinations) was 4.1% for Al concentration of 0.35 µg/ml. The detection limit (2σ) was 0.03 µg/ml. In Table 2 results regarding Al speciation in aqueous solutions following the procedure outlined in the experimental section are reported. Water samples were prepared as described in literature [5]. Experimental data agree very well with those reported by other Authors [5] for both the ferron colorimetric method and Nuclear Magnetic Resonance technique. Speciation did not change diluting the solution with degassed water as previously reported.

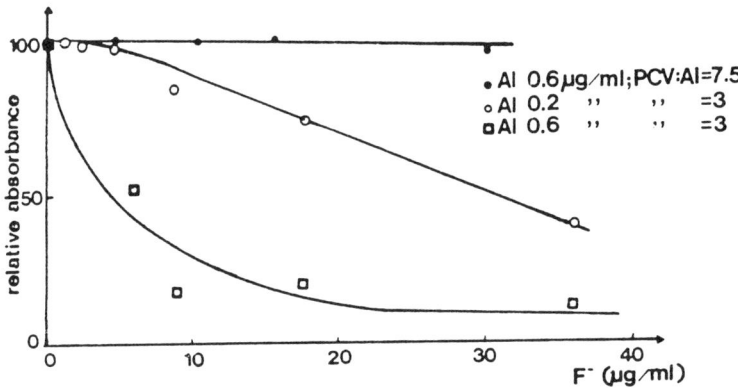

FIGURE 4.   *Fluoride-PCV Competition in Complexing Al.*

Results from PCV colorimetric method application

Similar tests performed on geothermal brine showed a decrease in sensibility (molar absorptivity = $2*10^4$ $l*mol^{-1}*cm^{-1}$) with an extension of linear working range up to 0.6 µg/ml of monomeric Al.

This method was used as a reference for all the other reported methodologies.

4.3 Adsorption on Alumina

Results regarding Al speciation in batch on alumina bed in deionized water and synthetic brine are reported in Table 5. No selectivity was found for monomeric or polymeric Al in water at different flow rates. Retention yield of monomeric Al increased at the lowest monomer initial concentration while polymeric Al was more retained at the highest polymer initial concentrations. Retention yield for total Al was greater at high polymer content. The difference between monomer and total Al retention for the solution with r = 0 (100% monomer) can be explained assuming the existence of catalytic phenomena (not further studied) altering the results of the speciation. No differences were observed dealing with synthetic brine; monomeric Al was quantitatively retained by alumina but a large polymer fraction ($\approx$ 80%) was adsorbed too.

The lack of selectivity makes alumina not suitable for Al speciation.

Origin of the solution	Flow rate (ml/min)	r	n	Monomer retained (%)	Polymer retained (%)	Total Al retained (%)
CO_2 free water	2	0 1 2	3 3 3	63 ± 7 75 ± 7 90 ± 10	-- 39 ± 16 87 ± 10	39 ± 13 62 ± 8 88 ± 10
CO_2 free water	3.5	0 1 2	1 1 1	61 74 71	-- 34 90	16 66 86
syntetic brine	2	0 1 2	3 2 2	100 ± 1 100 100	-- 71 85	89 ± 1 90 89

n: sample size.
r: $[OH^-]/[Al]$.

TABLE 5.   Speciation Tests on Alumina (Al Conc. = 1.8 µg/ml).

4.4 Filtration and Ultrafiltration

Table 6 shows that Al was not retained by ultrafiltration through membranes 10,000 dalton nominal weight cut-off.

The retention of a small fraction of polymeric Al by filtration through 0.01 µm porosity membranes is probably due to absorption on the material of the membrane (cellulose acetate) rather than to the presence of high molecular weight polymer (not verified in the ultrafiltration runs).

Tests on synthetic brine showed a non selective retention of Al, due to the already mentioned absorption phenomena, that in this case seem to be more evident.

Origin of the solution	Porosity (nm)	Mol.weight cut off (dalton)	r	n	Monomer retained (%)	Polymer retained (%)
CO_2 free water	10	250,000	0 1 2	2 2 2	2 1 3	-- 19 11
CO_2 free water	1.6	10,000	0 1 2	1 1 1	1 5 5	-- 0 4
syntetic brine	1.6	10,000	0 1 2	1 1 1	12 20 14	-- 17 36

n: sample size.
r: $[OH^-]/[Al]$.

TABLE 6.   Speciation Tests by Filtration and Ultrafiltration (Al Conc. = 1.8 µg/ml).

Filtration and ultrafiltration runs support the supposition that the polymeric fraction of Al is mainly composed with "Al_{13}" form [4,5,7,8] with molecular weight about 1,000.

The speciation might be successfully obtained by ultrafiltration through 500 dalton molecular weight cut off membranes. The latter would require a complex apparatus, hardly suitable for in field operations.

4.5 8-HQ Complexation - MIBK Extraction Method

Table 7 reports results concerning Al speciation in CO_2 free deionized water. The method was not selective being monomer and polymer extracted together. Total Al extraction was not quantitative, reaching about 70% in every case, and was not dependent on the monomer/polymer ratio. Monomeric Al was more retained at higher monomer/polymer ratios, while a great amount of polymer ($\approx$ 85%) was always extracted.

r	n	Al					
		Before extraction (µg/ml)			Extracted (%)		
		Total	Monomer	Polymer	Total	Monomer	Polymer
0	2	0.66	0.66	--	68-73	71-74	--
1	2	0.66	0.43	0.23	74-75	67-67	87-91
2	2	0.66	0.16	0.50	71-71	25-31	84-85

n: sample size.
r: $[OH^-]/[Al]$.

TABLE 7.   Speciation Tests by Chelation with 8-HQ and Extraction with MIBK.

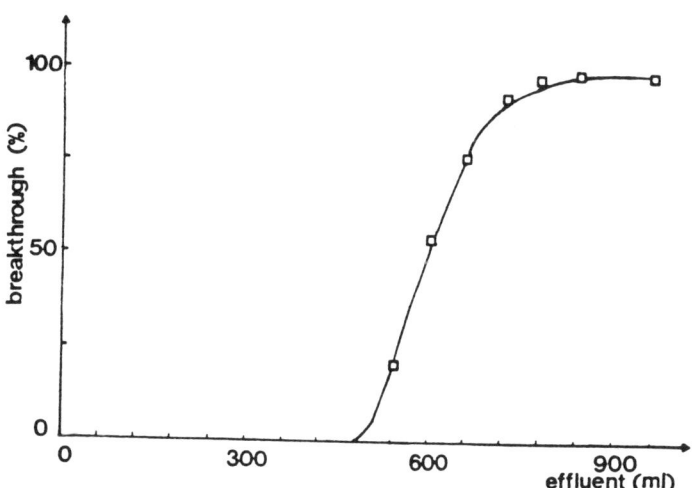

FIGURE 5.   Breakthrough Profile of an Al Solution (0.87 µg/ml) in the Dowex Column. Flow rate: 10 ml/min.).

4.6 Ion Exchange - Complexation Method

Figure 5 shows the breakthrough profile for a 0.87 μg/ml solution of Al at 10 ml/min. flow rate. Results show that the absorbance of the eluate remains constant and close to the blank value for 500 ml of collected solution and, then, it increases to the initial value of the solution before passing the column. A capacity of 48 μeq of Al was calculated. Taking: Ca = 70 μg/ml; Na = 35 μg/ml; K = 20 μg/ml and NH_4^+ = 30 μg/ml as the concentration of the major components in the used geothermal brine, a total capacity of 1.83 meq/g of resin is obtained (theoretical value: 2.12 meq/g). Using 100 ml aliquots of geothermal fluid sample, Dowex column can be used five times before regeneration.

As long as speciation tests are concerned, Table 8 reports the Al retention yield as a function of flow rate through the column. Geothermal brine at initial Al concentration of 0.2 μg/ml (55% monomeric Al) was used. The expected value of 45% of retention (that was the actual polymer content) is obtained with a flow rate of about 10 ml/min. Slower flow rates make both the time of analysis longer and poor the recovery, because of the partial complexation of the polymer by the PCV; on the other hand, higher flow rate (17 ml/min.) does not let the polymer equilibrate with the resin. For this reason, 10 ml/min. was chosen as the best flow rate giving good speciation efficiency and the shortest time of analysis.

No influence on Al speciation was observed if fluoride was present at concentration up to 15 μg/ml, as resulted from the spectrophotometric characteristics of the Al-PCV solution both before and after passage on the resin. The latter agrees with the statement of the possibility of determining Al monomer (including organic and fluoro complexes), by the PCV colorimetric method [22].

TABLE 8.   Al Retention Yield on Cationic Resin Dowex 50W-x12 as a Function of Flow Rate After PCV Complexation (Al Conc. = 0.2 μg/ml).

Flow Rate (ml/min)	Linear Velocity (cm/min)	Al Retained (%)
0.4	1.4	31
1.0	3.5	38
1.7	6	44
9.9	35	45
10.5	37	45
17.0	60	29

Origin of the solution	r	n	Al					
			Initial conc. (µg/ml)			Retained (% ± σ)		
			total	mon.	pol.	total	mon.	pol.
syntetic brine	1	3	0.43	0.28	0.15	35.7±1	--	101±2
real geother. brine	0	3	0.12	0.12	--	2.7±4	2.7±4	--
	2	2	0.60	0.12	0.48	81	4	100

n: sample size.
r: $[OH^-]/[Al]$.

TABLE 9.   Speciation Tests on Dowex 50W-x12, Na^+ Form After PCV Complexation.

In Table 9 results regarding a speciation test in synthetic and real geothermal brine are reported. It is evident the good selectivity of Dowex 50W-x12 for the retention of polymer and the good reproducibility of the results. In particular, the retention of the polymer was quite complete (101 ± 2%) while monomeric Al was not significantly retained (2.8 ± 3.8%). Concentration values for polymeric Al are corrected for the concentration of naturally occurring Al in the brine, while retention yield on the resin is calculated after elution of the element with 1M $HCl/0.1M$ HNO_3 and analysis by ICP-AES.

Silica gel supported 8-HQ	
This work (µeq/g)	Reference[14] (µeq/g)
42 ± 5 *	50 *
40 ± 6 **	--
XAD-7 supported 8-HQ	
This work (meq/g)	Reference[15] (meq/g)
1.5 ± 0.1 **	--
--	2.9 ± 0.1 *
1.5 ***	--

* with respect to Cu
** with respect to Al
*** with respect to Al after treatment with sulphide (3 mg/l)

TABLE 10.   Capacity of the Two Different Supported 8-HQ.

4.7 Chelation with Supported 8-HQ

Table 10 reports the capacity values of silica gel and XAD-7 supported 8-HQ. The capacity of XAD-7 supported 8-HQ (1.5 meq/g) is about 40 folds higher than the other material (40 μeq/g). further experiments on XAD-7 8-HQ showed that its capacity was not affected by treatment for 2 hours with a solution containing 3 mg/ml of sulphide. Sulphide concentration affected only the sulphide. Sulphide concentration affected only the kinetics of Al retention, being that chelation rate is higher in their presence. After 10 minutes of stirring, Al retention was about 100% with sulphide addition against a value of less than 50% without sulphide. As already said (see experimental), further experiments were performed employing silica gel supported 8-HQ that is easier to prepare in stable form than XAD-7 8-HQ.

The influence of washing the exchanger with different solutions before use was evaluated in terms of adsorption kinetics of Al in geothermal brine and capacity. Al retention is faster and the capacity is higher if the exchanger is previously washed with concentrated HCl: in fact, at the same equilibrating time, at least 20% of Al is retained and the capacity is about 1.5 folds greater with respect to an exchanger washed with 5 M HCl. The worst results were obtained using the 1M HCl/0.1M HNO_3 solution: this fact was not further investigated.

No variation of Al retention in geothermal brine was observed neither as a function of pH in the range 4- 7.5 nor as a function of sample temperature in the range 4°C-75 °C (only a little increase of retention rate at higher temperatures was observed). Working with 100 ml samples, a total retention can be obtained after 12 hours of stirring with 8 g of exchanger washed with concentrated HCl and then with deionized water until neutral reaction.

Al can be eluted from the exchanger, after filtration through 0.45 μm filter (larger pore size are also suitable), with 1M HCl/0.1M HNO_3 acid mixture. Two fractions of 5 ml each of eluent are enough to ensure a total recovery of retained Al when kept in contact with the exchanger for five minutes at least. In Table 11 the results regarding Al speciation in synthetic brine are reported. The retention of monomeric and polymeric Al was strongly dependent on their initial concentration; in particular, the retention of the monomer decreases and the retention of polymer increases at higher [OH-]/[Al] ratios. The amount of total Al retained was nearly constant. As in the case of alumina test, a partial polymerization of the monomer while stirring the solution in presence of 8-HQ would occur. This fact is evident in the first row of the Table 11 where retention of total Al and monomeric Al was not the same, although the solution contains 100% Al monomer.

r	n	Al					
		Monomer		Polymer		Total	
		Initial (µg/ml)	Retained (%)	Initial (µg/ml)	Retained (%)	Initial (µg/ml)	Retained (%)
0	2	2.30	90	--	--	2.30	76
1	2	1.56	82	0.84	56	2.40	74
2	2	0.65	46	1.75	87	2.40	76

n: sample size.
r: $[OH^-]/[Al]$.

TABLE 11. Speciation Tests on Silica Gel Supported 8-HQ.

We can conclude that even if less affinity for polymer does exist, there is not a sufficient selectivity to use silica gel supported 8-HQ for Al speciation in water or geothermal brine. Tests performed complexing Al with PCV before the treatment with the exchanger showed no variation in Al retention, because of the much greater stability of Al-8 HQ complexes with respect to Al-PCV and Al-OH ones.

A lot of column runs on both selected exchangers were previously performed in order to valuate the possibility of speciating Al on the basis of different contact time between the exchanger and Al solution.

Also, a faster exchanging kinetics was observed for the monomeric Al forms in comparison with polymeric ones, the separation efficiency was greatly depending on the initial composition of the solution to be speciate in terms of monomer/polymer ratio.

On the other hand, the flow rate affected both the speciation efficiency and the Al (monomer/polymer) retention yield making the results evaluation very difficult.

4.8 Proposed Method

On the basis of the obtained results, a method for Al speciation in geothermal brine was proposed. The method consists of a preliminary treatment with PCV at pH 6-6.2 to complex all the monomeric Al followed by the retention of the polymer on a column filled with Dowex 50W-x12, Na^+ form.

If necessary, the monomer is preconcentrated in batch on silica gel supported 8-HQ before the ICP-AES instrumental determination is made. The batch treatment supported with silica gel was applied to another sample aliquot to determine total Al. The difference between the values obtained in the two instrumental determinations (after elution with 1M HCl/0.1 M HNO_3 from the 8-HQ exchanger) allows the Al speciation to be performed.

This procedure may usefully be applied to Al concentrations not directly detectable by the PCV colorimetric method (D.L. = 0.075 μg/ml).

In case of higher concentrations, Al monomer can be determined by the PCV colorimetric method, while total Al can be directly determined either by ICP-AES or by the colorimetric method after depolymerization by acidification.

The block diagrams describing the two methods are reported in Figure 6.

The proposed procedure was applied to the speciation of Al in synthetic brine (Table 12) and in real geothermal brines (Table 13) collected in Larderello geothermic fields.

5. CONCLUSIONS

Among the tested methodologies, only the PCV colorimetric method and the complexation-ion exchange method proved to be successful for Al speciation in geothermal brines. Ultrafiltration requires a complex instrumentation to be appropriately used in field operations. Retention on alumina or supported 8-HQ, solvent extraction in MIBK after complexation with 8-HQ are not selective techniques if the aim is the Al speciation.

The proposed method requires complexation of monomeric Al with PCV and retention of the polymer onto a strong cationic exchanger followed by Al preconcentration on supported 8-HQ before ICP-AES analysis.

The main advantage of this method is in obtaining quite good preconcentration factors for both total and monomeric Al with high monomeric-polymeric Al separation efficiency.

SPECIATION OF ALUMINUM IN GEOTHERMAL BRINES

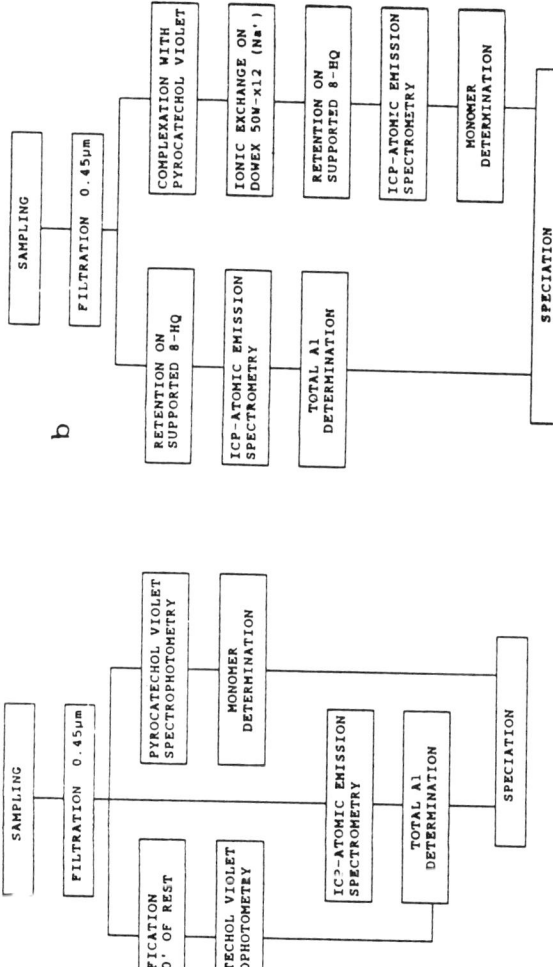

FIGURE 6. Flow Sheet of the two Proposed Analytical Procedure.
  a: Al concentration More than 0.37 µg/ml (5 Folds
     the D.L. of the PCV Colorimetric Method);
  b: Al Concentration Less Than 0.37 µg/ml.

r	Total Al initial concent. (μg/ml)	Dowex column		Silica gel supported 8-HQ	
		Final conc. (μg/ml)	Retained (%)	Final conc. (μg/ml)	Retained (%)
1	0.40	0.24	40	0.02	92
	0.42	0.26	38	0.04	85

r: $[OH^-]/[Al]$.

TABLE 12. Results of Speciation Tests in Synthetic Fluid Obtained with the Proposed Procedure (Monomer: 60%).

Origin of the sample	Al		
	Total (μg/ml)	Monomeric (μg/ml)	Monomeric (%)
brine	0.17	0.17	100
hot spring water	0.13	0.10	77

TABLE 13. Al Speciation in Real Sample Collected in the Larderello Geothermal Fields.

The accuracy was checked by comparison with the PCV colorimetric method and by using standard solutions with known polymeric and monomeric Al content.

Detection limit, considering a preconcentration factor of 10 and an instrumental detection limit of 0.03 μ.ml of the ICP-AES technique, is 3 ng/ml.

The simplicity of the tool, the short time required for the operations (about 15 minutes, excluding stirring time with supported 8-HQ) make this method very interesting for in field operation on geothermal sample.

REFERENCES

1. Seip, H.M., L. Muller, and A. Naas. "Aluminum Speciation: Comparison of Two Spectrophotometric Analytical Methods and Observed Concentration in Some Acidic Aquatic System in Southern Norway," *Water, Air and Soil Pollution*, Vol. 23, (9184) pp. 81-95.

2. Driscoll, C.T., J.P. Baker, J.J. Bisogni, and C.L. Scholfield. "Acid Precipitation: Geological Aspects," *Ann. Arbor. Sci.* (1983) pp. 55-75.

3. Watson, J.C. "Round Robin Evaluation of Methods for Analysis of Geothermal Brines," in "Geothermal Scaling and Corrosion." *ASTM STP* 717 (Casper, L.A. and T.R. Pinchback, 1980) pp. 236-258.

4. Bertsch, P.M., W.J. Layton, and R.I. Barnhisel. "Speciation of Hydroxy-Aluminum Solutions by Wet Chemical and Aluminum-27 NMR Methods," *Soil Sci. Soc. Amer. J.*, Vol. 50, (1986) p. 1449.

5. Bertsch, P.M., G.W. Thomas and R.I. Barnhisel. "Characterization of Hydroxy-Aluminum Solutions by Aluminum-27 NMR," *Soil Sci. Soc. Amer. J.*, Vol. 50, (1986) p. 825.

6. Buffle, J., N. Parthasarathy, and W. Haerdi. "Importance of Speciation Methods in Analytical Control of Water Treatment Processes with Application to Fluoride Removal from Waste Waters," *Water Res.*, Vol. 19, (1985) pp. 7-23.

7. Akitt, W. and A. Farthing. "Aluminum-27 NMR. Part 2. Gel Permeation Chromatography," *J. Chem. Soc. Dalton Trans.*, (1981) pp. 1606-1608.

8. Akitt, W. and A. Farthing. "Aluminum-27 NMR. Part 4. Hydrolysis Using Sodium Carbonate," *J. Chem. Soc. Dalton Trans.*, (1981) pp. 1617-1623.

9. Barnes, R.B., "The Determination of Specific Forms of Aluminum in Natural Water," *Chemical Geology*, Vol. 15, (1975) pp. 177-191.

10. LaZerte, B.D. "Forms of Aqueous Aluminum in Acidified Catchments of Central Ontario: A Methodological Analysis," *Can. J. Fish Aquat. Sci.*, Vol. 41, (1984) p. 766.

11. Driscoll, C.T., J.P. Baker, J.J. Bisogni and C.L. Schofield, "Effect of Aluminum Speciation on Fish in Dilute Acidified Waters," *Nature*, Vol. 284, (1980) pp. 161-164.

12. Wilson, A.D. and G.A. Sergeant. "The Colorimetric Determination of Aluminum in Minerals by Pyrocatechol Violet," *Analyst*, Vol. 88, (1963) p. 109.

13. Rogeberg, E.J. and A. Henrisksen. "An Automatic Method for Fractionation and Determination of Aluminum Species in Fresh-Waters," *Vatten*, Vol. 41, (1985) pp. 48-53.

14. Hill, J.M. "The Preparation of 8-Hydroxyquinoline Substituted Silica Gel for the Chelation Chromatografy of Some Trace Metals," *J. Chromatografy*, Vol. 76, (1973) p. 455.

15. Janak, K. and J. Janak. "Preparation and Properties of a Styrene-Ethylene Dimethacrylate Copolimer-based Chelating Ion Exchanger with Bonded 8-Hydroxyquinoline," *Collection Czechoslovak Chem. Commun.*, Vol. 51, (1986) p. 657.

16. Janak, K. and J. Janak. "Chelating Ion Exchangers with Bonded 8-Quinolinol on a Glycidyl Methacrylate Gel," *Collection Czechoslovak Chem. Commun.*, Vol. 48, (1983) p. 2352.

17. Sarzanini, C., E. Mentasti, V. Porta, and M.C. Gennaro. "Comparison of Anion-Exchange Method for Preconcentration of Trace Aluminum," *Anal. Chem.*, Vol. 59, (1987) pp. 484-486.

18. Martell, A.E. and M. Calvin. *Chemistry of the Metal Chelate Compounds*, (Prentice-Hall, Inc., New York, 1952) p. 17.

19. Meites, L. *Handbook of Analytical Chemistry*, (McGraw-Hill Book Company, Inc., 1963), pp. 3-107.

20. Dougan, W.K. and A.L. Wilson. "The Absorptiometric Determination of Aluminum. A Comparison of Some Chromogenic Reagents and the Development of an Improved Method," *Analyst*, Vol. 99, (1974) pp. 413-430.

21. Andelman, J.B. and J.R. Miller. "Impact of Acid Rain on Aluminum Species in Streams," pp. 19-25.

22. Jones, P., L. Ebdon, and T. Williams. "Determination of Trace Amount of Aluminum by Ion Chromatografy with Fluorescence Detection," *Analyst*, Vol. 113, (1988) p. 641.

DISCUSSION OF:
SPECIATION OF ALUMINUM IN GEOTHERMAL BRINES: COMPARISON OF DIFFERENT METHODOLOGIES

Frank J. Millero
Rosenthal School of Marine and Atmospheric Science, University of Miami

These workers report on the different methods that can be used to determine the speciation of aluminum in brines. In recent years, there has been a lot of interest in the speciation of aluminum in natural waters, due to the toxicity of Al^{3+}. In fresh waters with low alkalinity (HCO_3^-), acid rain lowers the pH enough to convert hydrolyzed aluminum to the ionic form:

$$Al(OH)_3 + 3\ H^+ \longrightarrow Al^{3+} + 3\ H_2O$$

Ionic interaction models are used to determine the various forms of aluminum at a given pH.

The authors have used the ligand pyro-catechol violet (PCV) to complex monomeric aluminum and used various separation techni-ques to determine total or polymeric Al. They made their measurements as a function of pH, temperature and ionic strength to completely characterize the separation techniques. The complexed monomeric aluminum is not related on a cationic resin; thus, it is possible to separate the monomeric and polymeric frac-tions.

The proposed analytical technique has two separate procedures based upon the level of Al. Both start after filtration through a 0.45 μm filter. At high levels of Al, the sample is treated with PCV to complex the monomeric Al, which is determined by spec-troscopic techniques. The total Al is determined after acidification, also with PCV. At low concentrations of Al, the total Al is determined after retention on silica gel-supported 8-HQ. Elution is made with 1 M HCl/0.1 M HNO_3 and the total Al is measured by ICP atomic emission spectrometry. The monomer is complexed with PCV and passed through a Dowex column to retain the polymer. The procedure appears to work very well and provides a reliable method to determine monomeric and polymeric Al in natural waters.

Some questions I have are given below:

1. The addition of CO_2 caused the monomeric Al to increase by 30%. Is this due to the formation of $AlCO_3^+$ complexes?

2. I feel that the use of molar concentration units would be more useful, especially when examining the competition effects of PCV and F for Al (Figure 4).

3. No discussion of the ICP-AES instrumentation is given. A short discussion of this technique would be helpful.

SPECIATION OF TIN IN SEDIMENTS OF ARCACHON BAY (FRANCE)

M. Astruc, R. Lavigne, R. Pinel, F. Leguille, and V. Desauziers
Laboratoire de Chimie Analytique, Universite de Pau, France

P. Quevauviller and O. Donard
Groupe d'Océanaographie Physicochimique, Laboratoire de Photophysique et Photochimie Moléculaire, Université de Bordeaux I, France

1. INTRODUCTION

Arcachon Bay, close to Bordeaux in the South West Atlantic Ocean coast of France (Figure 1), is the first place in the world where the impact of tributyltin pollution on the marine environment has been advocated [1]. In this 155 km^2 Bay did exist a very important oyster industry, producing about 12000 T/year of Crassostrea gigas. The same Bay, where tidal flushing is rather poor, is also a major mooring and sailing place where up to 15000 pleasure crafts are present in Summer. In 1973, 75 and 76 spatfall and production were good, but from 1977 to 1981 spatfall was disastrous and serious "balling" of adult oysters occurred. As a consequence the oyster production went down with serious consequences on local employment.

From circumstantial evidence French authorities came to the conclusion that TBT-based antifouling paints were concerned. In January 1982 a ban was imposed on the use of these paints for pleasure crafts of 25 m or less except those with aluminum hulls. As early as Summer 1982 spatfall was good again and remained good in the following years. Oyster shell deformations became more slowly less important. Commercial oyster production came up again from 3000 T in 1981 to 15000 T in 1985 [2]. Monitoring of total tin and organotin compounds in both sea water and oysters [3] during the same period demonstrated a steady decrease of tin pollution of sea water and oysters. However some shell malformations remained.

It seems thus that the decontamination of Arcachon Bay - thus effective - is much slower than expected. Several reasons may be advocated. Among these may be a storage of organotin compounds in sediments and a slow release to overlying waters.

Such a possibility seems to arise from distribution studies of TBT between particulates and water in estuarine conditions [4]. Persistence of TBT in the marine environment is a subject of controversy and half-life time reported thus far vary between hours and months [5]. Arcachon Bay, submitted during 5-6 years to a heavy TBT pollution and protected from it during the last 8 years is obviously of prime interest. Therefore several sediment cores were taken in Arcachon Bay in 1988 - 1989 to try and evaluate the depth profiles of inorganic tin and organotin compounds together with some other parameters that may help to understand TBT behavior in these conditions.

This proved difficult as Arcachon Bay is submitted to intense dredging operations and several of the sediment cores studied have been found perturbated by dredging or disposal of dredged material.

The results presented hereafter have been obtained on one unperturbated core only. However all the conclusions apply as well to the other cores studied but for the effects of dredging operations.

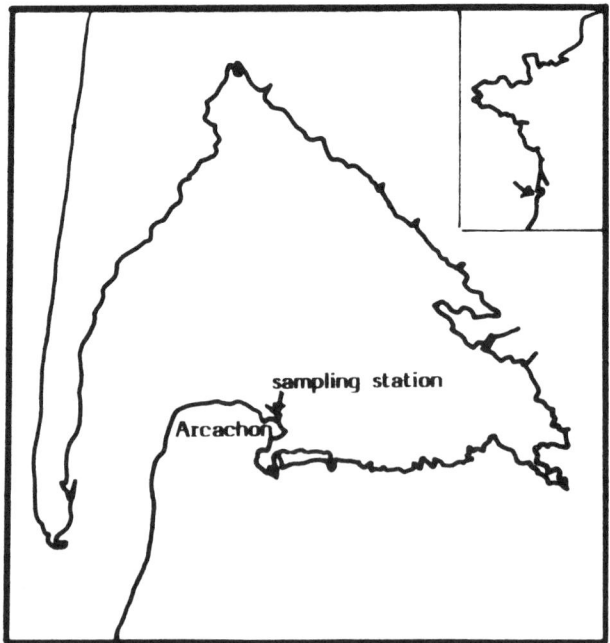

FIGURE 1. Location of Sampling Station.

2. MATERIAL AND METHODS

2.1 Sampling

A 50 cm sediment core was sampled in Arcachon marina (Figure 1) in a site that has not been perturbated during the last 15 years at least, using a hand operated stainless still sampler. The sediment core was then cut in 1 cm slices. Some of the slices were halved. Interstitial waters were extracted by centrifugation prior to immediate analysis. Sediment samples were prepared for analysis following two alternative paths: (1) air drying; crushing; 63 μm sieving; (2) wet sieving (63 μm); air drying. They were then kept at + 4°C in sealed polyethylene bags, awaiting analysis.

2.2 Analytical Techniques

Tin Speciation in Sediments

Tin speciation experiments in the sediment samples were run independently in Pau and Bordeaux using the same method [6,7] that has been cross verified previously [8]. 0.1 to 1 g of dry sediment is placed with 20 ml of concentrated acetic acid (Pour Analyses) in a 25 ml glass stoppered Pyrex flask on a magnetic stirrer for 4 hours. The flask is then centrifugated (4000 rpm, 3 min). Aliquots (0.1 - 2ml) of the supernatant solution are pipetted into a hydride generation reactor containing already c.a. 100 ml of ultrapure water (Millipore).

Tin Speciation in Interstitial Waters

Aliquots (1 ml) of interstitial waters are introduced in the hydride generation reactor and diluted to c.a. 100 ml with ultrapure water acidified by acetic acid (1 %).

Tin Speciation Analysis [6]

A 5 % $NaBH_4$ (for synthesis) in 1 % NaOH (Merck Suprapur) aqueous solution is added via a peristaltic pump (2.5 ml/min) to the Pyrex reactor. The $NaBH_4$ solution is injected during a variable time, depending on sample nature. Evolved hydrides are carried by the Helium flux (400 ml/min) to a glass GC column (2 g Chromosorb GAW-DMCS, 10 % OV 101) cooled in liquid nitrogen during 1 min. The liquid N_2 is then removed leaving the GC column to warm by contact with the atmosphere (3 min), then a gentle electric heating of the column is applied (up to 200 °C) during 3 min. Separated hydrides, flushed by Helium from the GC column, are introduced, after mixing with H_2 (200 ml/min) and O_2 (100 ml/min) in a quartz open-ended furnace, electrically heated to 1000°C, and situated in the light beam of a IL 151 Atomic Absorption Spectrometer (λ = 286.3 nm).

In these conditions a complete chromatographic analysis of inorganic, methylated and butylated (until TBT) species of tin is registered in 6 min (9 mn for the whole process). The detection limits in routine operation given in Table 1 could be seriously lowered if necessary by increasing aliquots introduced in the reactor, or replacing the Tin HCL by a Tin EDL.

Reproducibility is very satisfying (Table 1) in these routine conditions.

TABLE 1.   Analytical Performances of the Routine Hydride-GC-AA Tin Speciation Method.

SPECIES	TRIT	MMT	DMT	TMT	MBT	DBT	TBT
Absolute detection limit (ng)	0.2	0.13	0.07	0.09	0.09	0.40	0.24
Concentration detection limit* (ng.l^{-1})	2	1.3	0.7	0.9	0.9	4	2.4
Reproducibility %	<8	<13	<17	<18	<15	<15	<9

* with a 100 ml reaction flask

Determination of Total Tin

Solubilization of sediment is effected by a warm HNO_3/HF digestion in covered Teflon beakers. Total tin determination is performed by GFAAS following a procedure described previously [8].

3. RESULTS AND DISCUSSION

All the concentrations reported there after are expressed as nanograms of tin per g of dried sediment (ppb) or per liter of water (ppt). For sake of simplicity we will use systematically the abbreviations: MBT, DBT, TBT for respectively the mono-, ditributyltin cations.

3.1 Speciation of Tin in the Sediment

The depth profile of tin speciation in the sediment core is presented on Figure 2 together with a log of the core established from a radiography. The level of butyltin pollution being quite high the speciation method was operated at low sensitivity and no methyltin compounds were detected. Other investigations conducted on other cores taken in the same area evidenced the presence of various methyltin compounds at concentrations ranging from 0 to 10 $ng.g^{-1}$.

3.2 Total Tin

Total tin (Figure 3) is maximum in sediment layers where butyltin pollution is high but is not directly correlated to it. Only a small part of total inorganic tin is extractible by acetic acid (TRIT). Therefore a large proportion of total tin must be contained in detritic minerals solubilized only by the HNO_3/HF digestion.

3.3 Total Recoverable Inorganic Tin (TRIT)

TRIT is the fraction of inorganic tin that is extracted from sediment by acetic acid. It varies roughly (Figures 2 and 3) between 40 and 170 $ng.g^{-1}$ (i.e. a negligible part of total tin) with two well defined maxima at 0 and -10-11 cm depth. The latter coïncidates with a highly polluted layer. The former may indicate either an enhanced mobility or a production of inorganic tin in the top oxic layer by degradation of organotins, possibly by photolysis.

3.4 Butylated Species

The concentrations of TBT, DBT and MBT vary with depth. They are negligible at -50 cm and very low up to -30 cm. But

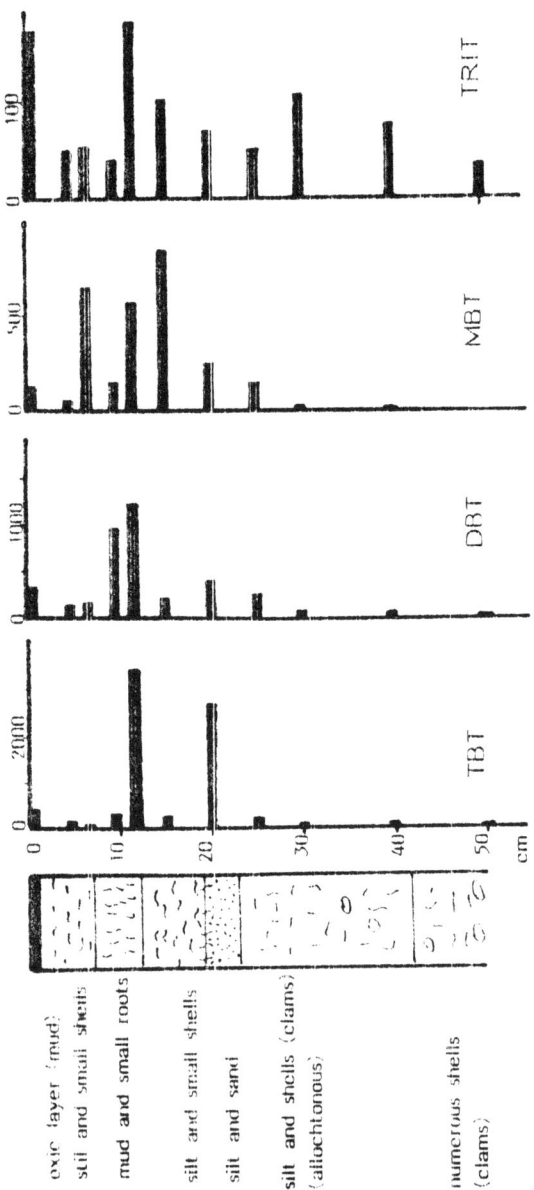

FIGURE 2. Tin Speciation Along the Sediment Core, Compared to a Radiographic Log. All Concentrations are Expressed as ng.Sn per g of dry Sediment.

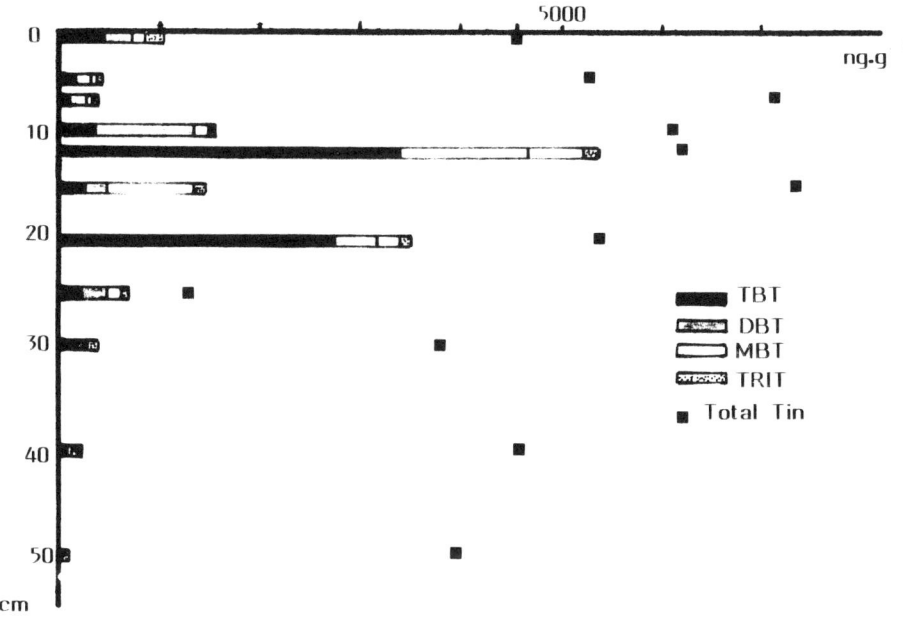

FIGURE 3. Tin Speciation Along the Core, Compared to Total Tin.

higher in the core they vary widely, TBT being always the major component, at concentrations reaching more than 3000 $ng.g^{-1}$ at -11 -12 cm - Two maximas are registered for all three butylated species, but if their positions are identical for TBT and DBT (-20 and -11 -12 cm) the MBT maxima are shifted towards the core top by approximately 5 cm (-15 and -6 -7 cm). One may notice as well that the surface concentrations (0-1 cm) of all three butylated species are higher than in the following layers (-4 -7 cm). The presence of two maxima is difficult to explain without a datation of the sediment layers. Perhaps it may be linked to the variations of the nature of sediment apparent in the log. However what is clear from these results is that TBT is not degraded rapidly in sediments as the period of highest water pollution by TBT in this Bay was

1979 -1981, i.e. 8 to 10 years ago. The five centimeter shift between the maxima of TBT/DBT and MBT may indicate a higher mobility of this species. A smaller difference of mobility may exist between TBT and DBT as the DBT maxima seem to be wider than the TBT ones. Analysis of interstitial waters (see after) is of interest on these points. In the top layer of the sediment the concentrations of all butylated compounds are higher than in somewhat deeper layers. These findings may indicate either that the partial ban on TBT paints is becoming less effective or that butyltins partially disappear from burying sediments (destruction or migration). There was no evidence of mixed butyl-methyl tin compounds in these sample nor in other cores studied in the same area.

3.5 Interstitial Waters

Total recoverable dissolved tin (TRDT) vary along the core (20-200 $ng.l^{-1}$) with a maximum in the top layers and another at -30 -40 cm depth, in layers deep enough to be older than TBT pollution (Figure 4). Tributyltin is absent. Mono and dibutyltin occur only in the top layers, at very low concentrations (7 $ng.l^{-1}$). These low concentrations were expected as this sediment core has been taken in an intertidal area where water fluxes may be important in the top layers.

The fact that MBT, to a lower extent DBT, and not TBT are present in interstitial waters of the top layers may help to explain the repartition of these compounds in the deeper layers of the sediment: MBT seems more mobile than DBT, and does not seem mobile at all in top layers. All of them lack mobility in deep layers. The ratios of dissolved ($ng.l^{-1}$) to particulate ($ng.g^{-1}$) MBT and DBT cannot be determined with any certainty from the results but only tentative values may be evaluated as 0.1 to 0.05 for MBT or 0.01 to 0.003 for DBT. There is a quite good relation between TRIT and dissolved inorganic tin (Figure 5)) with maxima at 0 and -10 cm depth, except in the -10 -20 cm region. From these data one may estimate the ratio of dissolved ($ng.l^{-1}$) to particulate ($ng.g^{-1}$) inorganic tin as varying between 1 (0 to -8 cm and -30 to -50 cm) and 0.05 (-20 cm). The lowest values of this ratio being obtained in the most polluted layers it may be supposed that the corresponding TRIT data are somewhat overestimated due to a very slight decomposition of organotins during the analytical procedure.

SPECIATION OF TIN IN SEDIMENTS OF A FRENCH BAY

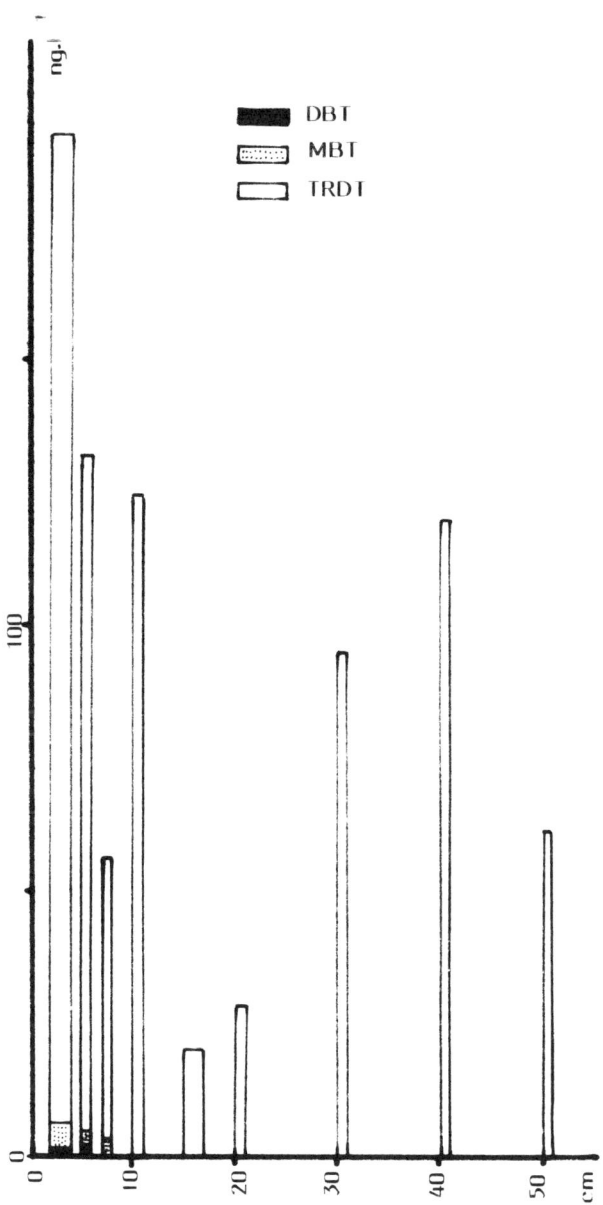

FIGURE 4. Tin Speciation in Interstitial Waters Along the Core.

272 SPECIATION AND COMPLEXATION IN NATURAL SYSTEMS

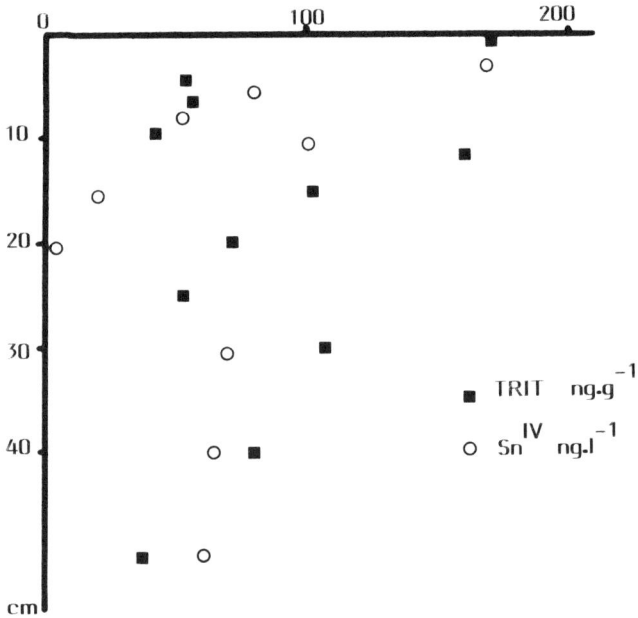

FIGURE 5. Comparison of Total Recoverable Inorganic Tin (TRIT) in Sediment and Inorganic Tin in Interstitial Waters Along the Core.

4. CONCLUSIONS

The first and most important conclusion concerns the stability of tributyltin in the intertidal sediment of Arcachon Bay studied. In deeply buried layers (10 - 20 cm) that have been polluted about eight to fifteen years ago the major component of organotin speciation is still tributyltin with minor contributions of the expected degradation products, dibutyltin and monobutyltin. In these anoxic conditions the half-life time of TBT must be expressed in years at least or rather in decades. Such a low degradation rate may perhaps be linked to inhibition of microbial activity at the high TBT concentrations found [5]. The mobility of the butylated tin species is low, decreases in the order MBT > DBT > TBT, and seems to be maximal in the top first centimeters.

Methylated tin species are not important components in the samples studied and methylbutyltin compounds have not been evidenced. Deeply buried sediments in Arcachon Bay may store large amounts of TBT whose release towards overlying waters could have disastrous consequences. However the mobility of TBT in these conditions seems to be low.

Systematic investigations along the French Atlantic Ocean coast should be undertaken to validate these findings on a more general basis.

Dredging operations with these polluted sediments should be considered with great care.

Thanks are due to M. Dupouts, Service Maritime Arcachon, MM. De Resseguer et Faugère, Dept. Gèologie et Océanographie, Université de Bordeaux for help in sampling.

REFERENCES

1. Alzieu, C., M. Heral, Y. Thibaud, M.J. Dardignac, and M. Feuillet. *Rev. Trav. Inst. Peches Marit.*, 45:101-116 (1982).

2. His, E. and R. Robert. Oceanis 13:317-335 (1987)

3. Alzieu, C., J. Sanjuan, J.P. Deltreil, and M. Borel., *Mar. Pollut. Bul.* 17:494 -498 (1986).

4. Randall, L.S. and J. Weber. *Sci. Total Environ.* 57:191 - 203 (1986).

5. Clark, E.A., R.M. Sterritt, and J.N. Lester. *Envorn. Sci. Technol.* 22:600-604 (1988).

6. Lavigne, R. "Tin Speciation in Waters, Sediments and Biological Matter by Hydride Generation, Gas Chromatography and Atomic Absorption Spectrometry," Thesis, University of Pau, France (1989).

7. Lavigne, R., F. Leguille, R. Pinel, and M. Astruc. Unpublished results (1988).

8. Astruc, A., R. Lavigne, V. Desauziers, R. Pinel, and M. Astruc. *Applied Organometal. Chem.* in press

9. Pinel, R., M.Z. Benabdallah, A. Astruc, and M. Astruc. *Anal. Chim. Acta* 181:187 - 193 (1986).

DISCUSSION OF:
SPECIATION OF TIN IN SEDIMENTS OF ARCACHON BAY (FRANCE)

M. Pettine and A. Puddu
Water Research Institute, CNR, Rome, Italy

INTRODUCTION

Tin is a widely used element. In the inorganic state it is present in natural waters in two main oxidation states (+2 and +4), whose relative weight is a function of the redox environment. The toxicity of inorganic tin species is generally very low [1]. This fact, together with the solubility limitations of tin(IV) hydroxide [2] made the environmental interest for tin quite negligible.

The increasing production of organotin compounds, due to their use in biocidal applications (wood preservation, marine paints) and in some industrial chemical processes (PVC stabilization, polyurethane catalysis), changed dramatically this picture in recent years [3].

The use as antifouling agents in marine paints or in cooling water systems represents the primary source of organotin for the aquatic environment, among the above different applications. To this aim tri-substituted organotin are the most used compounds.

Concern over the effects of organotin compounds first emerged in France, where the crash of a major oyster fishery in the Arcachon Bay was attributed to tributyltin compounds used in antifouling coatings of numerous recreational boats in that bay.

The relationship between TBT based paints and oyster malformations, demonstrated by Alzieu et al. [4], stimulated intensive scientific studies which better characterized these organometallic compounds, with reference to possible toxic effects, to bioaccumulation factors and to environmental distribution and behavior.

BACKGROUND

Tributyltin derivatives are generally a factor 10-100 times more toxic than the di-organo- or mono-organo-tin substituted for aquatic organisms [5]. Low Effect Concentration values were estimated to range from 0.002 to 0.02 µgTBT/l for gastropods, from 0.02 to 0.05 µgTBT/l for bivalves and from 0.09 to 0.14 µgTBT/l for crustaceans [6]. Toxic effects were also found at lower concentrations, of the order of 1 ngTBT/l, with the development of imposex in certain gastropods molluscs. The bioaccumulation factors range from approximately 1000 to above 10000 for macroorganisms and may reach much higher values (3000) for algae and bacteria [3].

Studies on the environmental distribution of organotin compounds indicate that contamination hazards can be expected in particular locations, such as harbors, small bays and coastal waters, characterized by intensive recreational boating and restricted water circulation, or by the outlets of industrial plants using organotin compounds as antifoulant for the protection of cooling systems. TBT concentrations in the range 2-500 ng/l have been reported from Chesapeake Bay [7] or San Diego Bay [8], with even higher concentrations in the surface microlayer, where organotin compounds tend to concentrate due to their limited aqueous solubility. Dibutyltin (DBT) or monobutyltin (MBT) concentrations were found generally at much lower concentrations with respect to TBT. As far as harbors in the Mediterranean area are concerned, Leghorn in Italy and Mersin in Turkey are the most contaminated sites with concentration levels up to 800-900 ngTBT/l [9].

On the basis of the information acquired on such chemicals, a provisional safe level of about 10 ngTBT/l has been indicated for the protection of seafood consumers and a similar value (10-20 ngTBT/l) has been suggested as an appropriate water quality standard for the protection of biota [9].

The reliability of the above limits is, however, questionable since there is a scarce knowledge of the fate of these chemicals. Literature information on the main physico-chemical properties of TBT compounds and its degradation products, such as aqueous solubility, octanol-water partition coefficients, is incomplete and sometimes contradictory. Tributyltin speciation and transfer pathways within the food chain need further studies [10, 11]. TBT persistence in aqueous solutions is still a subject of controversy.

The paper of Astruc and collaborators on the speciation of organo-tin compounds in the sediments of Arcachon Bay, subjected 10-15 years ago to a heavy TBT pollution, addresses the stability of TBT in the marine environment, providing new insights into TBT persistence to be considered when evaluating the environmental impact of organotin compounds.

FATE OF ORGANOTIN COMPOUNDS

TBT salts in natural waters produce an equilibrium mixture, composed of tributyltin chloride (TBTCl), tributyltin oxide (TBTOH), the aquo complex $TBTOH_2^+$ and tributyltin carbonato ($TBTCO_3^-$) species, whose distribution is influenced by pH, Cl^- and dissolved CO_2 concentration [10]. The above species interact with particulate matter showing a strong tendency to accumulate in sediments.

Reported k_{oc} values, calculated as the ratio between partition coefficients and fractional organic carbon content, are in the range $1.2 \cdot 10^4$ to $2 \cdot 10^5$ (12). These values, as well as the reported bioaccumulation factors, appear to be higher than what could be expected on the basis of n-octanol/water partition coefficient, suggesting an influence of polar interactions, other than hydrophobic ones, on sorption or bioconcentration phenomena [13].

Whether accumulation in sediments will occur in practice will depend on TBT degradation rate in water or sediment phase. The degradation process determines a stepwise loss of the organic groups, to which a progressive lowering of the potential aquatic toxicity corresponds. In natural waters, the main TBT degradation mechanisms involve ultraviolet irradiation (photolysis) and biological activity. Photolysis was shown to be possible, but the efficiency of such a process appears to be largely reduced by the limited sunlight penetration, providing half life time >89 days [14]. The microbial degradation seems to be the most important process for removing TBT in natural water conditions. TBT degradation rate by microbial activity results strongly dependent on its concentration. At nanomolar TBT levels the half life time appears to be relatively short in waters, with values in the range 2.5 to 15 days [8, 11, 14, 15]. In dark conditions or in cold waters half life values are longer and at high TBT concentrations they are in the order of months. In sediments or sediment-water mixtures similar high half life of months have been reported [16, 17]. Oxygenation conditions, moreover, were shown to influence the degradation rate in soils, the half life values increasing from 116 to 815 days when aerobic or anaerobic conditions were used [16]. Very little information is available on the degradation of organotin compounds in anaerobic sediments.

Astruc and collaborators in a sediment core collected in the Arcachon Bay found a very high TBT concentration in deeply buried layers (>3000 ng/g) and lower concentrations for the degradation products (>1000 ng/g for DBT and >5000 ng/g for MBT). On the basis of these results the Authors argue a very strong stability of TBT in anoxic sediments, suggesting the occurrence of inhibition phenomena at the high TBT concentrations measured, as already noted by other Authors [8] for degradation in waters. Some points need to be discussed with regard to this paper.

The half estimation in the order of years or decades is too vague and not supported by any reference data on TBT concentration in sediments before regulations, introduced in France to limit the TBT use, came into force.

The presence of two maxima found for TBT and DBT in the core profile is actually mysterious and the approximate 10 cm distance between TBT peaks is very hard to justify, even taking into account the reported mineralogical characteristics of the different layers or age of sediment which is lacking. Some more explanation is also required for the observed shift of MBT maxima from those of TBT and DBT, attributed to different mobility of such species. The analysis of interstitial waters does not help to understand the distribution of organotin compounds in the sediment. The upward shift of MBT maxima respect to TBT and DBT peaks does not appear to be related to a different diffusive flux of such species, since they are not present at all at >5-10 cm depth. The objective difficulty in understanding the observed organotin distribution could indicate that some other causes of disturbance, other than dredging operations, may have disturbed the core sample. The statement that the sediment core was taken in an intertidal area where water fluxes are important supports the above doubt. If this should be the case, the very low concentrations of MBT and DBT found in surface interstitial waters might be the result of dynamic exchange processes between bottom waters and surface sediment. Data on organotin concentrations in the bottom waters would be of interest to support this point.

The finding that "in the top layer of the sediment the concentration of all butylated compounds are higher than somewhat deeper layers" could suggest some degradation of organotin in the oxic layer at relatively low TBT concentrations. This degradation is also consistent with the high $Sn(IV)$ concentrations found in surface interstitial waters. The statement regarding the partial butyltin disappearing from burying sediments (destruction or migration) may perhaps include the above hypothesis, but it should be explained.

In sediment cores from San Diego Bay, presumably under a constant TBT input, Stang and Seligman [16] found total butyltin concentrations decreasing with depth, as well as the tributyltin fraction respect to total butyltin, and the ratio of monobutyltin to total butyltin increasing with depth. On the basis of this results and of degradation experiments in laboratory they suggested the occurrence of in-situ debutylation. The maximum concentration of total butyltin compounds in the above sediment core was 551 ng/g. This value is much lower than that found in most polluted Arcachon Bay sediments. Without doubt, the much larger TBT concentration in the latter Bay may have determined a higher stability of organotin, but the core profile distribution shown by Astruc and collaborators needs several factors to be considered. These include the rate of supply of TBT to the water, the degradation rate of TBT and its derivatives in water phase, the sedimentation rate, the partitioning to the sediment, the dilution of sediments by bioturbation and any degradation of debutylation processes occurring in-situ within the sediment.

The analytical techniques to perform tin speciation, based on conversion of tin compounds by reaction with sodium borohydride to the corresponding volatile hydride, their trapping on a cooled GC column and releasing upon warming, has been widely used, after the work of Braman and Tompkins [18] and Hodge et al. [19]. Thus, the possible overestimation of total recoverable inorganic tin (TRIT), attributed to decomposition of organotins during the analytical procedure needs some more explanation. A last point concerning the analytical procedure concerns the reasons for the choice of acetic acid to extract tin compounds from the sediment. The literature reports that many other solvents have been used to extract organotin from sediment or biological tissue, including dichloromethane, exane, ethylacetate, exane, ethylacetate, chloroform-methanol mixture, diethyl ether, benzene with 0.5% tropolone, methylisobutylketone. A standardization on this matter would be desirable to make the results from different laboratories comparable.

REFERENCES

1. Wong, P.T.S., Y.K. Chau, O. Kramer and G.A. Bengert. "Structure Toxicity Relationship of Tin Compounds on Algae," *Conc. J. Fish. Aquat.* Sci. 39 (1982).

2. Macchi, G. and M. Pettine. "Voltammetric Characterization and Chemical Behavior of Inorganic Tin in Natural Waters," *Environ. Sci. Technol.* 14 (7) (1980).

3. Laughlin, R.B., Jr. and O. Linden. "Tributyltin - Contemporary Environmental Issues," *Ambio* 16(5):252-256 (1987).

4. Alzieu, C., Y. Thibaud, M. Heral and B. Boutier. "Evaluation des Risques dus a l'Emploi des Peintures Anti-Salissures dans les Zones Conchylicoles," *Revue Trav. Inst. Pêch. Marit.* 44:301-348 (1980).

5. Walsch, G.E., L.L. McLaughlan, E.M. Lores, M.K. Louie and C.H. Deans. "Effects of Organotins on Growth and Survival of Two Marine Diatoms, Skeletonema Costatum and Thalassiosira Pseudonana," *Chemosfere* 14(3/4):383-392 (1985).

6. E.P.A. "Tributyltin Technical Support Document: Position Document 2/3," *PB88-161203*, Washington, C.C. (1987).

7. Hall, L.W., Jr. "Tributyltin Environmental Studies in Chesapeake Bay," *Mar. Pollut. Bull.* 19(9):431-438 (1988).

8. Seligman, P.F., A.O. Valkirs and R.F. Lee. "Degradation of Tributyltin in San Diego Bay, California, Waters," *Environ. Sci. Technol.* 20(12):1229-1235 (1986).

9. U.N.E.P. "Assessment of Organotin Compounds as Marine Pollutants and Proposed Measures for the Mediterranean," *UNEP(OCA)MED WG.1/7*, Athens (1988).

10. Laughlin, R.B., Jr. H.E. Guard and W.M. Coleman, III. "Tributyltin in Seawater: Speciation and Octanol-Water Partition Coefficient," *Environ. Sci. Technol.* 20:201-204 (1986).

11. Francois, R., F.T. Short and J.H. Weber. "Accumulation and Persistence of Tributyltin in Eelgrass (Zostera Marina) Tissue," *Environ. Sci. Technol.* 23(2):191-196 (1989).

12. Unger, M.A., W.G. MacIntyre and R.J. Huggett. "Equilibrium Sorption of Tributyltin Chloride by Chesapeake Bay Sediments," in *Proceedings of the Organotin Symposium, Oceans* (4), New York, pp. 1381-1385 (1987).

13. Laughlin, R.B. Jr., W. French and H.E. Guard. "Accumulation of Bis(Tributyltin) Oxide by the Marine Mussel Mytilus Edulis," *Environ. Sci. Technol.* 20(9):884-890 (1986).

14. Maguire, R.J., J.H. Carey and E.J. Hale. "Degradation of the Tri-n-butyltin Species in Water," *J. Agric. Food. Chem.* 31(5):1060-1065 (1983).

15. Seligman, P.F., A.O. Valkirs and R.F. Lee. "Degradation of Tributyltin in Marine and Estuarine Waters," in *Proc. Organotin Symp. Oceans 86 Conf. and Expos.*, Washington D.C., pp. 1189-1195 (1986).

16. Stang, P.M. and P.F. Seligman. "Distribution and Fate of Butyltin Compounds in the Sediment of San Diego Bay," in *Proc. Organotin Symp. Oceans 86 Conf. and Expos.*, Washington D.C. (1986).

17. Maguire, R.J. and R.J. Tkacz. "Degradation of the Tri-n-butyltin Species in Water and Sediment from Toronto Harbor," *J. Agricul. Food Chem.* 33:947-950 (1985).

18. Braman, R.S. and Tompkins, M.A. "Separation and Determination of Nanogram Amounts of Inorganic Tin and Methyltin Compounds in the Environment," *Anal. Chem.* 51(1):12-19 (1979).

19. Hodge, V.F., S.L. Seidel and E.D. Goldberg. "Determination of Tin (IV) and Organotin Compounds in Natural Waters, Coastal Sediments and Macroalgae by Atomic Absorptio Spectrometry," *Anal. Chem.* 51(8):1256-1259 (1979).

METAL COMPLEXATION BY WATER-SOLUBLE ORGANIC SUBSTANCES IN FOREST SOILS

A.T. Kuiters and W. Mulder
Department of Ecology and Ecotoxicology, Biological Laboratories, Free University, Amsterdam, The Netherlands

1. ABSTRACT

Freshly fallen leaves of oak and poplar trees were subjected to leaching under controlled laboratory conditions and the leachates were analyzed for soluble organic substances with metal-complexing abilities. By using a modified size-exclusion chromatographic technique, the soluble organics in the litter solutions were fractionated into distinct molecular weight groups and the metal-complexing capacity of each group was determined at the same time. Analyses of the leachates revealed the presence of two apparent molecular weight groups, with different leaching rates and metal-binding abilities. Group I compounds, with highest molecular weight and with polysaccharides and polyphenols as the presumably main components, had the ability to form strong metal-complexes. These compounds were only gradually released by leaching of the litter material. Group II, with low-molecular weight organic and phenolic acids presumably as the predominant compounds, had metal-complexing abilities as well, but with low to moderate strength. These compounds were rapidly leached from the leaf litter.

In contact with the humus layer of forest soils, these litter leachates promoted to solubilization of metal ions (Al, Fe, Cu, Pb, Zn and Mn). This was revealed by batch equilibrium experiments. For Al and Fe this was effectuated by strong complexants, presumably compounds of relatively high molecular weight (group I). The solubilization of Pb, Cu, Zn and Mn was correlated with the presence of low-molecular weight compounds (group II). The release of soluble organic substances from the top-layer of forest soils and the subsequent mobilization of metal ions by the percolating litter solutions may have serious consequences for the availability of (heavy) metals for uptake by the vegetation and facilitates the vertical transport of the complexed metals through the soil profile.

2. INTRODUCTION

The organic top layer of forest soils often act as a sink for heavy metals, deposited by air pollution from fossil fuel combustion, metal processing or other anthropogenic activities. The possible ecotoxicological effects of (heavy) metals is largely determined by the concentration and speciation in the soil solution. Metal ions are distributed between the solid phase and the soil solution by equilibrium processes. The chemical composition of the soil solution, especially the presence of soluble metal-complexing organic substances, influences these equilibria [1].

In forest soils, high amounts of soluble organic substances are produced in the organic top layers, resulting from the leaching and decomposition of plant residues, from root exudation or from microbial synthesis [2, 3, 4]. The metal-complexing abilities of the organic compounds are related to the presence of functional groups, i.e. carboxyl, phenolic hydroxyl or amine groups [3]. The soluble organics are transported to deeper soil layers by percolation of the soil solution and may bring large amounts of metal ions into solution by complexation. Consequently, complexed metal ions are more mobile in soils and are probably more available to plants.

Several investigators have demonstrated that leachates of forest leaf litter have the ability to solubilize Fe and Al [5, 6, 7, 8] and facilitate metal transport in the soil [2, 8, 9, 10]. There is still much uncertainty about the active components in litter leachates responsible for metal-complexation and subsequent mobilization. Some reports present evidence that polyphenols are the main component forming soluble organo-metal complexes [6, 11], while others ascribe this effect to the occurrence of low molecular-weight organic acids [2, 8, 10] or monomeric phenolic acids [12].

The aim of this study was to determine the metal-complexing capacity of leachates of forest leaf litter, how this property changes throughout the leaching process, and which molecular weight groups are responsible for the metal-complexation. Solubilization of metal ions (Al, Fe, Pb, Cu, Mn and Zn) by leachates of the top-litter layer during their vertical transport through the forest soil, was studied in batch equilibrium soil experiments.

3. METAL-COMPLEXING CAPACITY OF LEAF-LITTER LEACHATES

3.1 Leaching Procedure

Leaf litter from oak (*Quercus robur L.*) and poplar (*Populus nigra L.*) was collected during autumn leaf fall (October 1988) in a coastal nature dune reserve in The Netherlands. The leaching of litter material was simulated under laboratory conditions by leaching in triplicate a known amount of fresh litter (equal to 5 g dry wt) with 250 mL demineralized water. After standing in the dark for 20 h at 12°C (without shaking), the solution was separated from the litter material by filtration, followed by membrane filtration (Schleicher & Schuell, 0.45 µm). This procedure was repeated every 8-10 days during a period of seven weeks. During the days that the litter material was not leached, they were stored in open polyethylene pots in the dark at 12 °C. Per litter type, six leachates were collected, denoted by Q1-Q6 and P1-P6 for the oak and poplar leachates, respectively. The clear solutions were stored at 4 °C, prior to the chemical analyses. The leachates were analyzed for pH and dissolved organic carbon (DOC). Moreover, the metal-complexing capacity was determined by a modified size-exclusion chromatographic technique. The strength of the metal-complexes was determined by an ion-exchange technique.

3.2 The Size-Exclusion Chromatographic Technique

The metal complexing capacity of the litter solutions was studied using a modified size-exclusion chromatographic technique [13, 14]. A gel permeation column is equilibrated with a flowing buffer solution of known pH and ionic strength containing a known concentration of a metal of interest. A certain volume of the sample solution is injected and as it traverses the column it continuously binds metal ions from the column until it comes to equilibrium with the free metal ion concentration in the solution. Monitoring the metal content of the effluent from the column, peaks appear at the elution volumes of the molecular groups that form complexes with the metal ions. The total area of the metal peaks corresponds with the total amount of metal complexed by the sample solution. This is followed by a trough in metal concentration of the effluent of the same area as the total peak area, corresponding to the amount of metal withdrawn from the eluent to restore the equilibrium between gel adsorbed-metal and free metal concentration in the eluent.

This gel chromatographic technique was carried out with a Sephadex G-25 (Pharmacia Fine Chemicals) packed column, with an exclusion limit of 5000 Daltons. The column dimensions were 20 cm x 1.6 cm I.D., with a void volume of 12.5 mL and a total gel volume of 26.8 mL. The flow rate was approximately 1.0 mL min^{-1}. The eluent was a 0.01 M sodium acetate buffer with 31.5 μM Cu (2 ppm). As in most studies, investigating metal-complexing properties of organic substances, copper was used as the metal. It should be noted that the pH of the buffer system determines to a large extent the metal-complexing ability of the litter solution. Therefore, all samples were chromatographed at pH 4.0 and 6.0 to measure the metal-complexing abilities at two relevant pH values. Before applying the samples to the gel filtration column, cations as K, Ca and Mg were removed by a cation-exchanger. Otherwise, these cations may, by their high concentration in the leachates, desorb Cu from the column. After that, a 500 μL sample was mixed with 500 μL eluent (twice concentrated to avoid effects of a diluted eluent on the Cu- equilibrium in the column) and placed on top of the gel bed by injection from a sample loop. The effluent from the column was passed through an 280 nm UV-monitor and fractions of 2.5 mL were collected. After acidification with 100 μL 5 N HNO$_3$, the Cu concentration of the fractions was measured by flame atomic absorption spectrometry.

For studying the properties of the Sephadex column towards low-molecular weight acids and phenolic acids, citric acid and protocatechuic acid were applied as Standards. Blue dextran was used to determine the void volume (V_0). The elution parameter K_{av}, a relative measure of molecular size, was determined from the following equation:

$$K_{av} = (V_e - V_o)/(V_t - V_o)$$

Where V_e is the elution volume of the molecular size group or compound in question and V_t the total volume of the gel. Adsorption phenomena may interfere with size fractionation factors, resulting from complex formation and hydrophobic interactions. [15]. Especially aromatic compounds have an affinity to dextran gels. Previous elaborative experiments with the Sephadex G-25 gel (Kuiters & Mulder, in prep.) revealed that the adsorption phenomena were reduced to a minimum with a sodium acetate buffer with 0.01 M ionic strength.

3.3 The Ion-Exchange Technique

With the above described size-exclusion chromatographic technique, the total metal complexing capacity of the litter solutions, including weak and strong complexes, is measured. An ion-exchange equilibrium technique was used to determine the strength of the formed complexes. A column (10 cm x 1.0 cmI.D.) was packed with a weak acidic cation exchanger (Lewatit TP 207). The resin was brought into the Na(+)-saturated form. Leachate samples of 30 mL were spiked with 1 mL of a copper sulphate solution (1 mM). The samples were given the opportunity to form Cu-complexes by equilibration in the dark at 20 °C for at least 30 min and were then eluated through the column. When the copper-spiked sample solution traverses the column, the strong metal complexes remain intact but the weak complexes are split by the reactive sites of the resin (iminodiacetic acid groups) and the released copper ions are adsorbed to the resin. By monitoring the copper concentration of the effluent, the amount of strongly complexed copper can be determined.

3.4 Analytical Procedures

Dissolved organic carbon was measured by adding 180 μL of each sample to a small tin cup and acidified with 20 μL 0.1 N HCl to remove the inorganic carbonates. Glucose was used as standard. The content of the cups was evaporated at room temperature and analyzed for organic carbon by an elemental analyzer (Carlo Erba). K and Na were determined by flame emission spectrometry, whereas Ca, Mg, Fe, Mn and Zn were determined by flame atomic absorption spectrometry (Perkin Elmer 4000). Al, Cu and Pb were determined by graphite furnace atomic absorption spectrometry (Perkin Elmer 1100)

3.5 Results

Chromatograms of some of the litter leachates after separation on Sephadex G-25 are presented in Figure 1. Two distinct molecular weight groups were distinguished, denoted by group I and II, respectively. The peaks were rather broad, indicating a rather high dispersion of molecular weight of the compounds within each group. Group I was eluted at the void volume of the gel column and consisted of compounds with a relatively high molecular weight. Group II consisted of compounds with a K_{av}-value between 0.45-

2.00, and with low-molecular weight. Figure 1 also shows the elution volumes of the standards citric acid (K_{av} is 0.95) and protocatechuic acid (K_{av} is 2.00). Repeated leaching of the litter material resulted in a rapid decrease of group II, whereas group I was found even after repeated leaching.

TABLE 1. *Metal-complexing Capacity of the Successive Leachates of Oak and Poplar as Determined by the Modified Size-exclusion Chromatographic Technique at Two pH Conditions. Given are also the amounts of strongly complexed copper as determined by ion-exchange.*

Litter Type	Leachate	Metal Complexing Capacity (μmol Cu L^{-1})		Strongly Complexed (μmol Cu L^{-1})
		pH 4.0	pH 6.0	
oak	Q1	74	151	11
	Q2	98	332	26
	Q3	66	194	38
	Q4	60	128	34
	Q5	50	88	16
	Q6	22	64	11
poplar	P1	130	430	16
	P2	52	112	13
	P3	28	32	10
	P4	38	38	10
	P5	34	35	8
	P6	10	20	5

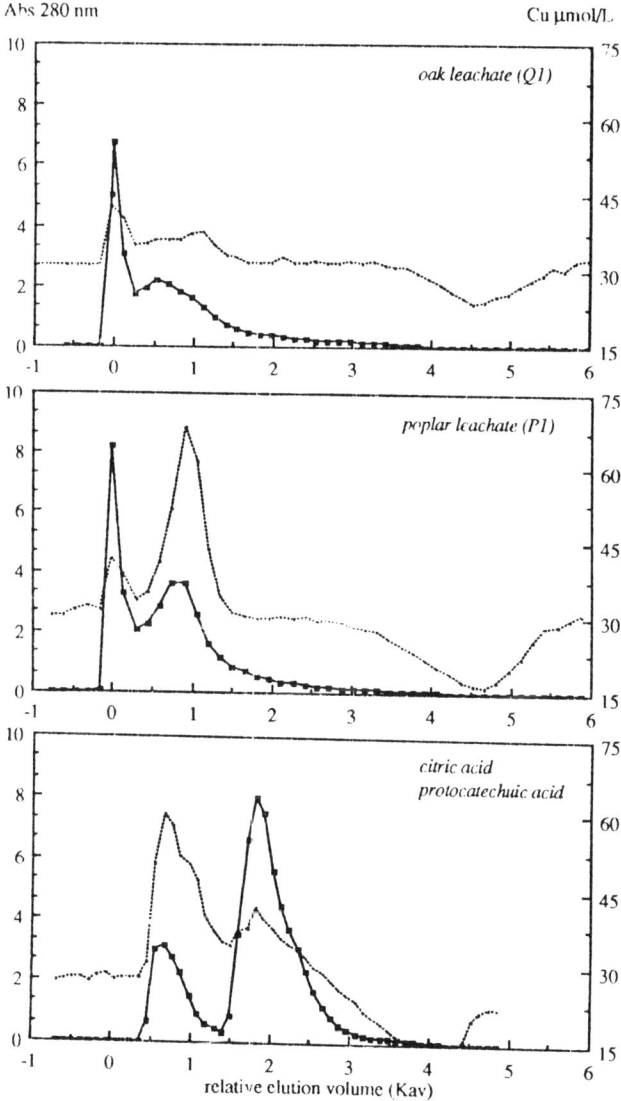

FIGURE 1. Chromatograms of Some of the Leaf Litter Leachates and a Standard Solution with Two Low-Molecular Weight Acids After Fractionation on a Sephadex G-25 Gel Filtration Column, Equilibrated with a Cu (31.5 µmol L^{-1}) Containing Eluent of pH 6. Given are the relative absorbance at 280 nm (solid line) and the Cu concentration of the effluent (dashed line) as function of the elution volume.

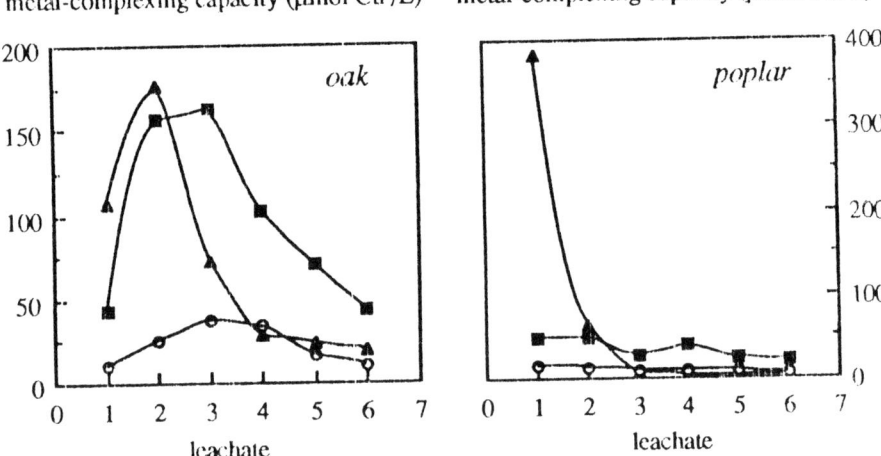

FIGURE 2. *Metal-complexing Capacity of Group I (High Mol wt; -■-) and Group II (Low Mol Wt;-△-) Compounds in the Successive Leachates of Oak and Poplar Leaf Litter. The Open Symbols (- o -) Indicate the Amount of Strongly Complexed Copper.*

Figure 2 shows the changes in metal-complexing capacity of the litter solutions during the course of the leaching procedure. The metal complexing capacity is presented per molecular weight group. Metal complexation by group II rapidly decreases during the leaching procedures, whereas group I compounds slightly changes (poplar) or even strongly increases (oak litter) after repeated leaching. This was also affirmed by the absorbances at 280 nm. The pH condition of the gel permeation column had a large impact on the metal complexing capacity of the leachates (see Table 1). Overall, the metal binding capacity at pH 6.0 was much higher than at pH 4.0. A comparison of the amount of strong complexes formed by the litter leachates with the total metal-complexing capacity makes clear that most of the metal-complexes had moderate to low strength as the main part was split by the ion-exchange column. From Figure 3, where strongly complexed copper is correlated with the metal-complexing capacity of the distinct molecular weight groups, it becomes clear that strong complexants occur mainly in group I.

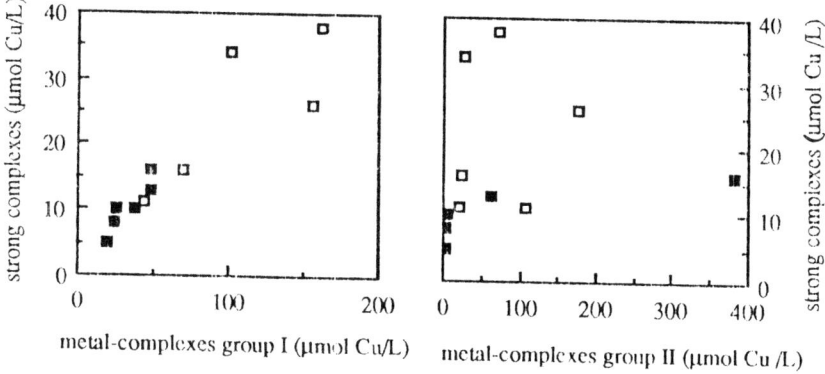

FIGURE 3. Diagram with the Amount of Strongly Complexed Copper by the Leachates of Oak (□) and Poplar (■) Leaf Litter Correlated with the Copper Complexed by Group I (left) and by Group II (right).

4. METAL-SOLUBILIZATION FROM FOREST SOILS BY THE LITTER LEACHATES

4.1 Batch Equilibrium Experiments

The ability of the litter leachates to desorb metals ions from forest soils was investigated in batch equilibrium experiments. Soil samples from the humus-layer (A_h-horizon, 0-8 cm depth, sandy calcareous dune soil) were collected under stands of oak and poplar trees (±30 years old) in November (1988) at the same location where the leaf litter material was sampled. These soils, denoted by Q-A_h (oak) and P-A_h, (poplar) respectively, were sieved (< 2 mm). Leachates of oak and poplar leaf litter were prepared in the same way as described before (repeatedly leaching every 8-10 days). Of both litter types, litter leachates were confronted with each of the two soils and the metal-desorbing ability was determined. This was done in two experiments. In the first experiment the first leachates of each litter type were used, denoted by Q1 (oak) and P1 (poplar).

In the second experiment this procedure was repeated, but with the third leachates of each litter type, denoted by Q3 and P3. The leachates were analyzed for pH, DOC and elemental composition (K, Na, Ca, Mg, Fe, Mn, Zn, Pb and Cu). The concentration of the predominant cations K, Na, Ca and Mg were summed and used as a measure for the ionic strength.

Batch equilibrium experiments were performed in triplicate in 250 mL polyethylene pots. A known amount of soil (equivalent to 10 g dry wt) was mixed with 25 mL litter leachate solution. The suspensions were briefly stirred and set away in the dark at 12 °C. As control, demineralized water was used. After 20 h, solid and solution phase were separated from each other by centrifugation (2500 rpm x 20 min) and pH was measured directly. The equilibrium solutions were membrane filtrated (Schleicher & Schuell, 0.45 -μm) and analyzed for DOC and metal concentrations (Al, Fe, Pb, Cu, Mn and Zn). Moreover, the metal-complexing capacity of the equilibrium solutions was determined by size-exclusion chromatography. All analytical procedures were the same as described before.

4.2 Results

Equilibration of the litter leachates with samples of the two forest soils resulted in higher metal concentrations in the equilibrium soil solution, compared to equilibration with demineralized water (Table 2). Al and Fe were mobilized most by leachates of oak, whereas Cu, Pb, Zn and Mn were especially mobilized by leachates of poplar. Pb and Cu were mobilized only by the first leachates. Compared to the Q-A_h soil, the equilibrium solution of the P-A_h soil had considerably lower metal concentrations and here the metal-mobilizing effects were relatively highest.

The pH of the equilibrium soil solution was only slightly changed by the litter leachates. Part of the dissolved organic-carbon (DOC) and the metal-complexing capacity of the litter leachates disappeared after contacting the soil, but the equilibrium solutions still had a considerably elevated concentration of DOC and metal-complexing capacity, compared to the equilibrium solutions with water.

The coefficients of correlation between metal concentrations and the chemical properties of the equilibrium soil solution (of both batch experiments) are presented in Table 3. It shows that the metals under study behaved as two groups. Fe and Al behaved similar and their equilibrium concentrations were slightly negatively

TABLE 2. *Metal Concentrations, pH, DOC and Ionic Strength of the Equilibrium Soil Solution of the Humus Layer of Two Forest Soils After Equilibration with the First Leachate from Oak (Q1) or Poplar (P1) Leaf Litter. Water Was Used as Control Treatment. In a Second Experiment the Third Leachate of Oak (Q3) and Poplar (P3) Was Used.*

Soil	Equilibration Solution	Al	Fe	Zn	Mn	Cu	Pb	pH	DOC (mg L^{-1})	Ionic Strength
		------(μmol L^{-1})------								
	Q1	1.1	0.5	0.5	4.6	0.13	0.03	5.6	169	0.002
	P1	0.6	0.4	7.2	7.6	0.35	0.09	6.3	496	0.010
Q-Ah soil	water	11.3	6.2	3.9	1.0	0.31	0.10	6.6	56	0.001
	Q1	14.1	10.5	3.5	3.4	0.64	0.16	6.4	149	0.003
	P1	10.2	8.1	5.9	15.3	0.77	0.40	6.2	360	0.007
P-Ah soil	water	3.0	2.0	2.2	0.7	0.31	0.05	7.6	35	0.002
	Q1	7.1	5.8	3.5	2.0	0.37	0.23	7.5	128	0.004
	P1	5.5	5.4	5.7	6.3	0.60	0.18	7.2	346	0.009
	Q3	1.1	1.1	0.2	1.5	0.01	0.01	5.7	81	0.001
	P3	0.5	0.6	1.2	0.4	0.01	0.02	6.9	49	0.001
Q-Ah	water	7.5	3.5	1.6	0.3	0.36	0.04	6.4	44	0.001
	Q3	12.9	7.8	1.9	1.2	0.34	0.10	6.7	104	0.002
	P3	11.3	6.5	3.3	1.1	0.31	0.06	7.1	103	0.002
P-Ah	water	3.2	1.3	1.0	0.1	0.26	0.01	7.7	23	0.002
	Q3	8.4	4.5	1.6	0.2	0.30	0.05	7.8	58	0.003
	P3	5.1	2.4	2.7	0.2	0.33	0.06	8.0	64	0.003

correlated with pH but were not correlated with DOC. The other group consisted of the metals Zn, Mn, Cu and Pb. Their equilibrium concentrations were positively correlated with DOC and to some extent also with metal-complexing capacity of the equilibrium soil solutions.

TABLE 3. *Coefficients of Correlation between Metal Concentrations and Several Chemical Properties of the Equilibrium Soil Solutions.*

Metal	pH	DOC	Ionic Strength	Metal-Complexing Capacity
Al	-0.61	0.38	-0.32	0.20
Fe	-0.67	0.49	-0.19	0.35
Zn	-0.40	0.78	0.52	0.62
Mn	-0.59	0.92	0.41	0.68
Cu	-0.52	0.79	0.43	0.59
Pb	-0.43	0.93	0.57	0.71

5. DISCUSSION

The analyses of the litter leachates of freshly fallen oak and poplar leaves revealed that considerable amounts of water-soluble organic substances with metal-complexing abilities are released by leaching. Fractionation of the soluble organics by size-exclusion chromatography resulted in two apparent molecular weight groups. Group I eluated at the void volume of the gel permeation column and contained compounds with high-molecular weight. These compounds were only gradually released from the litter material. Group II was composed of low-molecular weight compounds and was rapidly leached from the leaf material. This group was especially found in leaves of poplar. Both molecular weight groups had complexing abilities for copper. From the positive correlation between the amount of strongly complexed copper and the complexing capacity of group I, it may be concluded that organic substances with the ability to form strong metal-complexes occur especially among the high-molecular weight compounds.

Polysaccharides and polyphenols, with molecular weight of a several hundreds to a few thousands, are presumably the main component of group I. In fresh leaf litter, polysaccharides account for 40-50% of organic C [16], they are moderately hydrosoluble and are able to form stable complexes with metals [3]. By condensation reactions they are changed into the more stable fulvic acids during humification, which are also well known for their metal-complexing abilities [3]. Low-molecular weight organic acids and phenolic acids are the most likely compounds of group II. They are released in high amounts from freshly fallen deciduous litter [2, 4]. As most of the metal-complexes formed by this group of compounds were split by the weak acidic cation-exchanger it may be concluded that most of the metal-complexes formed by this group had moderate to low strength. This is in accordance with the stability constants of many commonly occurring organic and phenolic acids [17].

The batch equilibrium experiments demonstrated that the litter leachates, when brought into contact with the humus layer of forest soils, increase the concentration of Al, Fe, Cu, Pb, Mn and Zn in the equilibrium soil solution. The positive correlation found between the concentrations of the metals Cu, Pb, Zn and Mn and the concentration of soluble organic substances (DOC), indicate that these metals were solubilized mainly by complexation. The solubilization of these metals was highest in leachates with high concentrations of low-molecular weight compounds (poplar leachates). Although Al and Fe were also solubilized by the litter leachates, no correlation was found with DOC. This can be understood by assuming that the solubilization of these metal ions was realized mainly by strong complexants (group I). It is well known that Al and Fe are, as trivalent cations, strongly bound by mineral and organic soil particles [3] from which they can only be desorbed by strong complexants. The presence of strong organic complexants in the leachates was not correlated with DOC. Other factors that may influence the distribution of metal ions between the solid soil phase and the solution are pH and ionic strength of the equilibrium soil solution [1]. Although a weak correlation was found between Al and Fe and pH, this factor may factor may not explain the solubilization of Al and Fe as lower pH conditions coincided with the presence of strong complexants.

The ecological effects of (heavy) metals in forest soils are closely related to their concentration and speciation in the liquid soil phase. Finally, an increase in metal-solubility in forest soils by the presence of soluble organic substances has implications for the intensity of metal-cycling in forest ecosystems [18]. In this respect, detailed knowledge about the processes that influence the partitioning of metal ions between solid and solution phase of forest soils is of crucial importance.

ACKNOWLEDGMENTS

These investigations were made possible by the financial support from the Program Office for Fundamental Soil Research (PCBB), The Netherlands (project no. 8936).

REFERENCES

1. König, N., P. Baccini, and B. Ulrich, "Der Einfluss der Natülichen Substanzen auf die Metallverteilung Zwischen Boden und Bodenlösung in Einem Sauren Waldboden," *Z. Pflandzeneraehr. Bodenk.* 149:68-82 (1986).

2. Bruckert, S. "Influence des Composés Organiques Solubles sur la Pédogenèse en Milieu Acide. II. Expériences de Laboratoire Modalités des Agents Complexants," *Ann. Agron.* 21:725-757 (1970).

3. Schnitzer, M. and S. U. Khan, Eds. *Soil Organic Matter* (Amsterdam: Elsevier, 1978).

4. Kuiters, A.T. and H.M. Sarink. "Leaching of Phenolic Compounds from Leaf and Needle Litter of Several Deciduous and Coniferous Trees," *Soil Biol. Biochem.* 18:475-480 (1986).

5. Bloomfield, C. "Leaf Leachates as a Factor in Pedogenesis," *J. Sci. Food Agr.* 6:641-651 (1955).

6. Bloomfield, C. "The Possible Significance of Polyphenols in Soil Formation," *J. Sci Food Agric.* 8:389-392 (1957).

7. Malcolm, R.L. and R.J. McCracken. "Canopy Drip: a Source of Mobile Soil Organic Matter for Mobilization of Iron and Aluminum," *Soil Sci. Soc. Amer.Proc.* 32:834-838 (1968).

8. Pohlman, A.A. and J.G. McColl. "Soluble Organics from Forest Litter and Their Role in Metal Dissolution," *Soil Sci. Soc. Am. J.* 52:265-271 (1988).

9. Dawson, H.J., F.C. Ugolini, B.F. Hrutfiord, and J. Zachara. "Role of Soluble Organics in the Soil Processes of a Podzol, Central Cascades, Washington," *Soil Sci.* 126:290-296 (1978).

10. Pohlman, A.A. and J.G. McColl. "Kinetics of Metal Dissolution from Forest Soils by Soluble Organic Acids," *J. Environ. Qual.* 15: 86-92 (1986).

11. Davies, R.I. "Relation of Polyphenols to Decomposition of Organic Matter and to Pedogenetic Processes," *Soil Sci.* 111:80-85 (1971).

12. Vance, G.F., D.L. Mokma, and S.A. Boyd. "Phenolic Compounds in Soils of Hydrosequences and Developmental Sequences of Spodosols," *Soil Sci. Soc. Am. J.* 50:992-996 (1986).

13. Hummel, J.P. and W.J. Dreyer. "Measurement of Protein-Binding Phenomena by Gel Filtration," *Biochim. Biophys. Acta* 63: 530-533 (1962).

14. Mantoura, R.F.C. and J.P. Riley. "The Use of Gel Filtration in the Study of Metal Binding by Humic Acids and Related Compounds," *Acta Chim. Acta.* 78:193-200 (1975).

15. Sada, A., G. Di Pascale, and M.G. Cacace. "Salt Effects on Adsorption of Aromatic Compounds in Sephadex G-25 Chromatography," *J. Chromatogr.* 177:353-356 (1979).

16. Kögel, I., R. Hempfling, W. Zech, P.G. Hatcher, and H.R. Schulten. "Chemical Composition of the Organic Matter in Forest Soils. 1. Forest litter," *Soil Sci.* 146:124-136 (1988).

17. Martell, A.E. and R.M. Smith. *Critical Stability Constants.* Vol. 3. (New York: Plenum Press, 1977).

18. Bergkvist, B. "Soil Solution Chemistry and Metal Budgets of Spruce Forest Ecosystems in S. Sweden," *Water Air Soil Pollut.* 33: 131-154 (1987).

DISCUSSION OF:
METAL COMPLEXATION BY WATER-SOLUBLE ORGANIC SUBSTANCES IN FOREST SOILS

K. FYTIANOS
Environmental Pollution Control Laboratory, University of Thessaloniki, Thessaloniki, Greece

The aim of this interesting paper was to determine the metal-complexing capacity of leachates of forest leaf-litter, and to find out which kind of organic substances are responsible for the metal complexation. It also discusses the factors that influence this process. Finally, the solubilization of a number of metal ions by leachates of the litter layer was studied in batch equilibrium soil experiments.

This last fact may have serious consequences for the availability of metals for uptake by vegetation and vertical transport of the complexed metals through the soil profile. This work consists of two parts.

The first part describes the metal-complexing ability of two groups of organic substances leached from freshly fallen leaves of oak and poplar trees in a forest ecosystem. The sources of soluble organic substances are natural depositions of plant residues that reach the soil in the form of leaves and other organic debris.

The first group (I) consists of high molecular weight substances (polysaccharides and polyphenols) that are able to form strong metal-complexes. These compounds were only gradually released by leaching of the litter material. The second group (II) with low molecular weight organic and phenolic acids also has metal-complexing abilities, but with low to moderate strength. These compounds were rapidly leached from the leaf-litter. Soluble organics from the leaf-litter extracts were isolated into two homogeneous fractions, but the compounds of these groups were not identified and quantified. For the metal-complexing capacity at two pH values (pH 4.0 and 6.0) a modified size-exclusion chromatographic technique was used. The pH value determines, to a large extent, the metal-complexing ability of the litter solutions. The metal binding capacity at pH 6.0 was much high than at pH 4.0.

The second part of this paper describes the ability of these leachates to promote the solubilization of metal ions from forest soils. This was investigated in batch equilibrium experiments. The solubilization of Al and Fe was correlated with the presence of high-molecular weight compounds (group I) (leachates of oak) and of Pb, Cu, Zn and Mn with the presence of low-molecular weight compounds (group II) (leachates of poplar).

The metal complexing capacity and the solubilization from forest soils by organic substance leachates is too complicated, and for this reason a complete study is necessary to examine the kinetics of metal dissolution, and also, the different parameters which influence the whole process. Knowledge only of the total amount of metals does not indicate the nature of the various geochemical processes which may take place.

For this reason, the study of the partitioning of the metals into selective chemical fractions in the forest soils would be of great interest. This is the most efficient way to obtain detailed information about the origin, made of occurrence, biological and physicochemical availability, mobilization of metals and their transport media. The study of the effect of pH on metal solubilization from forest soils would also be very interesting. The prevailing soil acidification will increase the release and leachability of many elements in soil, those of anthropogenic as well as those of natural origin.

As it is known, in Central Europe the problem of metal release and cycling in forest ecosystems has also aroused considerable interest, due to the severe forest decline. An increased soil acidification could induce a nutrient deficiency (Mg) which seems to be a main factor in forest decline, which tends to become a widespread phenomenon in Europe. For that reason it is of great concern to study the loss of metals from the entire soil profile.

Finally, it is suggested that the results of this paper be compared and discussed to those reported in literature in order to obtain a complete idea about this serious problem.

THEORETICAL AND EXPERIMENTAL DRAWBACKS IN HEAVY METAL SPECIATION IN NATURAL WATERS

P. Papoff, M. Betti, and R. Fuoco
Dip. Chimica, (Univ.Pisa) - I.C.A.S. (CNR - c/o Dip. Chimica, Univ.Pisa) - Italia

1. INTRODUCTION

Metals in natural waters are present partly in soluble form, partly adsorbed or retained in suspended or sedimented solid matter. Operatively the soluble form is considered the one present in the liquid phase and not retained on a 0.45 μm membrane filter. The exchange of metals between different phases depends on the nature and size of the solids, and on some physico-chemical parameters intrinsic in each natural water as hydrodynamic conditions, temperature, dissolved oxygen, salt content and so on.

Studying the potential toxicity of an aqueous system, the metal species in all phases of the system which the organisms may enter in contact with, should be taken into account. Exchange mechanisms between phases have been extensively studied both in field and batch experiments [1-4].

As far as the soluble metals are concerned, they are shared between different molecular forms which may or may not be in mutual chemical equilibrium.

The total soluble metal concentration C_M of a given metal M can be expressed as:

$$C_M = [M] \cdot \underbrace{(1 + \beta \cdot [L_1] + ... + \beta \cdot [L_n])}_{\sum_i [ML_i]} + \sum_j [MX_j] + (C_M)_{adsorbed} \qquad (1)$$

Where [M] is the concentration of free metal ions, $\beta \cdot [L_i]$ is for the ligand L_i equal to:

$$K_1 \cdot [L_i] + K_2 \cdot [L_i]^2 + ... + K_n \cdot [L_i]^n$$

K_i is the overall conditional formation constant with the ligand L_i (the charge of the complexes, and the mixed complexes as well have been ignored for the sake of simplicity). $\Sigma[MX_j]$ are the metal containing compounds in non equilibrium with the free metal ions. $(C_M)_{adsorbed}$ is the concentration of metal adsorbed or included in the colloids which have passed through the 0.45 um membrane filter.

At present, the uptake mechanisms and the relevant toxicity of different species in which the metal may be shared in the aquatic environment, have not been well understood. As a matter of fact, the biological responses of organisms to trace elements are very complex [5]. For instance, the same element exerts different toxicities for different animals. As for the mechanism of toxicity, the most relevant is certainly the chemical inactivation of enzymes. All divalent transition metals and most electronegative metals are very active in this respect. They react very promptly with the amino, imino and sulphydryl groups of proteins; some of them (Cd, Hg) may compete with zinc and displace it in zinc-containing metalloenzymes [5].

In this respect the chemical approach, whenever the results have to be used for the biological interpretation of the toxicity of the system, must look at the biological target very carefully.

The aim of this communication is to show the quality of information chemists can give to biologists in understanding the potential toxic effects of heavy metals suspected to be present, under different molecular forms, in a water basin.

We will start with a short description of the basic biological problems and the kind of questions concerning the chemical domain for which chemists are expected to give answers.

2. THE BIOLOGICAL PROBLEM

Pollutants in an aqueous ecosystem affect the biotas' quality of life, according to mechanisms which are very much dependent on the nature and concentration of each pollutant, the exposure time and on the characteristics of the ecosystem as well; these characteristics undergo oscillation due to seasonal and daily meteorological effects. The main variables which need taking into account when studying pollutant toxic effects, as far as these variables influence the biological uptake mechanisms in natural waters, are: pH, temperature, dissolved oxygen content, ionic strength, molecular forms in which the toxic group is present, the presence of toxic or non-toxic foreign substances that may hinder or enhance the effect of the pollutant being considered. The calcium to magnesium [6] and the sodium to potassium concentration ratio is often considered. Owing to such a great number of variables, which are furthermore mutually dependent, biologists are forced to:

a) use simplified experimental models, so that some kind of information about the toxic activity of each pollutant, related to the used conditions can be obtained in the laboratory;

b) use some operative assumptions which allow much of the evidence inferred from the experiments performed as in point a), and concerning the mechanisms of uptake of the pollutant according to the biotic indicator used as a test, to be interpreted and extrapolated to real systems.

Experimental models are obtained by properly selecting for each pollutant of interest 1) the suitable biotic indicator; 2) the experimental laboratory conditions simulating at the best the natural hydrosystem characteristics; 3) the type of observable biological variable to be quantified (growth factor, nutrient uptake, etc.).

As far as the difficulties inherent to points 1), 2), 3) are or can be assumed to have been overcome, the biotic information obtained by laboratory experiments retains a sufficiently general meaning and can be cautiously managed for the interpretation of the pollutant behaviour in a real ecosystem, provided that this ecosystem has been sufficiently characterized.

It must be noticed that the use of biotic indicators resident on the field only provides time-integrated information about the overall effect due to physical, chemical and biological pollutants present during the exposure time of the indicator.

Furthermore, by changing the nature of the biotic indicator or some of the experimental conditions in laboratory tests, results may be obtained which present quite a large bias or appear, to some extent, contradictory. It is good practice in such circumstances to use *operative laboratory models* whereby the experimental conditions are fixed on the ground of agreements between laboratory leaders. These operative models may differ from each other according to the kind of natural water being considered: lake, sea, river, fresh-water and so on, in that different biotic indicators may be required.

The more the operative model is simplified compared with the complexity of the real system, i.e. the larger the difference between real and laboratory systems is, the lower the amount of information available for the extrapolation from laboratory to real system situation will be. It is a difficult decision that biologists, and not only biologists, are often called to take when studying pollution problems: 1) the use of simplified operative models which allow quite reproducible results to be reached, in spite of a loss of an unknown amount of correlation between real and laboratory systems; 2) the use of complex experimental models much closer in simulating the real system, whereby the recovery of this correlation is in principle by and largely uncertain since the effective reasons of the observed effects are uncertain.

The situation can further degenerate, whenever the analytical chemical steps as well, owing to the difficulties inherent to the measurement of very low concentration of metals in natural waters, require some simplification in the analysis to be adopted.

Biologists' operative assumptions may be summarized as follows:

1) When heavy metal is shared in natural waters between different phases, the fraction present in the water phase, which includes colloids with a size diameter lower then 0.45 μm, is considered as the only one directly involved in the biological uptake mechanisms. Suspended and sedimented matter are considered as a supply medium for heavy metal exchange. Exchange mechanisms generally reach steady or equilibrium conditions very slowly depending on steps which are kinetically or thermodynamically preferred [1-4, 7].

2) The uptake of heavy metals may occur through three main pathways: a) via the free metal ion, which can be adsorbed on the biological water/membrane interface and suitably transported inside the body through the membrane. The direction of mass transport depends on the difference of the chemical (or electrochemical) potential of the free metal ion at the external and internal membrane interfaces. All the complexed forms which are in chemical equilibrium with the free metal ion concur to its mass transport. The uptake rate may be mass transport or kinetically controlled whether the diffusion of the dissociation of the complexes in giving metal ion is the rate controlling process; b) by the displacement of the metal from an aqueous complex by means of a ligand group at the membrane interface; c) by the direct crossing of the biological membrane through a proteic channel or as a hydrophobic and liposoluble metal complex.

3) The toxic activity of a free metal ion depends on the nature of the natural water. In the case of fresh water this activity depends on the total amount of dissolved salt (hardness) as will be shown later. Evidences were found that the free ionic forms of several trace metals (e.g. copper, cadmium, lead and zinc) are the most toxic forms for phytoplankton, invertebrates and fish; complexed forms appear to be non-toxic, or at least are considerably less toxic than the free metal ion [8, 9];

4) in the case of different states of oxidation, as for Se, Cr, As oxyacids, the toxic activity very much depends on the oxidation state.

The uptake mechanism topic has been extensively discussed, among others, by Fraùsto da Silva and Williams [10]. The reader is kindly requested to refer to the original papers.

2.1 Short Comments on the Biological Postulates

As far as the uptake mechanisms previously mentioned are concerned, chemical models are not suited to predict any correlation between toxic effect and concentration of toxicant as a function of a given variable. Chemical models, nevertheless, may be useful to find an a posteriori quantitive explanation of phenomena. For instance, EPA has given a correlation between the maximum acceptable concentration (MAC) of several trace elements for aquatic life in fresh water and hardness, expressed as mg L^{-1} of Calcium Carbonate [11]. In this case no-competitive (equations 2 and 2a) and competitive (equations 3 and 3a) co-ion displacement equilibria can be considered:

$$S + M \rightleftharpoons SM \tag{2}$$

$$SL + M \rightleftharpoons SLM \tag{2a}$$

$$SCa + M \rightleftharpoons SM + Ca \tag{3}$$

$$SLCa + M \rightleftharpoons SLM + Ca \tag{3a}$$

where S and SL are respectively an adsorbing site and a ligand site on the membrane surface, M the heavy metal cation and Ca the co-ion in solution, respectively. (For the sake of simplicity, charges are omitted.)

According to equations (2) the toxicity effect can be related to the ratio:

$$[SM]/[S] = K_2[M] \cdot \gamma_M \quad \text{or} \quad [SLM]/[SL] = K_{2A}[M] \cdot \gamma_M$$

respectively: it will remain constant when increasing metal concentration (MAC in this case) provided that the product MAC $\cdot \gamma_M$ remains constant.

According to equations (3) the toxicity effect can be related to the ratio:

$$[SM]/[SCa] = K_3 [M] \cdot \gamma_M/[Ca] \simeq K_3 [M]/[Ca]$$

$$[SLM]/[SLCa] \simeq K_{3a} [M]/[Ca]$$

respectively: it will remain constant when increasing MAC provided that the ratio [M]/[Ca] will remain constant.

As a matter of fact, plotting the product MAC $\cdot \gamma_M$ vs. hardness (Figure 1b) a trend similar to that of MAC vs. hardness (Figure 1a) was obtained, as shown in Figure 1 for Cadmium. On the other hand, as shown in Figure 2, the ratio MAC/h (which is proportional to the ratio MAC/[Ca]) remains fairly constant for all the elements considered varying the hardness, showing that competitive substitutions according to equations (3) are predominant. Likely, these findings can be obtained whenever chemical equilibria are responsible for toxicity mechanism and the response of the biotic indicator used is not affected by lateral changing of other variables, related to the main variable considered, as it seems to be the case of the MAC behaviour considered above. The role of Ca is, according to biologists, to decrease the permeability of the membrane.

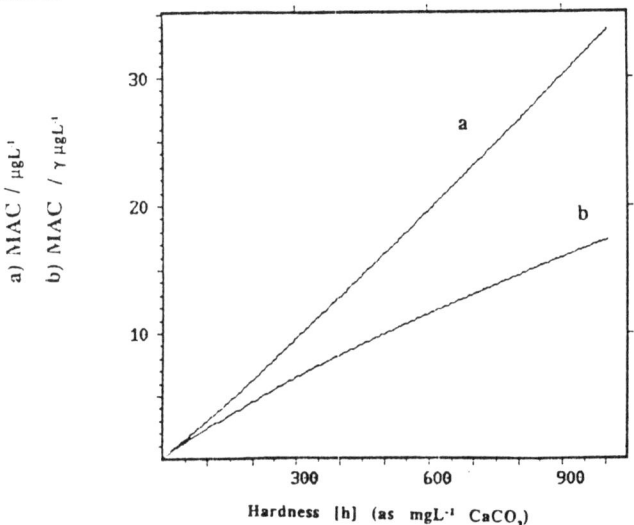

FIGURE 1. *Maximum Acceptable Concentration (MAC) vs. Hardness (mg/L $CaCO_3$) for Cadmium. a) According to the General Equation MAC = Exp(a(ln(hardness)±b)); b) MAC·γ_M.*

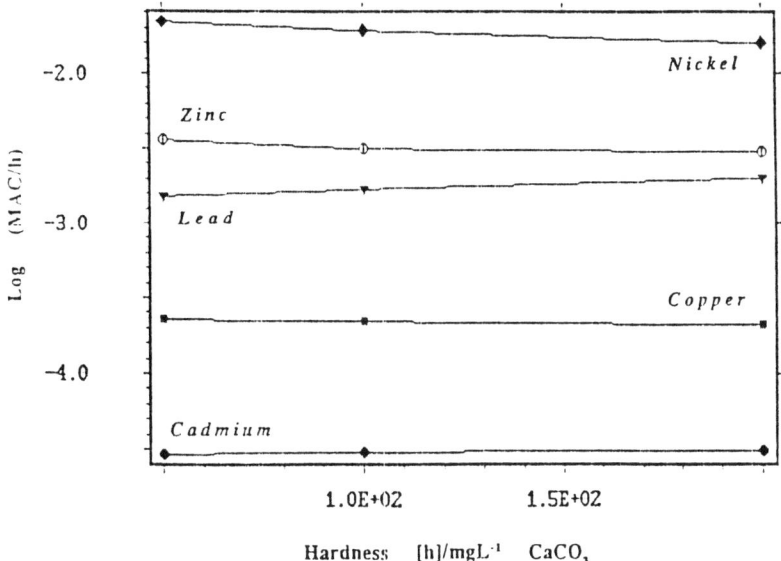

FIGURE 2. Log (MAC/h) vs. hardness for Cd, Pb, Cu, Zn, Ni; hardness (h) is expressed as mg/L $CaCO_3$.

As regards to chemical equilibrium at the interfaces, it must be pointed out that:
- the concentrations (activities) at the interfaces inside and outside the biological cells will be in true thermodynamic equilibrium, as assumed by biologists, provided that all the rearrangement mechanisms of the complexes are quick compared to the mass-transport ones towards and away from the interfaces;
- the metal ion mechanism uptake for any process type as summarized in Equations 2 and 3 will occur only via metal ion or via the direct contribution of other particles whether or not the reaction directly involving the free ion is kinetically much more preferred.

2.2 Conclusions Concerning the Biological Problem

The tolerance of a determined trace metal depends on parameters intrinsic to the biotic itself used as biotic indicator. These parameters may be different also for biotas belonging to the same group. In addition, nowadays biologists have no data on the effects of the different chemical forms of metals on the toxification and detoxification mechanisms [5].

Apart from this, in order to predict the individual poisoning potentiality of a given metal in a natural systems, the biologist should be but in principle interested in collecting information:
1) the concentration of the free metal ion in the aqueous phase;
2) the concentration of all the species in chemical equilibrium with the free metal ion;
3) the concentration of all the metal-containing species which do not dissociate up to free metal ion and are besides toxic (able to cross the membrane);
4) the presence of antagonistic or synergic foreign ions or molecules that may partake in the interface equilibrium.

In addition the biologist should be aware of the salt composition of the aqueous medium and what the rate determining step is, which is relevant to the most probable mechanism of uptake: the dissociation of complexes in giving the metal ion, the complex formation with the internal ligand L, the mass-transport through the membrane.

In the next section we will consider to what extent and which questions it is possible to give a correct chemical answer.

3. THE CHEMICAL PROBLEM

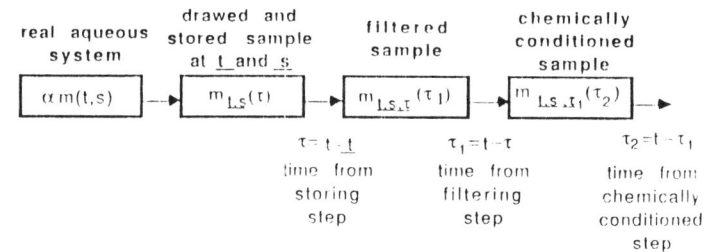

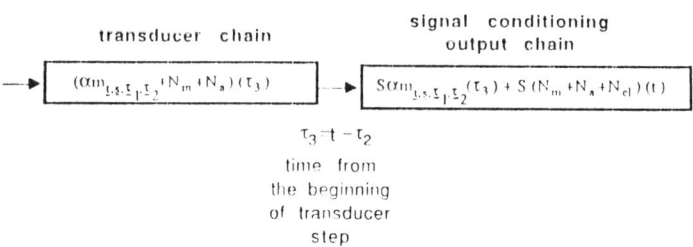

FIGURE 3. Skeletal Scheme Relevant to the Information Propagation When Characterizing an Aqueous System.

Let us consider the skeletal features of the model, operating under a given physical transducer channel, for the chemical characterization of a real aqueous system as shown in Figure 3. The relationship between the amount of chemical information $m(t,s)$ originally contained in the real system and the amount of information $y(t)$ measured depends on the reliability of each intermediate step. Clearly the goal in defining the best procedure for each step is to remove or to reduce to a known minimum extent its time-dependence and the inherent systematic and random errors. In this model the following points are worth commenting on.

1) *Correlation between $m(t,s)$ and $m_{t,s}$, the information contained in the sample drawn, at the time t in the point s:* It is well known that $m(t,s)$ can vary greatly depending on the time and the gathering point of the sample. When characterizing a specific aqueous system, a proper monitoring program involving different stations and depths of sampling, is required in order to take into account the effects due to vertical and horizontal movements of watermasses. Furthermore, the characteristic of effluents entering the water system, and hydrodynamic mixing conditions as well as the seasonal and daily effects must be considered. In this sense if $m(t,s)$ is confined to a given layer of the water column, to a given time interval and to given stations, it will contain less information, the lower the number of samples drawn.

2) *Loss of information during the phase separation, storage and sample chemical conditioning steps:* In each of these steps an unknown loss of information Δ_m can be caused by a non intentional addition or subtraction of analyte.

As for the sample preservation during the storage, the extent of exchange between sample solution and the container walls depends on the material of the container, the cleaning procedure used, the type and the concentration level of the elements in the liquid and solid phases and the nature of the sample as well. For instance, the presence of iron and organic substances in the sample may induce loss of heavy metals by coprecipitation/adsorption mechanisms while the presence of complexing agents may counteract this phenomenon.

In order to minimize spurious exchanges between dissolved/solid matter, it is generally good practice to separate the water phase from the solid phase as soon as possible and store the aqueous sample at a low temperature.

Owing to the extremely low concentration of the elements involved and to the complexity of the system, it should be born in mind that no physical methods for phase separation have been and can be suggested as essentially free from risks of stimulating in situ precipitation of flocculation processes, provoked by changes in pH, redox potential etc. These risks mainly depend on the nature and on the amount of suspended solid matter. Filtration is generally considered adequate for both particulate and dissolved matter analysis. Depending on the metal considered, different chemical stabilizers must be added to the filtered samples. After filtration, m' is the information related to the soluble compounds of the metal M.

3) *Loss of information due to the intrinsic feature of the physical channel - detector chain*: Once the information formerly m(t,s), has reached the transducer detector chain (which is mainly optical or electrochemical or chromatographic one), the y(t) signal at the final output after the signal conditioning step, it will have the form:

$$y(t) = S \cdot \alpha \cdot m'_{\underline{t},\underline{s},\,\underline{\tau}_1,\,\underline{\tau}_2}(\tau_3) + S\,[N_m + N_a + N_{el}]\,(t)$$

where S is the instrumental sensitivity for the particular molecular forms (see eq.1) which the transducer-detector chain is selective to; α, the operational selectivity factor, is the actual ratio between the concentration of these molecular forms and the total concentration C_M; N_m, N_a and N_{el} are respectively multiplicative, the additive and the actual electronic noises characteristic of the matrix nature of the sample and the nature of the transducer-device respectively.

The amount of information y(t) concerning a given metal in an aqueous system will be consequently related to a) the space point $\underline{s}$ and the time $\underline{t}$ of the draw sample; b) the sum of the errors (systematic and random) introduced from the drawing to the analysis; c) the times τ_i relevant to the start of each step.

The most frequently used techniques which may allow reliable results to be obtained in the best experimental conditions are summarized in Table 1 for different classes of compounds.

3.1 The Determination of the Total Concentration C_M

The origin of the water sample with its peculiar matrix effects affects total dissolved concentration determination.

Surfactants (normally present at 0.8-5 mgL^{-1} levels) as well as occasional traces of oils or grease may contribute in causing significant errors or a large decrease in sensitivity depending on the instrumental technique used. Surface-active compounds have been

TABLE 1. *Reliable Information Obtainable Through the Most Frequently Used Instrumental Techniques Under the Specified Conditions.*

Detectable Species

Technique	C_M	[M]	$\Sigma_i[ML_i]$	$\Sigma_j[MX_j]$	$C_{M_{ads}}$
Atomic Spectroscopy	X^a				
potentiometry		X^b			
PP		X^c	X^d	X^f	
DPASV		X^c			
Chromatography			X^g	X^g	
Ultracentrifugation					X

a) chemical pre-treatment for mineralization of volatile compounds and solvent extraction recommended;
b) not suitable at natural concentration
c) after matrix effect elimination by physico-chemical treatment;
d) by the E_p shift, in absence of matrix effects;
f) by i_p, provided that a linear plot is obtained by the standard addition method and the contribution of MX_j species is negligible at the applied E_{p1};
g) whenever derivatization and solvent extraction procedure prior to separation are available.

found by Hunter and Liss in sea water samples at a 0.4 mgL^{-1} concentration level (expressed as Triton X-100)[12]. These authors measured the surfactant concentration, using a polarographic maximum method based on recording the reduction wave of Hg(II) [13], and found that these surface-active compounds are highly resistant to ultraviolet photo-oxidation. In Atomic Absorption Sprectrometry (AAS), coupled with electrothermal vaporization and atomization steps, all the above classes of interferences may affect the mechanisms of vaporization and the yield of atom formation giving rise to negative errors. In Anodic Stripping Voltammetry (ASV), surface-active substances and oils affect the reduction-reoxidation mechanisms, while some colloids or soluble compounds present in the filtrate may be electroactive. In Potentiometry Stripping Analysis (PSA), where the oxidant step is chemically achieved, organic material appears to have a lower interference effect than in ASV [14].

Sample decomposition for total concentration determination therefore appears to be the recommended procedure on every occasion. Unfortunately decomposition procedures may give rise to systematic errors, through: 1) incomplete decomposition; 2) contamination by reagent, vessel material, work environment; 3) loss of analyte by volatilization, adsorption on-, reaction with- the vessel walls. Of the numerous decomposition methods that have been published, only a few, based on ultraviolet irradiation, meet the requirement of being both efficient and error-free [15-17]. In these methods a small amount of hydrogen peroxide, as a source of oxygen free-radicals, has to be added to the acidified sample. Depending on the organic content of the water [17] boiling from 1 to 4 hours, in order to destroy organic matter without producing perceptible errors, was found necessary.

With the photodigestion device proposed by Dorten et al. [18] the organic matter in water was reduced practically to zero at temperatures not above 65°C. In general the use of closed reflux apparatus is recommended to avoid the loss of volatile compounds during the initial step of the operation. It is likely that this risk still holds in part when ozone is flowing through the sample instead of when the H_2O_2 -UV procedure is used. Apart from this, ozone oxidation at pH 2 seems to be effective in destroying EDTA, APDC, humic acids and tannic acids at 3, 25, 25 and 100 mgL^{-1} levels respectively in less than 1 hour [19].

3.2 Bias Concerning C_M as Results from Interlaboratory Exercise

In order to get an idea about the size of bias in trace analyses, we will show the results obtained in some of several interlaboratory tests.

As a rule, one would expect that the selected laboratories accepting to participate in AQC tests consider themselves able to cope with the task and they will take special care to achieve the greatest accuracy and precision possible. In addition, it should be born in mind that each result in any AQC test is not less than the average of four replicates.

In the case of mercury after a 3-year trial, coordinated by the Water Research Center (WRC)(United Kingdom) in the framework of an Analytical Quality Control (AQC) [20-23], only 4 out of 9 laboratories gave acceptable values within a 10% bias at the concentration level of 0.36 ul/l., and 8 laboratories met this criterium at the 1.62-ug/L level [21]. Note that one laboratory which found good results at the 0.36-ug/L level failed the test at the 1.62-ug/L level. Having observed that the concentration of mercury in unpolluted natural waters does not exceed 0.01 ug/L, the question arises as to what the bias could be within AQC tests at concentration levels between 0.36 and 0.01 ug/L and how theses concentrations can be handled and determined reliably (see Figure 4)

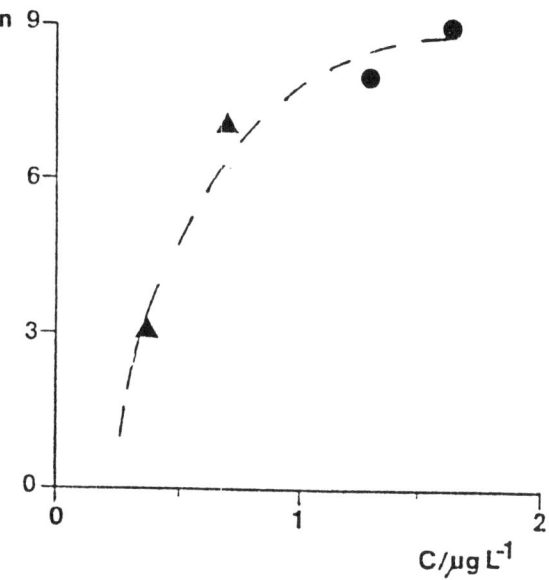

FIGURE 4. Correlation Between the Number of Laboratories (n) Which Obtained the Correct Value within the Accepted Bias for Mercury (C). (△) Standard solution; (●) Real samples (21).

Trace metals in filtered sea water considered in another calibration exercise [24], led to the following findings.
1. The average interlaboratory values for preacidified samples were, with respect to frozen samples, 43% higher for cadmium, 230% for copper, 23% for lead, 177% for zinc, with no statistical evidence for iron, manganese and nickel.
2. Large differences between the results: in Figure 5, the spread of the results relevant to the dissolved copper determination for preacidified sample is shown.
3. Very large relative standard deviations when the mean value of each laboratory is referred to the mean interlaboratory value: 87(59) for Cadmium, 78(115) for Copper, 70(95) for Lead, 30(38) for Zinc, 39(89) for Manganese, 80(65) for Nickel, and 70(112) for Iron (the values in parentheses refer to frozen).
4. Very poor spike recovery, which can be considered a measure of the accuracy reached by the participants: 36 to 96%, depending on the analyte.

314   SPECIATION AND COMPLEXATION IN NATURAL SYSTEMS

5. Significant improvements, although the above remarks still hold, under conditions where fewer participants have been selected and the number of individual analyses has been increased up to 100-130.

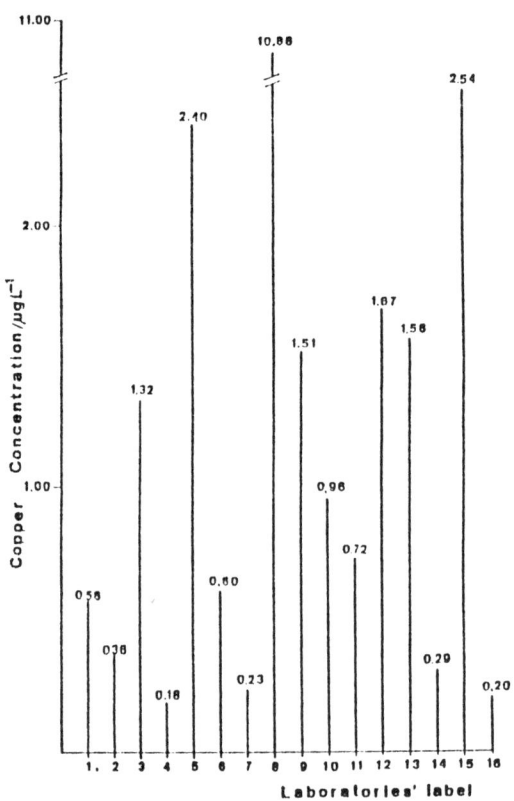

FIGURE 5.   *Interlaboratory Exercise Results Relevant to the Dissolved Copper Determination in Preacidified Samples. [25].*

Note that the above results refer mainly to samples filtered and aged, for which the bias relevant to the preceding steps has been eliminated and that total concentration C_M was measured.

3.3 Chemical Speciation of Single Species

Remarkable and praiseworthy endeavors have been made in the direction of chemical speciation in order to evaluate toxicological effects. Individual species determination is possible whenever analytical procedures do not involve element exchanges between the molecular forms during the separation and measuring time. This holds for organometallic compounds and for molecular species containing the element in different valence states. As far as we know, Chau was the first to explore the possibility of coupling chromatographic techniques to an Atomic Absorption detector in the speciation of organo-lead and organo-tin compounds. Derivatization through Grignard's reaction of the ionic forms, made the measurement of each distinct compound present in water, sediment and fish samples possible [25-27]. By successive improvements in the original procedure, very low detection limits, 0.1-0.3 ugL^{-1}, are now obtainable for organo-lead compounds and inorganic lead as well. Propylation, rather than butylation, is claimed to offer several advantages, such as a better recovery of dialkyllead species, a shorter analysis time and a better resolution of some peaks [28].

Using HPLC with an AA detector good separation of Sn(II), Sn(IV) and tributyltin was obtained at ugL^{-1} levels [29].

Differential pulsed electrochemical techniques coupled with selective organic phase extraction have been proposed by Colombini et al. [30] for consecutive determination of Me_4Pb, Et_4Pb, Me_3Pb^+, Et_3Pb^+, Me_2Pb^{+2} + Et_2PB^{+2}, inorganic Pb^2 in natural water samples. At ugL^{-1} levels, the bias in the determination of the above mixture was less than 8%, and typically 4%. Using differential pulse anodic stripping voltammetry Kenis and Zirino [31] were able to distinguish between inorganic tin and Bis(tris-n-butyltin) at ugL^{-1} levels. Cr(III)/Cr(VI) separation was obtained by Fuoco and Papoff using differential pulse polarography at ugL^{-1} levels [32].

3.4 Chemical Speciation of Classes of Compounds

When individual species determination is not possible, some operative and arbitrary classification has to be introduced.

Based on Batley and Florence's original scheme, the chemical groups of the Equation 1 have been classified into seven distinct groups according to their behavior in the following physico-chemical treatments: percolation through Chelex-100 column, ultraviolet irradiation and acid digestion [33, 34]. ASV was used for measuring concentrations prior to and after each treatment.

From a general viewpoint, it is a difficult, if not an impossible task, to directly follow the seven classes of compounds through eight different procedures, each potentially involving severe contaminations and large biases. If the previously mentioned interlaboratory tests concerning total concentration determination are used for comparative purposes, one may wonder what kind of discouraging results might be expected when different fractions of C_M have to be analyzed as well.

As for the use of ASV in the speciation scheme proposed by Batley and Florence for Cu, Pb and Cd [33, 34] apart from artifacts and experimental difficulties, the ASV peak currents (measured in the filtered sample acidified at pH 4.8, under a unique plating potential, -0.9 V) can not be unequivocally related to distinct classes of compounds with defined toxicological effects in the original sample. Instead, it would be advisable to use different plating potentials in order to test the quality of the results obtained, and, in the case of ill-defined anodic peaks, to use medium-exchange procedures [35, 36]). The medium-exchange procedure involves a deposition of the metals from the sample solution, followed by their stripping into a more suitable electrolyte. This approach is very efficient in minimizing during the oxidation step the effect of interferences which arise from various sample components. If the solvated metal ion is the main cause of the toxic effects [8, 9], Batley and Florence's speciation scheme fails to give sound information.

3.5 Conclusions Concerning Speciation by Electrochemical Measurements

Because of the generalized use of ASV in speciation measurements of non-digested real samples the features and the relevant sources of error peculiar to this technique are here summarized.

In principle, ASV is an unsurpassable tool for measuring very low total concentrations of analyte provided matrix effects can be overcome. However, when many species concur in giving M_{red} only one anodic peak is generally, but not always, observed.

At constant total concentration of reversible complexes depending on the relative concentration of each ML_i complex, a large variation of i_p may be caused by changes in the diffusion or kinetic mechanisms. That is, when the mass-transfer is the rate determining step, i_p value will depend on each D_{ML} according to the equation:

$$i_p \sim t_{pl} \left(a \sum_i D_{ML_i}^{2/3} \cdot C_{ML_i} + b \sum_i D_{ML_i} \cdot C_{ML_i} \right)$$

where i_p is the peak current; t_{pl} the pre-electrolysis time; D_i the diffusion coefficient.

When a kinetic mechanism is the rate determining step, the i_p value will depend on the dissociation rates of the ML_i species; for extremely low dissociation rates, i_p will be proportional to [M], the concentration of the free metal ion. Apart from in this latter case (which is extreme), the standard addition method will give the correct value of the total concentration of the reversible complexes provided that a) L_i concentrations are in excess; b) sufficient time elapsed between the addition and measurement in order to guarantee that equilibrium conditions are reached in the bulk of the solution; c) the instrumental sensitivity S be sufficiently high.

As for E_p (peak potential), when the formation rate at the electrode interface is for each ML_i complex, very large or very small compared to the metal oxidation rate, E_p will be relateable respectively to the polarographic $(E_{\frac{1}{2}})_c$ (half-wave potential in d.c. polarography of solvated ion) which should have been obtained for the same ligand-containing or ligand-free solution. Between these extreme conditions, E_p may assume any value [37].

In the absence of diagnostic tests to investigate kinetic contributions during reduction and/or oxidation steps at a very low concentration level, ASV is not a tool to be used for distinguishing between different classes of complexes in terms of their thermodynamic or kinetic effects. To some extent and under suitable conditions information may be obtained by using different E_{pl} [38]. However situations as shown in Figure 6, where the i_p varies irregularly with the plating potential, are representative of most real samples and are not easily interpretable. In addition, it may be observed that UV irradiation as used in this case and in many other speciation schemes is effective in destroying matrix effects only under thermic treatment and in presence of H_2O_2. In the above case, the total concentration of copper measured at -0.9V should be about 250% lower. Furthermore, speciation experiments performed at the natural pH of the water samples, may give rise to systematic errors owing to the tendency of metal ions to adsorb on the container walls. It has been claimed that adsorption of copper on teflon material can be noticed in standard addition procedures [42] for equilibration times higher than few hours. The question arises what would be the loss of analyte for adsorption from the collection of the sample to the analysis.

Criticisms on the reliability of procedures in solving speciation problems have been recently expressed by different authors [39, 40].

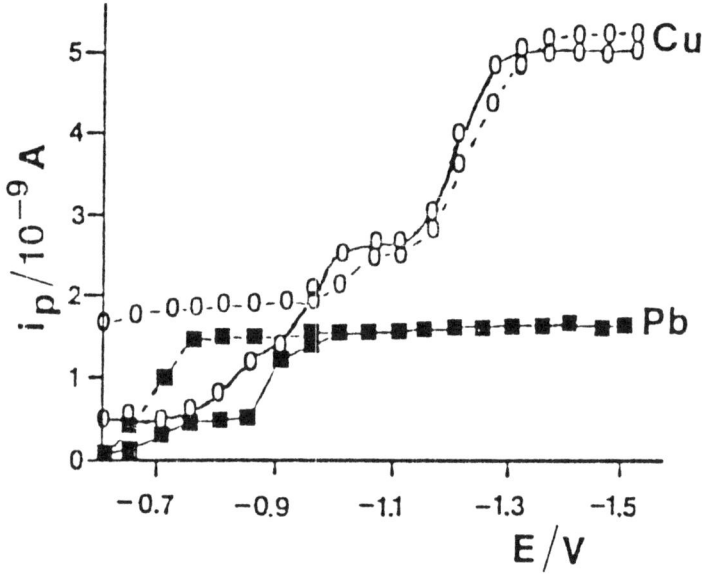

FIGURE 6. *Relationship Between Peak Current (i_p) and Plating Potential (E_p) for Copper (○) and Lead (■) in Filtered Baltic Seawater. Dotted Line Refers to the Results After UV Irradation [38].*

4. CONCLUSIONS

The poisoning effect of a soluble metal in natural waters depends on the ability of the biota to avoid toxification and to activate detoxification.

Owing to the complexity of the overall mechanisms, poisoning effects may vary from one biota to another even when they belong to the same family.

Nowadays biologists are far from having sufficiently delucidated the different aspects of poisoning mechanisms and most of them are inclined to think that free metal ions and hydrophobic metal-containing compounds are the species preferentially transported across a biological membrane. They are therefore interested but potentially to achieve information about metal speciation.

At present, the analytical procedures available, permit the measurements in the aqueous phase of: a) the total concentration of a metal, C_M, once destroyed the matrix effect; b) the concentration of any metal-containing compounds MX in non-equilibrium with the free metal ion, provided that they do not undergo decomposition during the time of analysis and that the detection limit of the method is lower than the actual concentration compound in the natural water.

The above procedures require great care and skilled operators as seen from the results obtained in several interlaboratory exercises.

As for the measurement of the whole concentration of the different complexes in equilibrium with the metal ion, the following points should be made:

- there are thousands of organic compounds present in natural waters of which the molecular composition and the relevant ability to bind the metal ion in complexed forms are largely unknown [41]. Consequently it is theoretically meaningless to define labile or non labile complexes which are not identified and which are not likely to be characterized in the near future. Apart from this all, the attempts made up to now to distinguish between labile and non-labile complexes in unpolluted waters when C_M is 100-1000 times lower than MAC, are useless since in these conditions the toxic threshold of the biotic indicator is far from being reached;

- the electrochemical determination of the whole concentration in the presence of matrix effect should be discouraged both for experimental and theoretical reasons. Both the chemical and biological information obtained in this way is quite poor.

Even though the biological information obtained in CM is quite poor, it is the all - apart from MX - that can be obtained with sufficient chemical accuracy.

ACKNOWLEDGMENTS

The authors wish to acknowledge Prof. V. Albergoni and Dr. S. Galassi for the helpful discussions on biotic indicators. Prof. F. Malatesta for the preparation of the figure relevant to the activity coefficients. The authors wish also to thank Mrs. E. Battaglia and Mrs. M. Cempini for the careful typewriting of this paper.

REFERENCES

1. Millward, G.E., and R.M. Moore. "The Adsorption of Cu, Mn and Zn by Iron Oxyhydroxide in Model Estuarine Solutions," *Water Res.* 16:981-985 (1982).

2. Salomons, W. "Adsorption Processes and Hydrodynamic Conditions in Estuaries," *Environ. Technol. Lett.*, 1:356-365 (1980).

3. Wander Weijen, C.H., M.J.H.L. Arnoldus, and C.J. Meurs. "Desorption of Metals from Suspended Material in the Rhyne Estuary," *Neth. J. Sea Res.*, 11:130-145 (1977).

4. Sholkovitz, E.R. "The Flocculation of Dissolved Fe, Mn, Al, Cu, Ni, Co, and Cd During Estuarine Mixing," *Earth. Planet. Sci. Lett.*, 41:77-86 (1978).

5. Albergoni V., and E. Piccinni. "Biological Response to Trace Metals and Their Biochemical Effects," in *Trace Element Speciation in Surface Waters and Its Ecological Implications*, G. G. Leppard, Ed. (New York: Plenum Press, Inc., 1983) p. 159.

6. Muller, H.G. "Acute Toxicity of Potassium Dichromate to Daphnia Magna as a Function of the Water Quality," *Bull. Environ. Contam. Toxicol.*, 25:113-117 (1980).

7. Duke, T.W., J.N. Willis, and D.A. Wolfe. "A Technique for Studying the Exchange of Trace Elements Between Estuarine Sediments and Water," *Limnol. Oceanogr.*, 13:541-541-548 (1968).

8. Sunda, W., and R.R.L. Guillard. "The Relationship Between Cupric Ionic Activity and the Toxicity of Copper to Phytoplankton," *J. Mar. Res.*, 34:511-520 (1976).

9. Gächter, R., J.S. Davis, and A. Mares. "Regulation of Copper Availability to Phytoplankton by Macromolecules in Lake Water," *Envir. Sci. Technol.*, 12:1416-1420 (1978).

10. Fraùsto da Silva, J.J.R., and R.J.P. Williams. "The Uptake of Elements by Biological System," in *Structure and Bonding, Dunitz,* J.D., P. Herumerich, K.J.A. Ibers, E.C.K. Jorgensen, J.B. Neilands, D. Reinen, and R.J. P. Williams Eds. (Berlin: Springer-Verlag, 1976) p. 67.

11. Environmental Protection Agency (Washington: *Water Quality Criteria, Fed. Reg.*, 45, (1980) p. 79318.

12. Hunter, K.A., and P.S. Liss. "Polarographic Measurements of Surface-Active Material in Natural Waters," *Water Res.*, 15:203-215.

13. Cosovic, B., V. Zutic, and Z. Kozarac. "The Surface-Active Substances in the Sea Surface Microlayer by Electrochemical Methods," Croat. Chem. Acta, 50:229-232 (1977).

14. Wang, J. "Stripping Analysis, Principles," *Instrumentation and Applications*, (VCH, Deerfield Beach, Fla., 1985) p. 112.

15. Batley, G.E., and Y.J. Farrar. "Irradiation Techniques for the Release of Bound Heavy Metals in Natural Water and Blood," *Anal. Chim. Acta*, 99:283-292 (1978).

16. Nygaard, D.D., and J.H. Lowry. "Sample Digestion Procedure for Simultaneous Determination of Arsenic, Antimony and Selenium by Inductively Coupled Argon Plasma Emission Spectrometry with Hydride Generation," *Anal. Chem.*, 54:803-807 (1982).

17. Mart, L., H.W. Nürnberg, and P. Valenta. "Prevention of Contamination and Other Accuracy Risks in Voltammetric Trace Metal Analysis of Natural Waters. III. Voltammetric Trace Analysis with a Multicell System Design for Clean Bench Working," *Z. Anal. Chem.*, 300:350-362 (1980).

18. Dorten, W., P. Valenta, and H.W. Nüberg. "New Photodigestion Device to Decompose Organic Matter in Water," *Z. Anal. Chem.*, 317: 264-272 (1984).

19. Clem, R.G. and A.T. Hodgson. "Ozone Oxidation of Organic Sequestering Agents in Water Prior to Determination of Trace Metals by Anodic Stripping Voltammetry," *Anal. Chem.*, 50:102-110 (1978).

20. Analytical Quality Control (Harmonised Monitoring) Committee, "Accuracy of Determination of Cadmium, Copper, Lead, Nickel and Zinc in River Waters," *Analyst* 110:1-10 (London), (1985).

21. Analytical Quality Control (Harmonised Monitoring) Committee, "Accuracy of Determination of Total Mercury in River Waters: Analytical Quality Control in the Harmonised Monitoring Scheme," *Analyst*, 110:103-111 (London), (1985).

22. Analytical Quality Control (Harmonised Monitoring) Committee, "Accuracy of Determination of Trace Concentrations of Dissolved Cadmium in River Waters: Analytical Quality Control in Harmonised Scheme," *Analyst* 110:247-252 (London), (1985).

23. Gardner, M.J., D.T.E. Hunt, and G. Topping. "Analytical Quality Control (AQC) for Monitoring Trace in the Coastal and Estuarine Environment," *Water Res. Tech.*, 18:35-41 (1986).

24. Bewers, J.M., J. Dalziel, P.A. Yeats, and J.L. Barron. "An Intercalibration for Trace Metals in Sea Waters," *Mar. Chem.*, 10: 173-193 (1981).

25. Chau, Y.K., P.T.S. Wong, and G.A. Bengert. "Determination of Methyltin(IV) and Tin(IV) Species in Water by Gas Chromatography/Atomic Absorption Spectrometry," *Anal. Chem.*, 54:246-249 (1982).

26. Chau, Y.K., P.T.S. Wong, and O. Kramar. "The Determination of Dialkyllead, Trialkyllead, Tetraalkyllead and Lead(II) Ions in Water by Chelation/Extraction and Gas Chromatography/Atomic Absorption Spectrometry," *Anal. Chim. Acta,* 146:211-217 (1983).

27. Chau, Y.K., P.T.S. Wong, G.A. Bengert, and J.L. Dunn. "Determination of Dialkyllead, Trialkyllead, Tetraalkyllead, and Lead(II) Compounds in Sediment and Biological Samples," *Anal. Chem.*, 56:271-274 (1984).

28. Radojevic, M., A. Allen, S. Rapsomanikis, and M.R. Harrison. "Propylation Technique for the Simultaneous Determination of Tetralkyllead and Ionic Alkyllead Species by Gas Chromatography/Atomic Absorption Spectrometry," *Anal. Chem.*, 56:658-661 (1986).

29. Ebdon, L.S., S.J. Hills, and P. Jones. "Speciation of Tin in Natural Waters Using Coupled High-Performance Liquid Chromatography-Flame Atomic Adsorption Spectrometry," *Analyst* (London), 110:515-518 (1985).

30. Colombini, M.P., R. Fuoco, and P. Papoff. "Electrochemical Speciation and Determination Organometallic Species in Natural Waters," *Sci. Tot. Environ.*, 37:61-70 (1984).

31. Kenis, P., and A. Zirino. "Quantitative Measurement of Tributyloxide in Sea Water by Differential Pulse Anodic Stripping Voltammetry," *Anal. Chem. Acta,* 149-157-165 (1983).

32. Fuoco, R., and P. Papoff. "Individual Detection and Determination of Cr(VI) and Cr(III) - Together with Cu, Pb, Cd, Ni and Zn - in Top Water at ppb Level," *Ann. Chem.*, 65:155-163 (Rome), (1985).

33. Batley, G.E., and T.M. Florence. "A Novel Scheme for the Classiciation of Heavy Metal Species in Natural Waters," *Anal. Lett.*, 9:379-388 (1986).

34. Batley, G.E., and T.M. Florence. "Determination of the Chemical Forms of Dissolved Cadmium, Lead and Copper in Sea Water," *Mar. Chem.*, 4:347-351 (1976).

35. Phillips, S.L. and I. Shain. "Application of Stripping Aanalysis to the Trace Determination of Tin," *Anal. Chem.*, 34: 262-265 (1964).

36. Ariel, M., U. Eisner, and S. Gottesfeld. "Trace Analysis by Anodic Stripping Voltammetry. II. The Method of Medium Exchange," *J. Electroanal. Chem.*, 7:307-312 (1964).

37. Betti, M., and P. Papoff. "Trace Elements: Data and Information in the Characterization of an Aqueous Ecosystem," *CRC Crit. Rev. Anal. Chem.*, 19(4):271-322 (1988).

38. Brügmann, L. "Electrochemical Speciation of Trace Metals in Sea Waters," *Sci. Tot. Environ.*, 37:41-60 (1984).

39. Batley, G.E. "The Current Status of Trace Element Speciation Studies in Natural Waters," *Trace Element Speciation in Surface Waters and its Ecological Implication,* G. G. Leppard, Ed. (New York: Plenum Prss, Inc. 1983) p. 17.

40. Campbell, P.G.C., and A. Tessier. "Current Status of Metal Speciation Studies," in *Metal Speciation, Separation and Recovery,* J.W. Patterson and R. Passino, Eds. (Chelsa: Lewis Publ. Inc. 1987) p. 201.

41. Hart, T.B. "Trace Metal Complexing Capacity on Natural Waters: A Review," *Environ. Technol. Lett.*, 2:95-118 (1981).

DISCUSSION OF:
THEORETICAL AND EXPERIMENTAL DRAWBACKS IN HEAVY METAL SPECIATION IN NATURAL WATERS

C. Haraldsson
Department of Analytical and Marine Chemistry
University of Göteborg and Chalmers University of Technology
S-412 96 Göteborg, Sweden

In the paper by P. Papoff, M. Betti and R. Fuoco some important aspects of trace metal speciation are discussed. The main points of the paper are:

A. A short discussion of physico chemical fundamentals.

B. A discussion of the biological problem that the results of the speciation studies should be applied to.

C. The main part of the paper addresses the many problems that are present when doing speciation studies.

In the part named "The biological problem" the many variables affecting the toxicity of a pollutant is presented. The authors point out that the variables pH, temperature, dissolved oxygen content, ionic strength and the molecular form of the toxic group are the important parameters to consider. The toxicity may also depend on other substances that may either hinder or enhance the effect of the pollutant considered. It is stressed that in this complicated situation the biologist will have to make simplifications order to get meaningful information. There is a part of this chapter that needs to be clarified i.e the part that discusses the influence of the activity factors.

The part named "The chemical problem" starts with a discussion of the problems involved in sampling and filtration and storage. It is pointed out that the data from a measurement strictly is valid only for the moment and gatheringpoint when the sample was collected. The authors does however not indicate the compromises possible in order to get meaningful data with a reasonable amount of resources. In many situations it is probably not necessary to consider daily variations of concentrations of toxic metals in an environmental monitoring program.

A significant part of the paper concerns the problems of determination of trace metals at the concentrations present in natural waters. The authors cites the results from intercalibration experiments to show that a large number of laboratories produces results that can be suspected to be unacceptable.

A number of instrumental methods useful for the determination of trace metals is listed. The authors does not mention the use of ICP-MS this technique is useful for the determination of total concentration in freshwaters and have further advantage in that sample decomposition not is needed for the determination of total dissolved metal concentrations. For the determination of total concentrations the authors concludes that sample decomposition always should be used. The problems present in sample decomposition is presented, and the conclusion is that only the methods based on ultraviolet irradiation is acceptable.

The speciation part of the paper starts with example of the few cases where identification and quantification individual species has been possible. This is followed by a presentation of the speciation methods in which the trace metals is divided into fractions with different chemical properties.

An overall comment is that the paper give a good review of the problems present in speciation studies, it would however be useful to see some more solutions to the problems presented.

PART IV: SORPTION ONTO SURFACES

PROTON AND METAL ION BINDING ON HUMIC SUBSTANCES

J.C.M. De Wit and W.H. van Riemsdijk
Department of Soil Science and Plant Nutrition, Wageningen Agricultural University, Wageningen, The Netherlands

L.K. Koopal
Laboratory of Physical and Colloid Chemistry, Wageningen Agricultural University, Wageningen, The Netherlands

1. ABSTRACT

The interaction of protons and metal ions with humic substances is analyzed. Specific attention is paid to effects of the site density, particle radius and ionic strength on the adsorption behavior. It is suggested to plot the proton adsorption isotherm, Θ_H, as a function pH_s, the proton activity at the plane of adsorption instead of the pH. Coinciding $\Theta_H(pH_s)$ curves at different ionic strength indicate that an appropriate double layer model has been used. Proton adsorption data on several humic acids have been analyzed.

For metal ion adsorption multicomponent (M,H) isotherm equations result. Adsorption isotherms for a monodentate positive surface complex are considerably different from those leading to a monodentate uncharged metal-surface complex. The adsorption isotherms for bidentate metal ion complexes formed on phthalic acid or salicylic acid-like surface structures are identical to the isotherms for the monodentate SOMOH complexes as long as the dissociation of the second surface groups of the phthalic acid and salicylic acid surface structures is negligible.

2. INTRODUCTION

Knowledge of the adsorption behavior of heavy metal ions in natural systems is of great importance for predicting the metal ion distribution and their bio-availability. The adsorption complex of natural systems is determined by the fixed charge surfaces of clay minerals and by the variably charged surfaces of metal (hydr-)-oxides and organic matter. Modeling of the adsorption onto variably charged surfaces, especially onto humic acids and fulvic acids, is the subject of this paper.

Variably charged surfaces have a pH-dependent surface charge caused by the protonation or deprotonation of surface groups. Of these surfaces, the oxides are well studied and exhibit an amphoteric behavior. Recently, Hiemstra et.al [1,2] have developed a multi-site complexation (MUSIC) model, in which the contributions of the various oxy and hydroxy groups at the different crystal planes to the overall surface charge of the oxide surfaces are taken into account explicitly. With the MUSIC model, it is possible to describe the proton titration isotherm with the priori predicted association constants [2].

The humic acids and the fulvic acids are considerably more complex than the oxides. Humic acids and fulvic acids are polydisperse mixtures of organic polyelectrolytes with a large variation, with respect to physical and chemical properties. In general, humic acids are larger and more cross-linked than the rather flexible fulvic acids. Due to the dissociation of functional groups, humic acids and fulvic acids are negatively charged in pH ranges relevant in natural systems. Important functional groups on humic acids and fulvic acids are the carboxylic, the phenolic and the carbonyl groups. According to several authors (for instance [3-5]) functional groups coordinated as in salicylic and phthalic acid, are important for the adsorption of metals like copper and aluminum.

The proton affinity constants of the functional groups are influenced strongly by the molecular structure and conformation of the organic colloid. Due to the existence of a large number of molecular structures in humic and fulvic acids, there will most probably be a continuous distribution of proton affinity constants for natural organic matter.

In order to obtain insight in the distribution of proton affinity constants, without knowing the entire structure of the molecules, several methods have been developed (see for an overview [6-9]). The most simple among these is the condensation approximation (CA) method [10], which relates to the affinity distribution to the first derivative of the adsorption isotherm. When these techniques are used, it should be realized that the degree of protonation of surface groups (Θ_H) is not only determined by the nature of the surface groups, but also by the surface charge density and by the ionic strength. As a consequence, even for an intrinsically homogeneous colloid, electrostatics introduce an "apparent heterogeneity" which results in ionic strength dependent apparent affinity distributions if the isotherms as such are analyzed [11]. In order to obtain the true or intrinsic affinity distribution, which reflects the chemistry of the surface, the electrostatics have to be taken into account explicitly. This can be done by adopting a double layer model and by applying this model to the adsorption data. If the double layer model is correct, the ionic strength dependency of the proton isotherms can be used to find an appropriate double layer model.

So far the emphasis was on proton adsorption, since in aqueous systems protons (and/or hydroxyl ions) are always present and the pH determines the state of the functional groups. If metal ions are present, it should be realized that in most cases, these ions compete with protons for adsorption sites so, for the description of metal ion adsorption a multicomponent approach and knowledge of the proton adsorption is necessary. In the second part of the paper, metal ion adsorption will be discussed and special emphasis will be given to complexation by salicylic and phthalic acid type of structures.

3. THE MONO-COMPONENT PROTON ADSORPTION ISOTHERM

3.1 Theoretical Framework

For a surface group of type i of a natural organic colloid, the protonation reaction can be written as:

$$S_iO^- + H_s^+ \rightleftharpoons S_iOH \qquad K_{i,H}^{int} \qquad (1)$$

The subscripts in Equation (1) symbolize the proton activity at the plane of adsorption should be used in the corresponding equilibrium relations, the superscript int, indicates that the defined $K_{i,H}^{int}$ is an intrinsic constant independent of the value of the surface charge. The activity of the proton at the plane of adsorption, H_s is given by:

$$H_s = H \cdot \exp\left(\frac{-F\Psi_s}{RT}\right) \qquad (2)$$

where H is the proton activity in the bulk solution, Ψ_s is the smeared out surface potential and F, R and T have their usual meaning. The assumption of a smeared out potential is plausible for a random heterogeneous surface [13,14].

We define the degree of protonation, $\Theta_{i,H}$:

$$\Theta_{i,H} = \frac{[S_iOH]}{[S_iO^-] + [S_iOH]} \quad (3)$$

Where $[S_iOH]$ and $[S_iO^-]$ are site densities, and the degree of dissociation α_i:

$$\alpha_i = 1 - \Theta_{i,H} \quad (4)$$

According to Healy and White [15]:

$$\Theta_{i,H} = \frac{K_{i,H}^{int} H_s}{1 + K_{i,H}^{int} H_s} \quad (5)$$

Equation (5) can also be written as:

$$\Theta_{i,H} = \frac{K_{i,H}^{app} H}{1 + K_{i,H}^{app} H} \quad (6)$$

with

$$K_{i,H}^{app} = K_{i,H}^{int} \cdot \exp\left(\frac{-F\Psi_s}{RT}\right) \quad (7)$$

The overall degree of protonation of the colloid is given by:

$$\Theta i = \sum_{i=1}^{n} f_i \Theta_i \quad (8)$$

where f_i is the fraction of the sites of type i with respect to the total numbers of sites.

The relation between $\Theta_{i,H}$ and H_s, Equation (5) is equivalent to a monocomponent Langmuir adsorption isotherm equation. It easily follows that $\log K_{i,H}^{int} = pH_s$ ($\Theta_{i,H} = 0.5$) = $pH_s(\alpha_i = 0.5)$. For a non-amphoteric negative surface, the $\Theta_{i,H}(pH)$ and $\Theta_{i,H}(pH_s)$ curves are descending, while the dissociation curves are ascending. Due to the effect of the surface potential the absolute value of the slope of the $\Theta_{i,H}(pH)$ curve is always smaller than that of the $\Theta_{i,H}(pH_s)$ curve. The same holds for the $\alpha_i(pH)$ curve, with respect to the $\alpha_s(pH)$ curve. Moreover, the slope of the overall $\Theta_i(pH)$ isotherm should be negative or zero and is always less steep than that of a homogeneous mono-component Langmuir isotherm.

The overall surface charge, σ_s is a function of the local surface charge or degree of protonation:

$$\sigma_s = \frac{N_s F}{N_{Av}} \sum_{i=1}^{n} f_i (1-\Theta_i) \quad (9)$$

where F and N_{Av} have their usual meaning and N_s is the site density (sites per unit area).

Several models have been proposed for the relation between the charge σ_s of the colloid and its electrostatic potential Ψ_s, relative to the potential in bulk solution. The ionic strength is an important parameter in these models. Here we will use relations derived from the Poisson-Boltzmann (PB) equation for spherical particles [16,17] and flat plates (see Eq. 18). For plates, the well known Gouy Chapman (GC) relation can be used whereas for spheres, use will be made of a numerical solution of the PB equation. The sphere and flat plate are seen in this study as limiting cases. The behavior of cylindrical or rod-shaped colloids will be somewhere in between these extremes.

On various mineral oxides, both proton and metal ion adsorption could be described well by adopting the GC model in combination with a Stern layer [1,2,13,19]. For organic materials with a less well-defined surface, the introduction of a Stern layer does not seem warranted.

With the help of the chosen DL model, the experimental $\Theta_{i,H}(pH)$ curves can be transformed into $\Theta_{i,H}(pH_s)$ curves. When the correct DL model is used, the $\Theta_{i,H}(pH_s)$ curves at different ionic strength coincide and one mastercurve is obtained.

3.2 Curvature and Ionic Strength Effects

The dependency of the surface potential Ψ_s on the surface charge σ_s, for a flat plate, and for a series of spheres is shown in Figure 1. Figure 1 clearly shows that for spheres larger than ca. 10 nm the course of Ψ_s is close to that of Ψ_s for a flat plate for all ionic strengths investigated. Hence, for most practical purposes, spheres larger than 10 nm can be considered as flat plates.

The size of humic substances depends on their molecular weight. According to Buffle [5] a radius of about 0.5 nm can be considered as a minimum value for fulvic acids. For humic acids, the gyration radius ranges from 1.5 to 15 nm, as the molecular weight increases from 2600 to 1.4 x 10^6 (Buffle [5], based on data from Cameron et al. [20]).

In Figure 2a and b the proton adsorption isotherms $\Theta_{i,H}(pH)$ for three ionic strengths are given for a homogeneous flat plate and for a sphere with a radius of 1 nm, respectively. Also included are the $\Theta_{i,H}(pH_s)$ curves, based on the correct double layer model, so that they show up as one mastercurve. The $\log K_H^{int}$ is for both cases the same, its actual value is irrelevant for the results.

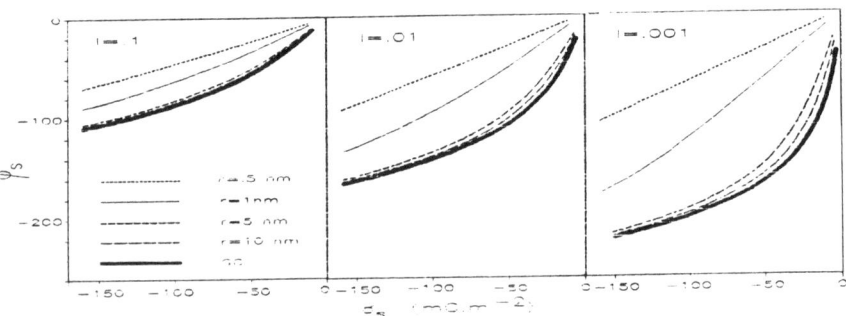

FIGURE 1. *The Surface Potential Ψ_s as a Function of the Surface Charge σ_s for a Flat Plate and a Series of Spheres.*

The $\Theta_{i,H}(pH)$ curves for the flat plate show a considerable ionic strength effect. The ionic strength dependency for the sphere is smaller because of the radial distribution of the electrostatical field. The $\Theta_{i,H}(pH_s)$ curves for the sphere and the flat plate are the same and equivalent to a monocomponent Langmuir isotherm.

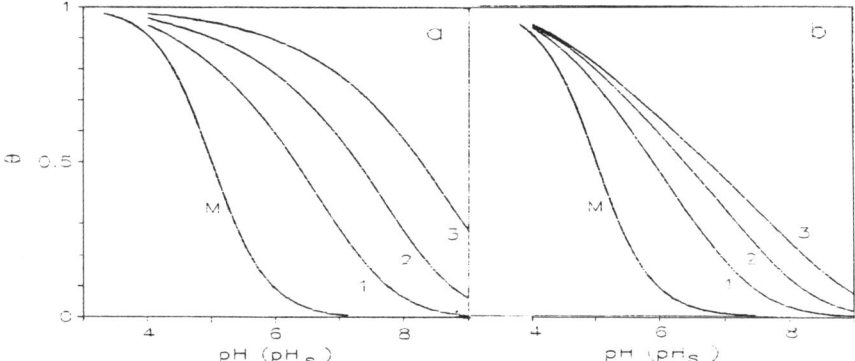

FIGURE 2. *Proton Adsorption Isotherms and the Corresponding Mastercurve, M, for a Homogeneous Flat Plate (a) and a Sphere with a Radius of 1 nm (b) with $\log K_H^{int} = 0.5$ and $N_s = 1$ (sites/nm^2). The Numbers Along the Curve Represent the Ionic Strength: 1, I=.1; 2, I=.01; 3, I=.001.*

3.3 Parameter Sensitivity

The site density and the radius or geometry of fulvic acids and humic acids proton adsorption data, are often unknown. Hence, it is important to establish the sensitivity of the double layer model for the site density and the radius. To check this effect, use is made of synthetic data, with known values of the parameters for two simple heterogeneous surfaces (see Figure 3a and b) at three different salt levels. Both surfaces contain two types of surface groups with $\log K_{1,H}^{int} = 3$ and $\log K_{2,H}^{int} = 5$, $f_1 = f_2 = 0.5$ and $N_s = 1$ (sites/nm^2). The $\Theta_{i,H}(pH_s)$ curves given in Figures 3 and 4 have been calculated, based on these synthetic data assuming several (incorrect) site densities, N_s^* and (incorrect) radii, r^*. In Figure 3, only the effect of an error in the site density on the $\Theta_{i,H}(pH_s)$ curves is shown, whereas in Figure 4, erroneous values for both the site density and the radius have been used. When the $\Theta_{i,H}(pH_s)$ curves coincide, the electrostatic effect is well predicted.

Figure 3 shows that the $\Theta_{i,H}(pH_s)$ curves of the small spheres are more sensitive to N_s than those of the flat plates. Further, from Figure 3 it follows that when the estimated site density is higher than the true value, the salt effect is over-compensated. At high Θ the $\Theta_{i,H}(pH_s)$ curves decrease more strongly than the monocomponent Langmuir equation. This is physically non-realistic.

When the site density is estimated too low, the $\Theta_{i,H}(pH_s)$ curves do not coincide. The lower the estimated site density, the higher the (incorrect) ionic strength dependency of $\Theta_{i,H}(PH_s)$ curves. If the site density is estimated to be infinitesimally small, the contribution of the surface potential to pH_s is negligible and the thus calculated (incorrect) $\Theta_{i,H}(pH_s)$ and the $\Theta_{i,H}(pH)$ curves coincide.

Similar effects are observed if a wrong value for r is chosen (see Figures 4a,c and e). For a radius too small, (Figures 4a,c) the compensation of the ionic strength effect is not large enough, while a too large radius (Figure 4e) leads to over-compensation.

From Figures 3 and 4a,c and e, it becomes clear that if either the site density or the radius is known, the freedom of choosing the value of the unknown parameter in order to obtain a mastercurve is very limited.

If neither the site density nor the radius is known, the situation is more complex. Figures 4b,d and f illustrate this. For several chosen values of r^*, N_s^* values can be found, which give $\Theta_{i,H}(pH_s)$ curves which, practically speaking, coincide. Therefore, it is advisable to measure a gyration radius or a hydrodynamic radius of the particles in addition to the proton adsorption data. A thus obtained radius narrows down the possible range of values in the double layer calculations. With, for instance, viscometry at a few pH values and/or ionic strength values, an impression is also gained about the rigidity of the particles.

4. CALCULATION OF $\Theta_{i,H}(pH_s)$ CURVES FROM EXPERIMENTAL PROTON ADSORPTION DATA FOR HUMIC AND FULVIC ACIDS

Unfortunately, for both humic acids and fulvic acids, only few experimental proton adsorption data sets at several ionic strengths are available in literature. In Figure 5, several of the available data sets [4,21-23] are replotted. Next to the experimental data also the $\Theta_{i,H}(pH_s)$ curves are plotted, which result in the best mastercurve. The Langmuir isotherm (L) is shown as a reference. In order to be able to calculate α for every datapoints of the dissociation curve, the total number of sites of the sample and the number of sites protonated for at least one of the datapoints, must be known. In our opinion, the uncertainties in determining these essential data may lead to an error in the α's of at least 5 to 10 percent. Considering the experimental errors and the small number of datapoints, the obtained result is satisfactory. The assumed values for r and N_s are shown in the figure caption.

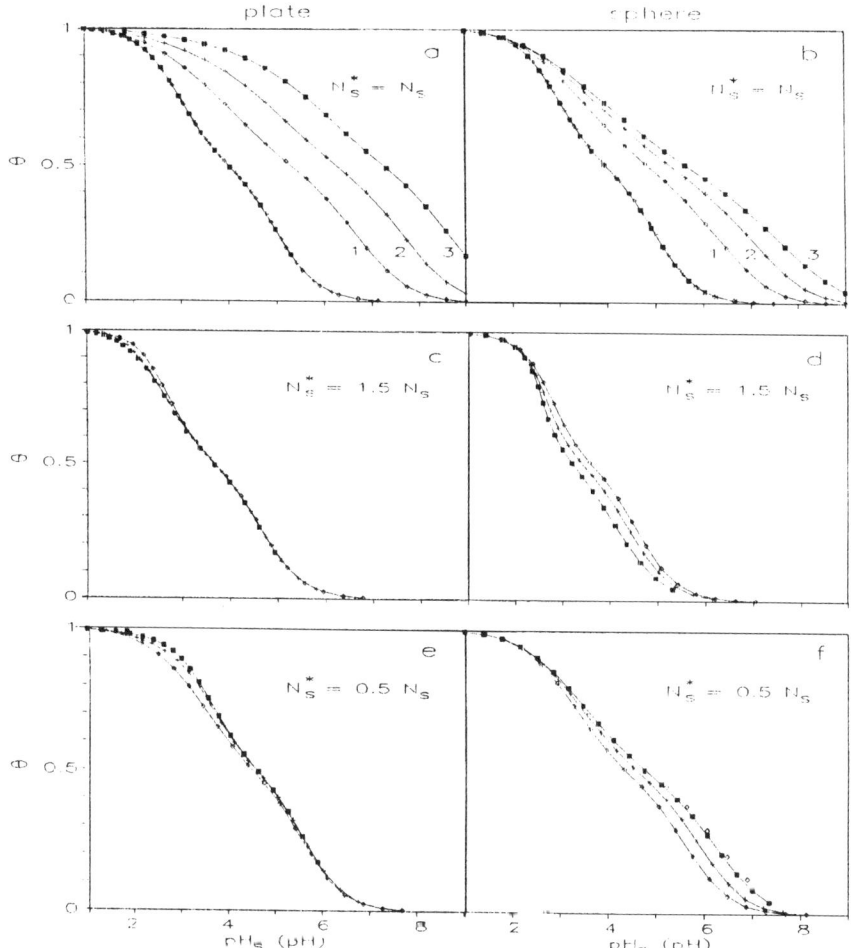

FIGURE 3. The Effect of the Site Density on the Acquisition of the Mastercurve from Synthetic Data for a (simple) Heterogeneous Flat Plate (a,c, and e) and a Sphere With a Radius of 1 nm (b,d and f). The synthetic results are obtained for $N_s = 1$ (sites/nm^2), $logK^{int}_{1,H} = 3$, $logK^{int}_{2,H} = 5$ and $f_1 = f_2 = 0.5$, at ionic strength values of .1(1),.01 (2) and .001M (3) Respectively.

The $\Theta_{j,H}(pH_s)$ curves have been calculated for three assumed values of N_s^*: Figure 3a and b: $N_s^* = 1.0$ (sites/nm^2)(the correct value), Figure 3c and d: $N_s^* = 1.5$ (sites/nm^2), Figure 3e and f: $N_s^* = 0.5$ (sites/nm^2).

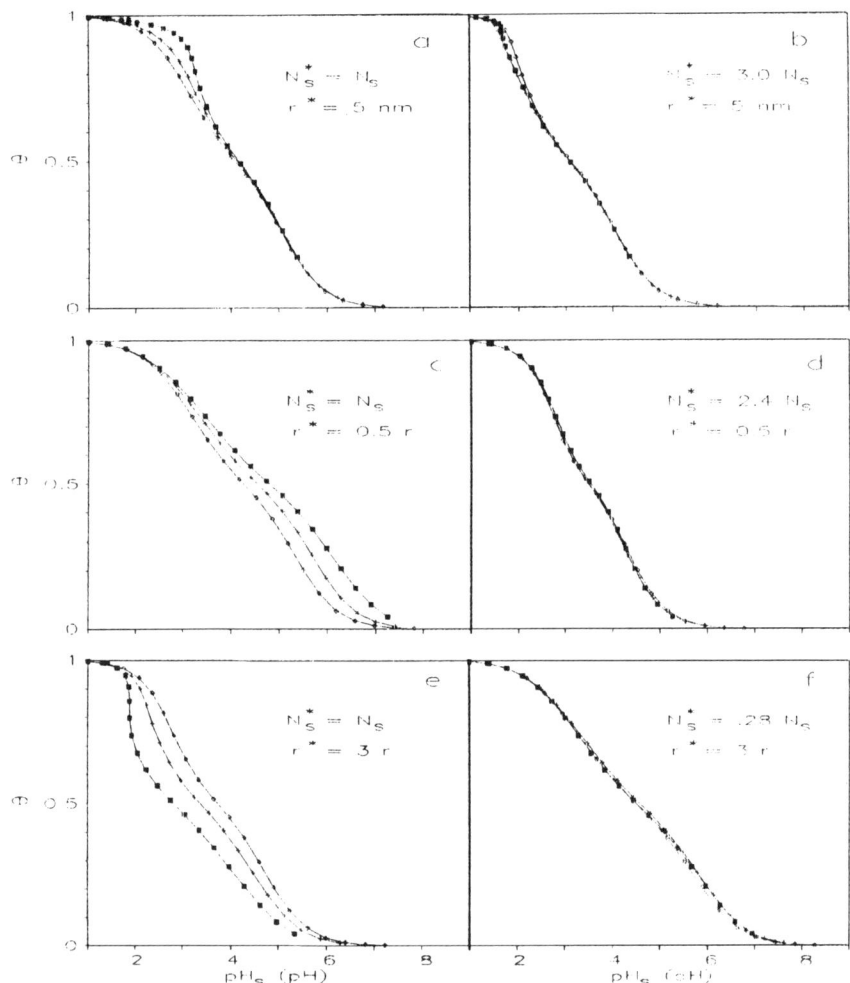

FIGURE 4. *The Combined Effect of the Radius and the Site Density on the Acquisition of the Mastercurve From Synthetic Data for a (Simple) Heterogeneous Flat Plate (a and b) and for a Sphere With a Radius of 1 nm (c-e).*
The Synthetic Results were obtained for $N_s = 1$ (sites/nm^2), $logK_{1,H}^{int} = 3$, $logK_{2,H}^{int} = 5$ and $f_1 = f_2 = 0.5$, at Ionic Strength Values of .001, 0.1 and .1 M (see Figs. 3a and b). The $\Theta_{i,H}$ (pH$_s$) Curves Have Been Calculated for:

Figure 4a,b: $r_* = 5$ nm; a: $N_{s*}^* = 1.0$ (sites/nm^2) b: $N_{s*}^* = 3.0$ (sites/nm^2)
Figure 4c,d: $r_* = .5$ nm; a: $N_{s*}^* = 1.0$ (sites/nm^2) d: $N_{s*}^* = 2.4$ (sites/nm^2)
Figure 4e,f: $r = 5$ nm; a: $N_s^* = 1.0$ (sites/nm^2) f: $N_s^* = .28$ (sites/nm^2)

The titration data of the humic substances presented in Figures 5a and b, exhibit a small ionic strength dependency. For both samples $\alpha_{i,H}(pH_s)$ curves have been obtained by assuming spherical particles with a radius of 1 nm. Because for such small spheres several combinations of the site density and the radius can be chosen (see Figure 4), coinciding $\Theta_{i,H}(pH_s)$ curves can also be obtained for other parameter values, as long as r < 3 nm. Assuming r > 3 nm does not lead to good results, while the application of the Gouy Chapman model was not successful at all. Due to swelling of the humic or fulvic acid, both their radius and site densities may change during the titration. This may explain the dispersion of the $\alpha(pH_s)$ curves at the endpoints of the isotherms for the different ionic strengths.

The "mastercurves" in Figures 5a and b are both less steep than the monocomponent Langmuir equation. This indicates that the materials studied are heterogeneous. For a further analysis of the heterogeneity, one of the methods discussed by Nederlof et al. [9] can be used.

The $\alpha(pH_s)$ curves for Figures 5c and d have been obtained by applying the GC model. The shape of the obtained mastercurves for the peat sample (Figure 5c) corresponds reasonably well with the monocomponent Langmuir equation with $\log K_H^{int} = 3.1$. Hence, in this case, adoption of a homogeneous surface model already leads to a reasonable description of the titration data.

The $\alpha(pH_s)$ curves for the humic acid data in Figure 5d are less steep than the monocomponent Langmuir equation. This indicates again, that the surface is considerably heterogeneous.

All proton titration data presented in Figure 5 are measured up to a pH = 6 for I = 0.1 M and up to pH = 8 for I = 0.001 M. Because the dissociation of the high affinity surface groups is suppressed by the dissociation of the low affinity groups, the contribution of the high affinity groups to the overall affinity is small in the pH ranges measured. Therefore, the existence of these groups cannot be obtained by the analysis of the presented proton adsorption isotherms, and the high affinity groups seem to be non-existent.

Several authors [4,21-24] have classified humic substances by using the so-called "Marinsky-plots" [4,21,24]. The Marinsky-plot can supposedly discriminate between permeable and nonpermeable macromolecules and between flexible and rigid linear polyelectrolytes. The application of Marinsky-plots as a classification method is, of course, only useful if the observed differences in physical and chemical behavior are uniquely caused by the degree of permeability or rigidity.

Figure 5 clearly shows that this is not the cause. In spite of their classification as permeable colloids [22,23], the datasets presented in Figure 5c,d can be described by adopting the GC relations for a non-permeable flat plate. Both humic substances characterized respectively as rigid (Figure 5a) and flexible (Figure 5b) impermeable linear electrolyte, can be described by assuming a rigid spherical particle. So, although Marinsky-plots can be helpful in visualizing differences in physical chemical behavior, they can not be used to characterize the nature of the colloid uniquely.

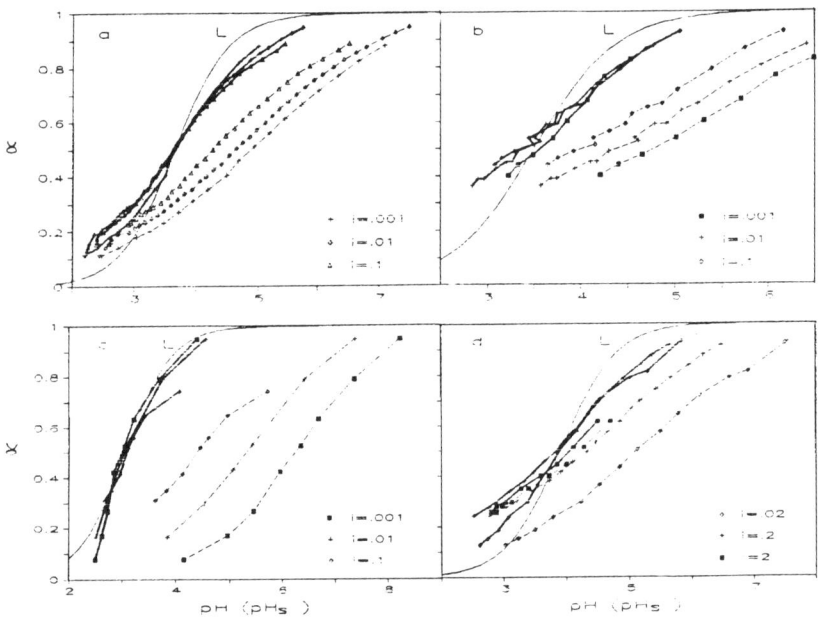

FIGURE 5. Experimental $\alpha(pH)$ Data (Dashed Curves) and the Best Coinciding $\alpha(pH_s)$ Curves for Humic Substances.

a. $\alpha(pH)$ Data for Armadale Horizons Bh Fulvic Acid [21]. $\alpha(pH_s)$ Curves Calculated with a Sphere with 1 nm, N_s = 0.7 (sites/nm^2).

b. $\alpha(pH)$ Data for Aquatic Humic Substances from a Stream at Lochard Forest (Central Region of Scotland) [4]. $\alpha(pH_s)$ Curves Calculated with a Sphere with 1 nm, N_s = 0.7 (sites/nm^2).

c. $\alpha(pH)$ Data for Sphagnum Pear [22]. $\alpha(pH_s)$ Curves Calculated with GC Model, N_s = 1 (sites/nm^2).

d. $\alpha(pH)$ Data for Humic Acid [23]. $\alpha(pH_s)$ Curves Calculated with GC Model, N_s = 0.4 (sites/nm^2).

For sake of reference also the monocomponent Langmuir Isotherm (L) is included when coinciding at α = 0.5 with the $\alpha(pH_s)$ curves.

5. METAL-ION ADSORPTION

5.1 Theoretical Framework

On analogy of proton adsorption, association reactions for adsorbing metal ions can be specified. In addition to Equation (1) for the proton, the association reaction for the adsorption of a divalent cation M^{2+} to a surface group of types i, can be given:

$$S_iO^- + M^{2+} \rightleftharpoons S_iOM^+ \qquad K_{i,M}^{int} \qquad (10)$$

Now both the fraction of surface groups is protonated, $\Theta_{i,H}$ and the fraction of the sites associated with metal ions, $\Theta_{i,M}$ can be represented by a multicomponent Langmuir equation:

$$\Theta_{i,H} = \frac{K_{i,H}^{int} H_s}{1 + K_{i,M}^{int} M_s + K_{i,H}^{int} H_s} \qquad (11)$$

and

$$\Theta_{i,H} = \frac{K_{i,H}^{int} M_s}{1 + K_{i,M}^{int} M_s + K_{i,H}^{int} H_s} \qquad (12)$$

where M_s is the metal activity in the plane of adsorption given by the product of the metal activity in the bulk solution and the Boltzmann factor for a divalent cation:

$$M_s = M \cdot \exp\left(\frac{-2F\Psi s}{RT}\right) \qquad (13)$$

Equations (11) and (12), are based on the assumption that protons and metal ions compete for the same sites. For multicomponent adsorption α_i is still defined according to Equation (3). The overall fractions $\Theta_{t,H}$ and $\Theta_{t,H}$ are given by the weighted sums of the $\Theta_{i,H}$ and the $\Theta_{i,M}$ for the individual groups (see Equation (8)).

The adsorption of both protons and metal ions affect the surface charge. According to Equation (10), high metal adsorption may even lead to a net positive surface charge. An increase of the metal adsorption will induce a further dissociation of the protonated surface groups and a release of protons from the surface. So for metal ion adsorption, an experimentally accessible proton-metal exchange ratio r_{ex} can be defined as:

$$r_{ex} = \frac{-\Delta \Gamma_H}{\Gamma_M} \quad ; \quad pH = \text{constant} \tag{14}$$

where $-\Delta \Gamma_H$ is the amount of protons, or hydroxyl ions required to maintain the pH constant and Γ_M, the amount of metal ions adsorbed.

The value of r_{ex} may change as a function of metal ion loading and pH. The change of the surface charge caused by the metal ion adsorption with respect to the proton adsorption isotherm can only be calculated if, in addition to the metal ion adsorption, r_{ex} is measured. Because the surface charge is essential for the calculation of the surface potential and therefore for pM_s, and pH_s, r_{ex} is an important parameter for an appropriate description of metal ion adsorption.

Due to the pH dependency of the metal ion adsorption, the acquisition of the intrinsic affinity distribution for the metal ion is far more complex than the acquisition of the intrinsic proton affinity distribution from a series of monocomponent proton adsorption isotherms. Even if a correct double layer model is used, the adsorption as a function of both pH_s and pM_s is no longer a mastercurve but a masterplane.

In the multicomponent Langmuir Equation (11-12), metal ions and protons compete for the same surface sites. If site competition is assumed to be non-existent, the metal ion adsorption can be described by applying a monocomponent Langmuir isotherm equation such as Equation (5), in which H_s is replaced by M_s. The $\Theta_{t,M}(pM_s)$ function is independent of pH_s and again we are dealing with a mastercurve in stead of a masterplane. The acquisition of the intrinsic affinity distribution for the metal ions is, in this case, similar to that of the intrinsic proton affinity distribution.

Note that also without site competition, metal ion adsorption is still influenced by proton adsorption via electrostatic effects. The $\Theta_{t,M}(pM_s)$ curves can only be obtained if Ψ_s and hence the surface speciation is known. This again stresses the necessity of determining

r_{ex} experimentally. In the remainder of this paper, only cases with site competition will be treated.

5.2 Model Calculations for Metal Ion Adsorption

In our opinion, data for proton titration isotherms at several ionic strength and data for the exchange ratios have to be available, in order to be able to analyze adsorption data for metal ions on humic substances rigorously. We have not been able to extract from literature a combination of proton adsorption data, metal adsorption data and exchange ratios. Therefore, we will restrict our analysis of metal ion adsorption to model calculations for homogeneous or simple heterogeneous surfaces.

The proton and metal ion affinity constants used are estimated from the complexation constants of protons and copper with salicyclic acid and with phthalic acid, as given by Martell and Smith [25]. The intrinsic affinity constants for surface groups and solution monomers are not necessarily identical to the tabulated values, but the latter give a first impression of the magnitude of the affinity constants. With all metal ion calculations presented, a GC double layer is used (flat plate).

5.3 Monodentate Metal Ion Adsorption

In the previous section, the multicomponent nature of metal ion adsorption was illustrated on the basis of complex formation according to Equation (10). However, this is not the only possibility for a monodentate complex. Instead of Equation (10), the following adsorption relation may also be assumed for the association reaction for a divalent metal ion:

$$S_iO^- + M^{2+} + H_2O = S_iOMOH + H^+ \qquad K^{int}_{i, MOH} \qquad (15)$$

On the basis of Equations (1) and (15), the following adsorption isotherm equation results:

$$\Theta_{i,MOH} = \frac{K^{int}_{i,MOH} M_s/H_s}{1 + K^{int}_{i,MOH} M_s/H_s + K^{int}_{i,H} H_s} \qquad (16)$$

In Figure 6, some model calculations for metal adsorption according to Equation (12) and Equation (16) onto a homogeneous surface with $N_s = 1$ (sites/nm^2) and $\log K_H^{int} = 3$ are presented. The metal affinity constants used are $\log K_M^{int} = 1.5$ (Equation (12)) and $\log K_{MOH}^{int} = -2.0$ (Equation (16)).

Figure 6 shows that the adsorption isotherms and the exchange ratio for the formation of the two complexes are completely different. The metal ion adsorption for the formation of SOM complexes is considerably lower and less pH-dependent than for the formation of SOMOH complexes. In both cases, the ion adsorption results in a further dissociation of the protonated surface groups. Therefore, the charge increase per metal ion adsorbed is not proportional with the metal ion valency z, but with $z-\tau_{ex}$. At pH = 4, an exchange ratio of about 1.3 is observed upon the formation of the SOM complex (see Figure 6b). So for every metal ion adsorbed 1.3 SOH groups dissociate. For the formation of the SOMOH complex $\tau_{ex} = 1.8$ (Figure 6d). This value is composed of 0.8 protons released, due to a further dissociation plus one proton resulting from the stoichiometry of Equation (16). The overall surface charge effect per adsorbed metal ion for the SOMOH complex (0.2) is thus smaller than for the SOM complex (0.7). Therefore, the formation of the SOMOH complex increases more strongly with increasing metal concentration, than that of the formation of a SOM complex.

An increase of the pH leads to a decrease in proton activity. Consequently, the proton becomes a weaker competitor. Therefore, at the same metal ion concentration, an increase of the pH results in a higher metal ion adsorption. This trend is enhanced by the fact that the degree of dissociation increases with pH, which leads to a more negatively-charged surface on which cations adsorb more easily. However, due to the larger degree of dissociation, the exchange ratio is smaller (Figure 6b and d) and the positive charge per metal ion added to the surface is much larger than at lower pH's. This charge effect suppresses further metal adsorption and counteracts the positive effects of an increase in pH.

At pH values where the SOH surface groups are almost completely dissociated, $\tau_{ex} \rightarrow 0$ and the $K_H^{int} H_s$ term in Equation (12) will be negligible. The surface charge and, therefore, also the surface potential will then be dependent on the metal ion concentration only. Hence, at high pH, the formation of SOM complexes will become independent of the pH. This effect is illustrated by the isotherms at pH = 6 and pH = 8. The pH effect between these curves is already very small.

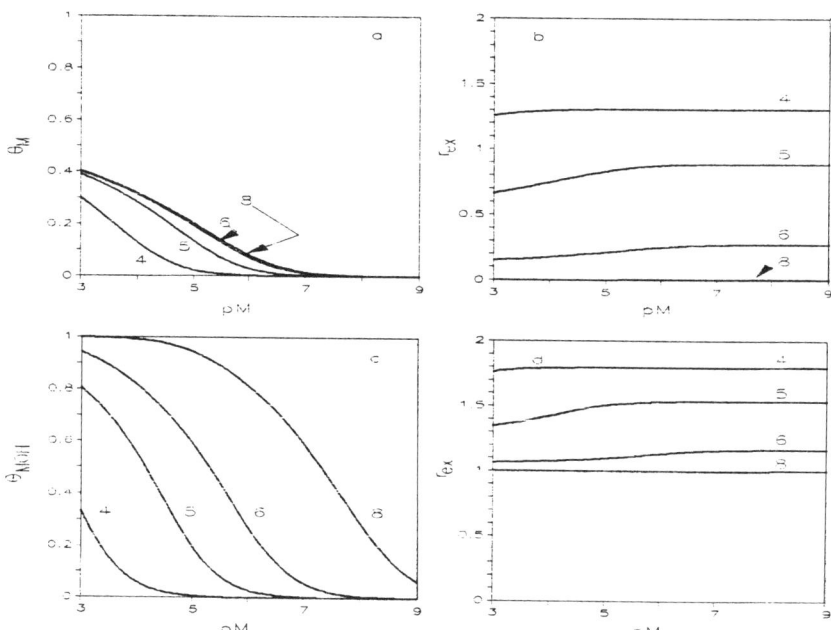

FIGURE 6. Calculated Adsorption Isotherms (a,c) and Metal Proton Exchange Ratios (b,d) for the Formation of SOM Complexes (a,b) and SOMOH complexes (c,d) for a Homogeneous Surface at pH Values 4,5,6 and 8. $logK_H^{int} = 3.0$; $N_s = 1$ $(sites/nm^2)$; *Figures a and b:* $logK_M^{int} = 1.5$; *Figures c and d:* $logK_{MOH}^{int} = -2.0$.

For the SOMOH complex, the $K_H^{int} H_s$ term in Equation (16) will also become negligible at high pH values. However, the metal ion adsorption will now become a function of M/H instead of M only. The metal ion adsorption isotherm will still exhibit a pH dependency. Note, that because the SOH surface groups are almost completely dissociated, the surface charge depends on the ratio M/H only. This means that the shape of the $\Theta_{MOH}(M)$ isotherms becomes identical for high pH values. Only their location on the pM axis is shifted.

5.4 Bidentate Metal Ion Adsorption on Salicylic Acid and Phthalic Acid-Like Surface Groups

According to Gamble et al. [3], Tipping et al. [4] and Buffle [5], phthalic acid and salicylic acid-type of structures of humic and fulvic acids play an important role in metal ion and adsorption. In those structures, two acid groups are in close proximity. The protonation of the functional groups of phthalic acid and salicylic acid-like structures, can be treated as two separate surface groups, each with only one reactive O^- site:

$$S_1O^- + H^+ \rightleftharpoons S_1OH \qquad K^{int}_{1,H} \qquad (17a)$$

$$S_2O^- + H^+ \rightleftharpoons S_2OH \qquad K^{int}_{2,H} \qquad (17b)$$

With each of these groups, the metal ion may react according to Equation (15) leading to the isotherm Equation (16). However, this may not be the most realistic complex.

We can also consider phthalic and salicylic-like structures each as one surface group with two reactive O^- sites which can be fully protonated in two steps:

$$SO_2^{2-} + H^+ \rightleftharpoons SO_2H^- \qquad K^{int}_{H1} \qquad (18a)$$

$$SO_2H^- + H^+ \rightleftharpoons SO_2H_2 \qquad K^{int}_{H2} \qquad (18b)$$

Logically, the site density while using Equation (17), should be taken twice as high as in the case of Equation (18). Furthermore, we will assume that $logK^{int}_{1,H}$ (17a) = $logK^{int}_{H1}$ (18a), $logK^{int}_{2,H}$ (17b) = $logK^{int}_{H2}$ (18b). In principle, the protonation of the surface according to respectively Equation (17) and Equation (18) are not equivalent. However, as long as $\Delta pK^{int}_H > 1$, the differences in proton association between both approaches are negligible. ΔpK^{int}_H is about 10 for salicylic acid and about 2.4 for phthalic acid.

From the solution chemistry of the salicylic and phthalic acid, it is known that for the complexation of metal ions with salicylic and phthalic acids, mostly bidentate metal complexes are formed. The corresponding formation of a bidentate surface complex SO_2M can be given by Equation (19):

$$SO_2^{2-} + M^{2+} \rightleftharpoons SO_2M \qquad K^{int}_{bi,M} \qquad (19)$$

The fractional coverage of the bidentate surface complex based on Equations (18) and (19) is given by:

$$\Theta_M = \frac{K_{bi,M}^{int} M_s}{1 + K_{H1}^{int} H_s + K_{H1}^{int} K_{H2}^{int} H_s^2 + K_{bi,M}^{int} M_s} \quad (20)$$

The same metal complex can also be formed through the reaction with the SO_2H^- group:

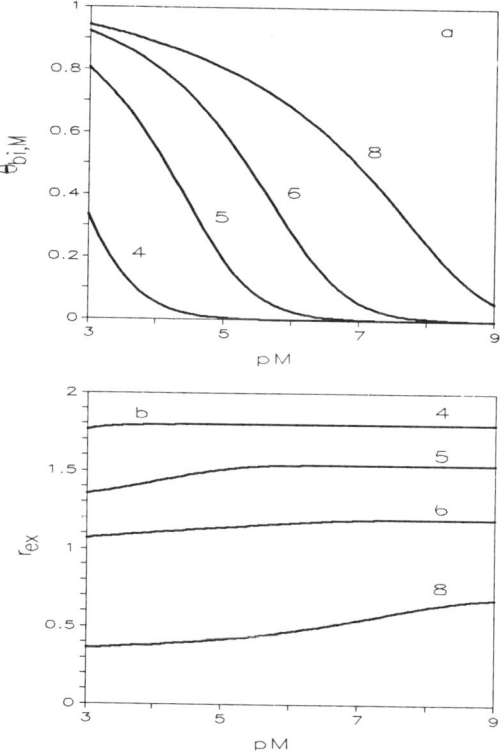

FIGURE 7. *The Calculated Adsorption Isotherms (a) and Metal Proton Exchange Ratios (b) for the Formation of SO_2M Complexes for a Surface with Phthalic Acid-Like Surface Groups Only at pH Values, 4,5,6 and 8.* $logK_{H1}^{int} = 5.4$, $logK_{H2}^{int} = 3$, $N_s = 1$ (sites/nm^2); $logK_{bi,M}^{int} = 3.4$.

348 SORPTION ONTO SURFACES

$$SO_2H^- + M^{2+} \rightleftharpoons SO_2M^+H^+ \quad \hat{K}^{int}_{bi,M} \qquad (21)$$

According to Equations (18a) and (19):

$$\log \hat{K}^{int}_{bi,M} = \log K^{int}_{bi,M} + \log K^{int}_H \qquad (22)$$

In Figure 7, the metal adsorption according to Equation (20), is calculated with $\log K^{int}_{H1} = 5.4$, $\log K^{int}_{H2} = 3$, $\log K^{int}_{bi,M} = 3.4$. The $\log K^{int}_H$ values are relevant for phthalic-like structures. The value of $\log K^{int}_{bi,M} = 3.4$ corresponds with a $\log \hat{K}^{int}_{bi,M} = -2.0$, which is equivalent with the $\log K^{int}_{MOH}$ value used in the calculations for Figures 6c and d.

The calculated adsorption isotherm and τ_{ex}-curves for the formation of the bidentate complexes at pH values 4, 5 and 6 (Figure 7), are identical with the curves for the formation of the SOMOH complexes presented in Figures 6c and d. The explanation of this phenomenon is simple. The fraction of the SO_2^{2-} groups is negligible up to pH = 6. The degree of dissociation of the second proton is completely determined by the formation of the bidentate complex. For every bidentate complex formed, 1 proton will be released (Equation (21)).

The stoichiometry with respect to M^{2+} and H^+ for the formation of a bidentate complex according to Equation (21), is equivalent to that of the formation of a SOMOH complex (Equation (15)). As long as the fraction of the SO_2^{2-} groups is small, no distinction can be made between bidentate complexes and monodentate SOMOH complexes, if corresponding affinity constants are used.

At pH values higher than 6, a notable fraction of the second group will dissociate. The τ_{ex} for the formation of the bidentate complexes now becomes smaller than τ_{ex} (SOMOH). Due to the larger charge effect per metal ion adsorbed, the metal adsorption for the bidentate complexes is smaller than for the monodentate SOMOH complexes.

At pH values where the SO_2H^- surface groups are almost completely dissociated, $\tau_{ex} \to 0$ and the formation of the bidentate metal complexes will become independent of the pH value. This situation is akin to the formation of the SOM complexes at high pH. When the SO_2H_2 and the SO_2H^- groups are fully dissociated, $K^{int}_{H1}H_s$ and the $K^{int}_{H1}K^{int}_{H2}H_s^2$ terms in Equation (20) are negligible. The surface charge and therefore, also the surface potential, will then become dependent on the metal ion concentration M only. Hence, the formation of SO_2M complexes will be independent of the pH at high pH values.

A $\Delta pK_H^{int} = 2.4$, as used in the calculations for the formation of the bidentate complexes (Figure 8) corresponds with the ΔpK_H^{int} of phthalic acid, but is small compared with $\Delta K_H^{int} = 10$ for salicylic acid. For a ΔK_H^{int} considerably larger than ΔK_H^{int} (phthalic acid), the contribution of the free SO_2^{2-} will be very small for most pH values. Therefore, up to high pH values, the formation of the SOMOH complexes will be identical to that of the SO_2M complexes.

6. CONCLUSIONS

The ionic strength dependency of proton adsorption isotherms of humic substances can be described reasonably well by using relations derived from the Poisson Boltzmann equation for spherical particles or flat plates.

On heterogeneous surfaces, the dissociation of the higher proton affinity groups is suppressed, due to the charge development. A significant degree of dissociation of such groups will occur at pH values considerably higher than expected on the magnitude of their intrinsic affinity constants only. For instance, if the proton adsorption isotherm of such a surface is measured at I = 0.1 up to pH = 6.0 - 7.0, the functional groups with a $\log K_H^{int} > 5.0$ hardly contribute to the overall proton adsorption isotherm. Therefore, those groups seem to be non-existent. Note that this is the case for the second site of salicylic acid and phthalic acid structures. This illustrates that in order to be able to obtain the intrinsic affinity constants for high proton affinity sites from $\Theta(pH_s)$ mastercurves proton adsorption isotherms have to be measured up to high pH values.

For the description of metal ion adsorption, a better insight is obtained when electrostatics are taken into account explicitly. In order to calculate the surface potential, the proton adsorption isotherm and the proton/metal exchange ratio, τ_{ex}, have to be determined in addition to the metal ion adsorption (at constant pH).

The proton and metal ion stoichiometry of the formation of bidentate metal ion surface complexes, for instance, with phthalic acid-like structures, is identical to that of the formation of a monodentate SOMOH complex as long as the dissociation of the second functional group is negligible. For phthalic acid-like surface groups this was the case up to pH = 6. The larger ΔpK_H^{int}, the larger the pH range for which both cases remain identical.

On the basis of titration experiments, no distinction can be made between the formation of bidentate and monobendate metal ion complexes at low pH values, if further information about the structure of the surface is absent.

ACKNOWLEDGMENTS

We want to thank Dr. A. De Keizer for making available the algorithm for the numerical solutions of the Poisson Boltzmann equation.
This work was partially funded by the European Comity Environmental Research Programme on Soil Quality under contract number EV4V-0100-NL(GDF), and partially funded by the Netherlands Integrated Soil Research Programme under contract number PCBB 8948.

REFERENCES

1. Hiemstra, T., W.H. Van Riemsdijk, and G.H. Bolt. "Multi-Site Proton Adsorption Modeling at the Solid/Solution Interface of (Hydr) Oxides: A New Approach. I: Model Description and Evaluation of Intrinsic Reaction Constants," *J. Colloid Interface Sci. in press,* (1989).

2. Hiemstra, T., J.C.M. de Wit, and W.H. Van Riemsdijk. "Multi-Site Proton Adsorption Modeling at the Solid/Solution Interface of (Hydr) Oxides: A New Approach.I: Application to Various Important (Hydr) Oxides," *J. Colloid Interface Sci. in press,* (1989).

3. Gamble, D.S., A.W. Underdown, and C.H. Langford. "Copper (II) Titration of Fulvic Acids Ligand Sites with Theoretical, Potentiometric, and Spectrophotometric Analysis," *Anal. Chem.*, 52:1901-1908 (1980).

4. Tipping, E., C.A. Backes, and M.A. Hurley. "The Complexation of Protons, Aluminum and Calcium by Aquatic Humic Substances: A Model Incorporating Binding-Site Heterogeneity and Macroionic Effects," *Wat. Res.*, 22(5):597-611 (1988).

5. Buffle, J. "Complexation Reactions in Aquatic Systems, An Analytical Approach," *Chichester:Ellis Horwood Limited,* (1988).

6. House, W.A. "Adsorption on Heterogeneous Surfaces," in *Colloid Science Specialist Report 4,* Everett, D.H. Ed. Chapter 1, (London: The Royal Society of Chemistry) (1983).

7. Jaroniec, M. and P. Brauer. "Recent Progress in Determination of Energetic Heterogeneity of Solids from Adsorption Data," *Surface Sci. Rep.,* 6:65-117 (1986).

8. Nederlof, M.M., W.H. van Riemsdijk, and L.K. Koopal. "Methods to Determine Affinity Distributions for Metal Ion Binding in Heterogeneous Systems." In *Heavy Metals in the Hydrological Cycle,* Astruc, M. and J.N. Lester Eds. (London: Selper Ltd., 1988) p. 361.

9. Nederlof, M.M., W.H. Van Riemsdijk, and L.K. Koopal. "Determination of Adsorption Affinity Distributions: A Comparison of Various Methods Related to Local Isotherms Distributions," *Submitted to J. Colloid Interface Sci.,* (1989).

10. Harris, L.B. "Adsorption on a Patchwise Heterogeneous Surface: Mathematical Analysis of the Step-Function Approximation to the Local Isotherm," *Surface Sci.,* 10:129-145 (1968).

11. Van Riemsdijk, W.H., L.K. Koopal and J.C.M. De Wit. "Heterogeneity and Electrolyte Adsorption: Intrinsic and Electrostatic Effects," *Netherlands J. Agricultural Sci.,* 35:241-257 (1987).

12. De Wit, J.C.M., W.H. Van Riemsdijk and, L.K. Koopal. "Proton and Metal Ion Binding on Homogeneous and Heterogeneous Colloids," in *Heavy Metals in the Hydrological Cycle,* Astruc, M. and J.N. Lester Eds. (London: Selper Ltd., 1988) p. 369.

13. Van Riemsdijk, W.H., G.H. Bolt, L.K. Koopal, and J. Blaakmeer. "Electrolute Adsorption on Heterogeneous Surfaces: Adsorption Models," *J. Colloid Interface Sci.,* 109:219-228 (1986).

14. Koopal, L.K. and W.H. Van Riemsdijk. "Electrosorption on Random and Patchwise Heterogeneous Surfaces: Electrical Double Layer Effects," *J. Colloid Interface Sci.,* (1989).

15. Healy, T.W. and L.R. White. "Ionizable Surface Group Models of Aqueous Interfaces," *Adv. Colloid Interface Sci.*, 9:303-345 (1978).

16. Loeb, A.L., J. Th. G. Overbeek, and P.H. Wiersema. "The Electrical Double Layer Around a Spherical Colloid Particle," (Cambridge: M.I.T. Press, 1961).

17. Stigter, D. "Functional Representation of Properties of the Electrical Double Layer Around a Spherical Colloid Particle," *J. Electroanal. Chem.*, 37:61-64 (1972).

18. Overbeek, J. Th. G. "Electrochemistry of the Double Layer," in *Colloid Science I* (Amsterdam: Elsevier Publishing Company), Chapter 4 (1952).

19. Van Riemsdijk, W.H., J.C.M. De Wit, L.K. Koopal, and G.H. Bolt. "Metal Ion Adsorption on Heterogeneous Surfaces: Adsorption Models," *J. Colloid Interface Sci.*, 116(2):511-522 (1987).

20. Cameron, R.S., B.K. Thornton, R.S. Swift, and A.M. Posner. "Molecular Weight and Shape of Humic Acid from Sedimentation and Diffusion Measurements on Fractionated Extracts," *J. Soil Sci.*, 23:394-408 (1972).

21. Ephraim, J., S. Alegnet, A. Mathuthu, M. Bicking, R.L. Malcolm, and J.A. Marinsky. "A Unified Physicochemical Description of the Protonation and Metal Ion Complexation Equilibria of Natural Organic Acids (Humic and Fulvic Acids) 2. Influence of Polyelectrolyte Properties and Functional Group Heterogeneity on the Protonation Equilibria of Fulvic Acid," *Envir. Sci. Technol.*, 20:354-366 (1986).

22. Marinsky, J.A., A. Wolf, and K. Bunzl. "The Binding of Lead (II), Copper (II), Cadmium (II), Zinc (II) and Calcium (II) to Soil Organic Matter," *Talanta*, 27:461-468 (1980).

23. Marinsky, J.A., S. Gupta, and P. Schindler. "The Interaction of Cu (II) Ion with Humic Acid," *J. Colloid Interface Sci.*, 89:401-411 (1982).

24. Marinsky, J.A. and J. Ephraim. "A Unified Physiochemical Description of the Protonation and Metal Ion Complexation Equilibria of Natural Organic Acids (Humic and Fulvic Acids). 1. Analysis of the Influence of Polyelectrolyte Properties on Protonation Equilibria in Ionic Media: Fundamental Concepts," *Envir. Sci. Technol.*, 20:349-354 (1986).

25. Martell, A.E. and R.M. Smith. "Critical Stability Constant, Volume 3: Other Organic Ligands." (*New York: Plenum Press* 1977).

DISCUSSION OF:
PROTON AND METAL ION BINDING ON HUMIC SUBSTANCES

Virginia L. Cunningham
Smith Kline & French Laboratories, King of Prussia, Pennsylvania

As described in the paper, humic and fulvic acids are polydisperse mixtures of organic polyelectrolytes, which exist as negatively charged species in the pH range of most natural waters. Some additional discussion of the nature of humic substances and their relevant functionalities may serve as an aid in elucidating the underlying organic chemical considerations on which this paper is based.

Figure 1 illustrates a typical humic acid structure. The molecule contains some regions which consist predominantly of hydrophobic moieties and other regions containing high concentrations of polar, hydrophilic functional groups. This disparity in physical properties in the two regions gives rise to a three-dimensional structure for these materials where the polar, hydrophilic groups are hydrated to the greatest extent possible, while the hydrophobic groups form what might be visualized as interior regions comparatively protected from the surrounding

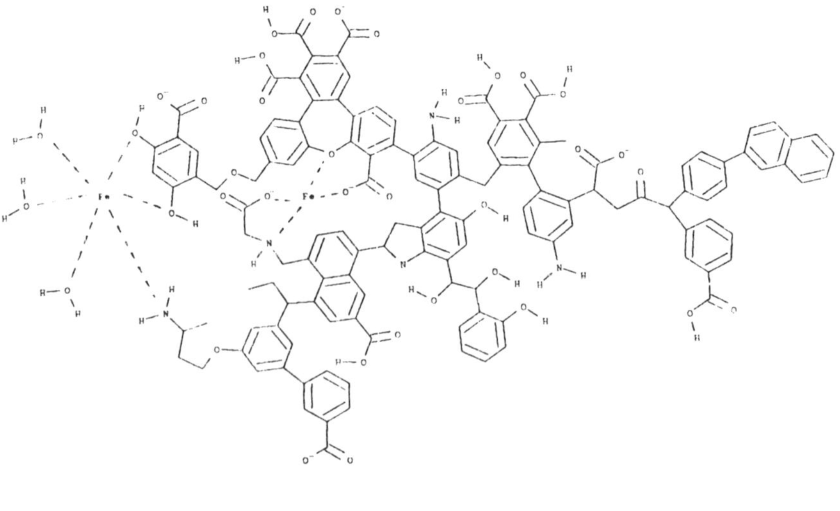

FIGURE 1.   Typical Humic Acid Structure.

aqueous medium. Thus, the surface of the particle will contain a high proportion of the polar functional groups, predominantly carboxylic acid and phenolic groups.

For the purposes of proton and metal ion complexation phenomena, these groups can be seen to consist of at least three types: carboxylic acid groups adjacent to other carboxylic acid groups or to hydroxyl (type I), the second carboxylic acid groups adjacent to an already dissociated carboxylic acid group (type II), and hydroxyl groups adjacent to carboxylic acid groups (type III). Phthalic acid (*ortho* isomer), shown in Figure 2, and salicylic acid, shown in Figure 3, are therefore good model compounds for proton and metal ion binding studies, since they contain the three types of acidic functionality present in humic substances in general.

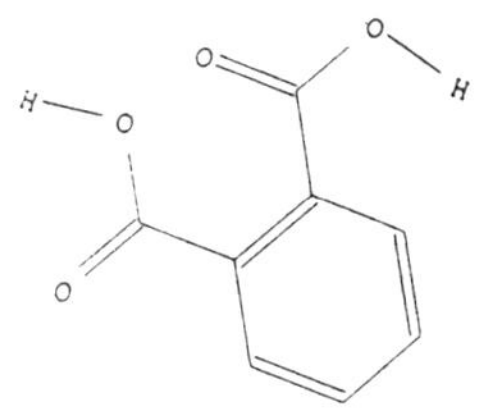

FIGURE 2.   o-Phthalic Acid.

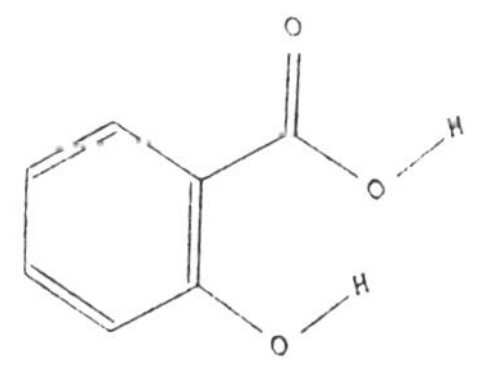

FIGURE 3.   Salicylic Acid.

The different acidities of the functional groups come about primarily because of the stabilizing influence of neighboring groups once the first proton has dissociated. For example, the pK_1 of o-phthalic acid is 2.89, similar to the pK_1 of salicylic acid. Both of these groups might be considered as type I, and have the least degree of proton affinity. However, the second carboxylic acid group in *o*-phthalic acid has a pK_2 of 5.51, indicating a higher proton affinity, while the pK_2 of salicylic acid, arising from dissociation of the hydroxyl proton is 13.4. The higher proton affinities of the second functional groups in *o*-phthalic acid and in salicylic acid can be attributed to hydrogen-bonding between the anion arising from the first dissociation and the proton in the second functional group. This is illustrated for *o*-phthalic acid in Figure 4 and for salicylic acid in Figure 5.

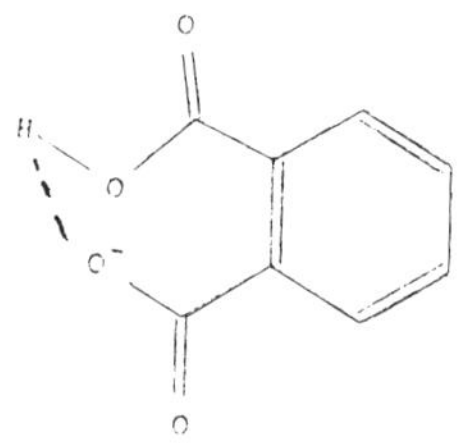

FIGURE 4.   *o-Phthalic Acid Anion.*

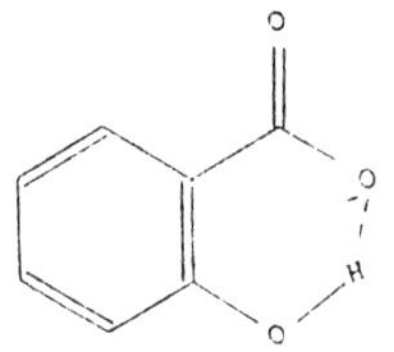

FIGURE 5.   *Salicylic Acid Anion.*

These hydrogen-bonding interactions account for the observations by the authors that the functional groups with pKs > 5.0 hardly contribute to the overall proton adsorption isotherm.

In the case of metal ion complexation, however, a somewhat different situation arises. The authors assume that protons and metal ions compete for the same sites, and that the adsorption of both protons and metal ions will affect the surface charge, possibly leading to a net positive surface charge. In addition, an increase in the adsorption of the metal ions will induce a further dissociation of the protonated surface groups and a release of protons from the surface. Evidence of this can be seen readily in the case of the pK_2 of salicylic acid. When only proton affinity is considered, that of the hydroxyl group is very high; however, in the presence of metal ions, the proton affinity becomes significantly less, while the metal ion affinity becomes much greater. For example, the pK of the aluminum ion complex with salicylic acid, shown in Figure 6, is 2.9 [1]. This accounts for the inability of a proton adsorption model alone to adequately predict metal ion adsorption and for the need to determine the proton-metal exchange ratio. It also explains why pH values up to 6, the degree of dissociation of the second proton is completely determined by the formation of the bidentate metal complex.

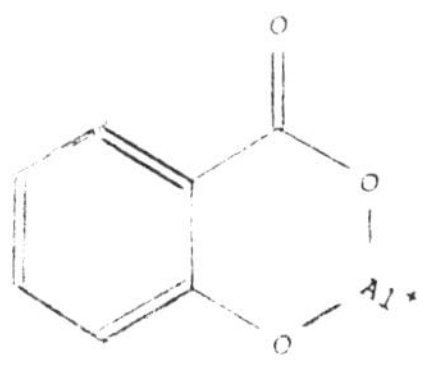

FIGURE 6. *Aluminum Complex with Salicylic Acid.*

REFERENCES

1. Tipping, E., C.A. Backes, and M.A. Hurley, "The Complexation of Protons, Aluminum and Calcium by Aquatic Humic Substances: A Model Incorporating Binding-Site Heterogeneity and Macroionic Effects," *Wat. Res.*, 22(5):597-611 (1988).

POLYELECTROLYTIC METAL IONS SEQUESTRANTS

L. *Campanella, V. Crescenzi, M. Dentini, C. Fabiani*
ENEA, Department TIB, Rome, Italy

F. *Mazzei, A.I. Nero Scheffino*
Department of Chemistry, University "La Sapienza" Rome, Italy

1. INTRODUCTION

Toxic metals can be removed from wastewaters by different, traditional procedures such as the precipitation of oxides, hydroxides, sulphites or carbonates, ion exchange, adsorption and electrodeposition or membrane processes.

These procedures often present a number of problems, including: secondary polluting effects (reagents added in strong excess), high operating costs (basically due to the commercial value of the reagents), need of tailoring the operational workout to the complexity of the waters to be treated and need of disposing of the resultant toxic sludges. Cheaper, non-polluting methods can be based on absorption on activated coal or on synthetic materials. However, the former can be considered only for the removal of trace contaminants while the latter, though of high absorbing capacity, cannot always be regenerated.

More recently, procedures for the removal of heavy metal ions from aqueous media based on the use of polyelectrolytes, either synthetic or natural, have been the subject of active investigation. These species, in fact, may exhibit at the same time a very high metal binding capacity---often accompanied by complex precipitation---and the ability to quantitatively flocculate colloidal, suspended particles. Polyelectrolytes may also be used in immobilized form, i.e. entrapped in water insoluble, stable polymeric membranes.

In the attempt of making such procedures both efficient and economically attractive, different polyelectrolytes have been recently considered including biopolymeric derivatives (e.g. starch derivatives).

Additional studies are needed however in order to better elucidate fundamental aspects connected with both the equilibrium and the kinetics of heavy metal ions binding by said species.

In our laboratories a research has been recently started on the possible use as metal ions sequestrants of maleic acid 1:1 copolymers as well as of ionic polysaccharides from non pathogenic bacteria. Motivations for the choice of these polymers may be briefly summarized as follows.

A number of ionic polysaccharides from microbial cultures, that is from renewable sources, have the advantage of being easily obtainable in pure form -- some are already commercially available -- and of being biodegradable/biocompatible.

A screening of microbial non-pathogenic polysaccharides in terms of their multivalent counterions binding under different working conditions appears therefore worthwhile.

Quite obviously, in order to scale up eventually, processes based on said biopolymers in the field of water decontamination, will have to substantially reduced their production costs.

Maleic acid copolymers, on the other hand, while not biodegradable can be prepared from relatively inexpensive raw materials, and present a high hydrolytic stability making them usable under experimental conditions prohibitive for the biopolymers.

We wish to report here a few original results obtained studying as chelating agents towards Cu(II) and Cr(III) ions the following polyelectrolytes:

1. a 1:1 maleic acid-ethyl vinyl ether copolymer;
2. a 1:1 maleic acid-acrylic acid copolymer; and
3. the exocellular polysaccharide extracted from cultures of Rhizobium trifolii strain TA-1 (Fig. 1).

Most experiments have been carried out using said species in dilute aqueous solutions. Polyelectrolytes (1) and (3) have been also entrapped in polyvinyl alcohol matrices and the resulting films studied for their Cu(II) ions absorbing capacity (Table 1).

2. RESULTS AND DISCUSSION

2.1 Potentiometric and Equilibrium Dialysis Data

Using the Cr(III) ions selective electrode described in the Experimental Part it has been possible to evaluate the percentage of such ions bound in dilute aqueous solution by each of the three polyelectrolytes considered (polymer concentration 6 meq/L, pH 5) upon addition of increasing amounts of $Cr(NO_3)_3$. The equivalent salt concentration had to be kept lower than that of the polyelectrolytes in order to avoid the onset of precipitation.

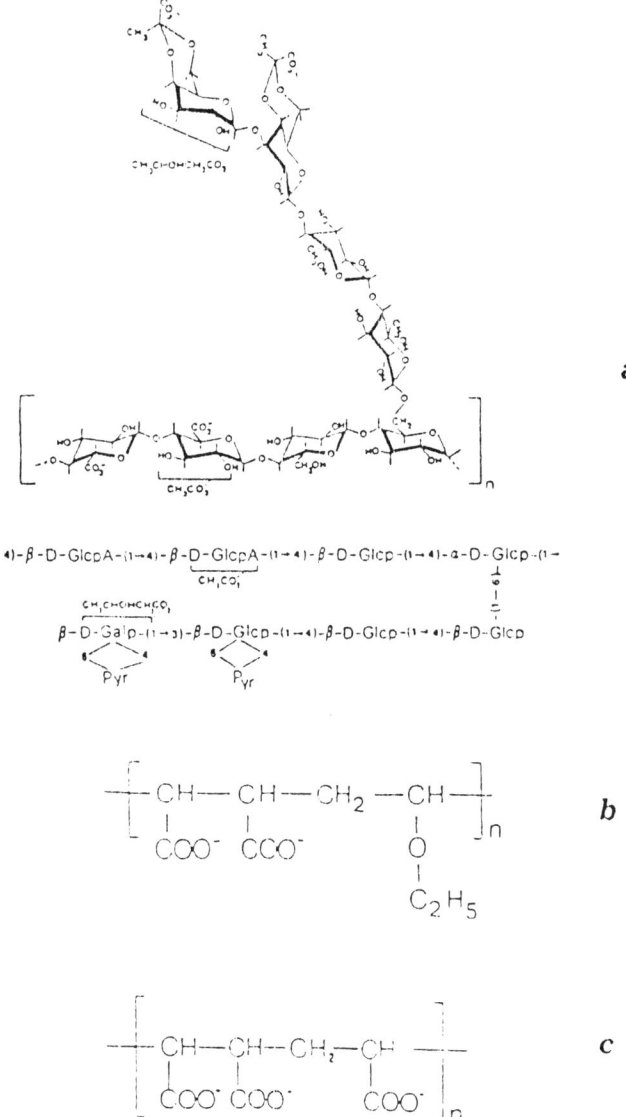

FIGURE 1. Schematic Drawings of the Repeating Units of: (a) the Exocellular Polysaccharide Form <u>Rhizobium Trifolii</u> Strain TA-1; (b) the Maleic Acid-Ethyl Vinyl Ether Copolymer (MAEVE): the Maleic Acid-Acrylic Acid Copolymer (MAAA).

TABLE 1.   Recovery of Metals by Membrane Technology: Examples of Industrial Applications.

Original Solutions	Metals	Membrane process
Iron and Steel	Fe(II), Fe(III)	Reverse osmosis, ultrafiltration
Galvanic industries	Mn(II), Cu(II) Cd(II), Pb(II) Cr(III), Cr(VI)	Reverse osmosis, electrodialysis, Donnan dialysis
Nuclear Wastes	Uranium Strontium (90) Cesium (137)	Donnan Dialysis, electrodialysis
Textile, tannery sludges	Zn, Cr(III)	Reverse osmosis, Donnan
Miscellaneous (photographic, minerary wastes)	Cu, Ni, Co, Cd, Ag, Au, An	Liquid membranes, Donnan dialysis electrodialysis

The results indicate that the fraction of Cr(III) ions bound is in all cases higher than 80%, reaching ca. 98% in the case of the MAEVE copolymer. The latter is therefore a very good sequenstrant which, in addition, offers the advantage of giving rise to phase separation at higher Cr(III) concentrations than the other two polyelectrolytes studied.

Similar results have been collected studying the interaction of Cu(II) ions with MAEVE, MAAA, and with the TA-1 polysaccharide, respectively, by means of a Cu(II) selective electrode. In this case, equilibrium dialysis experiments have also been performed in 50 MN $NaClO_4$ at 25°C: the results are summarized in Figure 2.

From these data it appears that the affinity for Cu(II) ions follows the order MAAA > MAEVE >> TA-1. Precipitation threshold Cu(II) concentrations follow just the same order.

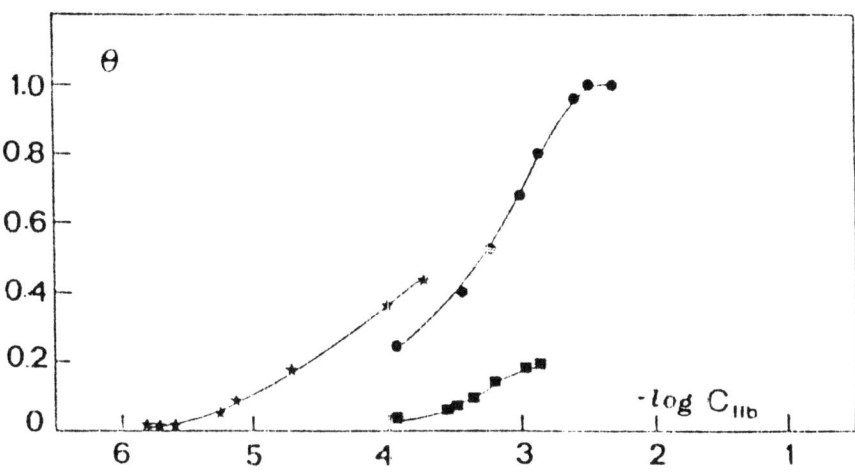

FIGURE 2. *Cu(II) Ions Binding Isotherms (25°C) in MAAA, MAEVE, and TA-1 Solutions (Solvent: 59 mM $NaClO_4$; Polymer Concentration: 2.4 meq/L.* Θ *Is the Equivalent Concentration Ratio (Bound Cu(II)/Polymer) as determined from Equilibrium Dialysis Experiments, and Cf Is the Concentration of Free Cu(II) Ions. The arrow indicates conditions for which precipitation takes place in the Cu(II)-MAAA system.*

2.2 UV Spectral Data

As already observed in the case of different maleic acid copolymers, interaction in dilute aqueous solution of Cu(II) ions with polycarboxylates considered in this work leads to typical spectral perturbations in the UV region, in particular: (a) a change of the spectrum in the 195-200 nm range, i.e. of the internal ligand transition band characteristic of bound (di) carboxylic groups; (b) the development of a strong electron transfer (ET) band between 245 and 265 nm due to electron transition between the central metal ion and the electronic system centered on the ligands.

In the 245-260 nm region the phenomenon is particularly evident and entails both a red-shift and a strengthening of the ET band upon increasing the stoichiometric ratio between added Cu(II) ions and polyelectrolyte in solution. This is clearly demonstrated by the results shown in Figs. 3-4 relative to MAEVE, MAAA, and to TA-1.

Optical density data, calculated in each case at the appropriate ET maximum wavelength, are reported in Fig. 5 as a function of the fraction of bound Cu(II) ions derived by means of equilibrium dialysis measurements.

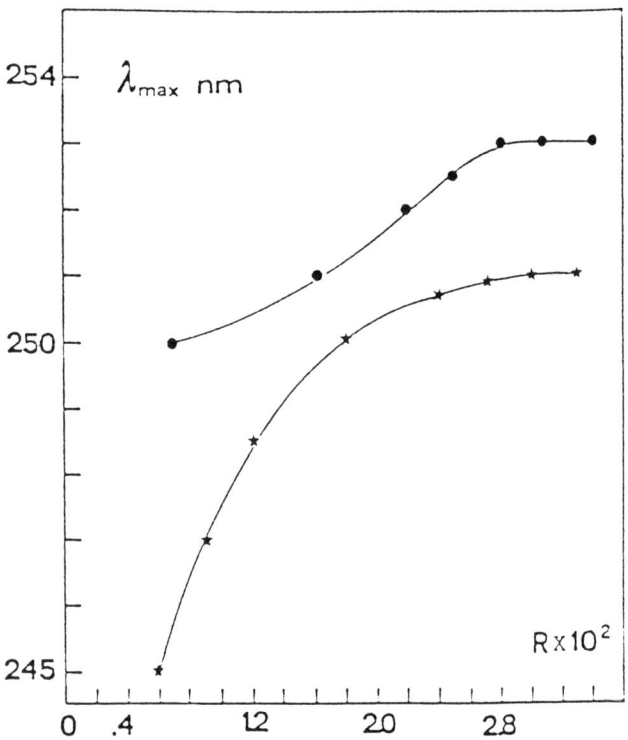

FIGURE 3. Wavelength of Maximum Absorption of the Electron Transfer Band Recorded for Solutions Containing Cu(II) Ions and MAEVE or MAAA. Polymer Concentration; 10 meq/L R Is the Stoichiometric Ratio Between Added $Cu(ClO_4)_2$ and Polymer Concentrations.

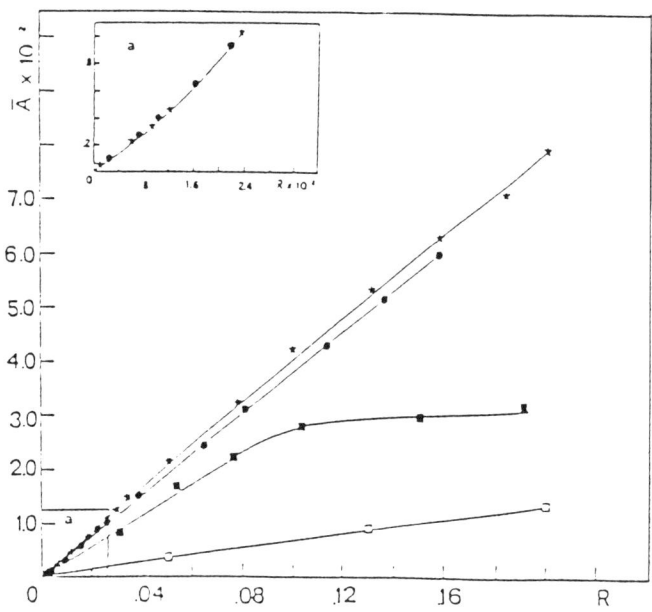

FIGURE 4. *Dependence on R (See Fig. 3) of Corrected Absorbance A for MAA, MAEVE, and TA-1 in 50 mM $NaClO_4$ in the Charge Transfer Band Region. Polymer Concentration: 7 meq/L (in the insert: 10 meq/L.).*

Mode and extent of Cu(II) ions interaction with the three polyelectrolyte studied thus result different: in particular, it is possible to conclude, in line with information derived from potentiometric and dialysis experiments, that the strongest interaction -- with an efficient penetration of carboxylate ligands in the inner coordination sphere of the counterions -- is achieved in the case of the MAAA copolymer.

2.3 Calorimetric Data

The results of the isothermal (25°C) calorimetry experiments are reported in Figure 6 in the form of a plot of the enthalpy of polyelectrolyte-$Cu(ClO_4)_2$ mix data as a function of the stoichiometric concentration ratio R.

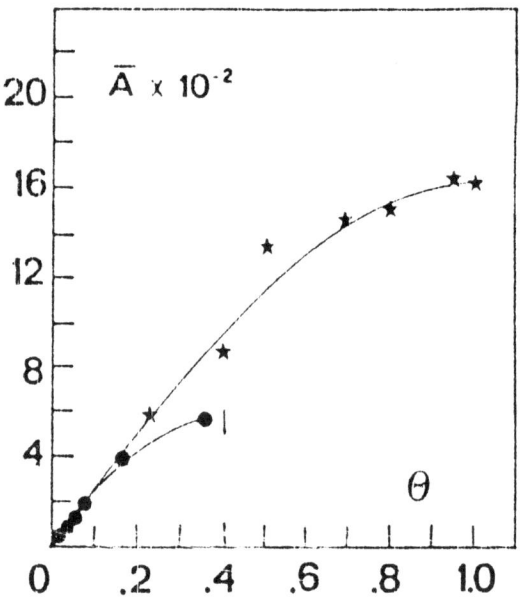

FIGURE 5. Dependence on Θ (See Fig. 2) of Corrected Absorbance A, Normalized by the Concentration of Bound Cu(II) ions, for Solutions of MAEVE and MAAA in 50 mM $NaClO_4$. Polymer Concentration: 2.4 meq/L.

The slopes of the resulting straight lines lead to estimate for the enthalpy of interaction -- expressed in kJ per equivalent of Cu(II) bound -- the following values: 6.2 (MAAA); 2.4(MAEVE). Clearly Cu(II) ions binding onto the polycarboxylate chains in dilute aqueous solution must be a process driven by the entropy, as already experienced with a number of different polyelectrolytes. In such a process, in fact, a relatively large number of water molecules gain degrees of freedom once disengaged by the hydration sheaths of closely interacting species. This phenomenon would be particularly important in the case of the MAAA copolymer whose chains exhibit a remarkably high fixed charges density.

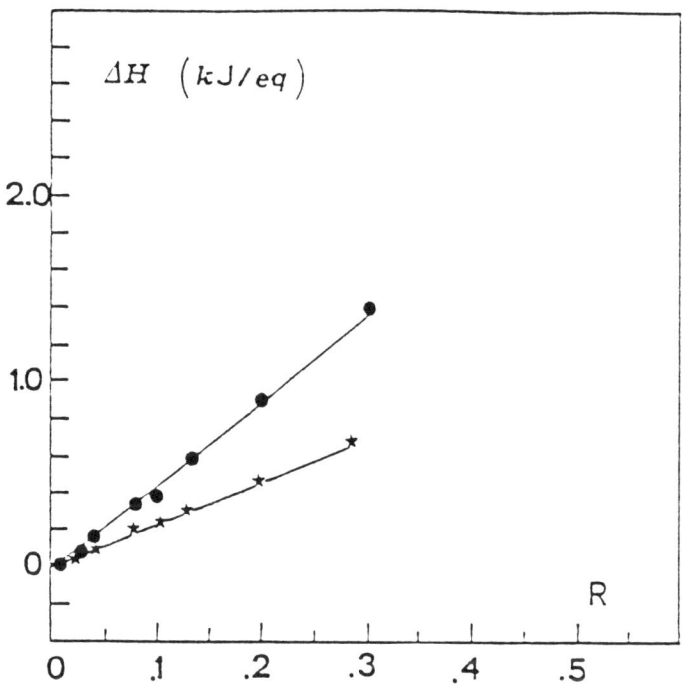

FIGURE 6. *Heat of Mixing (25°C) Aqueous Solutions of $Cu(ClO_4)_2$ and of MAAA, or of MAEVE as a function of the Stoichiometric Concentration Ratio R (See Fig. 3).*

2.4 Membrane Filtration Data

Membranes of polyvinyl alcohol with entrapped either TA-1 or MAEVE polyelectrolytes were tested for their ability to absorb Cu(II) ions from 1 mM $CuCl_2$ Solutions (see Experimental). The results show that TA-1 containing films can almost quantitatively remove Cu(II) ions for low elution volumes while films with MAEVE do exhibit a lower affinity but a distinctly higher integral capacity (Table 2) (Figure 7a-7b).

(a)

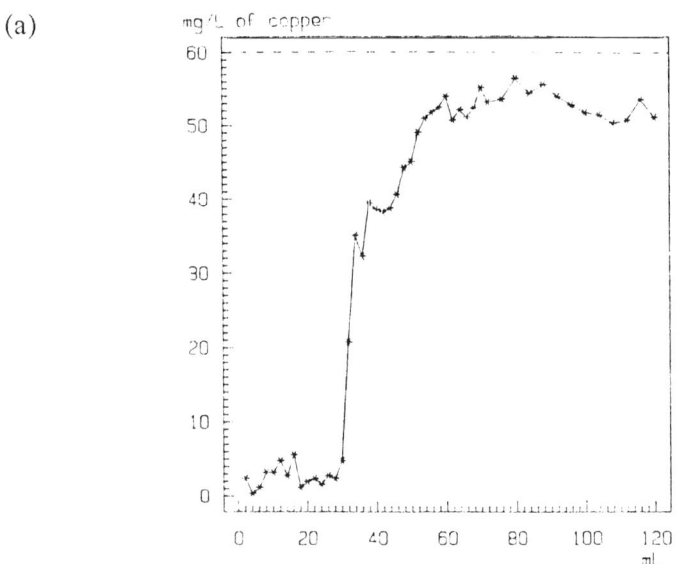

(b)

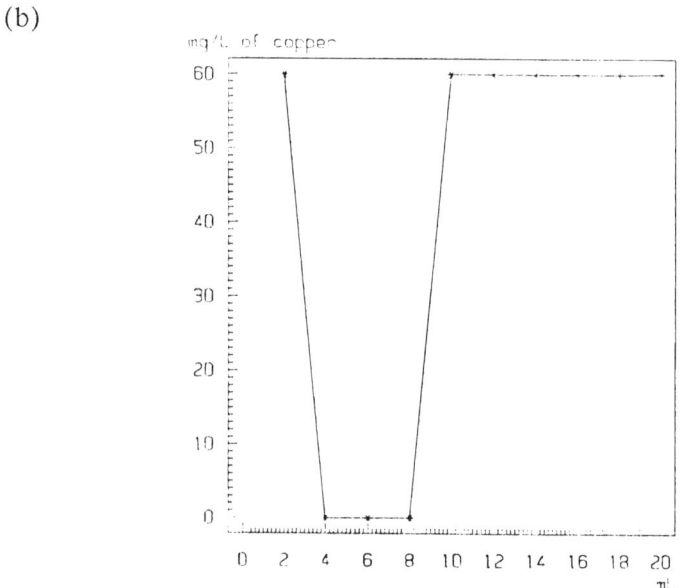

FIGURE 7. *Penetration Curves of Copper Ions (Initial Concentration 60 mg/L) on Polysaccharide membrane Columns.*
(a) MAEVE
(b) TA-1-Na

3. CONCLUDING REMARKS

The extent of Cr(III) and Cu(II) ions complexation by the three polyelectrolytes considered is very high, particularly so in the case of the MAAA copolymer. This is consistently shown by potentiometric, equilibrium dialysis, spectroscopic and calorimetric data pointing out that the main driving force is entopic in nature in all cases.

TABLE 2.   Behavior of Filtering Membranes of Polysaccharides to Copper Solutions.

Passed Volume (mL)	TA-1-Na (mg/L of copper)	MAEVE (mg/L of copper)
2	60	5.7
4	0	5.7
6	0	4.3
8	0	4.3
10	60	6.2
12	60	7.1
14	60	7.5
16	60	8.3
18	60	10.2
20	60	12.6
22	60	13.5
24	60	14.2
26	60	16.1
28	60	17.6
30	60	19.3

The MAAA copolymer, however, is sensitive to relatively high added trivalent or divalent salt concentrations yielding phase separation more easily than the MAEVE copolymer or the TA-1 polysaccharide. MAAA solutions might therefore be used for the removal of Cr(III) and/or Cu(II) ions from aqueous systems if it is desired to collect the resulting polymeric complexes in the form of a precipitate.

The MAEVE copolymer, on the other hand, might be useful in processes in which complexed metals ions have to remain in solution as well as in process based on ion-exchange, chelating membranes.

The polysaccharide studied (TA-1) does not seem to offer any advantageous feature as far as its possible uses in metal ions removal is considered. In the context, other ionic bacterial polysaccharides will have quite naturally to be studied: probably, however, whole bacterial cell cultures which in addition to the exocellular species exhibit other metal ions binding cellular polymers may provide better and more economic systems.

4. EXPERIMENTAL

4.1 Samples

The exocellular polysaccharide from Rhizobium trifolii strain TA-1, hereafter indicated as TA-1, was a sample kindly provided by Prof. L.P.T.M. Zevenhuizen of the University of Wageningen, The Netherlands. It has been purified by dialysis in aqueous concentrated NaCl (to remove possible traces of contaminating heavy metals), followed by extensive dialysis against distilled water, and finally freeze-dried. Elemental analysis, constituent sugar analysis, and potentiometric titration data are in complete agreement with the repeating unit structure depicted in Figure 1. The average molecular weight of our TA-1 sample is 1.3 x 10 (6) Dalton (light scattering data).

The copolymers of maleic acid with ethyl vinyl ether (MAEVE) and with acrylic acid (MAAA) have been synthesized in the laboratory of prof. M. Aglietto of the University of Pisa, Italy. Samples have been purified following the procedure outlined above for TA-1. On the basis of potentiometric titration data the equivalent weights of the two copolymers result in good agreement with the theoretical values for the repeating unit structures depicted in Figure 1.

The intrinsic viscosities of the copolymers in 50 mM $NaClO_4$ and at 25°C are: 240 (MAEVE), and 140 (MAAA) mL/g. All other reagents used were analytical grade.

4.2 Instrumentation

UV spectra were recorded using a Varian Cary 219 spectrophotometer. Absorption spectra in the UV region were recorded at 25°C for each polycarboxylate solution in the presence of increasing amounts of $Cu(ClO_4)_2$ at a constant ionic strength (0.05 M $NaClO_4$). Aliquots of $Cu(ClO_4)_2$ solutions were added directly into the polymer solution in the quartz cuvettes under stirring, and the readings collected at equilibrium.

Calorimetric measurements have been made using an isothermal LKB Model 107000 batch instrument following a procedure described in details elsewhere.

Potentiometric data were collected using an Orion Model potentiometric apparatus equipped with Cr(III) and Cu(II) electrodes: these were used in conjunction with reference calomel electrode.

Viscosity measurements were performed using a Schotte-Geraete automatic viscometer equipped with a water thermostat. Films containing either TA-1 or MAEVE were prepared as follows: 70 mg of polyvinyl alcohol and 100 mg of polysaccharide are dissolved in 2 mL of NaOH and 1 mL of ethanol; the solution is heated for few minutes so that full solubilization is favored; then the solution is boiled for 30 s to eliminate the alcohol excess. Then the solution is stratified in a Petri plate, letting the solvent to be evaporated for about 36 hours and the obtained membrane is washed with water and stored; it is used as a stationary phase in column (Figure 8).

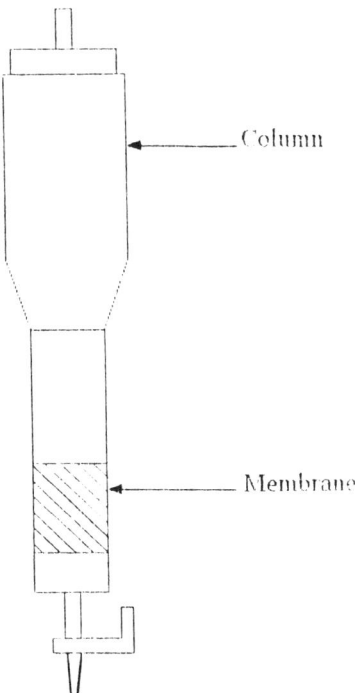

FIGURE 8. *Membrane Operating in a Column Bed.*

ACKNOWLEDGMENTS

This work has been carried out with financial support of ENEA, Department TIB, Rome.

REFERENCES

1. P.A. Sandford, I.W. Cottrel and D.J. Petit, "Pure & Appl. Chem.", 56, 879 (1984).

2. L. Kenne, and B. Lindberg in *The Polysaccharide*, G.O. Aspinali, Ed. Vol. 2, (Academic New York, 1983), p. 287.

3. V. Crescenzi, M. Dentini and I.C.M. Des., *Carbohydr. Res.*, 160, 283 (1987).

4. V. Crescenzi, M. Dentini and T. Coviello in *Industrial Polysaccharides*, S.S. Stivala, V. Crescenzi, I.C.M. Des., Eds., (Gordon & Breach, New York, 1987), p. 68.

5. C. Villiers, C. Braud, M. Vert, E. Chieleni, and M. Marchetti, *Eur. Polym. J.*, 14, 211 (1987).

6. S. Paoletti and F. Delben, *Eur. Polym. J.*, 11, 561 (1975).

7. S, Paoletti, A. Cesaro, A. Ciana, F. Delben, G. Manzini and V. Crescenzi, in *Solution Properties of Polysaccharides*, D.A. Brant, ED. *A.C.S. Symposium Series No. 150*, American Chemical Society, Washington, (1981), p. 379.

8. V. Crescenzi, F. Delben, S. Paoletti and J. Skerjanc, *J. Phys. Chem.*, 78, 607 (1974).

9. M. Dentini, T. Coviello, W. Burchard and V. Crescenzi, *Macromolecules*, 21, 3312, (1988).

DISCUSSION OF:
POLYELECTROLYTIC METAL IONS SEQUESTRANTS

Arup K. Sengupta
Lehigh University Environmental Engineering, Bethlehem,
Pennsylvania 18015

Basic Premises of the Paper

Polyelectrolytes with coordinating functionalities have the ability to sequester toxic metal cations in the aqueous phases through Lewis acid-base type interactions. The paper discusses the following three polyelectrolytes in relation to their ability to complex and sequester Cu(II) and Cr(III) ions:
  i) 1:1 maleic acid-ethyl vinyl ether (MAEVE) copolymer;
  ii) 1:1 maleic acid-acrylic acid (MAAA) copolymer; and
  iii) extracellular polysaccharide extracted from cultures of Rhizobium Trifolii strain TA-1.
Metal-sequestering ability of all the above-mentioned macromolecules/polyelectrolytes stem from their carboxylate functionalities which act as ion-exchange as well as coordination sites. Figure 1 shows a general structure of a Cu(II)-polymeric carboxylate complex which is representative of all the three macromolecules considered in this study. One of the key advantages of using extracellular polyelectrolyte is the fact that they are readily biodegradable.

Experiments were conducted at pH 5.0 to study the relative sequestration abilities of these three macromolecules for Cu(II) and Cr(III), and the salient observations can be summarized as follows:

a) Complexation of Cr(III) and Cu(II) ions by the three polyelectrolytes considered is fairly strong and the affinity sequence is MAAA > MAEVE >> TA-1. Precipitation threshold for Cu(II) follows the same order.

b) The colorimetric data suggest that the main driving force for the uptake of metals ions is primarily entropic in nature.

c) The TA-1 polysaccharide (biodegradable) studied offers less affinity toward metal ions compared to other synthetic macromolecules (MAAA and MAEVE). In view of the abundance of bacterial polysaccharides and their biodegradability, more work should be undertaken in this area.

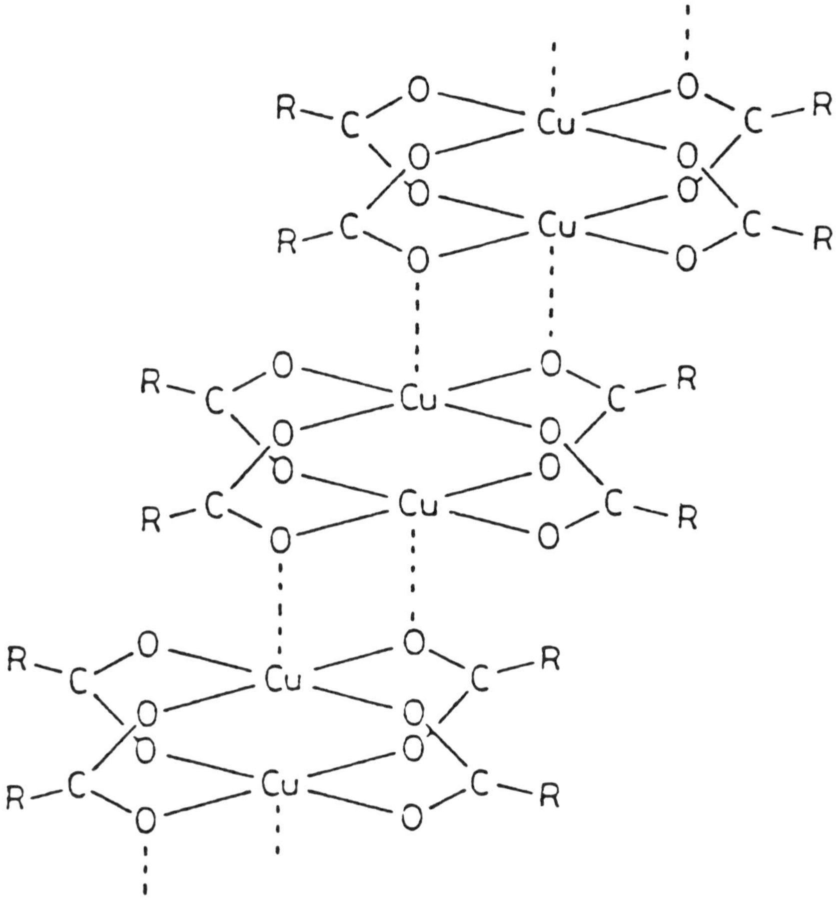

FIGURE 1. Schematic Structure of Polymeric Copper (II) Carboxylates.

General Comments and Related Problems

In general, all weak acids with carboxylate functionalities show fairly high affinities towards commonly encountered toxic metal cations, such as, Cu(II), Ni(II), Cd(II), Pb(II) etc. Using bacteria-derived macromolecules with -COOH functionalities may be an economic way to sequester particularly trace levels toxic metals present in the aqueous phase. However, the following considerations are important for such an application:

(a) The experimental data were collected at pH 5.0 where significant metal sequestering capacity was observed for all three polyelectrolytes. However, our work with polymeric ion-exchange resins with carboxylate functionalities indicate that such metal uptake capacities are strongly dependent on the pH and concentration of other ligands present in the aqueous phase. At more acidic pH and in the presence of inorganic and organic ligands, the metal uptake capacities of these polyelectrolytes may be reduced significantly.

(b) Rate of metal ion uptake by these polyelectrolytes is important and likely to be fast enough for labile complexes. However, for metal ions like Cr(III) and Co(III), which are known to form inert complexes, such an uptake may be extremely slow. This should be apparent from the comparison of the second-order rate constants for water exchange in the primary solvation shell of metal ions as shown below [1]:

Metal Ion	Rate Constant (mol^{-1} sec^{-1})
Cu^{2+}	9×10^8
Ni^{2+}	2.9×10^3
Zn^{2+}	5×10^6
Cd^{2+}	9×10^7
Cr^{3+}	3.6×10^{-5}
Co^{3+}	$< 10^2$

(c) Biodegradability is considered to be a favorable property for the bacteria-derived polyelectrolytes over other synthetic ones. However, sequestering toxic metal ions will affect such biodegradability and that should be looked into.

REFERENCES

1. Francois M.M. Morel, *Principles of Aquatic Chemistry*, (Wiley-Interscience, New York, 1983).

CHEMICAL SPECIATION OF HEAVY METALS IN SOILS FOLLOWING LAND APPLICATION OF CONDITIONED BIOLOGICAL SLUDGES AND RAW PIG MANURE

M. Baldi
Department of Hydrology and Environmental Engineering, University of Pavia, Italy

M.C. Negri
ECODECO-FERTILVITA Giussago, Italy

A.G. Capodaglio
Department of Civil Engineering, Marquette University, Milwaukee, Wisconsin

1. INTRODUCTION

The utilization of biological sludges from civil and industrial wastewater treatment processes as organic soil conditioners is nowadays a common disposal practice in several countries, as traditional sludge management practices are replaced by resource recovery and conservation alternatives.

Fertirrigation is also rapidly becoming a widely accepted and adopted method of disposal for untreated organic wastewaters, among the others, raw pig manure is often disposed of in this way, as it contains high quantities of macronutrients, and in particular nitrogen, necessary for crops.

For both disposal methods it is usually required that these wastes be tested for phytotoxicity, and metals and pathogens content prior to clearing for this form of disposal, and that criteria set by a regulatory agency be followed, to avoid negative effects on the crops and public health.

Both disposal procedures have several advantages for both the generator of the waste and the final user (i.e. the farmer on whose land the waste is disposed of), among which are relatively low operation costs, as compared to other "treatment" methods, and the recycling of organic matter, which is especially beneficial and demanded in alluvial soils subject to an intensive agriculture, as in the case of the Italian Po River valley, where this study took place. On the other hand, these procedures have a few drawbacks, one of

378 SORPTION ONTO SURFACES

them being the fact that they contain, and may transfer into the soil a certain amount of heavy metals.

Actually, the uncertainty about the fate of heavy metals in an agricultural soil matrix, and its effect on crops and the food chain, currently represents the main obstacle to a larger diffusion of these disposal practices. There is, in fact, a potential niche of unexploited market for them, that can be fully developed in the presence of a carefully planned distribution program, which should also take into account the results of agrichemical research.

The purpose of this study is to investigate the fate of some heavy metals introduced into the soil matrix by the addition of sludges and raw manure. The results show that the short-term consequences of these additions are virtually negligible. These results will be subsequently verified in the long period (5-10 years).

2. PRELIMINARY INVESTIGATION

At the beginning of this study, a series of analyses was conducted over a large number of sludge samples (about 100) collected at the FERTILVITA facilities, a subsidiary of ECODECO, in Corteolona near Pavia. These sludges, due for final disposal on agricultural soil, had undergone previous homogenization and treatment with ammonia solutions to increase their nitrogen content, and constitute the fertilizer produced and marketed by the company.

The analyses were aimed at the determination of the average heavy metals content of the sludges: the results obtained for zinc, nickel and copper will be presented herein.

Heavy metal concentrations in the "final product" are fairly constant over time, with an average content of 1650 mg/kg for zinc, 80.5 mg/kg for nickel and 551 mg/kg for copper respectively. All concentrations are referred to the dry weight of the sludge.

All soils for which these disposal practices are contemplated, are sampled and periodically monitored, as mandated by the disposal permit temporarily granted by the governing authority, Regione Lombardia. Among the parameters that are determined in laboratory tests are: heavy metals content, cationic exchange capacity (CSC) and pH. On these results will ultimately depend the acceptance of the disposal practice and its modalities.

Five hundred samples of soil were analyzed to determine these parameters. The average heavy metals concentrations, CSC and pH in the soils under investigation are reported in Table 1. All samples have been collected according to the regulations from the Regione Lombardia, which require the representative sample to be obtained by a mixture from 25 samples collected over 5 hectares.

TABLE 1. Physical-Chemical Characteristics of Soils Samples.

Parameter	Concentration or value
Zn	61 mg/kg dry matter
Ni	26" "
Cu	18" "
CSC	13.5 mE/100 g
pH	6.36

3. HEAVY METALS IN AGRICULTURAL SOIL

Considering that, on the average, the soil quantity in the upper layer influenced by agricultural practices is about 4×10^6 kg/ha, and making the reasonable assumption that any quantity of heavy metals introduced into the soil mass will be quantitatively held within this superficial layer (root zone), even in the case of disposal according to the maximum rate and sludge metal concentrations allowed by the current regulations of Regione Lombardia, a total increase in heavy metals content in the soil ranging from less than 2 to almost 8% could be expected at most, as indicated in Table 2.

Furthermore, considering also the large variability that heavy metals concentrations usually show even between contiguous sampling sites, it may be expected that any significative variation in their concentrations, positively identifiable with these resource conservation practices, would, at most, be detected after 4-5 years of continuous, repeated applications.

TABLE 2. Expected Heavy Metal Concentrations Increase in Agricultural Soil Following Application of Sludges Respectively with Maximum Allowed and Average Heavy Metals Content.

Metals	Max. allowed content	Average sludge content
Zn	+ 7.0 %	+ 3.8 %
Ni	+ 1.6 %	+ 0.5 %
Cu	+ 7.9 %	+ 4.3 %

However, the total heavy metals content of a soil, although of primary importance for both disposal practices and crop compatibility, is only one of the facets of the problem.

From an environmental and agronomic perspective, in fact, the chemical species in which metal ions are present is also important, since they might be differently bioavailable to the crops depending on the specific form or species in which they exist. Since both sludges and pig manure are extremely rich in organic matter content, it may be reasonably assumed that the metals contained therein are more readily bioavailable than those contained, for example, in mainly inorganic wastes.

Disposal procedures for these ultimate "product types" by direct recycle on agricultural land should be considered valid alternatives to other applicable treatments until their effects are negligible not only considering the total heavy metals increase in the soil, that should indeed be limited, but also while consistent quantitative variations of the most available species are also insignificant. In other words, these disposal practices are valid and applicable until the soil retains its capacity to inhibit metals mobility.

The objective of the present study is to determine the bioavailable fraction of some heavy metals in soils conditioned with sludge residuals or subject to fertirrigation, as compared with untreated soils of the same characteristics.

For this purpose, a separation of the metals in the following fractions, which are the more readily available for crops, has been attempted:

Water Soluble: With no doubt this is the most readily bioavailable fraction. Usually, only extremely low quantities of heavy metals can be extracted with H_2O, so that they can be quantitatively determined only when the most sophisticated analytical techniques are used. Heavy metals bound to the humic and fulvic fraction with the strongest acid functions, and therefore with higher ionization capacity, shift in solution in water. This can be demonstrated by acidifying the aqueous solution with HNO_3 (2%) that will cause an organic colloid to precipitate.

Exchangeable: This fraction is constituted by those metals loosely bound to the active sites, which behave as cation exchange resins. The active sites are saturated by using concentrated saline solutions, such as 1 M $MgCl_2$ [1], $BaCl_2$ [2], NH_4OAc [3, 4] and KNO_3 [5, 6]. It should be remembered that the possibility to form complexes with $CH3COO^-$ or Cl^- allows the transfer in solution of metals that are not necessarily bound with these active sites [7, 8, 4].

Adsorbed: This fraction consists of those ions that are more strongly bound to active sites and that are extractable by generation of soluble floururate complexes [9].

Bound to the Organic Fraction: The metals that are bound to the organic fraction are those that form complexes mainly with the nonsoluble fraction of humic acids in the above mentioned solutions. For an analytical quantification of this fraction, two procedures can be employed: by oxidation with H_2O and HNO_3 [2], or by extraction with the chelating solutions DTPA [10] or $Na_4P_2O_7$ [11] or NaOH [12]. In the first case, sulphures are determined at the same time. The latter procedure is used to extract the weakest humic acids, that would not otherwise go in solution. In fact, extraction with 0.1 M DTPA solution also allows carbonate extraction.

4. MATERIALS AND METHODS

A parcel of agricultural land, with an area of about 1600 m^2, was selected for this experiment. The soil type is a sandy loam typical of an Italian alluvial plain.

The parcel was than divided into six equal, well separated subportions, as illustrated in Figure 1. Four of these were subject to "treatment" with sludges and raw pig manure (A,E and B,D respectively), while the remaining two (C,F) were left at their natural condition, as control. This apportionment was based on a random choice.

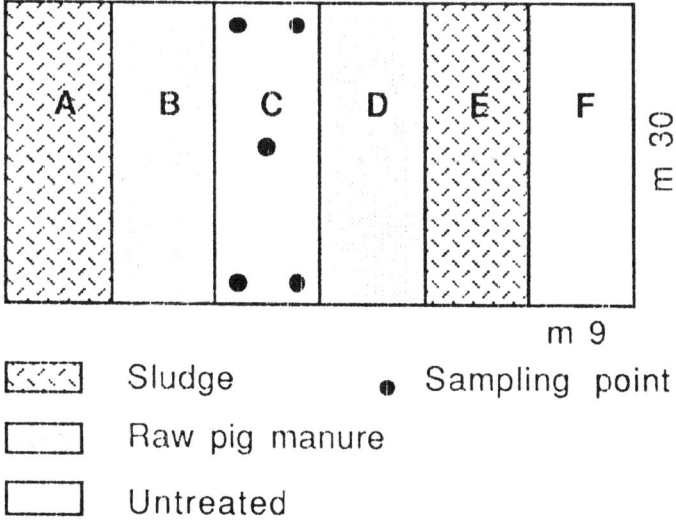

FIGURE 1. *Experimental Site.*

Those parcels subject to conditioning or fertirrigation received an amount of waste sufficient to provide the necessary dosages of macronutrients (especially N) for corn crops. In particular, parcels A and E, subject to sludge conditioning, had 3 tons of dry matter/ha added, while the addition of pig manure to parcels B and D reached 4.8 tons/ha.

On the experimental site corn was grown according to standard agricultural practices, including chopping and dispersion on the field of the nonedible portions of the plants. Five soil samples were then collected for each of the six subparcels, as shown in Figure 1 for subparcel C, and thoroughly mixed to obtain a representative, average sample for each subparcel.

The samples, thus homogenized, were passed through a 20 mesh sieve, and subsequently analyzed to determine the content of heavy metals of concern and, in addition, pH and CSC.

The analytical procedure for quantitative determination of the metals was the following: First, the soil matrix was mineralized with $HNO_3+HCl+HF$ in teflon beakers, then the solution obtained was analyzed with an atomic adsorption spectrophotometer Perkin-Elmer 5000. Aliquots of samples were then subject to chemical speciation procedures, until all bioavailable species had been determined.

To analytically determine bioavailability, several different chemical speciation procedures can be employed: methods based both on subsequent extractions with the same reactive and sequential extractions with different reactives have been developed and are reported in literature [13, 14, 15].

The latter seem to be the most promising procedures, although it should be noted that there is not, yet, a well defined protocol for the choice of a reactive sequence rather than another. This is due to the fact that any available reactive will, in practice, interfere to a certain degree with all the other chemical species, while bringing into solution the one for which it is being specifically employed. Minimization of this interference should be achieved on a case by case basis.

Three different speciation schemes have been adopted here, which are similar to those proposed by Latterell et al. [16], Stover et al. [9] and Legret et al. [4] respectively. The original procedures have been integrated and modified, where possible, as to allow for all of them the same separation procedure. The schemes adopted are described in Table 3.

In each extraction, a solid/liquid ratio of 1:20 was maintained. An aliquot of 2.50 g of soil was introduced in a teflon container to which 50 cc of bidistilled water was also added. The containers were then sealed and put in a rotative shaker for 72 hours, after

TABLE 3. *Summary of Extraction Procedures Adopted.*

Fractions	Procedure		
Soluble	H_2O	H_2O	H_2O
Exchangeable	KNO_3	KNO_3	$BaCl_2$
Adsorbed		KF	
Organically Bound	DTPA	$Na_4P_2O_7$	$H_2O_2+HNO_3+AcONH_4$
Available Carbonate	EDTA	EDTA	$AcOH+AcON_a$
Sulfide	HNO_3	HNO_3	$H_2O_2+HNO_3+AcONH_4$
Oxide Bound		$NH2OH.HCl+AcOH+AcONa$	
Residual		$HNO_3+HCl+HF$	

which period the solid phase was separated from the aqueous extract by means of centrifugation at 6000 rpm. The aqueous solution, collected in glass probes, were analyzed with atomic adsorption spectrophotometry with electrotermic atomization. The results thus obtained are reported in Figure 2, where they have been arranged to evidentially support the reproducibility of the analytical results, for the same sample size.

The solid phases, separated by centrifugation, were reintroduced in teflon containers and treated with 50 cc of 1 M KNO_3 solution for the first and second series, and 1 M $BaCl_2$ solution for the third one.

The mixtures were, again, kept in a shaker for 72 hours and then separated by centrifugation. The analytical results relative to this fraction, relative to the KNO_3 and $BaCl_2$ extractions, are presented in Figure 3 and 4.

All three solid phases were then treated with 50 cc aliquots of a 0.5 M KF solution that was brought to a 6.5 pH by addition of fluoridric acid. The results of this latter step are shown in Figure 5.

The solid phases coming from the third extraction were treated with a 0.1 M DTPA (diethilenetriaminepenthacetic acid) solution, a 0.1 M $Na_4P_2O_7$ solution and a 1 M CH_3COONa solution for the first, second and third series respectively. The third solution was buffered to pH = 5 by addition of acetic acid.

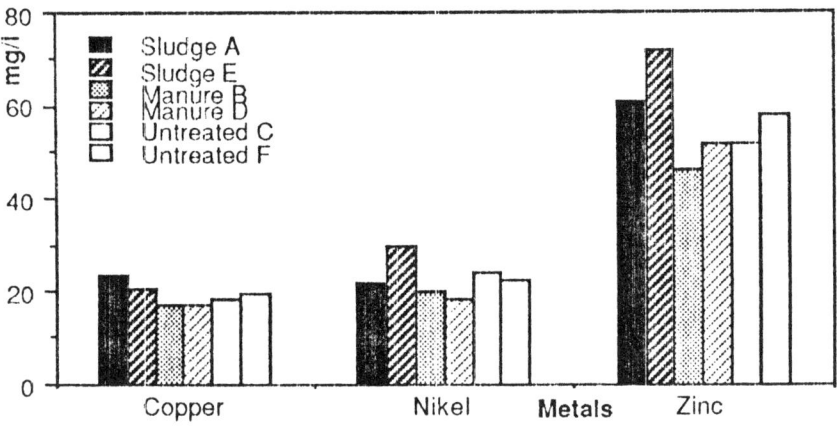

FIGURE 2.   Soluble Fractions Extracted with Water.

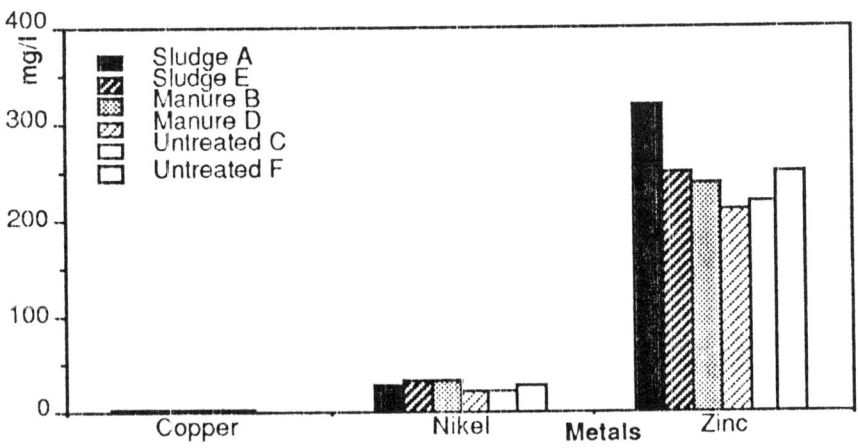

FIGURE 3.   Exchangeable Fraction Desorbed with KNO_3 Solution.

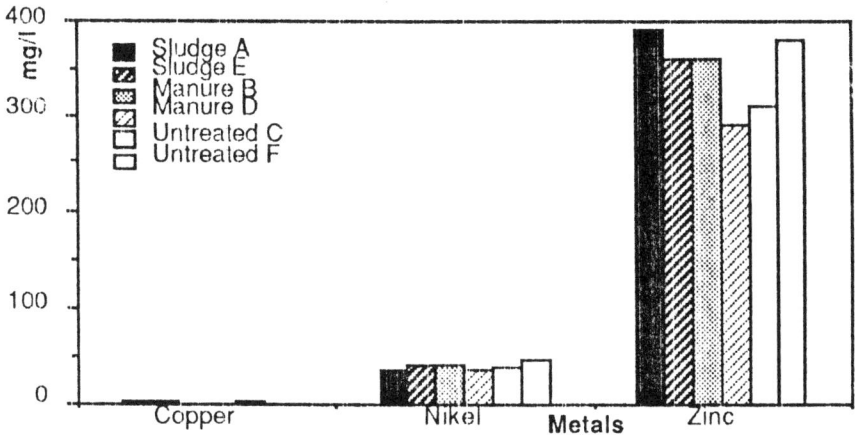

FIGURE 4. Exchangeable Fraction Desorbed with $BaCl_2$ *Solution*.

The DTPA solution partially extracts carbonates, unlike the $Na_4P_2O_7$ solution that extracts almost exclusively those metals bound to the organic matter. The AcONa/AcOH solution at pH = 5, instead, dissolves carbonates. This extraction in the third sequence with H_2O_2 and 1 M HNO_3 as oxidants will solubilize the sulphures and the metals bound to the organic phase and would simultaneously dissolve carbonates, given the low pH values.

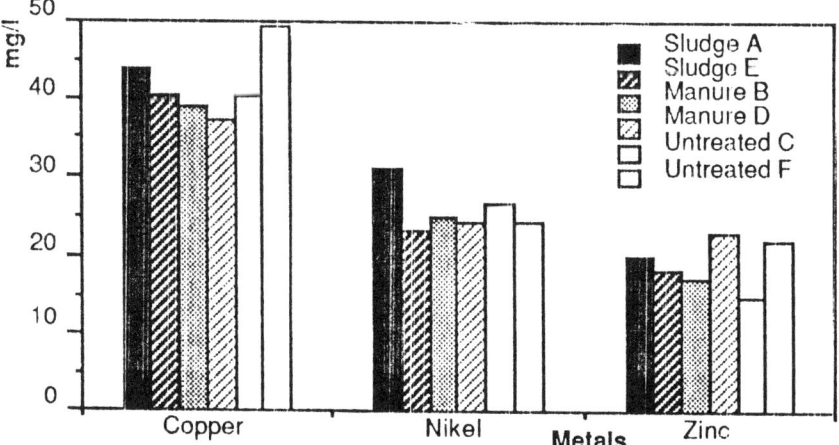

FIGURE 5.   Adsorbed Fraction Soluble in KF Solution at pH 6.5.

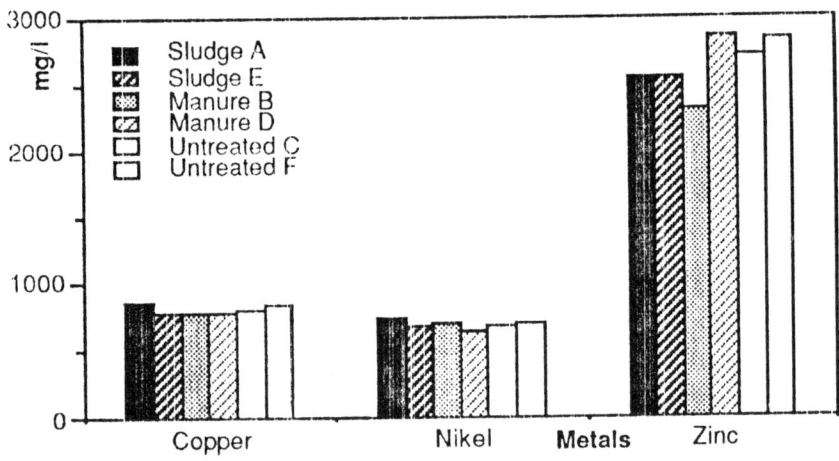

FIGURE 6. Organically bound and Carbonate-Available Fractions Soluble in DTPA.

The experimental results of the forth extraction with DTPA are represented graphically in Figure 6.

All the compounds employed in the preparation of the necessary solution for these experiments were high purity "Suprapur Merck" products. The high degree of purity was necessary to avoid gross contamination of the soil matrix by the highly concentrated solution employed by the procedures .

5. RESULTS AND DISCUSSION

The results from the first extraction step with bidistilled water, which is common to all separation sequences, reported in Table 4, show a certain uniformity of the average concentration data obtained for the three metals. Copper has been detected, in fact, with an average concentration of about 20 ppm, nickel content is about 30 ppm, and zinc, although more mobile, reaches the highest value at 60 ppm.

TABLE 4. *Results and Subsequent Averages from the First Extraction Step.*

Parcel	Zn			Cu			Ni		
Sludge A	62±14			22±4			24±8		
		62±10			22±4			26±]	
Sludge E	62±9			19±2			28±6		
Manure B	53±7			19±2			24±6		
		51±6	60±14		19±2	20±3		23±4	27±7
Manure D	50±7			18±2			22±3		
Untreated C	54±8			19±3			26±10		
		68±20			21±4			30±9	
Untreated F	81±20			22±3			33±10		

The results show that the soil metal content can be considered statistically unchanged both within the single subparticle and each treatment subset, as proven by an analysis of the data variance. Table 4 shows the subsequent means and variances of the results obtained for the three metal concentrations, for each land parcel, determined over the total sample number. Furthermore, analysis of these data shows that all treatment schemes (including the blank) are statistically equivalent as far as elution of those three metals is considered, and seem to indicate that nothing particularly significant occurred as a consequence of either of the soil treatment procedures considered.

The second extraction, with 1 M KNO_3 solution (first series) shows that there are considerable amounts of zinc available for cationic exchange, as the concentrations measured in the solutions analyzed range from 200 to 250 mg/l. Nickel is, by far, less available, and quantities in the solution are the same as measured during the first extraction with water. Finally, copper available for cationic exchange is present in extremely low quantities, or about 1/10 than those previously extracted.

Similar results have been obtained by extraction with a 1 M $BaCl_2$ solution (second series). In this case, copper concentrations remained unchanged in the range 1-3 ug/l, while nickel and zinc concentrations have been detected at levels which double those measured with KNO_3. This could be due to formation of stable chlorine complexes, that could allow ion extraction independently from the cationic exchange mechanism.

Nickel and copper levels were, however, brought to comparable levels by the subsequent extraction step, common to all series, with a 0.5 M KF solution, conducted at pH=6.5. Here, ions adsorbed on the active sites, but not subject to a simple cationic exchange mechanism, were finally extracted, since their mobilization requires an additional weak complexing capacity, such as that provided by the fluoride ion.

From the data just obtained, it can be seen that the availability of zinc is extremely poor, and that its concentration is reduced by an order of magnitude with respect to that obtained by cation exchange. Nickel maintains an elution capacity compared with that already observed with KNO_3 and $BaCl_2$. Copper, however, is extracted at levels ranging from 10 to 20 times those previously measured, therefore, at the level of 40-50 ug/l.

This latter result highlights the different conditions in which the metals considered in this study can be observed as a consequence of their very same nature: zinc is bound to active sites that are easily saturated with concentrated saline solutions, nickel is almost equally distributed among those active sites and others with stronger adsorbing capacity. Copper seems to be bound almost exclusively to this latter type of active sites, since it is practically left uneluted by KNO_3 and $BaCl_2$ solutions.

The fourth extraction step was designed to verify the distribution of heavy metals among those fractions bound to the organic phase and those in carbonate form.

The first series of samples, treated with DTPA, allowed a preliminary quantification of both forms of metals in the soil. The second series, subject to extraction with $Na_4P_2O_7$ solution, allowed determination of the organic phase bound fraction, while the third series was treated with CH_3COOH/CH_3COONa, at pH = 5, to measure the quantities present in carbonate form.

The results show that regardless of the extraction procedure followed for the determination, the major part of metals (Zn, Ni and Cu) is bound to either the organic or carbonate fractions. The DTPA extraction is, obviously, the one yielding the greatest concentrations in the liquid phase, the remaining two series bring a few subsequent considerations: zinc is equally distributed between the humic fraction of the soil and the carbonates, copper is tendentiously bound more frequently to the organic phase, while nickel behaves exactly inversely.

As an example, Figure 7 shows the average efficiency achieved in the different extraction steps for zinc, nickel and copper using the first extraction method, in samples from land parcels A and F (Sludge application); Figure 8 is the same for raw pig manure fertirrigation.

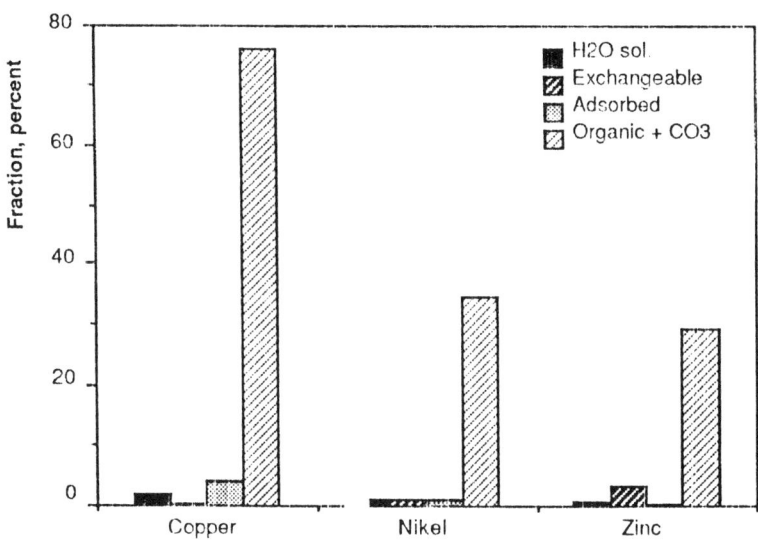

FIGURE 7. Extraction Percentages in Sludge Amended Soils.

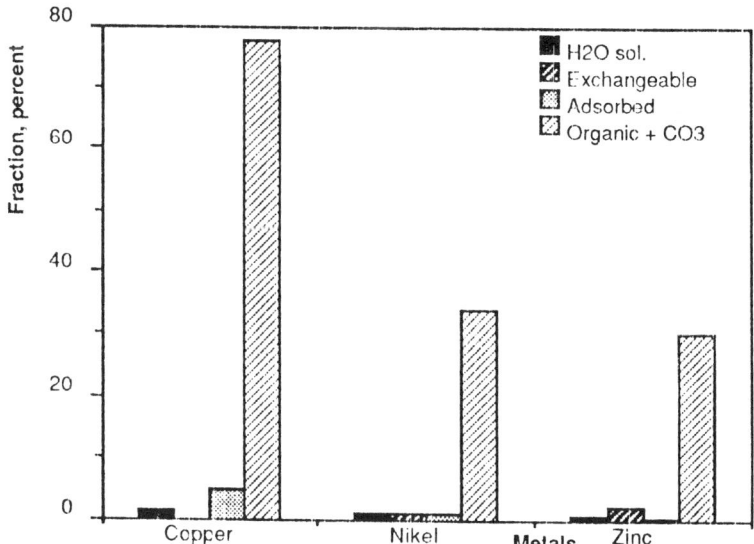

FIGURE 8. Extraction Percentages in Soils Treated with Raw Manure.

6. CONCLUSIONS

An analysis of the data obtained and herein presented shows that the three separation techniques adopted in this study yield results that are, although with slight differences, comparable. In particular, the 1 M $BaCl_2$ solution proved to be much more effective than the 1 M KNO_3 solution. Apart from this consideration, all the analytical procedures lead to analogous final considerations.

The three heavy metals considered in this study dissociate into the various chemical species depending on their chemical nature, and independently from sludge or raw manure soil treatments. This is proved by the fact that the intrinsic metal concentration variance from the untreated parcels samples is often comparable with that observed in the remainder of the treated samples. This implies that, at least after a whole year of treatment, not only was there no increase in the total quantity of metals in the agricultural soil layer, but even the most easily metabolized species did not show any significant differences. Any small difference observed, on the other hand, could not be univocally attributable to the disposal of the wastes.

The present study will continue with two main objectives: first, the speciation according to the three procedures illustrated will be extended to other heavy metals of interest. The treatment of the parcels will continue according to the same modalities just exposed, and soil samples will be collected and analyzed with the same techniques in about 2 to 5 years and again in 10 years, to verify the medium and long-term effects of the disposal procedures under study.

REFERENCES

1. Hickey, M.G. and J.A. Kittrick. "Chemical Partitioning of Cadmium, Copper, Nickel and Zinc in Soils and Sediments Containing High Levels of Heavy Metals," *J. Environ. Qual.*, 13: 372, (1984).

2. Legret, M., D. Demare, and P. Marchandise. "Speciation of Heavy Metals in Sewage Sludges," in *Proc. Int. Conf. Heavy Metals in Environment*. Vol. 1. 350, (CEP Consultants Ltd., Edinburgh, 1983).

3. Foerstner, U., W. Calmano, K. Conradt, H. Jaksch, C. Schimkus, and J. Schoer. "Chemical Speciation of Heavy Metals in Solid Waste Materials (Sewage Sludge, Mining Wastes, Dredged Materials, Polluted Sediments) by Sequential Extraction," in *Proc. Int. Conf. Heavy Metals in Environment,* (CEP Consultants Ltd., Edinburgh, 1981) pp. 698,

4. Steinhillber, P. and F.C. Bosswell. "Fractionation and Characterization of Two Aerobic Sewage Sludges," *J. Environ. Qual.* 12:529, (1983).

5. Sposito, G., L.J. Lund, and A.C. Cheng. "Trace Metal Chemistry in Aridzone Fields Soils Amended with Sewage Sludge. I. Fractionation if Ni, Cu, Zn, Cd and Pb in Soil Phases," *Soil Sci. Soc. Am. J.*, 46:260, (1982).

6. Chang, A.C., A.L. Pope, J.E. Warnike, and E. Grgurevic. "Sequential Extraction of Soil Heavy Metals Following a Sludge Application," *J. Environ. Qual.*, 13:33, (1984).

7. Doyle, P.J., J.N. Lester, and R. Perry. "Survey of Literature and Experience on the Disposal of Sewage Sludge on Land," *Report DGR/480/60,* (UK Dept. of the Environment, London, 1978).

8. Tessier, A., P.G.C. Campbell, and M. Bissom. "Sequential Extraction Procedures for the Speciation of Particulate Trace Metals," *Anal. Chem.*, 51:844, (1981).

9. Stover, R.C., L.E. Sommers, and D.J. Silviera. "Evaluation of Metals in Wastewater Sludge," *J. Water Poll. Control Fed.*, 48:2165, (1976).

10. Petruzzelli, G., L. Lubrano, and G. Guidi. "The Effect of Sewage Sludge and Composts on the Extractability of Heavy Metals from Soil," *Environ. Technol. Lett.* 2:449, (1981).

11. Lake, D.L., P.W.W. Kirk, and J.N. Lester. "The Effects of Anaerobic Digestion on Heavy Metal Distribution in Sewage Sludge," *Water Pollut. Control.*, 84:549, (1985).

12. Emmerich, W.E., L.J. Lund, A.L. Page, and A.C. Chang. "Solid Phase Forms of Heavy Metals in Sewage Sludge-Treated Soil," *J. Environ. Qual.*, 11:178, (1982).

13. Berrow, M.L. and J. Webber. "Trace Elements in Sewage Sludges," *J. Sci. Food Agric.* 23:93, (1972).

14. Rudd, T., D.L. Lake, P.W.W. Kirk, and J.N. Lester. "Chemical Fractionation of Heavy Metals in Sewage Sludges," *2nd Report to Water Research Center*, Medmenham, U.K. (1985).

15. Jenkins, S.H. and J.S. Cooper. "The Solubility of Heavy Metal Hydroxides in Water, Sewage and Sewage Sludge. III. The Solubility of Metals Present in Digested Sewage Sludge," *Int. J. Air Water Pollut.* 8:695, (1964).

16. Latterell, J.J., R.H. Dowdy, and W.E. Larson. "Correlation of Extractable Metals and Metal Uptake of Snap Beans Grown on Soils Amended with Sewage Sludge," *J. Environ. Qual.* 7:435, (1978).

DISCUSSION OF:
CHEMICAL SPECIATION OF HEAVY METALS IN SOILS FOLLOWING LAND APPLICATION OF CONDITIONED BIOLOGICAL SLUDGES AND RAW PIG MANURE

Gianniantonio Petruzzelli
Istituto per la Chimica del Terreno, C.N.R., Pisa Italy

Sequential Chemical Extractions for Heavy Metal Speciation in Sewage Sludge-Treated Soils

Land application of sewage sludge could introduce great quantities of potentially harmful heavy metals into agricultural soils. Identifying chemical forms of these elements in soil is an increasingly important objective in evaluating possible environmental consequences of this practice.

Sequential extraction procedures are considered of great value for the identification of the chemical forms of heavy metals in soil because of the ability of the used reagents to interact with metals in different labile pools in soil from which these elements can enter the food-chain through plant roots or percolate downward to the ground water.

Different schemes of sequential extraction procedures have been developed in recent years (Lake et al. 1984). The diversity of the reagent used, and the different concentrations employed to extract specific metal forms make the comparison of results from different authors difficult. Nevertheless it is evident that total concentration of a metal in soil is too generic a parameter to determine its environmental behavior, mobility and availability for the nutritional process of plants. Mobility of an element in the soil is affected by its source (sewage sludge, raw pig manure, fertilizers...), the type of interactive process in the soil (sorption, desorption, precipitation), and mainly by the characteristics of soil involved (pH, amount of organic matter, type and content of clay, available surfaces etc.).

In any case, the chemistry of the metal in the soil solution plays the predominant role in the interactions with the different reagents used in sequential extraction procedures.

To obtain successful data with chemical extraction, during each extraction step the amounts of metal must be small, compared to the concentration of the used reagent, to limit the possibility of the extraction of one metal influencing the extractable amounts of the others. The characteristics of the soil on which sewage-sludges are used are of paramount importance; their great variability among soil types requires careful consideration before using any extractant. For example, among the different complexing agents commonly used, EDTA and DTPA provide an effective chelation of heavy metals only in a narrow range of pH and not in the whole pH range of agricultural soils, particularly in the case of metal-rich soils, where the concentrations of heavy metals my exceed the complexing capacity of the extractant. Since the extracted quantities are the equilibrium concentrations of heavy metals obtained by the contact of the solid-soil phase with the extracting solution, changing the chemical conditions of the reaction, according to the soil characteristics, the extracted amounts will also change.

In all cases the aim of a chemical test is to evaluate the transferability of the heavy metals from sewage sludge treated soil to plants. As previously stated, this process is, first of all, influenced by soil characteristics so that for the same quantity of heavy metals added with sludge, the uptake from plants is quite different according to the soil type.

In Table 1 some data from field experiments (Davis and Stark, 1981) are reported. Similar results were obtained also from green-house experiments with different metals after the addition of compost derived from the organic fraction of urban refuse (Petruzzelli et al., 1986). Moreover, a great influence on the availability of heavy metals in soil is determined by the growing plants. For the same levels of heavy metals in soil, some species tend to accumulate higher quantities than other crops. Lettuce, for example, is well known as an indicator species for Cd uptake while potato tubers are among the least sensitive to this metal in soil. Since a chemical test is not always able to evaluate the concentrations at which heavy metals exert their effects on the physiological mechanism of plant nutrition, a procedure for the combination of chemical and biological methods is often useful to improve the evaluation of environmental hazards from sewage sludge utilization on soil. (Petruzzeli et al. 1986).

Another important point to consider is the chemical form of each heavy metal in the utilized sludge which will greatly influence the mobility of these elements in soil, particularly in the first periods following sludge application. In fact, adsorbing sites of soil immobilize a certain amount of heavy metals which are in mobile chemical forms in sewage sludge, this process in turn produces the release of more mobile heavy metals from sludge.

If the adsorption properties of soil surfaces are greater than the corresponding sewage sludge, the mobility of heavy metals will be governed by the soil characteristics. However, if adsorption capacities in the sludge are higher than in the soil, only few heavy metals will be immobilized in the soil and the chemical properties of the sewage sludge greatly determine the mobility of the elements.

This is particularly true in the case of high addition rates of sludge to soil, when adsorption sites of the soil are readily saturated by metals, so that the remaining amounts of these elements will show a behavior greatly influenced by their chemical forms in sewage sludge.

It is, therefore, often necessary to extend characterization methods, such as sequential chemical extraction, also to original sewage sludge in order to obtain a more precise knowledge of the environmental behavior of heavy metals.

CONCLUDING REMARKS

After years of sewage sludge application to soils the most important implications concerning heavy metals are their long-term effects. These effects involve also the possible remobilization of metals due to the eventual lowering of pH, the increasing concentration of salts and complexing agents, and the change in redox conditions.

Sequential chemical extraction procedures are able to supply relevant information about the mobility and hence bio-availability during the years of sludge application and also after the termination of this practice.

Despite the clear usefulness of these procedures, there are some problems associated. The non-selectivity of reactions, which depend on a series of parameters such as the time of extractions, the ratio soil-solution, the adsorption and desorption phenomena during extraction, particularly in the case of ion-exchange interactions. Finally and most important, during each extraction step the sample is subjected to changes which alter its original chemical characteristics. It is, therefore, necessary to utilize a chemical fractionation scheme which at least reduces the possibilities of altering the original soil samples. This is particularly true in a heterogeneous system such as soil where thermodynamic models of speciation are not able to take into account the role of kinetically controlled processes, which are of predominant importance in the heavy metal biogeochemical cycles.

TABLE 1.   Cd Concentration, in Wheat Grain after Sewage Sludge Addition to Different Soils.

Cd Soil Concentrations $\mu g\ g^{-1}$	Cd Wheat Grains Concentrations $\mu g\ g^{-1}$		
	Calcareous	Sandy Loam	Clay
0.9	0.1	0.22	0.24
1.5	0.12	0.30	0.33
3	0.22	0.39	0.42
5	0.32	0.55	0.65

Rearranged from Davis and Stark 1981.

REFERENCES

1.   Lake, D.L., P.W.W. Kirk, and J.N. Lester. "Fractionation, Characterization, and Speciation of Heavy Metals in Sewage Sludge and Sludge-Amended Soils: A Review," *J. Environ. Qual.* 13:175-83 (1984).

2.   Petruzzelli, G., G.Guidi, and L. Lubrano. "Assessment of Bioavailable Heavy Metals in Compost-Treated Soils," In *Int. Conf. Chemicals in the Environment,* 772-778 (1986).

3.   Davis, R.D. and J.H. Stark. "Effects of Sewage Sludge on the Heavy Metal Content of Soils and Crops: Field Trials at Cassington and Royston," in *Characterization, Treatment and Use of Sludge,* Eds. P. L'Hermite and H. Ott. (Published for the Commission of the European Communities by D. Reidel, Dordrecht, 1981) pp. 687-698.

PART V: ION SEPARATION

A PROCESS TO IMPROVE THE REGENERATED EFFLUENT CONCENTRATION OF ION EXCHANGE

Xiang Liangkui
The Seventh Design and Research Institute of the State Commission of Machinery Industry, Xian, P.R. China

Peng Dangcong
Department of Environmental Engineering, Xian Institute of Metallurgy and Construction, Xian, P.R. China

1. INTRODUCTION

Wastewater from metal finishing and plating industries usually contains a lot of metals, such as copper, zinc, cadmium, nickel, chromium and so on. They not only lead to environmental pollution but also consume resources. Many methods have been developed to treat this kind of wastewater. Ion exchange is a good candidate. it can eliminate pollution and recover metal, simultaneously, water treated can be reused. In China, more than one thousand sets of ion exchange equipment have been installed since 1975. But some problems have arisen in use, especially management of regenerated effluent. The ideal way is to recycle them to plating bath. From the view of reuse, the concentration of regenerated effluent must exceed the concentration in plating bath, but the traditional regeneration process can not meet the requirement. For example, when weak caustic ion exchange is used to treat Chromic wastewater, NaOH concentration of regenerating liquid is only about 1.5 mol. and $C_r(VI)$ concentration of regenerated effluent is only 40 g/l, but Cr(VI) concentration in the electrolyzer is about 80g/l (Standard chromic Plating). In order to refill the electrolyzer, regenerated effluent is often evaporated. The installation of evaporators not only increases the capital costs, but also increase the operating expenses. Therefore, how to improve the regenerated effluent concentration of ion exchange has theoretical and practical meanings.

2. BACKGROUND

The regeneration is an important link of the ion exchange process. As we know, the principle of strong ion exchange is concentration effect and weak ion exchange will form weak electrolyte. The use of highly concentrated regenerating liquid is expected in consideration of reuse. But the concentration of regenerated effluent is not directly proportional to the concentration of regenerating influent used. Figure 1 is a typical regenerated effluent concentration diagram of Chromium. We can easily see that the concentration of regenerated effluent increases slowly with the one of regenerating influent used when the concentration of NaOH is over 1.5 mol. This is because:

(1) When the concentration of NaOH increases, the stickiness of liquid increases rapidly, so the ion diffusions along the liquid membrane outside resin and the hole within resin decrease.

(2) Vertical diffusion increases because of the difference of specific gravity between remains liquid in the bed and regenerating influent.

(3) Shrinkage of resin (about 10%) in regeneration results in a wrap effect which slows down the ion diffusion. In addition to those, specific consumption of regenerated liquid increases and Cr(VI) breakthrough is ahead of time because of the wrap effect when highly concentrated regenerating liquid is used.

Considering the factors above, 1.5 mol. NaOH is often used in the regeneration of Chromates. The volume of regenerating liquid is 2 times that of the resin bed and Cr(VI) concentration of regenerated effluent is about 40 g/l.

3. LIMIT REGENERATED EFFLUENT CONCENTRATION THEORY

In order to improve the regenerated effluent concentration, it is important to know how much the maximum regenerated effluent concentration in theory is under the assumptions as follows:

(1) absorbed water within the resin may be negligible (Figure 2)

(2) exchangeable groups being distributed evenly within the resin

(3) regenerated reaction taking place easily

(4) the bed of ion exchanger being thick enough

(5) ion diffusion in the regenerating liquid and regenerated effluent can be rapid enough, so thinning action may be negligible. We can get a maximum regenerated liquid concentration (C_w) within the ion exchanger:

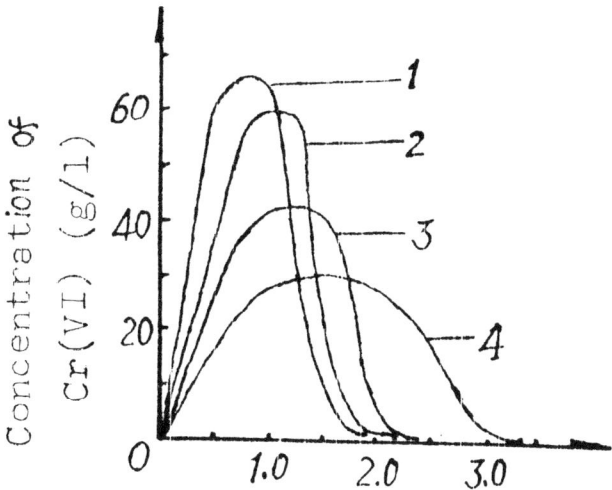

Regenerating Liquid Volume/Resin Bed Volume

FIGURE 1. Relationship between the regenerated effluent and regenerating liquid concentrations (type of resin: 710A; flow rate of regenerating liquid: 1m/h).

1--15% NaOH   2--9.5% NaOH
3--6% NaOH   4--4.5% NaOH

The concentration (C_o) outside the ion exchanger is the one regenerated effluent. Generally, Co is less than Cw because of diffusion. Under the limited condition, Co is equal to Cw. Therefore, we can consider that Cw is the theoretical limit regenerated effluent concentration (C_T).

According to the Equation (1), when weak caustic type 710A is used in the treatment of Chromate Wastewater (q is about 70g/l Cr (VI) and Σ is 45%), we have:

$$C_T = \frac{q}{\Sigma} = \frac{70}{5\%} = 156 \text{ (g/l)}$$

4. STEP BY STEP REGENERATION PROCESS

Based on the equation (1), we know that the maximum regenerated effluent concentration of Cr(VI) can reach about 156g/l when weak anion exchanger is used. But in fact, about 40 g/l Cr(VI) can be gained if the traditional regeneration process is taken. Although regenerated effluent concentration can be enhanced when highly concentrated regenerated influent is adopted, other disadvantages, such as breakthrough ahead of schedule due to the wrap effect, decrease of regeneration effectiveness and so on, will be brought on. In order to resolve this contradiction, we introduce a new process in which, first, saturated ion exchanger is regenerated by a 1.5 mol. regenerating liquid, we call it first step; then the regenerated effluent in which solid or highly concentrated liquid regenerating agent is added to 1.5 mol. again well be reused as regenerating influent in the next run, we call it second step. This is called step by step. The regenerated effluent concentration will be improved after several recyclings. Simultaneously, wrap can be reduced to the lowest limitation because lower regenerating influent concentration is used in each run. Therefore regeneration effectiveness is not reduced.

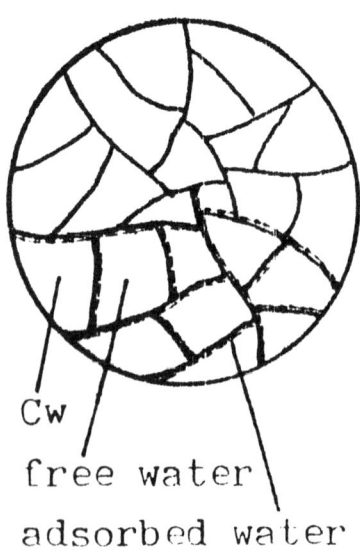

FIGURE 2.   *The Structure of Resin.*

5. EXPERIMENTAL RESULTS

In order to examine the process, experiments were made using regeneration of Chromates as a representative. Table 1 is the coefficients of resin used. Table 2 is the experimental operating conditions.

TABLE 1.   The Characteristics of Resin Used.

resin type	710A (weak anion)
Water ratio of resin	45%
Specific gravity	1.05
operating exchange volume per liter wet resin	70 g Cr (VI)

TABLE 2.   The Regeneration Coefficients.

diameter of column	25 mm
packed height of resin bed	600 mm
flow rate of regenerating influent	1 m/h
regenerating liquid flow direction	down flow with co-flow of exchange process

Figure 3 is the diagram of regenerated effluent concentration. In the first step, 6% NaOH and two times of concentration in the middle part (1 time of the resin bed volume) is about 40 g/l Cr (VI). In the second part, 1.5 mol. NaOH is added to the effluent from the first step and then used as regenerating influent. Simultaneously, fresh NaOH liquid whose concentration is 1.5 mol. and volume is 1 time of resin bed is added; we have the maximum and average re-generated effluent concentration in the middle part (1 time of the resin bed) are 80 g/l, 70 g/l Cr(VI) respectively and 102 g/l, 90 g/l respectively in the third step. The value is more than the concentration in the plating bath. The regeneration effectiveness in the step by step process is equal to the first step (more than 95%) because both have the same regenerating liquid concentration in fact.

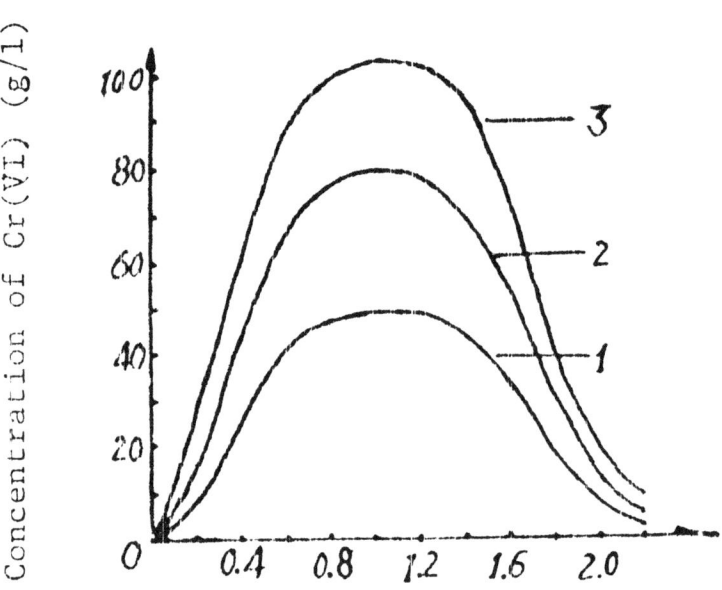

FIGURE 3. *Regenerated Effluent Concentrations Using the Step Process (Type of Resin: 710A; Flow Rate: 1 m/h).*

1--first step (6% NaOH)
2--second step (40 g/l Cr(VI) + 6% NaOH)
3--third step (70 g/l Cr(VI) + 6% NaOH)

The initial part in which Cr(VI) concentration is lower and final part in which concentration of Cr(VI) is lower also and alkali contained are often returned to raw wastewater. According to the principle of step by step process, we mix the two parts together and NaOH is added to it, then use it as regenerating influent also in the next run. In this way, the regenerated effluent concentration is improved and the consumption of regenerating liquid is reduced also.

Compare the results with the highly concentrated regeneration process introduced by Dawlowski, L. et. al. [1] in which 10 mol.

NaOH (45%) is used and the maximum concentration of regenerated effluent is only about 78 g/l Cr(VI), we can easily recognize that this process is superior to other processes.

6. CONCLUSIONS

(1) A limited regenerated effluent concentration formula is given. This formula assists in the selection of regenerating liquid concentration.

(2) A step by step regeneration process is introduced. According to this process, the regenerated effluent concentration can be increased greatly and the regeneration process can be operated in optimum conditions.

(3) Evaporator can be omitted when the process is used for the recovery of Chromates an other metals from plating wastewater. The regenerated effluent can be directly refilled to the plating bath.

(4) The initial and final parts cut from regenerated effluent can be reused as regenerating influent after solid or highly concentrated regenerating agent is added. The specific consumption of regenerating agent is reduced.

(5) This process can be used in other applications which the regenerated effluent concentration is expected to improve.

ACKNOWLEDGMENTS

Jin Qiting, Professor of Xian Institute of Metallurgy and Construction, who examined the manuscript and gave beneficial advice and Ms. Jin Ruihong who typed the paper are acknowledged.

REFERENCES

1. Pawlowski, L., et.al. "A New Method of Regeneration of Anion Exchangers Used For Recovering Chromates from Wastewater," *Water Research*, Vol. 15, (1981), pp. 1153-1156.

DISCUSSION OF:
A PROCESS TO IMPROVE THE REGENERATED EFFLUENT CONCENTRATION OF ION EXCHANGE

L. Liberti
Istituto Chimica Applicata, Facoltà di Ingegneria, Università di Bari, Italy.

D. Petruzzelli
Istituto di Ricerca sulle Acque, Consiglio Nazionale Ricerche, Bari, Italy.

INTRODUCTION

Ion exchange technology for metal removal and separation from industrial effluents offers the following advantages over conventional application based on chemical precipitation and/or solvent extraction:

-- Recovery of metals and water for eventual reuse.
-- High concentration of metals in the elution streams.
-- Relatively low cost.

Of course not all the above advantages would be simultaneously achieved in each application, but even one of them would justify the adoption of reference technology.

On the other hand the need to prevent environmental pollution and the raising cost of metals has resulted in a greater interest in conservative technologies for recovery and reuse of the valuable products resulting from wastewaters and/or side liquid streams from productive processes.

Table 1 reports a list of metals that can be efficiently removed and recovered by ion exchange, according to a recent review by Dillman [1].

Ion exchange is efficiently adopted, for example, in the viscose process to recover zinc from rinse waters of the fiber by means of an hydrogen form strong cation resin which is regenerated with the segregated spent acid bath from the process [2].

The most important copper recovery system based on ion exchange is associated with the rayon industry. Wastewaters from rayon production result in a copper-ammonia complex exceeding 100 ppm: By means of a carboxylic resin quantitative recycling of both chemicals in the productive process is obtained; the resin is

TABLE 1. Guide to Selective Sorption of Metals by Ion-Exchange Resins. (from ref.[1])

Metal (form)	Resin	Resin form	Influent pH
Antimony (antimonite)	SB	Cl	1—14
Antimony (antimonite)	WB	FB	4—10
Arsenic (arsenate, arsenite)	SB	Cl	1—14
Arsenic (arsenate, arsenite)	WB	FB	4—10
Barium (Ba^{++})	WA	H, Na	1—8
Cadmium (Cd^{++})	WA	H	3—8
Cadmium (anionic complex)	WB, SB	FB, Cl	6—12
Cesium (Cs^+)	WA	Na	7—14
Chromium (Cr^{+++})	SA	H	0—5
Chromium (chromate)	SB, WB	SO_4, Cl	0—14
Cobalt (Co^{++})	SA	NH_4	1—8
Cobalt (Co^{++})	WA	Na, H	5—9
Cobalt (Co^{++})	WB	FB, SO_4	4—10
Cobalt (anionic complex)	SB	Cl	strong acid
Copper (Cu^{++})	WA	H, Na	5—8
Copper (Cu^{++})	WB	FB, SO_4	3—10
Gold (anionic complex)	SB	OH	0—14
Iron ($FeCl_4^-$)	SB	Cl	2—12 M HCl
Iron (organic chelate)	WB	FB	3—8
Lead (Pb^{++})	WB	FB, SO_4	4—6
Lead (chloride complex)	SB	Cl	0—14
Mercury	WB	FB	5—14
Mercury (chloride complex)	SB	Cl	0—14
Molybdenum (molybdate)	WB	FB, SO_4	3—9
Nickel (Ni^{++})	WA	Na, NH_4	5—9
Nickel (Ni^{++})	WB	FB, SO_4	4—7
Platinum (anionic complex)	SB	Cl	0—14
Rhenium (anionic complex)	SB, WB	Cl, FB	0—14
Selenium (selenite)	WB	FB	2—6
Silver (Ag^+)	SA	H, Na	0—7
Silver (Ag^+)	WB	FB	5—9
Silver (anionic complex)	SB, WB	Cl, FB	0—14
Tungsten (tungstate)	SB, WB	Cl, FB	0—14
Vanadium (vanadate)	WB	FB, SO_4	2—8
Zinc (Zn^{++})	WA	Na	5—10
Zinc (Zn^{++})	WB	FB, SO_4	4—7
Zinc (Zn^{++})	SA	H	0—7

Note: WB, Weak-base anion exchanger. SB, Strong-base anion exchanger. WA, Weak-acid cation exchanger. SA, Strong-acid cation exchanger. FB, Free-base.

then regenerated with the same exhausted bath from rayon production line [2].

Cobalt and nickel, because of their high economic value and scarcity, are ideal metals for ion exchange recovery. Accordingly hydrometallurgical processes for metal recovery from ore are essentially based on weak cation carboxylic resins [3]. The advantage of using these latter resins resides essentially in their selectivity toward reference ions and good elution property during regeneration step by using cheap and commercially available chemicals.

Phosphonic resins have been used to keep the level of soluble lead below 50 ppb in several industrial process streams. The resin efficiently removes lead even in the presence of competitors in acidic dilute solutions [4]. High in the list of priority pollutants, mercury is present essentially in wastewaters from chlor-alkali plants. From these effluents colloidal mercury is oxidized by chlorination and removed by cation exchange where concentration down to about 100 ppb are reached; the final polishing step is carried out with non-functional adsorbent resins allowing for final effluent concentration in the range of few parts per billion [4, 5].

Generally speaking, decontamination of industrial liquid streams from toxic metals by ion exchange is feasible and may in most cases, be economical even with conventional resins. A useful discussion of the problems and related considerations in choosing an appropriate ion exchange technique in metal technology has been recently given by Anderson [7]. However, if complexing agents are present in the liquid effluent, such as, for example, in some spent tanning baths or plating rinse waters, strong difficulties in metal recovery and separation may arise.

In fact conventional ion exchangers in commercial use today for metal recovery and separation are "selective" rather than "specific" [8].

Specificity involves the retention of only a single ion, whereas selectivity is the relative affinity of the ion exchanger among a number of ions. Selective ion exchangers may remove preferentially one ion together with smaller amounts of other environmentally innocuous ions from the liquid-phase.

Accordingly the tendency is to synthesize "tailor made" specific reactive polymers for the retention of the ion of interest without interference from competitors. This is the goal resin manufacturers hope to reach in the near future: The task, however, is very hard.

It goes without saying that from a thermodynamic point of view the higher the free energy gain in moving a reaction toward products, the more difficult it would be, in terms of mass action, to reverse the equilibrium. In other words, the higher the affinity of a resin toward an ion, the higher the regenerant consumption to recover the exchanged species would be. Chelating resins, for example, present selective ligands for the metallic ions innested onto the polymer matrix. However, once retained, metals are not easily regenerated in most cases, unless a specific chemical able to form stronger adducts (chelate, complex) with the ion of interest is used. Eventually another chemical able to change speciation of the exchanged ion could be efficiently used as regenerant.

A strong effort to improve selectivity of conventional anion and cation resins is underway: Typical examples are represented by some conventional anion resins extremely selective to nitrates [9] or to chromates [10]. They do not solve the problem, however, because, even in this case the tendency is to improve selectivity not specificity of the resin.

Recent examples of specific exchangers (essentially aminophenol or polybenzoimidazole matrixed resins) have appeared in the literature [11, 12]. Reference resins show an extreme specificity toward ferric and cupric ions, for example, and, most important, are regenerated very easily by using cheap commercial chemicals, e.g., dilute acidic solutions.

This behavior is interpreted in terms of metal ion chelation when the free base amino groups, present onto the matrix, have solitary electron pairs available for chelation. During regeneration usually performed in acidic conditions, the resulting protonated quaternary ammonium ion would reject cations by Donnan exclusion thus leading to an immediate elution of the metal ions from the resin.

ION EXCHANGE IN PLATING INDUSTRY

The most extensive use of ion exchange in metal finishing is in the treatment of spent chromic baths and related rinse waters [3-5, 13-16]. In this ambit, the chromic acid baths are contaminated by the trivalent chromium formed by the reduction of the acid and other heavy metals (Al, Zn, Cu, Ni, Fe) which dissolve from materials being plated. As a result the quality of the baths deteriorates and a part of the same, still containing large amounts of chromic acid, has to be discharged.

Additionally, the materials being plated or finished must be rinsed after each step to improve the quality of the product. The feed rinse waters (often of high quality) are easily contaminated from the drag-out of the plating solutions: Ion exchange appears the most efficient unit operation to recover and reuse rinse waters.

Figure 1 reports the flow sheet of a typical process for chromic acid recovery with recycle of rinse waters from plating effluents [14-16]. The rinsing wastewaters together with spent baths, eventually filtered-out on a sand bed to remove suspended matter, are passed onto a strong cation exchanger, where metals are taken-up. The decationized stream is then percolated on a weak (eventually strong) anion resin for chromic acid exchange and recovery to the plating bath. Deionized water obtained, usually backed-up with fresh water to maintain residual chromium concentration $\leq$ 200 ppm, is then recycled to the rinsing water tank ("closed loop" system).

As already mentioned, retention of chromic acid onto anion exchange bed is possible by the use of strong basic [15, 16] and weak basic [4, 16] exchange resins. The first allows for a lower Cr(VI) leakage (<1mg/l), but requires larger amounts of caustic for its regeneration, whereas a weak base resin, although leading to higher Cr(VI) leakages, requires lower regenerant consumption. The use of weak base resins is accordingly preferred in the case of closed loop systems.

The critical benefit in using weak base resins, however, is the NaOH regenerant consumption which provides savings in the general economy of the process [16]. Chromate concentration in the regeneration eluates, in fact, strongly depends on the concentration of NaOH used during this operation. However, as reported in the paper under discussion, the concentration of chromates in the regeneration eluates does not increase appreciably beyond a certain limit. Typically, with weak basic resins regenerated by 4.5%; 6.5%; 9.5%; 15% NaOH solution, the corresponding maximum chromate concentration is respectively 30; 40; 55; 65; g/L, figures well below the 80 g Cr(VI)/L required for direct recycling in the chromic plating baths.

On the other hand for the conventional regeneration of strong anion resin a regeneration dosage in the range of 80-150 g NaOH/lr at 4% solution concentration is recommended [15].

In both cases further concentration by thermal evaporation of the regeneration eluate would be required in order to refill the spent plating bath (see Figure 1).

A breakthrough in the field has been recently introduced by Pawlowski [16] by using small amounts (say 0.2 bed volumes, BV) of an highly concentrated caustic solution (45% NaOH w/v) for regeneration of weak base resins. After the operation chromate concentration in the regeneration effluents resulted in the range of 150 g/L of Na_2CrO_4, directly reusable to refill spent plating baths.

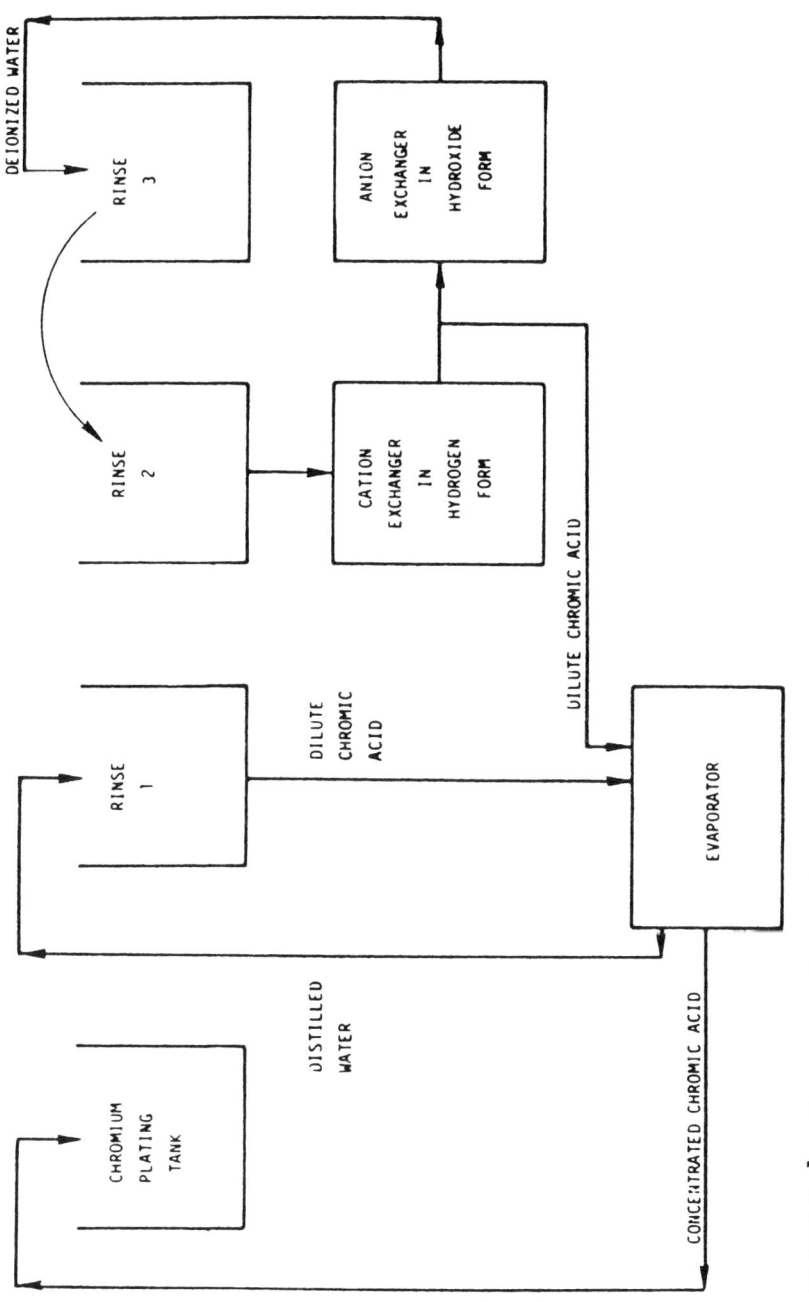

FIGURE 1 Recovery of chromic acid and the production of deonized water from chromic acid rinse waters. (from ref. 1)

The procedure, however, seems to be affected by the following limitations:
 a) Regenerant losses;
 b) Strong mechanical stress of the resin beads (swelling-shrinking) due to osmotic shocks.

THE STEP-BY-STEP REGENERATION TECHNIQUE

The paper under discussion presents a "step by step" regeneration procedure for the weak base anion resin used in a conventional ion exchange process for recovery and reuse of spent plating baths. The procedure allows for recycling of rinsing waters and recovery of chromates in the productive process without the need of additional thermal concentration.

According to this technique, during the initial step the resin is regenerated with 2 BV os 6% NaOH and the regeneration eluates are collected in two fractions of 1 BV each. The fraction corresponds to the central volume ("the heart" with the maximum elution peak of chromates). The second fraction is recomposed by mixing the "head" and "tail" volumes. After backing-up both fractions with fresh NaOH to restore 6% concentration, the same solutions are reused for the subsequent regeneration step in the following order:
 1) The hearth fraction;
 2) The head-tail fraction.

The procedure is then repeated for subsequent steps until the required minimum chromate concentration for refilling plating bath (80 g/L) is reached (see Figure 2). According to this procedure, the following advantages should be evidenced: a) Increasing chromate concentration in the spent regeneration eluates after subsequent steps; b) High regeneration efficiency in terms of equivalent of Cr(VI) recovered per equivalent of OH^- dosed; c) Acceptable osmotic shocks on the resin bead due to the quite low concentration of regenerant solution.

Disadvantages would be essentially centered on the following points: a) Chromate leakage from the resin would be quite high during exhaustion; b) Regenerant losses; Although the step by step regeneration procedure for ion exchange resins is not an innovation in the strictest sense [17, 18] the presented method looks quite interesting in general terms.

The following general remarks can be drawn onto the procedure as compared to Pawlowki's: a) Both procedures avoid thermal evaporation of regeneration eluates allowing for direct refilling of the plating baths; b) Xiang procedure calls for higher regenerant consumption (up to 150 g/L_r) respect to Pawlowski ($\approx$100 g/L_r); c) Pawlowski procedure calls for lower regenerant volume (0.2 BV), with a more complicated operativity.

Generally speaking, regeneration procedures allowing for economical savings are always welcomed in ion exchange technology where the chemical consumption during regeneration step results to be the weakest point. Although the presented step by step procedure looks quite attractive a better substantiation of the principles on which the authors base their thesis, together with a deeper insight into the operative details would be necessary for a definite judgement.

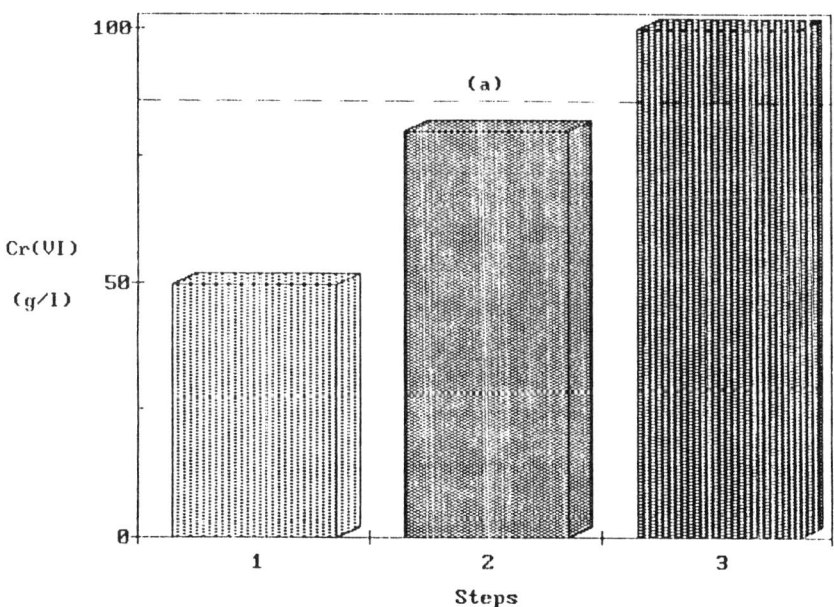

FIGURE 2.   Cr(VI) Build-Up in the Regeneration Eluate by the Xiang Step-by-Step Procedure. (a) Minimum Cr(VI) concentration for refilling plating bath (80 gCr(VI)/L).

REFERENCES

1. Dillman, T.R. "Metal Ion Removal and Recovery." in *Ion Exchange for Pollution Control*, C. Calmon, Ed. (CRC Press, Boca Raton, Fl, 1979).

2. Greig, J.A., B.P. Lindsay. "Copper Removal from Zinc Bearing Leach Liquor." M. Streat Ed. *Ion Exchange for Industry*, (Ellis Horwood, Chichester, UK, 1988).

3. Gold, H., G. Czuprina, R.D. Levy, C. Calmon. "Purifying Plating Baths by Chelate Ion Exchange." J.W.Patterson, R.Passino, Eds., *Metals Speciation Separation and Recovery*, (Lewis Pub. Co. Chelsea, MI, 1987).

4. Gold, H. "Metal Finishing Wastes." in *Ion Exchange for Pollution Control*, C. Calmon, Ed. (CRC Press, Boca Raton, FL, 1979).

5. Weiner, R. *Effluent Treatment in the Metal Finishing Industry*. (R. Draper Ltd., Teddington, UK, 1970).

6. Liberti, L. "Scambio Ionico," in *I Metalli Nelle Acque: Origine, Distribuzione, Metodi di Rimozione*, (Quad. IRSA, No. 71, 171, 1981).

7. Anderson, R.E. "Toxic Metals Decontamination: Does Ion Exchange Fit?" *Amber Hi Lites*, No. 181, (Rohm & Haas Co. Philadelphia, Summer 1987).

8. Calmon, C. "Specific and Chelate Exchangers New Functional Polimers for Water and Wastewater Treatment." *J.A.W.W.A.*73,12 (1981), pp. 652-56.

9. Ambrus, P. et. al. "Nitrate Removal by Selective Ion Exchange." (Rohm & Haas, Tech. Bull., 1988).

10. Sengupta, A.K., D. Clifford, and S. Subramonian. "Chromate Ion Exchange Process at Alkaline pH." *Wat. Res.* 20, 9 (1986), pp. 1177-84.

11. Hodgkin, J.H. and R. Eibl. "Ferric Ion Chelation by Aminophenol Resins." *React.Polym.* 4:285-91, (1986).

12. Chanda, M., K.F. Driscoll, and G.L. Rempel. "Polybenzimidazole Based Resins New Chelating Agents; Ferric Ion Selectivity of Resins." *React. Polym.* 9:277-84,(1988).

13. Thompson, J., J. Miller, and V. Miller. "Role of Ion Exchange in Treatment of Metal Finishing Wastes," *Plating*, East Orange, NJ, 58:809, (1971).

14. Bolto, B. and L. Pawlowski. "Reclamation of Wastewater Constituents by Ion Exchange." *Part I. Effl. Wat. Treat. J.* 23, 1 (1983), pp. 6-17 and 55-57.

15. Kunin, R. "Ion Exchange for Metal Product Finisher." *Product Finisher*, Cincinnati, OH, (April-June 1969).

16. Pawlowski, L., B. Klepacka, and R. Zalewski. "A New Method of Regeneration of Anion Exchangers Used for Recovering Chromates from Wastewaters." *Wat. Res.*, 15 (1981), pp. 1153-56.

17. Polta, R.C., R.W. DeFore, and W.K. Johnson. "Evaluation of Physico-Chemical Treatment at Rosemount." VA. E.P.A. Rep. #600/2-78-201, (Dec. 1978).

18. Liberti, L., N. Limoni, A. Lopez, R. Passino, and G. Boari. "The 10 m3/h RIM-NUT Demonstration Plant at West Bari for Removing and Recovering N and P from Wastewaters." *Wat. Res.*, 20 (1986), p. 735.

HEAVY METAL REMOVAL USING NATURAL ZEOLITE

M.D. Loizidou
Athens National Technical University, Greece

1. INTRODUCTION

The removal of heavy metals from industrial effluents is drawing an increased interest. It is well known that the presence of toxic metals in the environment is a potential health hazard. Measures must be taken to eliminate these toxic metals before they reach the final recipient, which will be a river, a lake or the sea.

Industrial effluents that contain high metal quantities, are those of the metal plating and metal finishing industries, mining industries, tanneries, etc.

From a study that has been carried out [1] for the Athens area, concerning the metal plating and metal finishing industries, it was found that around 150 small plants exist and produce effluents that contain very high metal concentrations, the type of the metal and its concentration, depending on the process taking place.

For the heavy metal removal from aqueous solutions or effluents, several physicochemical methods have been used. Such methods are precipitation, activated carbon adsorption, ion exchange, reverse osmosis, foam flotation techniques, and cementation [2,3]. The cost of such systems renders many times, the impossibility for treatment of effluents to the acceptable level.

As mentioned, the number of the electroplating and metal finishing industries in the Athens area is quite high, but the plants themselves are very small (most of them employing 5 to 10 people). This indicates that the units cannot provide funds for their effluent plants. Hence, the methods that must be employed have to be within their capabilities to cope with the problem.

Usually the treatment employed includes pH control and precipitation, which are low-cost methods but not always effective, since long settling times are required and the disposal of sludge produced must be handled. These needs have prompted us to study an effective and low-cost method for removing the toxic cations from the effluents, by using naturally occurring materials, the so-called zeolites. Zeolites have the ability to exchange cations, like the organic resins. For this particular work, the zeolite used was clinoptilolite of Greek origin. Very recently, it was found that extensive deposits of this mineral exist in the northern part of Greece.

In this study, several ion exchange systems were examined using this natural zeolite in order to evaluate its efficacy for the treatment of effluent containing heavy metals such as lead, cadmium, zinc, copper, nickel, iron and chromium.

From previous work [4,5,6] that has been carried out, it was found that natural clinoptilolite from Nevada, USA, is very selective for certain heavy metals. Also, some other researchers have shown that clinoptilolite of various origins, is capable of removing heavy metals. Though there is a good indication that clinoptilolite is quite selective for heavy metals, it is absolutely necessary to examine the properties of a natural zeolite in great detail, since the ion exchange selectivities and ion exchange capacities vary with geologic origin [7,8,9].

Since this is the first time that clinoptilolite of this particular origin was examined, it was necessary to carry out a detailed study.

2. THEORETICAL ASPECTS OF THE ION EXCHANGE PROCESS

During an ion exchange the zeolite, which in this case is clinoptilolite, can exchange its sodium with another cation present in the solution phase. The ion exchange process can be presented by the following equation, for a binary system:

$$Z_A B_C^{Z_B^+} + Z_B A_S^{Z_A^+} \rightleftharpoons Z_A B_S^{Z_B^+} + Z_B A_C^{Z_A^+} \tag{1}$$

where Z_A, Z_B are the valencies of the exchange cations A and B and the subscripts (C) and (S) refer to the exchanger and solution phases, respectively. The ion A, which is the initial ion in the solution, is called the counter ion.

Usually, the equilibrium properties of an ion exchange system are depicted by an isotherm. This is an equilibrium plot of the exchanging ion concentration in solution, against the corresponding equilibrium concentration of the same ion in the zeolite at constant temperature and solution concentration. The equilibrium concentrations are expressed as equivalent fractions A_S, A_C for the solution and exchanger phases, respectively. For a binary exchange, the equivalent fractions are defined by:

$$A_S = \frac{Z_A m_A}{Z_A m_A + Z_B m_B} \quad (2)$$

$$A_C = \frac{Z_A M_A}{Z_A M_A + Z_B M_B} \quad (3)$$

where m_A, m_B are the concentrations of ions A^{Z_A+}, B^{Z_B+} in solution, and M_B and M_B are the concentrations of the ions in the exchanger. From the isotherm plot it is quite easy to calculate the thermodynamic parameters, the equilibrium constant K_a and $\Delta G°$, the free energy change per equivalent associated with the reaction described by Equation (1). The thermodynamic parameters K_a and $\Delta G°$ obtained, refer to the ion exchange system as a whole, and therefore give an indication of the <u>overall</u> preference that a given zeolite displays for a particular ion. This is a rigorous approach and provided sufficient activity coefficient data are available for both phases for the system concerned. It can be used to predict the behavior of the system under widely differing conditions, such as external solution concentration. Various attempts have been made using data from the binary system, to predict the behavior of multicomponent systems but this was not very successful [10,11].

Actual information on the behavior of an ion exchange system at any value of A_s or A_C, can be given by the separation factor "a".

$$a = \frac{A_C \cdot m_B}{B_C \cdot m_B} \quad (4)$$

which can be calculated from an isotherm, as shown in Figure 1.

$$a = \frac{Z_A}{Z_B} \frac{\text{Area I}}{\text{Area II}}$$

$a > Z_A/Z_B$ exchanger selective for A^{Z_A+}
$a = Z_A/Z_B$ exchanger shows no preference.
$a < Z_A/Z_B$ exchanger is selective for B^{Z_B}.

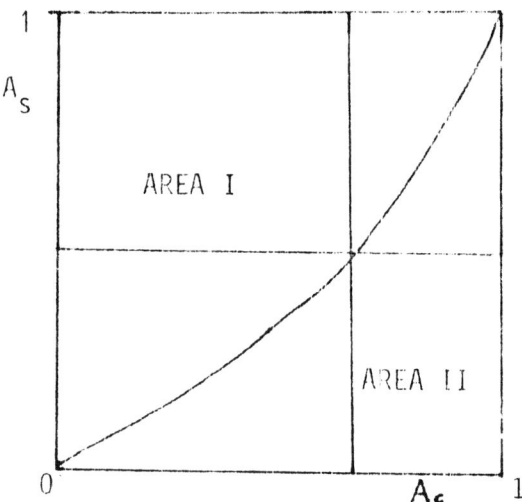

FIGURE 1. Ion Exchange Isotherm.

Also, very practical and useful information for an ion exchange process, can be given by studying the kinetic behavior of the systems. The kinetic tests indicate how an ion exchange system behaves with respect to time, as the ion exchange process is taking place. These tests can describe an ion exchange process under varying conditions. It is possible to examine the behavior of a zeolite towards competing cations at low external solution concentrations, which is usually the case of effluents. It must be emphasized that the kinetic tests describe the ion exchange process under very specific conditions, and the results obtained cannot be used to predict an ion exchange behavior as in the case of the thermodynamic parameters. However, the kinetic behavior of the ion exchange system is key information, regarding the treatment of industrial effluent using zeolites, which is the main objective of this work.

3. EXPERIMENTAL PROCEDURE

3.1 Zeolite Treatment and Analyses

The zeolite used was collected from a deposit in the northern part of Greece. Since no information existed about the deposits, an x-ray examination was carried out. It was found that this deposit was very rich in clinoptilolite. Also it was found that substantial quantities of calcite were present.

The natural zeolite was crushed and separated in various fractions. The 100-mesh fraction was chosen for this particular study. The zeolite was converted into a homoionic form by equilibrating it with sodium nitrate solution, the concentration being 1 mol/dm^3. The exchange was repeated several times (12), at the temperature of 298°K. Then the zeolite was washed several times with distilled water, dried and stored in a desiccator over saturated sodium chloride solution for two weeks before chemical analysis. The zeolite was analyzed using various methods which are described elsewhere [13,14,15].

3.2 Ion Exchange Experiments

The selectivity trends were measured by determining kinetic curves at 298°K, using various metal concentrations. Known quantities of the zeolite (sodium clinoptilolite) usually about 0.3 g, were equilibrated with aliquots of solutions containing the various metals.

In the initial study, a solution (50 ml) of 50 mgl^{-1}, for one particular metal cation originating from its nitrate salt, was equilibrated with the zeolite. The metals chosen were Pb^{+2}, Cd^{+2}, Zn^{+2}, Cu^{+2}, Ni^{+2}, Fe^{+3}, Cr^{+3}. The equilibrations were performed for the time intervals of 1, 5, 10, 15, 30, 60 minutes and 24 hours. Duplicates for each point were carried out. At the appropriate time, the separation of the two phases was done by vacuum filtration through a porous membrane. The initial solutions and the solutions after the exchange were analyzed for their metal content, using atomic absorption spectroscopy.

Next, 50 ml of a mixed solution of the seven metals studied previously were equilibrated with zeolite, the total metal concentrations being 700 mg/l and 100 mg/l for each metal. Similar experiments were performed for concentrations of 50 mg/l, 25 mg/l and 10 mg/l, for each of the seven metals involved in the study.

After the exchange, the zeolite phase was washed with distilled water, followed by $NaNO_3$ solution in order to examine whether the metal taken by the zeolite could be replaced again by the sodium ions. The regeneration step was done by equilibrating 20 ml of molar $NaNO_3$ solution with the zeolite for ten minutes, followed by filtration. The equilibration/filtration processes were repeated three times and the filtrate in each step was analyzed for the metal content. Then the zeolite structure was examined using x-ray diffraction.

Also, a set of experiments were performed using solutions of $NiSO_4$, $NiCl_2$, and $Ni(NO_3)_2$ at a concentration of 0.1 N. These experiments were carried out in order to investigate the effect of the different anions on the ion exchange process.

4. RESULTS

The results of the chemical analyses of the sodium clinoptilolite are given in Table 1. The oxide formula was calculated and found to be:

$$0.95Na_2O.0.0.4K_2O.0.21CaO.Al_2O_3.8.60SiO_2.6.81H_2O$$

The zeolite exchange capacity was also calculated and found to be 2.53 mequiv.g^{-1}. The percentage metal removal and the metal uptake expressed in mequiv.g^{-1}, with respect to time for the system involving one type of cations, are given in Figures 2,3 and Tables 2,3. Also, Table 4 gives the maximum exchange levels and the percentage of the theoretical exchange capacity of the zeolite that has been covered by the counter-ions 24 hours after equilibration.

The percentage metal removal for the mixed solutions having concentrations of 700 mgl^{-1}, 350 mgl^{-1}, 175 mgl^{-1}, and 70 mgl^{-1}, are shown in Figures 4 to 7 and Tables 5 to 8. Table 9 gives the results of the regeneration process. Figure 8 presents the selectivity of clinoptilolite towards nickel ions, when different anions are present.

5. DISCUSSION

5.1 Exchange Capacity

The natural zeolite used was found to have a Theoretical Exchange Capacity (T.E.C.) of 2.53 mequiv.g^{-1}. This capacity was based on its aluminum content, resulting from the chemical analysis that was carried out as shown in Table 1.

TABLE 1.   Analysis of Sodium Clinoptilolite.

SiO_2	H_2O	Al_2O_3	Na_2O	K_2O	CaO	Fe_2O_3	MgO	MnO
60.715	14.454	12.035	6.975	0.490	1.505	1.398	0.129	0.049

For synthetic zeolites, the exchange capacity can be safely based on the aluminum content under normal conditions. However, for this work natural clinoptilolite was used and it was therefore quite difficult to decide what was the true exchange capacity, since impurities were likely to be present. Thus, some of the aluminum present might not contribute to the ion exchange capacity.

If for the Na-CLI the exchange capacity is based on the sodium contents of the zeolite, its value becomes 2.07 mequiv.g^{-1}. This value is still higher, compared with the exchange capacities based on the sodium content and reported by several researchers [12,13]. For practical reasons it is found convenient to base the exchange capacity on the zeolite's aluminum content.

5.2 Ion Exchanges

Single Ion in Solution

In this work, various experiments were performed in order to investigate the behavior of sodium clinoptilolite towards the metals Pb^{+2}, Cd^{+2}, Zn^{+2}, Cu^{+2}, Ni^{+2}, Fe^{+3}, Cr^{+3}.

Examining first the ion exchange systems that contain only one type of metal cations of initial concentration 50 mg/l^{-1}, the following observations were made. Figures 2, 3 and Tables 2, 3 give the results of these ion exchange systems. It is clearly shown that the metal removal is a very fast process for all seven metals. From the very first minute most of the cations in the solution were taken by the zeolite in exchange with its sodium. The Pb/Na-CLI exchange system gives the higher metal removal, followed by Cd/Na-CLI, Zn/Na-CLI, Cu/Na-CLI, Ni/Na-CLI, Fe/Na-CLI and Cr/Na-CLI. In other words, the selectivity series at the equilibrium point which is considered to take place in 24 hours follows the pattern:

$$Pb > Cd > Zn > Cu > Ni > Fe > Cr$$

It is worth mentioning here, the significance of these kinetic tests in order to examine the behavior of the zeolite towards the metal removal. From an earlier study [4,5], it was found that the G values obtained for the system Pb/Na-CLI and Cd/Na-CLI were -3.81 KJ (g.equiv)$^{-1}$ and 2.26 KJ (g.equiv)$^{-1}$, respectively. The thermodynamic parameters so obtained, refer to the system as a whole and therefore give an indication of the overall preference that a given zeolite shows.

For this particular case, Na-CLI is more selective to lead than sodium, but the zeolite is more selective to sodium than cadmium. Referring again to Figure 2, it is shown that the zeolite is quite selective to cadmium at this particular external solution concentration, 50 mg/l^{-1}. This is in good agreement with the results obtained using data from an isotherm plot and calculating the separation factor "α", from the unnormalized data [15]. It is shown that for dilute solutions, the zeolite can be very selective to cadmium.

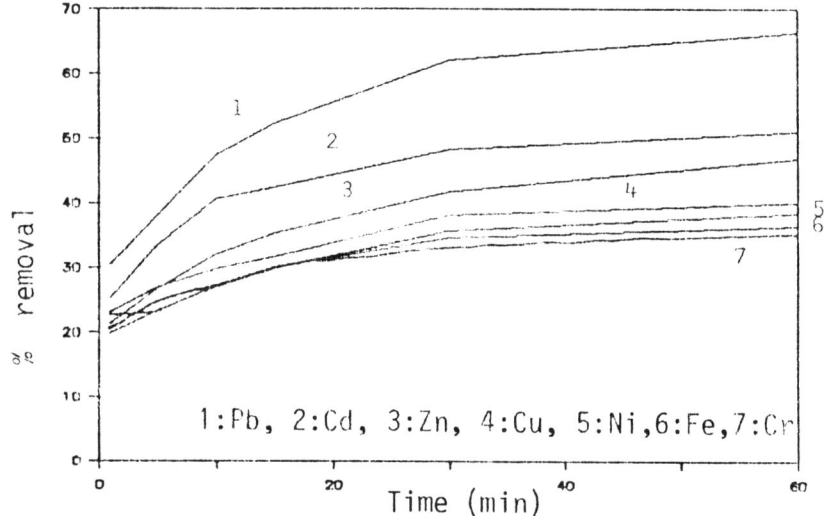

FIGURE 2. *Percentage of Metal Removal by Na-CLI for Single Ion Solutions (50 mg/l).*

TABLE 2. *Percentage Removal of Metal Cations by Clinoptilolite (Na-CLI) Single Ion Solutions of Initial Concentration 50mg/l.*

Exchange time (min)	Percentage of metal removed by the zeolite						
	Pb	Cd	Zn	Cu	Ni	Fe	Cr
1	30.49	25.19	21.33	23.09	19.84	22.71	20.57
5	37.88	33.26	26.56	26.88	23.19	23.18	24.64
10	47.51	40.69	32.09	29.83	27.06	27.21	27.18
15	52.31	42.45	35.27	31.71	29.95	30.09	30.06
30	62.05	48.37	41.74	38.16	35.78	34.64	33.24
60	66.39	50.99	46.87	40.09	38.43	36.36	35.17
1440	71.25	54.84	50.15	47.21	43.84	39.64	39.53

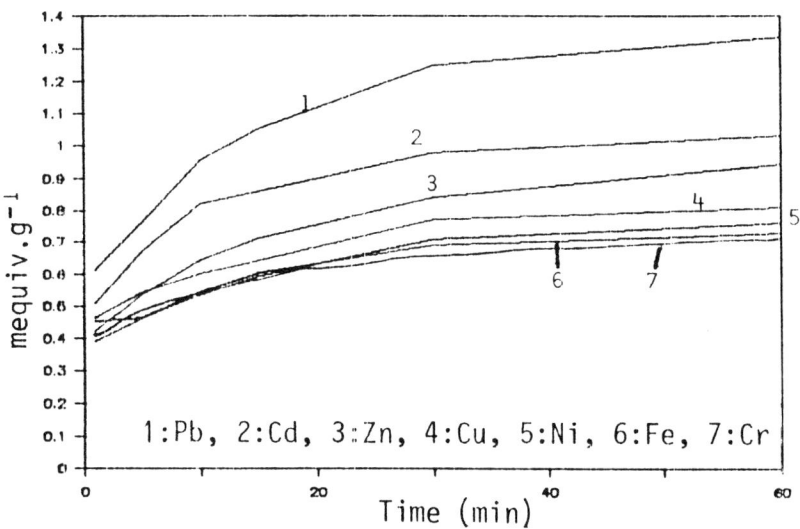

FIGURE 3. *Metal Uptake by Clinoptilolite.*

TABLE 3. *Metal Uptake by Clinoptilolite Expressed in Mequiv/g.*

Exchange time (min)	Percentage of metal removal by the zeolite						
	Pb	Cd	Zn	Cu	Ni	Fe	Cr
1	0.613	0.509	0.424	0.466	0.393	0.455	0.412
5	0.762	0.672	0.535	0.542	0.461	0.465	0.493
10	0.955	0.823	0.646	0.602	0.536	0.546	0.544
15	1.052	0.858	0.711	0.641	0.593	0.603	0.601
30	1.248	0.978	0.841	0.771	0.709	0.691	0.665
60	1.335	1.031	0.943	0.809	0.761	0.728	0.712
1440	1.433	1.109	1.011	0.952	0.868	0.795	0.793

Additional important information is obtained examining Figure 3 and Table 3. The results indicate that the metal uptake expressed in mequiv.g^{-1}, hence, indicating the actual exchange capacity of the zeolite for a particular ion, is much lower than the theoretical exchange capacity. After a 24 hour equilibration process it was shown that the metal taken by the zeolite, with the exception of lead, covered less than half of the theoretical exchange capacity of sodium clinoptilolite.

For the system Pb/Na-CLI, the maximum exchange capacity was 1.43 mequiv.g^{-1}, which corresponds to 56.6% of the theoretical exchange capacity, as shown in Table 4. The Cd/Na-CLI system covered 43.8%, Zn-Na-CLI 39.9%, Cu/Na-CLI 37.6%, Ni/Na-CLI 34.4%, Fe/Na-CLI 31.4%, and Cr/Na-CLI 31.3%.

Mixed Ion Solutions

The results obtained for these ion exchange systems are given in Figures 4-7 and Tables 5-8. In Figure 4, the percentage metal removal is given for the seven metals under examination and for an initial concentration of 100 mg/l, for each metal cation, the total concentration being 700 mg/l. The percentage of metal removed follows the series: Pb > Cd > Zn > Cu > Zn > Cu > Ni > Fe > Cr. This selectivity series follows the series observed in the case of the single type cation exchanges, although the concentrations of the external solution were much higher. At equilibration after 24 hours, 65.4% of the initial lead has been taken by the zeolite. It is worth noting that around 50% is exchanged in the very first minute. This is very useful when treating large amounts of effluents. Table 6 and Figure 5 present the results of the ion exchange systems for an external solution concentration of 50 mg/l for each metal cation, i.e. a total concentration of 350 mg/l. The selectivity series Pb > Cd > Zn > Cu > Ni > Cr > Fe, remained practically unchanged when the external solution concentrations were 175 mg/l and 70 mg/l (Figures 6,7, Tables 7,8). It is worth noting that the selectivity series for four metals based on the hydration energy, was reported [8] to be

$$Pb > Cd > Zn > Cu$$

Thus, copper with the largest hydration energy, 498.7 Kcal/g-ion, prefers the solution phase. The experimental results are in agreement with this selectivity series.

TABLE 4.   Theoretical and Actual Exchange Capacity of Na-CLI After a 24 Hour Exchange.

	Pb	Cd	Zn	Cu	Ni	Fe	Cr
Metal uptake in mequiv/g	1.433	1.109	1.01	0.952	0.868	0.795	0.793
% of T.E.C. taken by cations	56.6	43.8	39.9	37.6	34.3	31.4	31.3

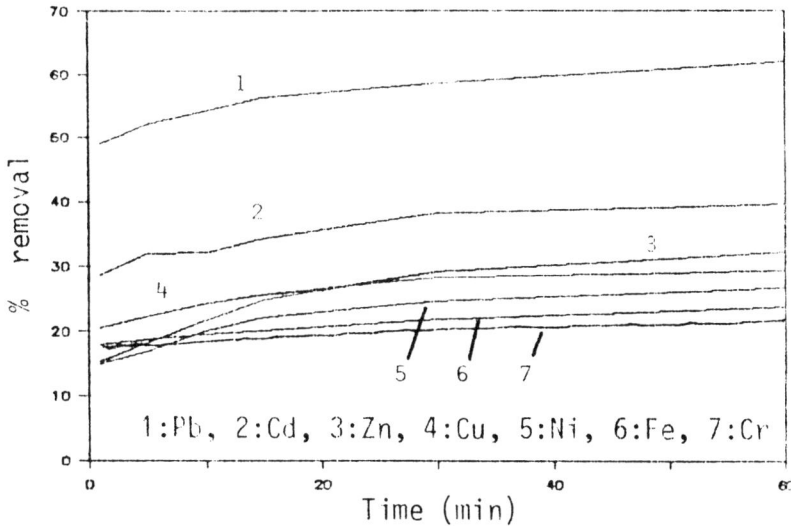

FIGURE 4. Percentage of Metal Removal from Mixed Metal Solution (7x100 mg/l).

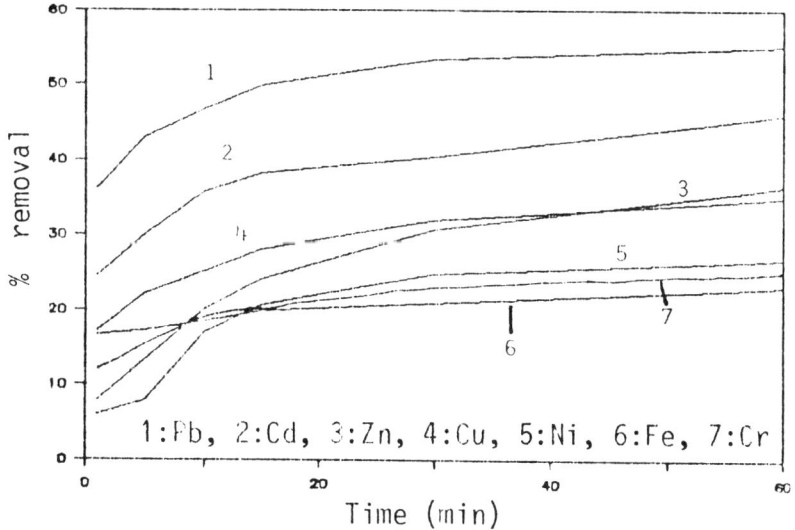

FIGURE 5. Percentage Metal Removal from Mixed Metal Solutions (7x50 mg/l).

TABLE 5. *Percentage Removal of Metal Cations by Clinoptilolite in Mixed Solutions of Initial Concentration 100mg/l of Each Metal.*

Exchange time (min)	Percentage of metal removed by the zeolite					
	Pb	Cd	Zn	Cu	Ni	Fe
1	48.97	28.55	15.48	20.62	15.09	18.01
5	52.11	31.97	18.22	22.29	16.94	18.84
10	54.31	32.12	21.83	24.3	20.13	19.64
15	56.45	34.28	24.89	25.62	22.12	20.09
30	58.72	38.22	29.26	28.41	24.72	21.87
60	62.01	39.71	32.41	29.62	26.82	23.96
1440	65.41	42.33	38.48	34.16	28.19	26.85

Regeneration of the Zeolite

After the metal uptake, the zeolite was examined for the reverse process, i.e. for the release of the metal and the uptake of the sodium ion, which was the initial ion in the zeolite. The results as shown in Table 9, indicate that significant quantities of the heavy metals are released into the solution, but further treatment is probably required to get more metals out of the zeolite. The reversibility of the process is very essential, since the zeolite is meant to be used for many cycles of uptake and metal release, in order to have an application for the effluent treatment. After the regeneration, the zeolite was examined by x-ray diffraction, which indicated that the structure remained unchanged.

Influence of the Anions

Some studies were carried out in order to evaluate the influence of the anions towards the metal removal by sodium clinoptilolite. The metal chosen was nickel, and the anions were Cl^-, NO_3^- and $SO_4^=$. The total external solution concentration was 0.1 N. The selectivity as shown in Figure 8, was higher for nickel when nitrate anions were present, followed by sulphate and chloride:

$$NO_3 > SO_4 > Cl$$

TABLE 6.   Percentage Removal of Metal Cations by Clinoptilolite in Mixed Solutions of Initial Concentration 50mg/l of Each Metal.

Exchange time (min)	Percentage of metal removed by the zeolite						
	Pb	Cd	Zn	Cu	Ni	Fe	Cr
1	36.31	24.66	7.99	17.27	6.07	16.76	11.93
5	43.09	30.07	13.33	22.17	7.94	17.21	15.48
10	46.73	35.76	20.09	25.22	17.06	18.51	19.27
15	49.89	38.26	24.14	28.11	20.71	19.88	20.39
30	53.33	40.35	30.75	32.02	24.81	20.99	23.14
60	55.14	45.99	36.51	35.02	26.82	23.11	25.08
1440	61.03	46.09	39.12	36.51	29.75	28.31	29.66

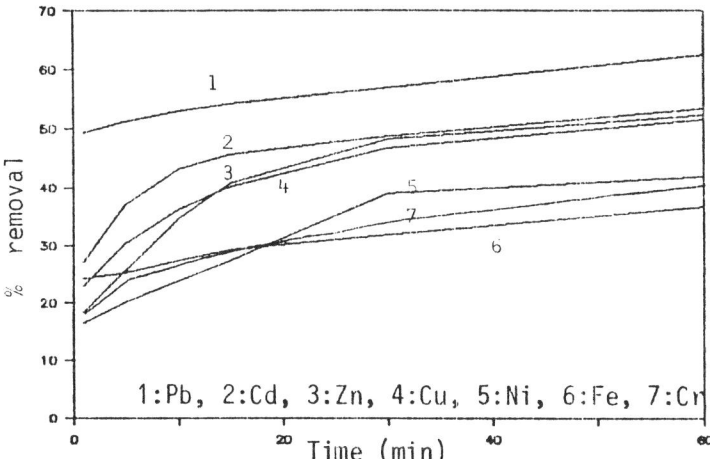

FIGURE 6.   Percentage Metal Removal from Mixed Metal Solutions (7x25 mg/l).

TABLE 7.   Percentage Removal of Metal Cations by Clinoptilolite in Mixed Solutions of Initial Concentration 25mg/l of Each Metal.

Exchange time (min)	Percentage of metal removed by the zeolite						
	Pb	Cd	Zn	Cu	Ni	Fe	Cr
1	49.43	27.08	18.43	22.94	16.57	24.32	18.32
5	51.23	37.14	25.61	30.35	20.11	25.25	24.01
10	53.01	43.21	34.61	36.28	23.95	27.41	26.56
15	54.31	45.69	40.86	40.15	27.45	29.33	29.28
30	56.92	48.79	48.23	46.72	38.89	31.91	34.15
60	62.46	53.47	52.42	51.55	41.92	36.64	40.42
1440	64.94	57.97	55.85	54.23	46.22	38.84	42.93

430 ION SEPARATION

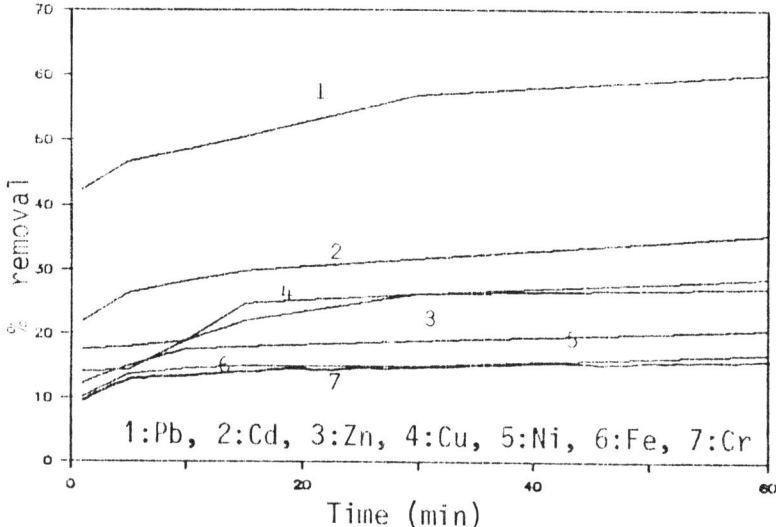

FIGURE 7. *Percentage Metal Removal from Mixed Metal Solutions (7x10 mg/l).*

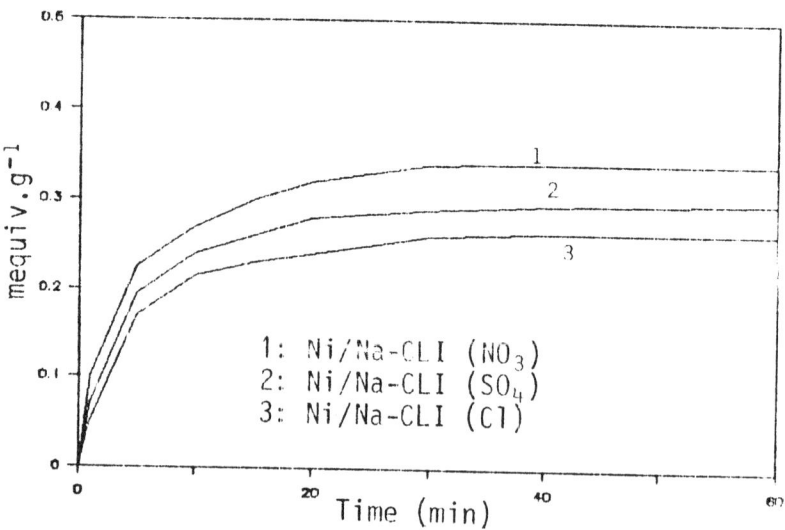

FIGURE 8. *NA-CLI Selectivity Graphs.*

TABLE 8. *Percentage Removal of Metal Cations by Clinoptilolite in Mixed Solutions of Initial Concentration 10 mg/l of Each Metal.*

Exchange time (min)	Percentage of metal removed by the zeolite						
	Pb	Cd	Zn	Cu	Ni	Fe	Cr
1	42.38	21.92	14.21	17.54	12.43	10.26	9.64
5	46.73	26.37	14.42	18.01	15.02	13.73	12.99
10	48.44	28.19	18.91	18.95	17.61	14.69	13.12
15	50.45	29.86	22.01	24.76	17.99	14.97	14.21
30	56.73	31.73	26.23	26.23	18.89	15.01	14.81
60	60.06	35.35	27.15	28.63	20.74	16.89	16.03
1440	62.91	38.57	29.51	31.07	22.91	21.14	20.91

TABLE 9. *Regeneration of Sodium Clinoptilolite.*

Initial metal concentration equilibrated in the zeolite for 24 h	Metal	% of metal released			% of the metal released from the zeolite
		First reg	Sec.reg	Third reg	
100	Pb	11.93	9.41	7.13	28.47
50		18.52	12.29	8.22	39.03
25		22.95	19.63	17.44	60.01
10		31.58	26.04	19.45	77.08
100	Cd	2.21	1.83	1.37	5.41
50		3.61	3.11	2.37	9.08
25		7.71	4.81	4.56	17.08
10		18.11	13.24	12.69	44.04
100	Zn	2.94	2.28	1.49	6.71
50		3.93	2.73	2.12	8.84
25		5.06	2.67	2.52	10.26
10		13.63	6.41	5.48	25.51
100	Cu	2.11	1.75	1.56	5.42
50		3.52	2.08	1.54	7.13
25		5.29	3.22	2.23	10.74
10		13.62	12.16	6.34	32.12
100	Ni	8.91	6.11	4.31	19.31
50		8.73	6.37	5.08	20.26
25		14.17	12.43	10.66	37.25
10		21.24	17.38	14.99	53.62
100	Fe	8.36	5.73	5.08	19.16
50		13.74	11.12	10.68	35.54
25		13.75	11.37	11.16	36.28
10		15.89	14.53	12.42	42.84
100	Cr	20.99	18.17	14.85	53.96
50		28.79	26.46	14.44	69.71
25		34.47	32.85	22.71	90.03
10		35.82	33.34	24.09	93.25

6. CONCLUDING REMARKS

The experimental results indicate that clinoptilolite has quite high ion exchange capacity, compared with clinoptilolite from different origins.

The metal removal percentage is quite satisfactory and the selectivity series for the various cations does not change when the external solution concentration changes. The actual exchange capacity of the zeolite is much lower compared with the theoretical one, indicating that for these particular experimental conditions, the entering cations do not occupy all the available positions in the crystal. This partial ion exchange is very common in natural zeolites. The regeneration process indicates that significant amounts of the metal taken, could be released in an exchange with sodium.

REFERENCES

1. Loizidou, M.D. "Heavy Metals in the Industrial Effluents in the Greater Athens Area," *Ministry of the Environment and Public Works* (1985).

2. Patterson, J.W. *Wastewater Treatment Technology,* (Ann Arbor Science Publisher, Inc. 1978).

3. Patterson, J.W. and R.A. Minear. "Physical-Chemical Methods of Heavy Metals Removal," in *Proceedings of the International Conference on Heavy Metals in the Aquatic Environment*, (Pergamon Press 1975).

4. Loizidou, M.D. and R.P. Townsend. "Ion Exchange of Natural Clinoptilolite, Ferrierite and Mordenite: Part 2 Lead-Sodium and Lead-Ammonium Equilibria," *Zeolites*, Vol. 7, (March, 1987) p. 153.

5. Loizidou, M.D. and R.P. Townsend. "Exchange of Cadmium Into the Sodium and Ammonium Forms of the Natural Zeolites, Clinoptilolite, Mordenite and Ferrierite," *J. Chem. Soc. Dalton Trans.*, (1987) p. 1911.

6. Sakellarides, P.O., M.D. Loizidou, and K.J. Haralambous. "Heavy Metal and Ammonium Removal From Effluents Using Natural Zeolites," *Balkan Scientific Conference in Environmental Protection in the Balkans, Varna,* (Sept. 20-23, 1988).

7. Loukatos, A., P.O. Sakellarides, and M.D. Loizidou. *11th Greek Conference on Chemistry,* The Association of Greek Chemists, Athens, (1986) p. 80.

8. Semmens, M.J. and M. Seyfarth. "The Selectivity of Clinoptilolite for Certain Heavy Metals," *Intern. Conference on Natural Zeolites*, Tuscon, Arizona (1976).

9. Barrer, R.M., R. Papadopoulos, and L.V.C. Rees. *J. Inorg. Nucl. Chem.*, (1967) p. 29.

10. Barrer, R.M. and J. Klinowski. *J. Chem. Soc. Faraday*, 1, 70, 2362 (1974).

11. Golden, C.T. and G.J. Robert. *J. Chem. Eng.*, V. 26, pp. 366-367 (1981).

12. Townsend, R.P. and M.D. Loizidou. "Ion Exchange Properties of Natural Clinoptilolite, Ferrierite and Mordenite: 1. Sodium-Ammonium Equilibria," *Zeolites*, Vol. 4, (April, 1984), p. 191.

13. Barrer, R.M. and R.P. Townsend. *J. Chem. Soc. Faraday Trans.*, 1, 72, 661, (1976).

14. Barrer, R.M. and R.P. Townsend. *J. Chem. Soc. Faraday Trans.* 1, 72, 2650 (1976).

15. Loizidou, M.D. "Ion Exchange of Lead and Cadmium With the Sodium and Ammonium Forms of Some Natural Zeolites," Ph.D. Thesis, The City University London, (1982).

DISCUSSION OF:
HEAVY METAL REMOVAL USING NATURAL ZEOLITE

Robert H. Rosset
Laboratoire de Chimie Analytique, Ecole Superieure de Physique et de Chimie de Paris, 10 rue Vauquelin, 75005 Paris, France

The use of a natural zeolite for the heavy metal removal from aqueous effluents may be an interesting way, due to:

- Economic considerations as zeolite is less expensive than organic ion exchanger

- Good permeability of the columns and pressure resistance so that high flow rate may be obtained.

A zeolite existing in the northern part of Greece, CLINOPTILOLITE (which will be abbreviated as CL), has been studied for this application.

It must be mentioned that only fundamental studies have been done, and this is probably the main drawback of this study. A column study with breakthrough fronts would be of value because it is the only way to decide if a material may be used for the treatment of real-life effluents.

The calculated exchange capacity of CL has been found to be 2.53 meq g^{-1}, which is a high enough value compared to organic ion exchangers as there is limited swelling of the material. The experimental value for sodium exchange is 2.07 meq g^{-1}.

Selectivity Order

This order is the following:

$$Pb > Cd > Zn > Cu > Ni > Fe > Cr$$

I consider that this experimental order is more accurate and significant than the order deduced from $\Delta G°$ values. It is difficult to admit that Cd (II) may be less preferred than sodium.

Heavy Metal Removal

The kinetic exchange curve would have to be related with column breakthrough experiments. If $\approx 70\%$ removal for lead is an interesting value, $\approx 35\%$ for chromium is perhaps insufficient regarding the chromium toxicity and the needs for a quantitative removal.

It is important to notice that exchange is rapid at the beginning. For instance, 49% of lead is removed in one minute and this percentage increases to 62% after one hour.

Regeneration

The regeneration study would need some comments on the feasibility of the process for effluents treatment. The author has used 1 mol L^{-1} $NaNO_3$ solution and a 20 mL volume for a 0.3 g of zeolite. An important practical problem would be:

- Is it economically possible to replace the zeolite without regenerating it in the case of diluted effluents?

- If regeneration is needed, one can calculate that with three regenerations for lead (e.g. at initial concentration of 10 mg L^{-1}), there is a dilution of the solution: the concentration which is initially of 10 mg L^{-1} becomes 6.4 mg L^{-1}.

CONCLUSIONS

CL has indeed interesting properties. However, breakthrough columns experiments are necessary to reach a conclusion on the economical interest of the proposed process.

METAL ION SEPARATIONS FROM HAZARDOUS WASTE STREAMS BY IMPREGNATED CERAMIC MEMBRANES

J. Yi, R. Ferreira and L.L. Tavlarides
Department of Chemical Engineering and Materials Science, Syracuse University, Syracuse, New York 13244

1. ABSTRACT

The objective of the research is to determine the feasibility of an advanced separation technology to selectively extract metal ions from aqueous waste streams by ceramic membranes impregnated with organic chelation acids. The mechanism of coupled transport provides selective extraction of the metal ions into the organic phase as organometallic complexes which then diffuse across the porous membrane and are stripped on the receiving streamside by adjusting the pH of the receiving stream. To provide experimental information to test the concept, the rotating diffusion cell, RDC, is employed with on-line UV-visible spectrophotometry to provide flux data under well characterized hydrodynamic mass transfer conditions.

Meaningful flux data have been obtained using the RDC to test the concept using a synthetic sample to mimic a dilute copper aqueous waste stream. The trial membrane system was an α-alumina/silica membrane (0.2 cm thick, 49-55 μm average pore size, and 41.7 volume % porosity) impregnated with a 100%, 2-M [Cu^{++}] in sulfate solution of pH 5, copper metal flux values of 1.17 x 10^{-5} and 4.82 x 10^{5} gmol/cm^2-hr were obtained with disk rotating speeds of 200 and 350 rpm respectively. These results are reproducible and prove the validity of the experimental technique and the concept of metal ion separation via impregnated ceramic membranes.

Membrane modules of shell and tube configurations are also envisioned feasible for multiple metal ion separation systems. Assuming that adequate fluxes can be obtained, the use of several membrane modules arranged in an appropriate sequence with proper choice of chelation acids will selectively remove metals from hazardous multiple metal ion streams. This capability of producing concentrated specific metal streams will permit the recovery and recycle of many metal ions.

2. INTRODUCTION

Disposal of toxic and hazardous metal ions in aqueous waste streams is a significant industrial waste problem. Waste streams from the electronics, electroplating, and photographic industries contain metal ions such as copper, nickel, zinc, chromium(II), chromium (IV), cadmium, aluminum, silver and gold, amongst others in various aqueous solutions such as sulfates, chlorides, fluoroborates and cyanides. Compositions of typical plating solutions are listed in Table 1.

Next to precious metals, strategic non-ferrous metals are the most important for recycling and removal [1,2,3,]. The non-ferrous component of municipal waste in the United States is estimated to be about 2%, with aluminum the predominant metal. Non-ferrous metals other than aluminum amount to approximately 500,000 tons per annum. Nickel, for example, is one of the principle non-ferrous metals of economic significance, yet only about 10% is recycled [4].

The predominant existing method of treatment of these waste streams is the precipitation of the metal ions in the form of hydroxides or carbonates [5], which results in a sludge that is dumped on hazardous waste sites. Other recovery methods recently considered include evaporation, reverse osmosis, ion exchange [6], electrolytic metal recovery, and solvent extraction. These processes have met with various degrees of success.

The use of liquid ion exchange chelation molecules to selectively extract metal ions by on site treatment of these waste streams is currently receiving substantial interest [7-12]. This interest stems from an awareness of the advances made in the development of commercial extractants for metal ion separations in the hydrometallurgical industries [13,14]. Metal cations react with the organic acids and acid chelating agents to form neutral complexes that are preferentially dissolved by the organic phase:

$$M^{+n} + \overline{nHR} \rightleftharpoons \overline{MR_n} + nH^+ \qquad (1)$$

Here the overbar denotes species in the organic phase. The above equation describes a cation exchange reaction wherein hydrogen ions are exchanged for the metal cation, so the degree of extraction of metal ions depends on the pH of the aqueous phase. The forward and reverse steps represent the extraction and stripping steps respectively.

The promise of such systems for treatment of these wastes is exemplified in recent works. Knocke et al. [8] indicated that

Table 1. Composition of Plating Solutions*

Metal	Purpose	Plating Solutions	Composition of Plating Solutions
Copper	Printed circuit boards, undercoat in decorative finishes	Plain cyanide Rochelle cyanide Copper sulfate Copper Fluoroborate	$CuCN, NaCN, Na_2CO_3$ $CuCN, NaCN, Na_2CO_3, NaOH$ Rochelle salt $CuSO_4 \cdot 5H_2O, H_2SO_4$ $Cu(BF_4)_2$
Nickel	Bright coating under thin Cr electroplate for decorative, corrosive- and wear-resistance purposes	Watts Sulfamate Fluoroborate Chloride	$NiSO_4, NiCL_2, Ni, H_3BO_3$ $NiCl_2, (NiSO_3NH_2)_2, Ni, H_3BO_3$ $NiCl_2, Ni(BF_4)_2, Ni, H_3BO_3$ $NiCl_2, Ni, H_3BO_3$
Chromium	Decorative or industrial finishes	Chromic acid	H_2CrO_4, H_2SO_4 or H_2SO_4F-
Zinc	Protect iron and steel against corrosion	Cyanide Noncyanide baths	$Zn(Cn)_2, NaOH, NaCN, Na_2S_5$ or Na_2S_4, $Zn_2P_2O_7, Na$ citrate, EDTA
Cadmium	Corrosion protection	Cyanide Fluoroborate	$CdO, Cd, NaCN, NaOH, Na_2CO_3$ $CdBF_4, Cd, NH_4CN, H_3BO_3$
Lead or Lead-tin Alloys	Improves solderabillity, coating properties and performance of steels	Fluoroborate	Pb, HBF_4, H_3BO_3, glue, resorcinal, gelatin, hydroquinone
Tin or Tin Alloys	Improve solderability, corrosion protection, antifriction properties	Sulfate Fluoroborate Halide	$SnSO_4, H_2SO_4$, gelatin, B-naphthol Gelatin, B-naphthol, Sn, HGR_4 $H_3BO_3, SNCl_2$ or SnF_2
Gold	Engineering (switches, semiconductors), decorative	Cyanide Acid	$AuCN, KCN, K_2CO_3, K_2HPO_4$, alloy metals, $Au(CN)_3, CN-$, citrates
Iron	Rare: for electroformed parts, dies, and cylinder liners	Chloride Sulfate/Chloride Fluoroborate	$FeCl_2, CaCl_2$ $FeCl_2, NH_4Cl, FeSO_4$ $Fe(BF_4)_2, NaCl, H_3BO_3$

*Data from Jacobsen, Kurt, and Laska, Richard, "Advanced Treatment Methods for Electroplating Wastes." *Pollution Engineering*, October, 1977.

Source: Industrial Pollution Control: Issues and Techniques. Van Nostrand Reinhold, 1981.

thenoyltrifluoroacetone and 8-hydroxy quinoline are promising extractants for separating Cu, Ni, Zn, Cr and Cd by liquid-liquid extraction in plating waste streams. Clevenger and Novak [9] attempted to find extractants that would selectively extract a single metal from a plating sludge that contained cadmium, chromium, copper, nickel and zinc by sequential liquid-liquid extraction. They found that acetylacetone and thenoyltrifluoroacetone selectively extracted copper and sodium diethyl dithiocarbomate removed all the metals from the sludge. However, reagent activity loss reduces the promise of this system. McDonald and Bajiva [7] demonstrated that tricaprylylamine was capable of extracting 97 wt% of hexavalent chromium from metal plating waste water and that the extractant could be reused. Comparisons of liquid-liquid extractions with evaporation and resin ion exchange suggest that liquid extraction could be the most cost effective if the process water can be recycled. In another study, Brookes [10] conducted sequential solvent extractions of multicomponent metal wastes of iron, zinc, copper, nickel and chromium employing di-2-(ethylhexyl) phosphoric acid, LIX622 (a hydroxy oxime), lauric acid and acetyl acetonate. The results are promising in that selective separation could be obtained by varying the pH to be compatible with the extraction and stripping range of the given chelation agent, and that trivalent chromium was separable with acetyl acetonate.

The technology of ceramic mineral membranes has advanced rapidly in recent years to a state where they can have significant potential for facilitated or coupled transport of chemicals and metal ions in separation processes. Mineral membranes appeared in 1980 with the marketing of the CARBOSEP [15] membrane based on the concept of a macroporous support (inside diameter 6mm) made of carbon, coated with a thin microporous layer of zirconia. These CARBOSEP membranes can work in the entire pH range (0 to 14) at temperatures up to 300°C, pressure gradients as high as 15atm, and are inert in all media. Since the success of the CARBOSEP membranes in separating uranium isotopes, new ceramic membranes such as MEMRALOX, DERAFLO, DYCACERAM, CERAM-FILTRE using alumina, silicon carbide and mixed minerals have been developed [16].

The above multilayered ceramic membranes have been manufactured in planar or cylindrical geometry using sol gel dipping, coextrusion of two powders of different grain sizes, coating the primary substance with a fine layer using the slip casting technique and even spraying of a painting on a cylindrical mandrel. These preparative techniques result in the preparation of a microporous membrane layer on top of a macroporous support. The significant advantages of these multilayered ceramic membranes include stability in organic media and at temperatures up to 500°C; good behavior in the whole pH range, reliability and long lifetime; neither creep nor deformation during cooling; and high permeability. Accordingly, these membranes, impregnated with suitable chelation agents, have great promise for selective removal of metal ions.

3. Coupled Transport of Metal Ions in Ceramic Membranes

Immobilized liquid membranes have received considerable interest for possible hydrometallurgical application [17]. In this process, solutions of the liquid ion exchange molecules are impregnated into the pores of a solid matrix support, typically polymeric membranes. A conceptual example is shown in Figure 1 where the forward reaction occurs on the feed side and the reverse reaction occurs on the receiving side. In the pores of the membrane, the chelated metal ion is transported to the receiving side and regenerated ion exchange molecules diffuse in the opposite direction to renew the process. The solution can be fixed in the pores due to capillary forces by properly selecting the membrane pore dimension and porosity.

The model to calculate flux through the membranes assumes that the copper ions experience a series of resistances which include transport through the aqueous feed solution, complexation reaction at the membrane/feed solution interface, diffusion across the membrane, stripping reaction at the membrane/strip solution interface, and transport through the aqueous strip solution. These steps are shown in Figure 2. It can be shown that for the extraction reaction

$$Cu^{++} + 2\overline{HR} \rightleftharpoons \overline{CuR_2} + 2H^+$$

occurring on either side of the membrane/film interface, that a plausible reaction mechanism yields an expression for the reaction rate at these interfaces which can be approximated by a simple first order rate law of the form

$$\text{rate} = k_f^* [Cu_{++}] - k_r^* [\overline{CuR_R}] . \qquad (2)$$

where k_f^* and k_r^* are pseudo rate coefficients. Assuming steady state, the flux equations for metal ion transport in Figure 2, J_S may be written as,

$$J_S = K_W \Delta C \qquad (3)$$

where

$$\Delta C = \frac{C_o}{K_1} - \frac{C_5}{K_2} \qquad (4)$$

$$\frac{1}{K_W} = \frac{\delta_1}{D_1 K_1} + \frac{1}{\varepsilon k_1} + \frac{\delta_m}{D^e_m \varepsilon} + \frac{1}{\varepsilon k_2} + \frac{\delta_2}{D_2 K_2} \qquad (5)$$

and

$$K_1 = \frac{k-1}{k_1} = \frac{\bar{C}_1}{C_2} \qquad K_2 = \frac{k-2}{k_2} = \frac{C_4}{\bar{C}_3} \qquad (6)$$

C is the metal ion concentration, $\bar{C}$ is the metal ion complex concentration in the organic extractant, D is the diffusivity D^e_m is the effective diffusivity in the membrane phase, ε is the porosity, δ is the boundary layer thickness, and δ_m is the membrane thickness.

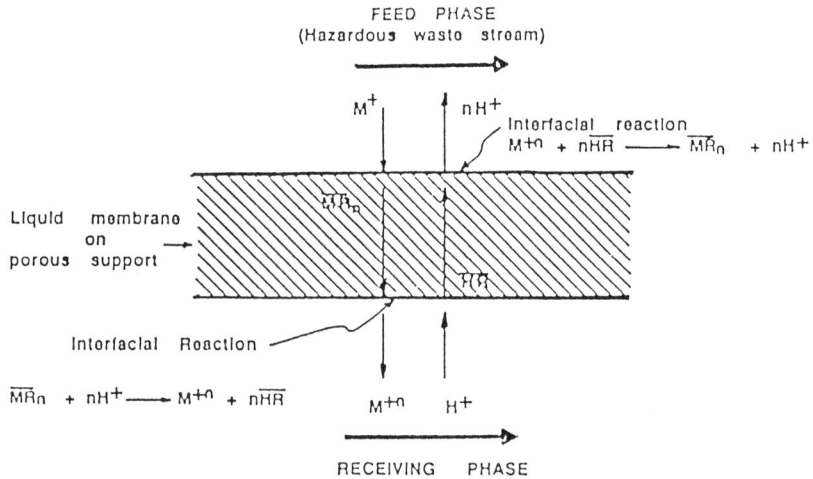

Figure 1. Coupled Transport of a Metal Ion in Ceramic Membrane

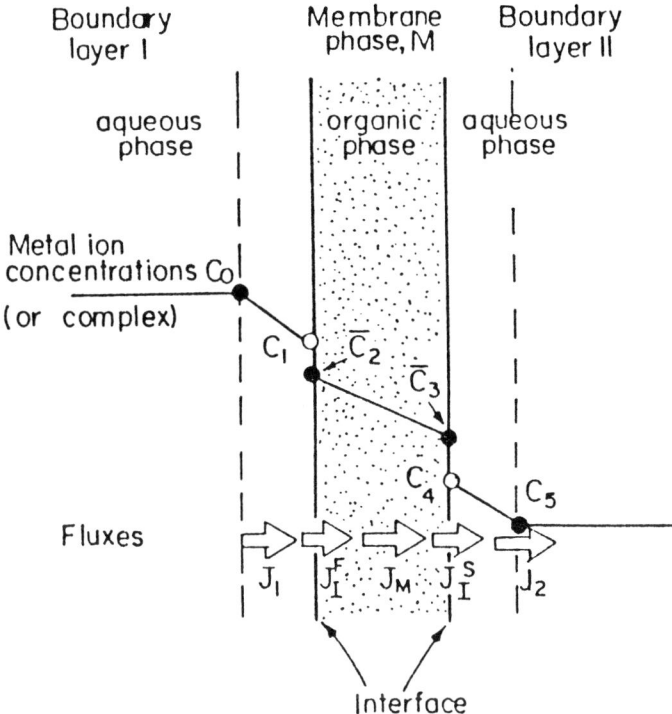

Figure 2. Schematic Representation of Metal Ion Extraction Flux using ICM

Equations (3)-(6) allow one to determine the parameters from designed tracer experiments, since the hydrodynamics and interfacial area for the RDC apparatus are well-defined and the film thickness is given by

$$\delta = 0.643 \, D^{1/3} \, \nu^{1/6} \, \omega^{-1/2} \tag{7}$$

where δ is the boundary layer thickness, D is the diffusivity coefficient, ν is the kinematic viscosity and ω is the rotating speed.

4. Experimental Apparatus and Technique

The rotating diffusion cell (RDC) is employed to obtain overall mass transfer rates and kinetic rate coefficients for the metal ion coupled transport across membranes. The RDC is one of the few examples where an exact solution of the hydrodynamic equations can be obtained.

For this study, the RDC has been suitably modified to incorporate an impregnated ceramic membrane, which brings the aqueous phase and strip phase into contact. Figure 3 is a schematic of the entire apparatus and shows that instantaneous monitoring of the internal fluid is possible using an on-line UV-visible spectrophotometer. Also, a pH unit and controller will permit maintenance of constant pH during the flux experiments. The system is isothermally controlled, a N_2 purge prevents oxidation of the solutions, and a controller is used to maintain constant rpm.

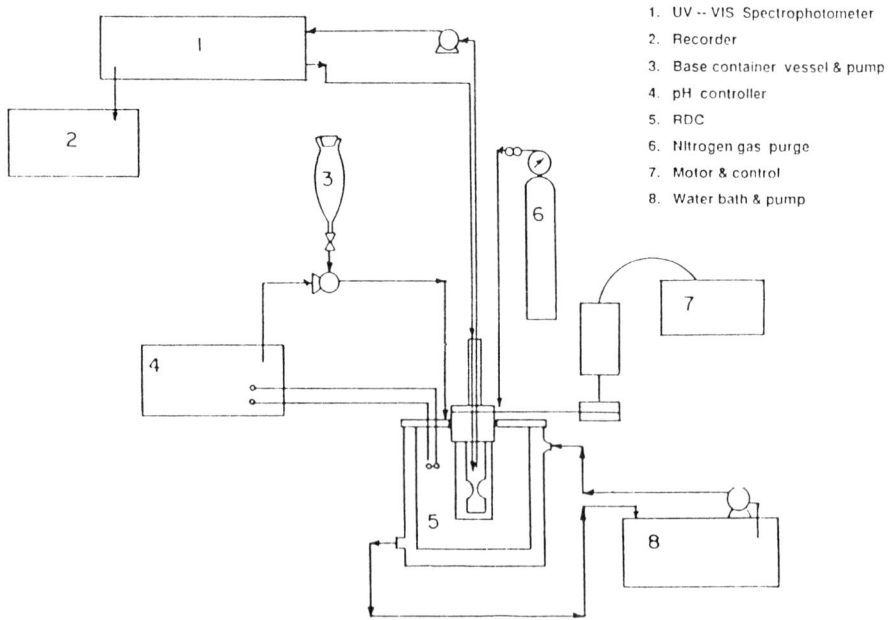

Figure 3. Rotating Diffusion Cell Set-up.

Figure 4 is a detailed design of the rotating diffusion cell (RDC). It is comprised of a thermostated beaker outside compartment and a glass internal compartment. A teflon internal baffle is located in the center of the RDC and is stationary during its operation, while the internal compartment rotates giving the necessary hydrodynamics. A teflon pulley block supports the internal compartment as well as the wheel bearing used to rotate it. The internal compartment has an approximate volume of 100 cc while the outer compartment has volume of 400 cc.

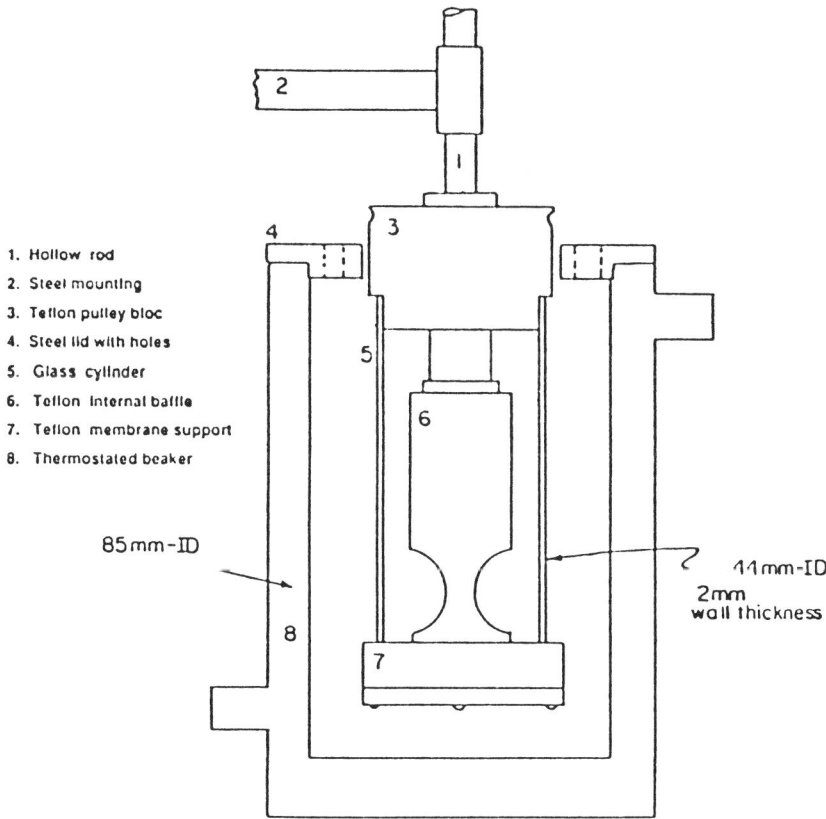

1. Hollow rod
2. Steel mounting
3. Teflon pulley bloc
4. Steel lid with holes
5. Glass cylinder
6. Teflon internal baffle
7. Teflon membrane support
8. Thermostated beaker

Figure 4. Rotating Diffusion Cell.

Analytical measurement of the copper transferred into the stripping solution is made in two ways. First, the stripping solution is directly measured by UV visible spectrophotometry without destroying the sample by circulating a portion of this stream constantly through a flow cell in the spectrophotometer and back into the RDC. In addition, a pH control unit for the feed solutions automatically adds and monitors the hydroxide ions fed to the feed solution to maintain constant pH. The hydrogen ions thus buffered are stoichiometrically related to the copper ions transported across the membrane. This technique provides a second independent method to compute the copper flux through the membrane by either measuring the copper concentration change in the stripping solution or monitoring the number of moles of hydroxide added to the feed solution.

A membrane impregnation apparatus was designed, constructed and commissioned to uniformly impregnate the ceramic membranes with the organic phase solutions of chelation acid with or without the carrier solvent. The apparatus is suitable for membrane impregnation by direct contacting of untreated membranes or treated membranes.

In this research the dilute metal ion aqueous phase is simulated with a copper sulfate solution of varying concentrations between 0.003 mol/l to 0.3 mol/l. The organic mixture for ceramic membrane impregnation is LIX84*, 2-hydroxy-5-nonylacetophenone oxime, in Kermac-400/500†, a 91% aliphatic and 7.9% aromatic kerosine. LIX84 and Kermac-400/500 are commercially available and have been widely used in the hydrometallurgical industry.

Ceramic membranes have been secured which cover a wide range of porosity and average pore diameters. They are composed mostly of α-alumina/silica with average pore sizes of 4.5 μm to 110 μm and porosites of 36 to 57 volume percent.

5. Preliminary Results

The calibration data between the copper concentration and the absorbance by UV-spectrophotometry shows a good linear relationship as shown in Figure 5 and permits direct measurement of flux. It should be noted that although the absorbances at 0 ppm concentration are changed intentionally by varying the set point voltage, the slopes of both curves on Figure 5 are the same. This result confirms that the copper flux through the impregnated membrane can be measured independently on the initial absorbance

---

*   Trademark of Henkel Corp.
†   Trademark of Kerr-McGee Corp.

simply by monitoring the absorbance change with time. The copper flux into the stripping solution, J_S, can be given as

$$J_S = \frac{V_S \, dC_S}{A_S \, dt} \tag{8}$$

where V_S is the volume of stripping solution, A_S is the contact area of the extractant between the membrane and the stripping solution, and C_S is the copper concentration in the stripping solution. Figure 6 shows that the relationship between the absorbance, A_{abs} and the copper concentration, C_S, can be expressed as

$$A_{abs} = \alpha \, C_s + \beta \tag{9}$$

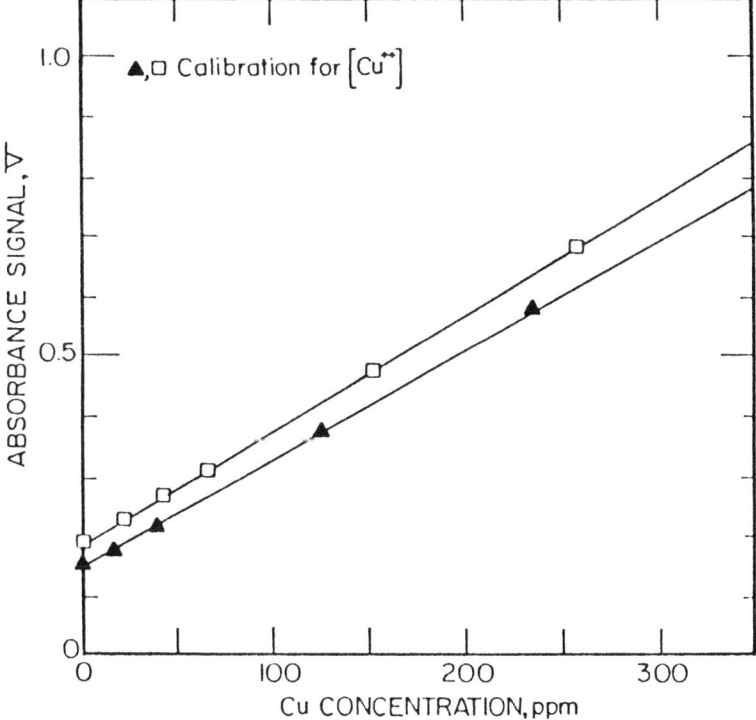

Figure 5. Calibration Line for the Copper Solution.

448   ION SEPARATION

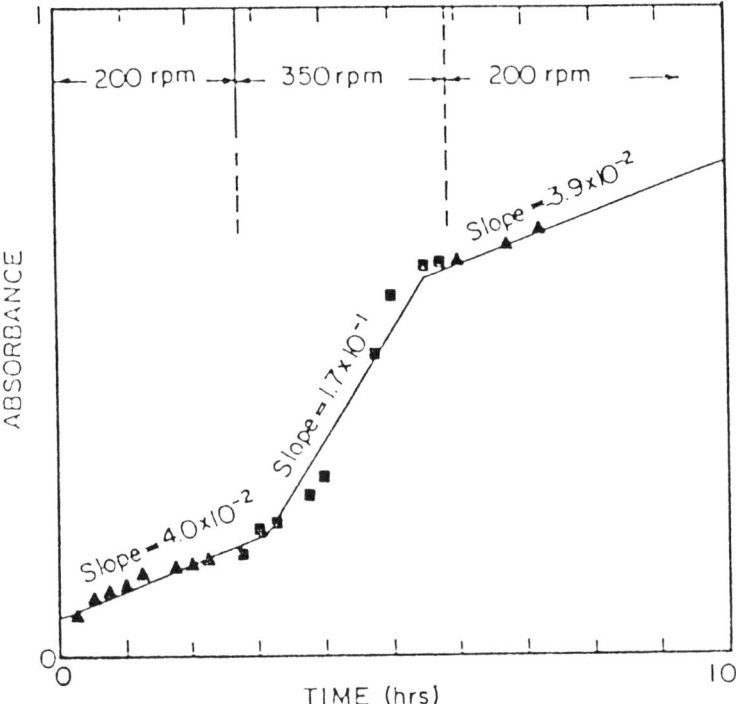

Figure 6. Preliminary Copper Extraction Experimental Result Using Impregnated Inorganic Membrane.

Where α is the slope, and β is the intercept. Accordingly, equation (8) may be written as

$$J_S = \left(\frac{V_S}{A_S}\right) \left(\frac{1}{\alpha}\right) \left(\frac{dA_{abs}}{dt}\right) \quad (10)$$

Equation (10) implies that the flux can be measured from the monitor of the absorbance change with time. Also, when this change with time (dA_{abs}/dt) is constant, the steady state flux is measured.

Preliminary flux experiments show that the RDC is suitable to investigate and to characterize various impregnated ceramic membranes for the metal ion extraction process as shown in Figure 6. Absorbances by UV spectrophotometer are directly monitored with time. For this experiment, 0.2 wt% Cu^{++} solution, 100% LIX84, and 3.06N sulfuric acid were used as feed, impregnating phase, and stripping solution, respectively. The inorganic membrane has the porosity of 41.7 volume % and a pore size range of 49־55µm. The acidity of feed was maintained at pH = 5. The rotating speeds were changed from 200 rpm to 350 rpm, and, subsequently, reduced to 200 rpm to investigate the reproducibility of RDC apparatus and the capability to obtain reliable flux data.

The slopes in Figure 6 are related to the copper ion flux transported through the impregnated membrane by equation (10). As the rotating speed increases, the flux increased due to the decrease of the diffusion resistance through the boundary layer as expected from equation (3)-(7). When the rotating speed is changed from 350 rpm to 200 rpm, the flux decreases. Furthermore, the two slopes for the flux experiments at 200 rpm are parallel. This implies that RDC apparatus is feasible to investigate and to characterize the impregnated membranes for the metal ion extraction process.

6. Conclusions and Future Work

The preliminary experimental results were reproducible and proved the validity of the experimental technique and the concept of metal ion separation via impregnated ceramic membranes. Also, the RDC allows one to obtain overall mass transfer rates and kinetic rate coefficients for the metal ion transport through the impregnated membranes.

A detailed study is in progress to define capabilities of the ceramic membrane system for the copper and nickel metal ion systems.

ACKNOWLEDGMENTS

The financial support provided by the New York State Center for Hazardous Waste Management under grant number 150-W012A for this project is gratefully acknowledged.

NOTATION

A_{abs} Absorbance

A_S The contact area of the extractant

C Metal ion concentration

$\bar{C}$ Metal ion complex concentration in the organic extractant

D Diffusivity

D^e_m Effective diffusivity in the membrane phase

J_s Flux

k Lumped parameter consisting of reaction rate coefficients and concentrations of the other reacting species

δ Boundary layer thickness

δ_m Membrane thickness

ε Porosity of a membrane

ω Rotating speed

υ Kinematic viscosity

REFERENCES

1. Spendlore, M.J., Bureau of Mines Information Circular 8711, (1976).

2. U.S. Department of Commerce, TIS PB245924, (1975).

3. Hanna, H.S. and C. Rampacek. "Fine Particles Processing," Ed. P. Somasundaran, *A.I.M.E.*, New York, NY (1980).

4. U.S. Department of Commerce, U.S. Office of Technology Assessment, Washington, D.C. PB80-102619, (September, 1979).

5. Lee, E., W. Strangio, and B. Lim. *Proceedings of the 41st Industrial Waste Conference, Purdue University, West Lafayette, IN*, Ed. John H. Bell, (Lewis Publishers, Inc., Chelsea, MI, May 13-15, 1986), pp. 652-658.

6. Dejak, M and T. Nadeau. *Hazardous Waste and Hazardous Materials*, 4, N3, (1987), 261-271.

7. McDonald, C.W. and R.S. Bajwa. *Separ. Sci.*, 12 (4) (1977), 435-445.

8. Knocke, W. R., T. Clevenger, M. M. Ghosh, and J. T. Novak. *Proceedings of the 33rd Industrial Waste Conference, Purdue University, West Lafayette, IN*, Ed. J. M. Bell, (Ann Arbor Science Publishers, Inc., Ann Arbor, MI, May 9-11, 1978). pp. 413-426.

9. Clevenger, T. E. and J. T. Novak, *Journal WPCF*, 55 (7), (1983) 984-989.

10. Brooks, C. S. *Proceedings of the 41st Industrial Waste Conference, Purdue Univ. West Lafayette, IN*, (Lewis Publishers, Inc., Chelsea, MI, May 13-15, 1986) pp. 642-651.

11. Davis, M. L., M. Chang, D. Copedge, and M. Strong. *Proc. of the 42nd Industrial Waste Conf., Purdue Univ. West Lafayette, IN*, Ed. J. M. Bell, (Lewis Publishers, Inc., Chelsea, MI, May 12-14, 1987) pp. 803-807.

12. Brooks, C. S., Ibid. pp. 847-852.

13. Tavlarides, L. L., J. H. Bae, and C. K. Lee. *Sep. Sci. and Tech.*, 22, (2,3), pp. 581-617, (1987).

14. Murthy, T. K. S., K. S. Koppiker, and C. K. Gupta. in *Recent Developments in Separation Science*, Vol. III, Ed. N. N. Li and J. D. Navaratil, (1986), pp. 1-44.

15. Cacciola, A. and P. Leung. EPO 040.282, Publ. (Nov. 25, 1981).

16. Charpin, J., P. Bergez, F. Valin, H. Barmier, A. Maurel, and J. M. Martinet. in *High Tech. Ceramics*, Ed. P. Vincenzini, (Elsevier Science Pub., B. V. Amsterdam, 1987), pp. 2211-2223.

17. Kim, K., *J. Mem. Sci.*, 21, 5-10, (1984).

DISCUSSION OF:
METAL ION SEPARATIONS FROM HAZARDOUS WASTE STREAMS BY IMPREGNATED CERAMIC MEMBRANES

Peter C. Nwogu
Department of Materials Engineering, AT&T Network Systems
Norcross, Georgia 30071

INTRODUCTION

In the last decades, few types of environmental pollution control technology have been introduced in the market. Each new advanced control technology has a burden to prove distinctive qualities in order to displace the existing commercial available technology for the same purpose [1, 2].

The objective of this paper is to discuss few Best Available Technology Economically Achieveable (BAT) i.e. electro-chemical method, ion exchange method, cementation method in comparison to the Impregnated Ceramic Membrane method under laboratory evaluation. A copper-bearing waste stream will be used for discussions [3, 4, 5, 6].

In the United States, there was a public concern and outrage over the consequences of heavy metals in the environment. Toxic metals were found to cause problems to Public Owned Treatment Works (POTW), runoff, and surface waters [2, 3]. In November 1982, the Environmental Pollution Agency (EPA) addressed the issue of the presence of aqueous complexing agents bearing toxic metal ions and are discharged into the environment. In July 15, 1983, the USEPA developed copper pretreatment standards for new and existing metal finishing sources. Two types of permits were developed. The National Pollution Discharge Elimination Sources (NPDES) and the Public Owned Treatment Works (POTW) [7, 8].

This development was an effort to regulate industries that discharge effluents bearing copper and other heavy metals into the environment. Industries such as printed circuit boards, telecommunication equipment manufacturers to name a few were affected [9].

By the nature of the copper process operation, waste stream effluent process contains organic complexing agents such as Ethylene Diamine Tetra Acetic Acid (EDTA) or other amino acids [3, 6].

It became the law of the land in July 1983 that pretreatment standards must be adopted and enforced, in order to meet the goal of the elimination permit. Affected industries must meet an average monthly which shall not exceed 2.07 mg/liter copper.

Industries such as Printed Circuit Board, may not exceed pretreatment standards of 4.5 mg/liter, maximum for any one day and 2.7 mg/liter for the average of four consecutive monitoring days [6, 10].

However, municipal or counties public owned treatment works have the option to require stringent standards for indirect pretreatment effluent dischargers i.e. a POTW permit may require less than 0.1 mg/liter copper per any given day monitored [4, 6].

Since 1983 changes have been made in the United States to specifically address the copper-forming point source category. The USEPA made these recent provisions in the Code of Federal Regulations (40 CFR 125.30 through 125.32). These liberal provisions allow industries to meet effluent limitations for toxic metals under flexible techniques of technology. An example was the introduction of effluent Limitation Standards, with the best available technology economically achievable (BAT). The unit of measurement of pollutants were modified under these provisions.

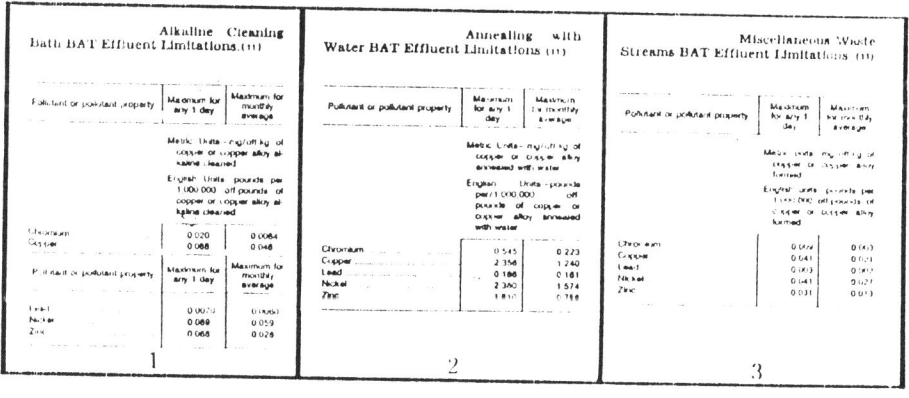

TABLES 1, 2 and 3. BAT Effluent Limitations As Were Provided and Became Effective in August, 1986 [9, 11].

The agency also established pretreatment standards Tables 4 and 5 for existing sources (PSES), for similar copper process operations [8, 11]

Annealing with Water PSES.[11]

Pollutant or pollutant property	Maximum for any 1 day	Maximum for monthly average
	Metric units—mg/off-kg of copper or copper alloy annealed with water	
	English units—pounds per 1,000,000 off pounds of copper or copper alloy annealed with water	
Chromium	0.545	0.223
Copper	2.356	1.240
Lead	0.188	0.161
Nickel	2.380	1.574
Zinc	1.810	0.756
TTO	0.808	0.421
Oil and grease [1]	24.800	14.880

[1] For alternate monitoring.

Alkaline Cleaning Rinse PSES.[11]

Pollutant or pollutant property	Maximum for any 1 day	Maximum for monthly average
	Metric units—mg/off-kg of copper or copper alloy alkaline cleaned	
	English units—pounds per 1,000,000-off pounds of copper or copper alloy alkaline cleaned	
Chromium	1.854	0.758
Copper	8.008	4.214
Lead	0.632	0.547
Nickel	8.090	5.351
Zinc	6.152	2.570
TTO	2.739	1.432

TABLE 4.   Annealing with Water PSES.

TABLE 5.   Alkaline Rinse PSES.

COPPER ION SOURCES

The most common source of metal ions into the environment are industrial plating and electro-depositing baths solutions. Copper articles are manufactured by deposition from solution of copper as copper sulfate [1, 3].

Table 6 shows a list of common industries for copper bath chemicals in the electroplating process, where by-copper is deposited.

COOPERATIVE EFFORTS

Joint efforts are continuing between academicians, industries, government agencies, and private citizens to combat and win the war against waste stream environmental pollution.[1,2,3,10] As a result, different techniques are continuously developed to pretreat industrial waste streams to the minimization of waste stream volumes, prior to discharge to the environment, and are continuously under these cooperative efforts [4, 5, 6].

Operation	Typical bath constituents and proportions		Principal forms of copper in bath
	Compound	Concentration, mg/liter	
Plating	CuCN NaCN Na₂CO₃ KNaC₄H₄O₆·4H₂O NaOH to pH 12.6	26 35 30 45	NaCu(CN)₂ Na₂Cu(CN)₃ Na₃Cu(CN)₄
Plating	CuCN NaCN HCN NaCNS NaOH KOH	75 45 g 57 g 9 g 15.0 21.0	NaCu(CN)₂ Na₂Cu(CN)₃ Na₃Cu(CN)₄
Plating[b]	CuCN NaCN NaCNS NaOH	75 84 9.4 19	NaCu(CN)₂ Na₂Cu(CN)₃ Na₃Cu(CN)₄
Plating	Copper Pyrophosphate Oxalate Nitrate Ammonia pH	22-38 150-250 15-30 5-10 1-3 8.2-8.8	K₆Cu(P₂O₇)₂
Electro-deposition and plating	CuSO₄ 5H₂O H₂SO₄	150-250 45-110	CuSO₄
Electro-deposition	Cu(BF₄) HBF₄ H₃BO₃ pH 1.2-1.7	224-448 15-30 15-30	Cu(BF₄)₂

[a] Taken from Reference 9.
[b] Plating solution used in this study for cyanide plating solution tests.

TABLE 6.   Industrial Plating and Electro-Depositing Baths.

IMPREGNATED CERAMIC MEMBRANE METHOD

The Impregnated Ceramic Membrane technique (ICM) has potential for the recovery of reusable dragout. Dragouts are obtained in electroless copper baths. Therefore, Figure 1 is a schematic system where such a system may be applied [1].

The ICM application may serve as a unit to treat alkaline cleaning baths or any miscellaneous waste stream. Some of the components of the ICM, such as the UV-VIS spectrophotometer may serve as an on-line monitoring of the copper ion concentration in the rinse tank. A high concentration of copper ion is a reflection of high adsorbency, which may indicate poor plating depending on the copper plating solution.

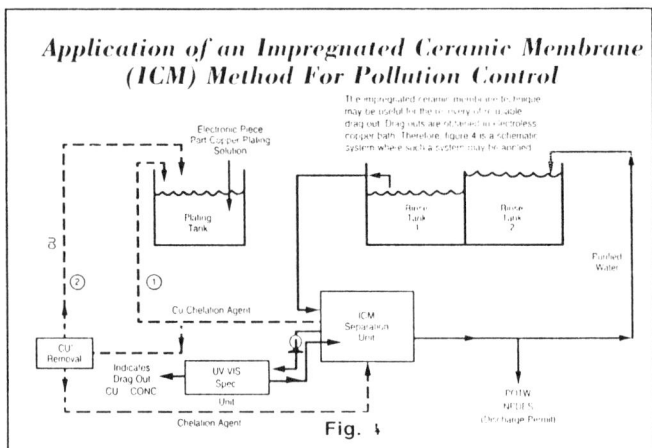

FIGURE 1. *Application of an Impregnated Ceramic Membrane (ICM) Method for Pollution Control.*

ION EXCHANGE METHOD

The ion exchange techniques over recent years have served as an economic method to remove copper or other precious metals from complexed solution. The problem is that the resin to remove copper has to be a powerful complexing agent in order to selectively remove the complexed copper from the copper complexing compound in the electroless copper effluent [7].

Figure 2 shows a schematic for an ion exchange technique [9].

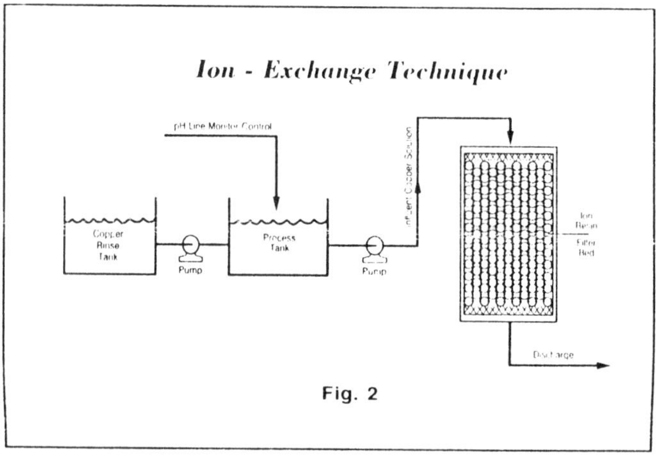

FIGURE 2. *Ion-Exchange Technique.*

ELECTROCHEMICAL METHOD SEPARATION

Commercial techniques such as the electrochemical method are currently used in plating industries to selectively remove copper metal ion from solution. The solution contains copper in a complexed form and are separated by electrochemical displacement.

The electrochemical displacement technique involves pH adjustment in the influent copper solution. This technique operates with a pH of 4 under room temperature. This process can handle a capacity flow rate of 5 gallons per minute. The process can also function under the presence of EDTA copper solution, with a high load up to 60 milligrams per liter of CU-EDTA, depending upon the system design [5].

During the process, copper is removed by a filter media. The copper can be recovered and the effluent waste stream is discharged [5, 6].

CEMENTATION METHOD

Brass mill process industries have used the Cementation Method for metal recovery. During the process of cementation, shredded iron is used for separating metal irons from waste streams bearing copper. The waste stream is an acid solution, and has a pH of approximately 4.0. An electromotive force is applied to the process to create a contact between the metal-bearing solution with the shredded iron metals. The process can operate with a feed solution of 20 grams per liter copper (20 g/l Cu). During the process, copper is recovered as cement copper [5]. The iron is in dissolution, but later removed by the lime precipitation method. The Cementation Method may also be applied to treat waste streams containing cyanide solution, thereby, separating metals such as copper, cadmium, lead and mercury [5, 6].

DISCUSSIONS

Metal ions are currently separated from waste streams by suitable techniques depending on the particular characteristics of the waste streams. The compatibility of the chemistry of the metal ion to be separated in the presence of oxidizing and reducing agents in a particular waste stream, have an affect on the efficiency of any method used to separate metal ion from the waste stream.

The ICM unit has the potential to be cost-effective as an on-line copper metal ion separation. Such an on-line process may serve for both effluent limitation and copper plating solution monitoring, respectively. The ICM unit may serve as an on-line process to reduce chemical purchase and recover reusable dragouts.

Experience has shown that in the electronic piece part copper plating process, recovery and reuse pays off the cost of equipment.

The ICM unit is different from other metal ion separation techniques, because the UV-VIS spectrometer serves as a monitoring device for the copper plating solution concentration. In other words, it indicates the on-line monitoring of the copper ion in the rinse tank and copper plating solution.

During cleaning in the plating process, oxides are removed from the surface. Spent solutions resulting from sulfuric acid cleaning of copper may be saturated with copper sulfate, which suggests that the application of the ICM unit to such a waste stream may be feasible. However, whether the ICM process performance will improve by operating either at a decreasing or increasing copper metal ion concentration is to be evaluated, even though early results indicate that treatment performance of ICM shows promising at a low influent of copper metal ion.

The presence of strong chelation agents tends to influence the separation of metal ions from waste streams. Therefore, it would be useful information to evaluate the ICM process performance under the presence of chelation agents such as Cyanacid and EDTA. These two materials have a strong association with plating baths. Such data will aid to determine areas for improving the efficiency of the ICM unit.

The problems which affect the existing available technology for metal ion separation from waste streams includes loading capacity, high cost operation, and the presence of other destructive impurities. Examples of such processes that encounter such problems are ion exchange, electrodeposition and cementation. However, the effect of these problems vary with the particular operation.

Some of the factors that may influence the ICM unit's proven performance may include porosity, concentration of organic load, the pH, the cell path, and the hand path wavelength. These factors need to be determined as to what extent they affect ICM efficiency in terms of removal, cost and durability. The commerciability of the ICM unit and its chance to displace other currently available technologies will depend upon future laboratory results.

CONCLUSIONS

In order to appraise the ICM method in relation to other BAT's for metal ion separation from waste streams, further studies on some of the recognized parameter for evaluation need to be obtained. Such parameters include pH variation, the presence of cyanaid and EDTA in the solutions, the presence of other metal ions in the solutions, and the effect of high concentration of metal ion in the influent solution.

The interesting thing about ion exchange, the electrodeposition and cementation techniques respectively, have to do with the regeneration of the resins, recover baths and reuse. The copper recovered may be exchanged for credit to the basic supplier of copper bars. The cementation process is used in the brass mill for recovery of metals. These processes are cost-effective and they solve effluent disposal problems.

Therefore, the ICM method will have to compete under these criteria in order to be classified under Best Available Technology, economically achievable for effluent limitations control.

ACKNOWLEDGMENTS

Gratitude is expressed to the CAPS LTD., of the Second International Symposium on Metals Speciation, Separation and Recovery, and the engineering managers of the AT&T Network Systems Norcross, Georgia.

REFERENCES

1. Yi, J., R. Ferzeira, and L.L. Tavlarides. "The Metal Ion Separation from Waste Streams by Impregnated Ceramic Membranes," *Second International Symposium on Metal Speciation, Separation and Recovery*, Rome Italy (1989).

2. Swalheim, D.A. "Rinsing Recycle and Recovery of Plating Efficiency," an AES, Inc. sponsored Lecture, one of Educational Publication, (1983), pp. 1-16.

3. Nwogu, P.C. "Cuprous Ion Versus Cupsic Ion Treatability in Biological Trickling Filter Systems," Masters Thesis, University of Arkansas, Fayetteville, Arkansas (1983).

4. Spearot, R.M. and J.V. Peck "Recovery Process for Complexed Copper-Bearing Rinse Waters." *Environmental Progress*, Vol. 3, No. 2, pp. 124-128 (1984).

5. Dean, G., L. Bosqui and H. Lanouette "Removing Heavy Metals from Waste Water," *Environmental Science & Technology*, Vol. 6, No. 6, (1972), pp. 518-524.

6. Lanouette, H. *Heavy Metals Removal*, Chemical Engineering Deskbook Issue I, (October 17, 1977), pp.73-80.

7. Williams, S.K. and Clifford Risley, Jr. "Treatment and Reuse in a Metal Finishing Job Shop," USEPA, EPA-670/2-74-042 National Environmental Research Center, Cincinnati, Ohio, (July 1974), pp. 1-18

8. Federal Register, "Toxic Pollutant Effluent Standards," 38,173,24342 (1973).

9. U.S.Environmental Protection Agency 40 CFR 468, "The Copper Forming Categorical Standards," Vol. 48,36942 (August 15, 1983).

10. McCabe, L.J. and J.M. Symons "Trace Metals in Water Supplies, Survey of Community Water Supply System," *J. Amer. Water Works Assoc.*, 62:670-681 (1970).

11. U.S. Environmental Protection Agency 40 CFR Ch. 1, Section 467.11, (July 1, 1986), pp.631-651.

PART VI: SOILS CONTAMINATION AND DECONTAMINATION

PARTITIONING OF HEAVY METALS INTO SELECTIVE CHEMICAL FRACTIONS IN SEDIMENTS FROM RIVERS IN NORTHERN GREECE

V. Samanidou and K. Fytianos
Environmental Pollution Control Laboratory, University of Thessaloniki, Thessaloniki, Greece

1. ABSTRACT

A fine-step sequential extraction technique was used to determine the chemical association of heavy metals (Pb, Cd, Cu, Fe, Mn, Zn, Cr) with major sedimentary phases (exchangeable cations, easily reducible compounds, organic sulfidic phases, carbonates and residual components) in samples from rivers in northern Greece (Axios and Aliakmon).

From the obtained data it can be seen that the surplus of metal contaminants introduced into the aquatic system from anthropogenic sources usually exists in relatively unstable chemical forms. A high proportion of the elements studied remains in the residual fraction for the Axios river and in the organic fraction for the Aliakmon. Most of the non-residual portion is bound to ferromanganese oxides and to organic matter.

2. INTRODUCTION

Environmental pollution is clearly demonstrated by the high concentrations of heavy metals in water and aquatic sediments in highly industrialized and densely populated areas,. Most of the sediments are enriched with heavy metals from municipal and industrial wastes and surface run-off. The chemistry of contaminated sediments is of a complex nature since their heavy metal content varies significantly [1,2].

Dredged material may contain appreciably high concentrations of heavy metals, however the question arises as to what percentage of the total concentration of heavy metals will be available for release to the water over a long period of time [3,4]. Knowledge of the total amount of metals does not indicate the risk of toxicity or the nature of the various geochemical processes which may be taking place [5-7]. Moreover, such studies have not yielded information regarding the partitioning of trace elements between the various components of the sediments and their potential to affect water quality under different environmental conditions [5,8].

The scope of this work was to investigate the chemistry of sediments and the abundance of trace metals within each chemical fraction. Various forms and associations of heavy metals, such as the exchangeable form, carbonate mineral phase, easily reducible form (Fe-Mn oxides), interactions with organic and sulfide fractions and presence in the lattice structure (residual) of sediments, were studied. This is the most efficient way to obtain detailed information about the origin, mode of occurrence, biological and physicochemical availability, mobilization of metals, and their transport media [9].

3. MATERIALS AND METHODS

Samples of surface sediment were collected at 3 month intervals from the Axios and Aliakmon rivers, northern Greece, over a 1-year period (1986-1987). Two sampling stations were established on each river, the distance between them being ~ 70 and 10 km respectively for the Axios and Aliakmon rivers. The following heavy metals were determined by AAS: Pb, Cd, Cu, Fe, Mn, Zn and Cr. The distribution of these metals was examined in the different sediment phases. The speciation of heavy metals in sediments was examined using a sequential extraction method in order to obtain the concentration of metals bound in cation exchange positions, carbonates, Fe-Mn oxides, organic matter, sulfides and crystal lattice [10-14]. The sequential extraction method followed is shown in Figure 1.

For the determination of heavy metals associated (chelated or adsorbed) with humic and fulvic acids, the sediments were treated with 0.1 N NaOH for 10 h. Leaching with 0.5 N HX1 has also been used for the evaluation of the pollution of the areas examined [15]. This extraction mainly removes the "anthropogenic" trace element fraction from the sediment.

4. RESULTS AND DISCUSSION

Comparing the total concentrations in sediments between the Axios and Aliakmon rivers, we conclude that metal pollution in sediments of the Axios river is considerably higher. The Axios river, which flows into the Thermaikos Gulf, is characterized by high levels of heavy metals. These pollutants, which flow into the river, originate in Yugoslavia as domestic effluents and industrial wastes [2]. Concerning the total heavy metal analysis in the Axios sediments, a significantly higher concentration was observed at station 2 (near the Greek/Yugoslavian border) than at station 1 (near the estuaries). This increase is probably due to sedimentation caused by coagulation or coprecipitation of the heavy metals.

No significant differences in heavy metal concentrations were observed in the sediments from the two stations on the Aliakmon River due to the short distance between them (< 10 km). The Aliakmon is the longest river in Greece (350 km in length) and originates in the Vernon mountains. Its water quality is affected by domestic effluents and industrial wastes. The water discharge of the river is variable, because it depends on the operation of a hydro-electric power plant.

The partitioning range of heavy metals extracted by different chemical extractants for the four sampling stations on the Axios and Aliakmon rivers is presented in Tables 1-4. The values (average from 1 year's measurements) are given by the ratio of the extracted amount to the total amount of metals (%).

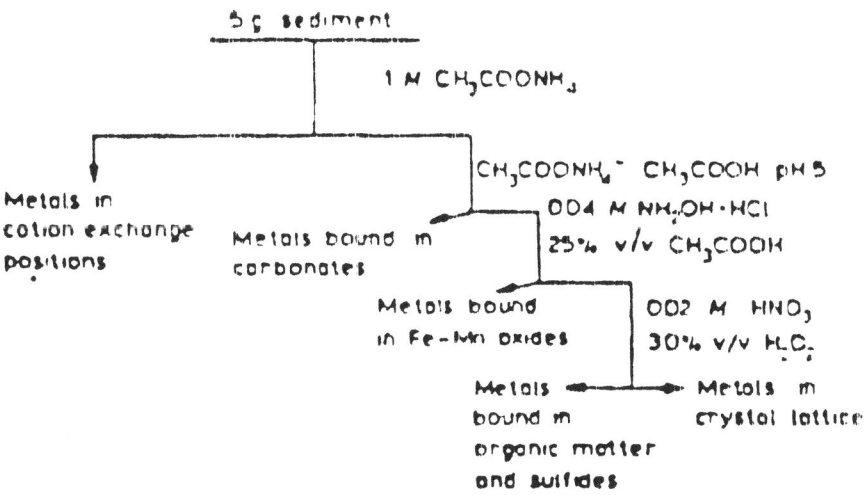

FIGURE 1.

TABLE 1. *Partitioning Range (%) of Heavy Metals Extracted by Different Chemical Extractants from Axios River Sediment (Station 1, Estuaries)*

	Pb	Zn	Cu	Mn	Fe	Cr	Cd
Cation-Exchangeable	1.16	0.99	0.98	5.57	0.19	0.19	9.59
Carbonates	13.26	9.08	2.19	21.94	0.96	0.63	35.53
Fe-Mn Hydrous Oxides	23.82	27.08	6.01	27.55	18.31	9.69	22.16
Organic Sulfides	33.13	29.12	42.98	23.99	37.28	30.70	10.77
Residual	28.63	33.71	47.84	20.95	43.26	58.79	22.13
Total ($\mu g/g$) or (mg/g)*	153.71	220.05	45.71	814.57	15.64*	166.97	13.52
"Anthropogenic" (non-lithogenous phases)	63.80	37.05	44.24	44.52	15.47	22.50	56.45
Humic and Fulvic Acids	9.98	10.92	13.59	0.69	0.84	1.34	4.88

TABLE 2. *Partitioning Range (%) of Heavy Metals Extracted by Different Chemical Extractants from Axios River Sediment (Station 2, Greek/Yugoslavian Border).*

	Pb	Zn	Cu	Mn	Fe	Cr	Cd
Cation-Exchangeable	0.96	0.76	0.87	7.24	0.14	0.52	10.79
Carbonates	15.54	8.36	1.47	19.64	1.60	0.52	24.89
Fe-Mn Hydrous Oxides	25.50	28.41	3.08	27.36	18.81	8.52	29.60
Organic Sulfides	29.80	32.03	44.65	22.44	37.06	32.51	19.05
Residual	28.20	30.44	49.69	23.32	42.39	57.93	15.67
Total ($\mu g/g$) or (mg/g)*	216.92	249.38	46.91	983.63	17.37*	193.48	17.59
"Anthropogenic"	72.47	44.80	57.52	43.33	15.05	28.25	62.14
Humic and Fulvic Acids	4.24	6.69	17.32	0.58	1.05	1.60	3.49

In most cases the sum of the sequential extractions of trace elements was close (± 10%) to the "total" metal concentrations.

Based on this selective leaching procedure, it has been possible to define the percentages of the various elements which are associated with certain fractions of the sediments, and also the distribution patterns of each of these in the sediments of the Axios

and Aliakmon rivers. The main conclusions which can be drawn from the chemical fractionation of the sediments are the following.
 (i) The heavy metals exhibited distinct distribution trends. The distribution patterns of the total metals were similar in some cases. However, on separating the metals into different chemical fractions,

TABLE 3. *Partitioning Range (%) of Heavy Metals Extracted by Different Chemical Extractants from Aliakmon River Sediment (Station 1, Estuaries).*

	Pb	Zn	Cu	Mn	Fe	Cr	Cd
Cation-Exchangeable	0.23	1.76	0.96	3.95	0.23	0.22	1.03
Carbonates	24.65	4.21	2.11	23.33	3.65	0.65	15.25
Fe-Mn Hydrous Oxides	19.72	13.65	4.67	32.67	31.61	2.87	41.43
Organic Sulfides	29.55	50.36	73.98	31.23	40.05	69.88	24.80
Residual	25.85	30.02	18.28	8.82	24.46	26.38	17.49
Total ($\mu g/g$) or (mg/g)*	156.19	80.79	36.77	503.81	6.59*	178.79	1.71
"Anthropogenic"	23.73	21.99	35.30	35.14	10.06	13.34	46.47
Humic and Fulvic Acids	4.89	13.30	6.69	0.35	0.50	0.80	9.07

TABLE 4. *Partitioning Range (%) of Heavy Metals Extracted by Different Chemical Extractants from Aliakmon River Sediments (Station 2, Greek/Yugoslavian Border).*

	Pb	Zn	Cu	Mn	Fe	Cr	Cd
Cation-Exchangeable	0.21	1.75	1.15	3.26	0.14	0.23	1.22
Carbonates	24.68	3.80	2.25	23.15	0.65	0.69	13.54
Fe-Mn Hydrous Oxides	16.09	13.37	4.97	32.91	26.92	3.54	42.82
Organic Sulfides	35.40	52.06	70.85	30.69	46.39	66.90	24.10
Residual	23.62	29.02	20.78	9.99	25.90	28.64	18.32
Total ($\mu g/g$) or (mg/g)*	94.86	77.94	42.76	533.97	7.03*	212.02	2.03
"Anthropogenic"	38.88	22.04	34.79	32.65	10.06	10.92	63.42
Humic and Fulvic Acids	11.16	9.51	8.96	0.61	0.42	0.41	7.16

the distribution patterns of the metals in the separate fractions differed considerably. This stresses the importance of knowing the form of trace metals in the environment. The cation exchangeable fraction, which represents the elements adsorbed on the sediments, was comparatively limited except for that of Cd in the River Axios.

(ii) Hydrous Mn and Fe oxides in sediments are generally strong scavenging agents for heavy metal ions [16, 18]. Manganese oxides in Axios sediments were found to be readily dissolved by hydroxylamine hydrochloride solution., leaving the major part of Fe oxides in the residue.

(iii) More than 50% of the Cd was associated with the carbonate and reducible fraction. It therefore appears that the Cd is incorporated within Fe and Mn oxides and/or with carbonates.

(iv) In both rivers Cu and Cr responded similarly to this treatment. A relatively small percentage of the total concentration was removed.

The hydrogen peroxide treatment should remove mainly sulfides and organic matter and have only a minor effect on the silicate lattice. The distributions of metals as organic complexes in sediments were very sensitive indicators of contamination. Of the metals examined in this study, Cu was the major component in this fraction. This is probably an indication of the well-known close association of Cu with organic matter. The residue remaining after the hydroxylamine hydrochloride, acetic acid, and the hydrogen peroxide leaches, could be expected to consist almost entirely of detrital silicate minerals and organic matter, such as humic material which is resistant to peroxide attack. This residual solid should contain mainly primary and secondary minerals which may hold trace metals within their crystal structure. These metals would not be expected to be released in solution over a reasonable time span under the conditions normally existing in nature.

The treatment of the residue with nitric-perchloric acid mixture, shows that for the Axios river the metals Fe, Cr, Cu and Mn were concentrated mainly in this fraction of the sediment. By contrast, in the Aliakmon river, most of the examined metals seemed to be mainly bound to organic matter and to sulfide minerals. The amount of trace metal in the residue decreases in the order
Cr > Cu > Fe > Zn > Pb > Mn > Cd for the Axios, and
Zn > Cr > Fe > Pb > Cu > Cd > Mn for the Aliakmon.

Most of the Cr in the sediments in the Axios was associated with the residual fraction. Therefore, it appears that in these sediments Cr is probably terrestrially derived.

Copper was contained largely in the organic and sulfide phase in the Aliakmon river, which has 70% of the total copper. About 50% of the total copper was found in the residual fraction of the Axios river, which indicates that copper is probably bound to terrestrially derived materials.

Approximately 35% of the fe was incorporated in the organic and sulfide phase. The percentage of Fe in this fraction was, however, variable. The variability probably results from competition between Fe organic complexes and hydrous Fe oxide forms. This situation is complicated because hydrous Fe oxides themselves can complex with organics, especially humic substances in sediments [19,20]. About 43% of the total Fe was found in the residual phase of the Axios sediment but only 25% in the Aliakmon sediment. Approximately 14% of the iron was associated with the reducible phase, although this showed considerable variation. An average of ~30% of the total Mn was associated with the reducible fraction probably in the form of Mn oxides. The Mn in the residual fraction arises from precipitation of Mn minerals.

Regarding the respective concentration in the sediment, the following list of relative levels was found: Fe > Mn > Zn > Pb > Cr > Cu > Cd for the Axios and Fe > Mn > Cr > Pb > Zn > Cu > Cd for the Aliakmon. Iron was shown to have the highest concentration factor.

Comparing the partitioning range of heavy metals between the two stations on the Axios we can conclude that there is no significant variation in the different fractions. The same results are also observed for the Aliakmon river. However, there is a large difference in the partitioning range of heavy metals in the fractions between the two rivers. The fractions generally vary from higher to lower concentrations for the metals Cr, Fe, Cu and Zn in the following order: Residual fraction > organic fraction > easily reducible fraction > carbonates fraction > cation exchangeable fraction for the Axios, and organic fraction > residual fraction > easily reducible fraction > carbonates fraction > cation exchangeable fraction, for the Aliakmon. The anthropogenic heavy metals in sediments follow the order: Pb > Cd > Cu > Mn > Zn > Cr > Fe > in the Axios and Cd > Pb > Cu > Mn > Zn > Cr > Fe > in the Aliakmon.

5. CONCLUSIONS

From the results of the present investigation the following conclusions can be drawn:
- concerning the total concentrations in sediment, the Axios river is considerably more polluted than the Aliakmon River, especially with the metals Fe, Mn, Zn, Pb and Cd;

- iron has been shown to have the highest concentration factor;
- the heavy metal concentrations in the different phases obtained by sequential extraction show distinct distribution trends;
- the two rivers show a large difference in the partitioning range of heavy metals in the fractions, indicating a different origin and transport media of the metals in the two rivers.

REFERENCES

1. Rapin, F., et al., "Heavy Metals in Marine Sediment Phases Determined by Sequential Chemical Extraction and their Interaction with Interstitial Water." *Environ. Technol Lett.* 4 (1983),pp. 387-396.

2. Chester, R. and M. Hughes, "A Chemical Technique for the Separation of Ferro-Manganese Minerals, Carbonate Minerals and Absorbed Trace Elements from Pelagic Sediments." *Chem. Geol.*, 2 (1967),pp. 249-262.

3. Gibbs, R.J., "Mechanisms of Trace Metal Transport in Rivers." *Science*, 180 (1973),pp. 71-73.

4. Gibbs, R., "Transport Phases of Transition Metals in the Amazon and Yukon rivers." *Geol. Soc. Am. Bull.* 88 (1977),pp. 829-843.

5. Salomons, W., "Sediments and Water Quality." *Environ. Technol. Lett.*, 6 (1985), pp. 315-326.

6. Presley, B.J., et al. Early Diagenesis in a Reducing Fjord. Saanich Inlet. British Columbia II. "Trace Element Distribution in Interstitial Water and Sediment." *Geochim. Cosmochim*, Acta. (1972), pp.1073-1090.

7. Aoyama, J., and Y. Urakami, "Local Distribution and Partial Extraction of Heavy Metals in Bottom Sediments of an Estuary." *Environ. Pollut.*, Ser. B. 4 (1982),pp. 27-43.

8. Guy, R., C. Chakrabari, et al., "An Evaluation of Extraction Techniques for the Fractionation of Cu and Pb in Model Sediment Systems." *Water Es.*, 12 (1978), pp. 21-24.

9. Weltè, B., et al., "Etude des Differents Methodes de Spèciation des Mètaux Lourds Dans les Sèdiments. 1 Etude Bibliographique." *Environ. Technol. Lett.*, 4 (1983), pp. 79-88.

10. Tessier, A., P. Campbell, et al., "Sequential Extraction Procedure for the Speciation of Particulate Trace Metals." *Anal. Chem.*, 51 (1979), PP. 844-850.

11. Etcheber, H., A. Bourg, et al., "Critical Aspects of Selective Extractions of Trace Metals from Estuarine Suspended matter. Fe and Mn Hydroxides and Organic Matter Interactions." *Int. Conf. Heavy Metals in the Environment*, Heidelberg, CEP Consultants, Edinburgh, (Sept., 1983), pp. 1200-1203.

12. Calmano, W., "Chemical Extraction of Heavy Metals in Polluted River Sediments in Central Europe." *Sci. Total Environ.*, 28 (1983), pp. 77-90.

13. Rapin, F., U. Förstner, et al., "Etude de la Répartition des Métaux Lourds Dans les Sédiments Superficiels de la Baie des Anges (Méditerranée, France) par Spéciation Chimique." *Vles Journ. Estud. Pollut.* Cannes, CIESM, (1982), pp. 107-114.

14. Gupta, S.K. and K.Y. Chen, "Partitioning of Trace Metals in Selective Chemical Fractions of Nearshore Sediments." *Environ. Lett.*, 10(2) (1975), pp. 129-158.

15. Agemaina, H. and A. Chau, "Evaluation of Extraction Techniques for the Determination of Metals in Aquatic Sediments." *Analyst*, 101 (1976), pp. 761-767.

16. Schör, J., "Iron-Oxo-Hydroxides and Their Significance to the Behaviour of Heavy Metals in Estuaries." *Environ. Technol. Lett.*, 6 (1985), pp. 189-202.

17. Gadde R. Rao and H.A. Laitinen, "Adsorption by Hydrous Iron and Manganese Oxides." *Anal. Chem.*, 46 (1974), pp. 2022-2026.

18. Chao, T.T., "Selective Dissolution of Mn Oxides from Soils and Sediments with Acidified Hydroxylamine Hydrochloride." *Soil Sci. Soc. Am., Proc.*, 36 (1972), pp. 764-768.

19. Smith, J.D. and P.J. Milne, "Determination of Fe in Suspended Matter and Sediments of the Yana River Estuary, and the Distribution of Cu, Pb, Zn and Mn in the Sediments." *Aust. J. Mar. Freshwater Res.*, 30 (1979), pp. 731-739.

20. Nembrini, G.P., et al., "Speciation of Fe and Mn in a Sediment Core of the Baie de Villefrance (Mediterranean Sea, France)." *Environ. Technol. Lett.*, 3 (1982), pp. 545-552.

21. Fytianos,K., V. Samanidiou, and T. Angelidis, "Comparative Study of Heavy Metals Pollution in Various Rivers and Lakes of Northern Greece." *Ambio*, 15(1986), pp. 42-44.

DISCUSSION OF: PARTITIONING OF HEAVY METALS INTO SELECTIVE CHEMICAL FRACTIONS IN SEDIMENTS FROM RIVERS IN NORTHERN GREECE

Alexander P. Mathews
Kansas State University, Manhattan, Kansas

INTRODUCTION

Heavy metal contamination of rivers may occur through natural or anthropogenic sources. If background concentration levels can be established, the analysis of river sediments can provide good indications of the quality of the river water and potential impact on the aquatic ecosystem. The most important aspects of heavy metal contamination are the hazards to aquatic organisms from bioaccumulation of toxic metals, and to human beings from the consumption of the contaminated organisms.

The analysis of total metal concentration in sediments provides an indication of gross pollution levels. However, to determine the source of contamination and to assess the environmental impact, it is necessary to know the types of metal species present in the sediment. Metals in unpolluted rivers are mostly present in the crystalline structures of detrital minerals, whereas, heavy metals of anthropogenic origin are largely associated with organic matter. In addition, the chemical form of the metal in the sediments will affect the remobilization of the metal from the sediment into the overlying waters and subsequent transfer through the ecosystem.

The paper by Samanidou and Fytianos fills the void in this important area of metal species characterization for sediments from Axios and Aliakmon rivers in Northern Greece. The discussion of this paper will be devoted to specific aspects of this work and suggestions for additional future work.

DISCUSSION

Surface sediment samples were analyzed for total metal concentrations and the metal species were characterized using the sequential extraction procedure of Tessier, et al. [1] and using atomic absorption spectroscopy. The data presented are average values, and provide baseline information on gross pollution levels. Additional information on sediment particle size distribution and organic content could be developed for better comparative analysis of the data between the two rivers and the sampling stations. The AAS analysis for some of the metals may be complemented by the use of X-ray fluorescence techniques [2].

The sampling stations for Axios river are located approximately 70 km apart and show differences in the sediment concentrations for some metals. The effect of sampling station location and associated water quality on the precipitation of dissolved minerals such as Fe and Mn should be considered. Metals such as Cu and Cd are transported without appreciable change between Stations 1 and 2. Water quality parameters such as pH, ionic strength, etc., would be useful in this regard to elucidate mechanisms that may be operative. It is not clear from the analysis why the concentration of Pb decreases from Station 2 to Station 1 for Axios river, whereas the opposite is true for Aliakmon river.

The distribution patterns for Cu are consistent with data in the literature indicating that it is mostly complexed by organic compounds. Some Cu is also present in the residual fraction. Lead is found to be evenly distributed between the various phases except the cation-exchangeable fraction. The higher concentration of Pb in the carbonate fraction for Aliakmon river could be due to the higher concentration of carbonates in the sediment. High concentrations of Pb and Cu in the organic fractions indicates industrial and municipal discharges as potential sources of contamination of the river.

The conclusions from this study indicate the level of heavy metal contamination for Axios and Aliakmon rivers. axios river appears to be somewhat more heavily polluted than Aliakmon river with some metals, and there is a difference in the distribution of metal species between the two rivers. These conclusions have to be tempered to some extent to take into account the effect of variables such as sediment particle size distribution [3], organic matter content [4], water quality variations at the location, and seasonal variations [3]. Future research should address some of these concerns.

REFERENCES

1. Tessier, A., P.G.C. Campbell, and M. Bison, "Sequential Extraction Procedure for the Speciation of Particulate Metals," *Anal. Chem.* 51:844-851 (1979).

2. Rauret, G., R. Rubio, J.F. Lopez-Sanchez, and E. Casassas. "Determination and Speciation of Copper and Lead in Sediments of a Mediterranean River (River Tenes, Catalonia, Spain)," *Wat. Res.*, 22:449-455 (1988).

3. Moriarty, F. and H.M. Hanson. "Heavy Metals in Sediments of the River Ecclesbourne, Derbyshire," *Wat. Res.*, 22:475-480 (1988).

4. Duzzin, B., B. Pavoni, and R. Donazzolo. "Macroinvertebrate Communities and Sediments as Pollution Indicators for Heavy Metals in the River Adige (Italy)," *Wat. Res.*, 22:1353-1363 (1988).

CHEMICAL DECONTAMINATION OF DREDGED MATERIALS, SLUDGES, COMBUSTION RESIDUES, SOILS AND OTHER MATERIALS CONTAMINATED WITH HEAVY METALS

German Müller
Institute für Sedimentforschung der Universität Heidelberg, Heidelberg, W. Germany

1. INTRODUCTION

Fine-grained sediments ("muds") in rivers, lakes and estuaries are increasingly recognized as both a sink and a source of heavy metals and organic pollutants in aquatic systems. In order to keep shipping channels and harbour basins open, extensive amounts of sediments must be removed. For example, from Rotterdam Harbour more than 20 million m^3 of sediments are dredged annually, of which about 10 million m^3 are more or less contaminated with heavy metals. In about 200 locations elsewhere in the Netherlands, highly contaminated sediments have been found in channels, rivers and other water areas with quantities varying from 1,000 up to 100,000 m^3 [1].

In the fairways and harbour basins of the port of Hamburg (W. Germany) about 2.5 million m^3 of mud and sand have to be dredged per year [2].

In recent years great environmental problems have arisen in the disposal and storage of dredged materials heavily contaminated with heavy metals. Especially the high cadmium concentrations that occur in many of these materials do not allow uncontrolled disposal or even use in agriculture. Dumping into the sea, often practiced in the past, does not offer a solution, representing, rather, a highly questionable transport of the contaminated material (and the problem itself!) into other compartments of the ecosystem that are equally in need of protection.

Similar problems as with dredged materials exist with metal-polluted sewage sludge, fly ash, residues from combustion and pyrolysis and - last not least - with soils that have been exposed to heavy metal emissions or have been treated with contaminated materials.

According to the long-term perspectives in waste treatment by incineration inorganic residues enriched in heavy metals will become dominant in the future.

2. TREATMENT TECHNIQUES

Only recently van Gemert et al. [1] have summarized the basic strategies for the treatment of contaminated residues such as dredged sediments. Two categories of techniques ('A' and 'B') can be distinguished (Figure 1):

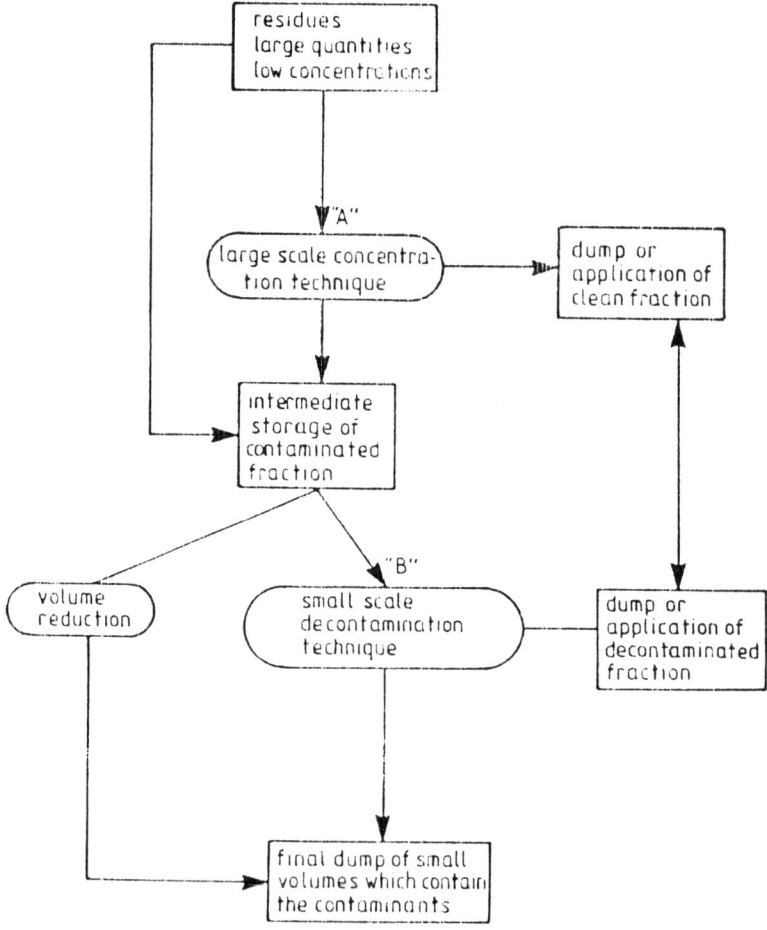

FIGURE 1.   Scenario for the Treatment of Large Quantities of Residue [1].

A: Large-scale concentration techniques. They are characterized by large-scale applicabilities, low costs per unit of residue to be treated and low sensitivity to variations in circumstances.

B: Decontamination or concentration techniques for relatively small-scale operations. They are generally suited for the treatment of residues which contain higher concentrations. Operation costs per unit to be treated are higher and they require specific experiences of operators.

In many cases B-techniques are not feasible for the treatment of large quantities of residue. If, however it is possible to preconcentrate the contaminants into concentrated flows of smaller quantities by using A-techniques, it may be feasible to treat these concentrated flows by B-techniques.

Among the A-techniques (Classification-flotation-high gradient magnetic separation) only classification is applicable to dredged sediments or soils contaminated with heavy metals.

In both sediments and soils heavy metals are mainly bound to the fine particles fraction, thus a classification will separate a sediment or soil into and uncontaminated coarse-grained sand fraction and a fine-grained mud (or sludge) fraction (consisting of silt-and clay-sized particles) in which the heavy metals are concentrated.

Figure 2 shows the mass balance and distribution of the heavy metals Pb, Cd, An and Cu by mechanical processing ('classification') of dredged sludge from the port of Hamburg in the pilot plant 'METHA' [3]. The fine particles are separated from the sand fraction by a hydrocyclone with subsequent washing of the underflow in an elutriator.

The final products are 'pure' sand which makes up 57.3% of the original sludge and 42.7% of fine sludge in which the heavy metals are enriched by a factor of about 2. It should be mentioned, however, that dredged materials from other localities contain lesser sand (often less than 10%). In such cases a classification would not be very effective and lead to very high costs.

Among the 'B'-techniques considered by the authors (1) to be applicable to dredged sediments contaminated with heavy metals (NaOCl-leaching/ion exchange/acid leaching) NaOCl-leaching (4) is restricted to mercury if present in the sludge in a form oxidizable with hypochlorite, i.e. as a sulphide or an organic compound.

Where ion exchange is concerned, three mutually independent developments run parallel.

a). selective cation exchange through existing, complex-forming cation exchangers [5].

b). Development of new selective ion exchangers.

c). Strongly basic anion exchangers [6, 7].

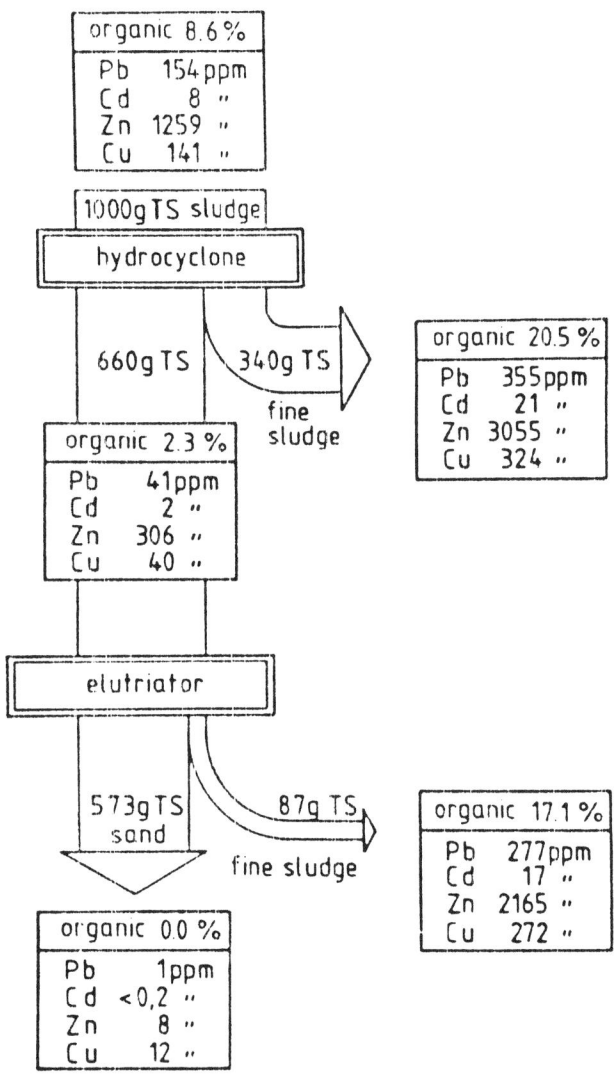

FIGURE 2. *Hamburg Harbour: Mass Balance and Distribution of Heavy Metals After Hydrocyclone and Elutriator Classification [3]. TS = Trockensubstanz = dry material.*

ad a) the reviewers came to the conclusion that 'the process appears to involve extremely long contact times, the required times may even be in the order of 1 week. In practice, the process will consequently turn out to be difficult to realize on a continuous basis. Due to the large supply of calcium, the resin's selectivity will not be high enough, thus necessitating the presence of large excess of resin with regard to the sludge.

ad b) The comment is more an outlook to the future: 'at least some years of research are expected to be required before a result, with practical applicability, may be realized.'

ad c) More promising are results from laboratory experiments obtained with strongly basic anion exchangers after acid extraction. Metals may be mobilized through Ph reduction. When the chloride concentration is sufficiently high in the liquid phase, metals such as Cd, Zn and Cu show a tendency to dissolve as negatively charged chloride complexes. In this form the metals may be immobilized to a strongly basic anion exchanger, which is in chloride form.

In the case of sufficiently low concentrations of chloride, the chloride complexes mentioned will be unstable and the metal ions will be converted to the cation form. For this reason, the anion exchanger once charged may be regenerated through washing with water. After treatment the ion exchanger may, in principle, be used again.

Research has shown that the effect of mobilization by pH reduction is strongly influenced by ventilation the sediment. Extraction only, has minimal effect, even when pH =1. Ventilating of the suspension greatly improves the extraction result, probably as a result of oxidation. A more or less optimal result is obtained through ventilating at pH = 4 and subsequently extracting at pH = 3.

In the review article of van Gemert [1] 'acid leaching' refers to the method developed by the author [8-10] designated as'chemische Entgiftung"/"chemical decontamination', in which acid leaching is only the first step in a process leading to 'clean' end products with respect to heavy metals.

3. CHEMICAL DECONTAMINATION: A CONCEPT FOR THE FINAL DISPOSAL OF MATERIALS CONTAMINATED WITH HEAVY METALS

Our method (Figure 3) is based on A) an extraction of heavy metals by treatment of the sediments, soils or other materials with acids, B) a removal of the solvents and repeated washing of the

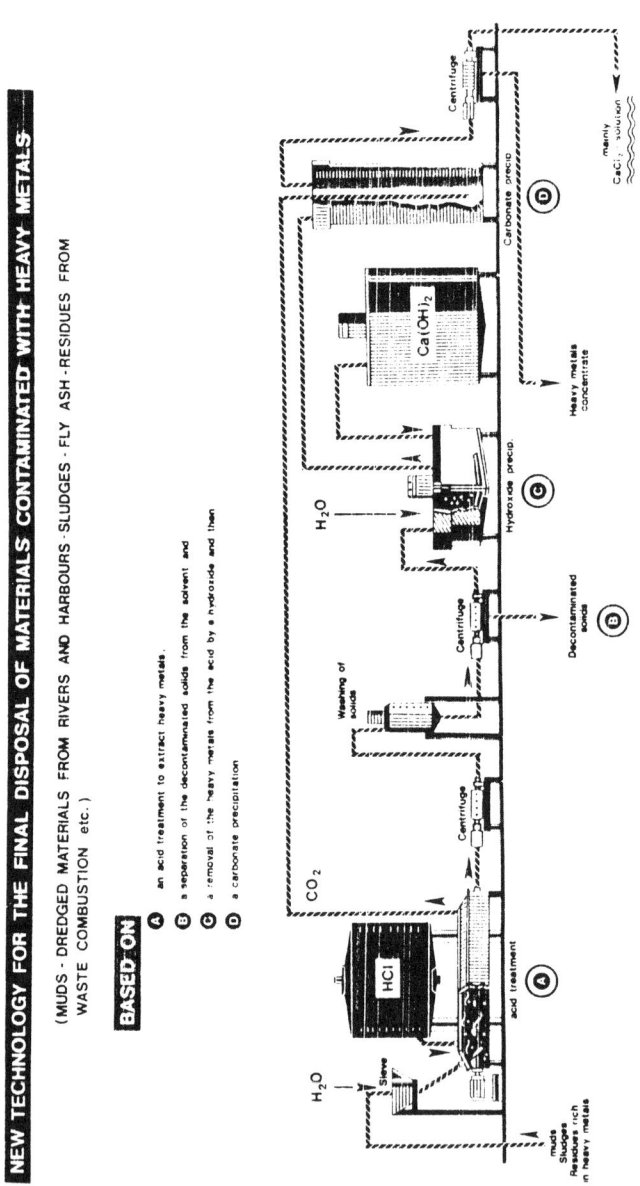

FIGURE 3. Simplified Model of a Chemical Decontamination Plant. A further pH-based Step in Separating iron hydroxides from other - 'valuable' - metal hydroxides may be installed in the hydroxide precipitation phase 'C'.

solids, C) a precipitation of the heavy metals in the acidic solution with calcium hydroxide and D) a subsequent elimination of minor concentrations of cadmium and other heavy metals still in (the now alkaline) solution by a carbonate precipitation (introduction of CO_2) during which cadmium and other heavy metals are incorporated into the calcite lattice.

3.1 Extraction by Acid Treatment - Selection of Suitable Acids

A) Extraction experiments with various acids and acid combinations were carried out on highly contaminated sediments from the Neckar and Rhine Rivers, both heavily polluted with heavy metals. Aqua regia and hydrochloric acid had a similarly high extraction rate as could be achieved with a hydrofluoric acid/nitric acid mixture. Sulfuric acid showed the least favorable results; for nitric acid, values for extracted iron, chromium, nickel and cobalt were clearly lower than those for aqua regia or hydrochloric acid.

As the introduction of nitrate appears undesirable for ecological reasons, hydrochloric acid seems to be the most suitable acid for decontamination. During acid treatment, not only do heavy metals go into solution, but carbonates normally present in the sediments dissolve and the carbon dioxide hereby produced can be applied in the final carbonate precipitation step. During the dissolution process, acid (30% HCl) is added continuously and a pH value of 0.5 must be maintained. The entire reaction occurs at room temperature and lasts only 10-15 minutes.

3.2 Separation of the Decontaminated Solids from the Solvents

B) The separation of the solids from the acidic solution is carried out by centrifuging, however, other techniques (settling tanks, band filter press etc.) are also thinkable. After repeated washing the now decontaminated solids may be used for various purposes, p.ex. brick making, cement admixtures, light building materials, landfill material, improvement of soil structure in agriculture etc.

3.3 Precipitation of Heavy Metals as Hydroxides

C) After solids have been separated the acidic filtrate is neutralized with calcium carbonate, calcium oxide or calcium hydroxide and calcium hydroxide is further added until a pH of 10 has been achieved. During this step, dissolved metals are precipitated nearly quantitatively as insoluble hydroxides. Up to a pH of 4 preferentially iron hydroxides precipitate and can be removed from the system.

3.4 Subsequent Precipitation of Remaining Minor Concentrations of Cadmium and Other Heavy Metals

D) A nearly complete elimination of traces of heavy metals, especially those of cadmium, is accomplished by introducing carbon dioxide into the basic hydroxide suspension. The precipitation of calcium carbonate (calcite) occurs, whereby practically all cadmium traces are coprecipitated and thus are bound in the calcite lattice. This preferred bonding is based on the nearly identical ionic radii of calcium and cadmium and the identical structure of both isomorphic trigonal carbonates $CaCO_3$ (calcite) and $CdCo_3$ (otavite) (Figure 4).

The apparent distribution coefficient of cadmium between calcium carbonate (as calcite) and solution was experimentally determined [11] and a constant value of about 20 was observed throughout the process of calcite formation in the system $CaO-H_2O-CO_2$. Lower values were found for the apparent distribution coefficient when varying amounts of chloride ions were added to the solution since cadmium ions form complexes with chloride ions [12].

The high distribution coefficient leads to a rapid decrease of cadmium concentrations in solution when calcite precipitation is initiated. Figure 5 shows the results of experiments [11]: For all Cd^{++} concentrations applied, cadmium ions had been removed almost completely from the solution before 10 per cent of calcium ions had precipitated as calcite.

Cations with ionic radii < 1 A° e.g. zinc, copper, cobalt and nickel (Figure 4) are also preferentially bound in the calcite lattice, leading to a further strong decrease of the already low concentrations of these metals in the alkaline solution.

The $CaCl_2$-rich filtrate separated after the hydroxide and subsequent calcite precipitation contains extremely low concentrations of heavy metals.

Table 1 shows results of analyses from large-scale experiments carried out with heavily polluted dredged materials from the Neckar River (SW-Germany) and sewage sludge from the city of Mannheim (SW-Germany). With respect to heavy metals the filtrates have drinking water quality.

The precipitated heavy metal concentrates - chiefly hydroxides - are separated from the filtrate by continuous filtration. The purified filtrate discharged after neutralization contains approx. 3% $CaCl_2$.

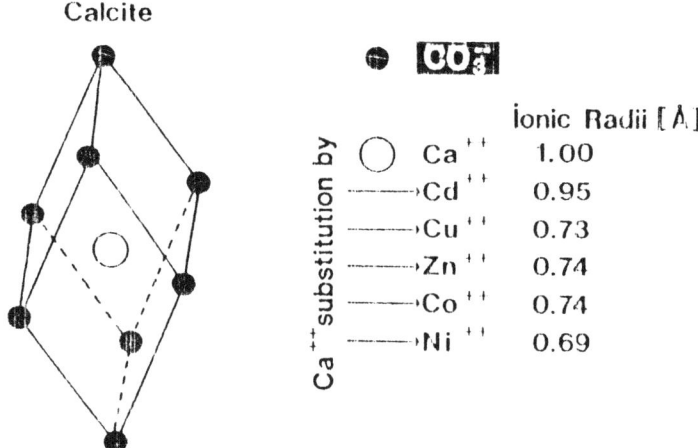

FIGURE 4. *Substitution of Ca^{++} in the calcite lattice by other divalent cations with ionic radii < 1 $A°$.*

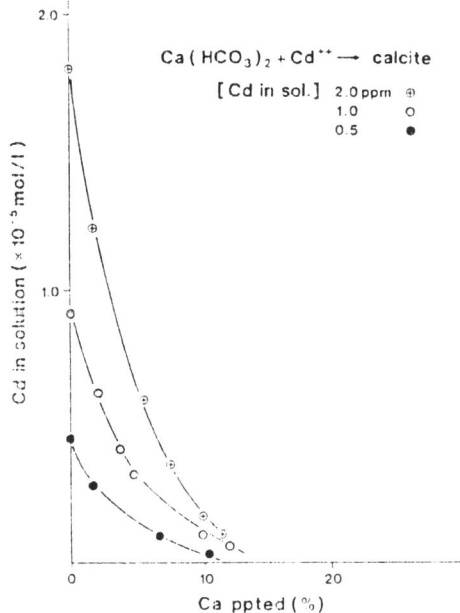

FIGURE 5. *Decrease of the Cadmium Content of Parent Calcium Bicarbonate Solution from Which Calcite is Precipitated [11].*

TABLE 1. Heavy Metal Concentration in Filtrates After Hydroxide and Carbonate Precipitation.

	Dredged Material Neckar River	Sewage Sludge Mannheim
Fe	<0.01	<0.008
Mn	<0.050	0.006
Cu	0.033	0.015
Pb	<0.005	0.28
Zn	0.007	0.04
Cd	0.001	0.04
Cr	<0.005	0.004
Ni	0.013	0.430
Co	<0.001	0.055
Hg	<0.0001	n.d.

all concentrations in ppm

4. Advantages of Chemical Decontamination - Practical Realization

The advantages of a chemical decontamination are obvious: it leads to a total recycling by which the materials obtained can be used or reused for various purposes. The decontaminated solids are suitable raw materials for brick or cement production or may be used as landfill material; the heavy metal concentrate poor in Fe can be reprocessed and reused. Only small volumes of Fe-rich precipitates obtained during hydroxide precipitation at a low pH have to be dumped.

In summary, Figure 6 shows a scenario specific for the chemical decontamination of materials enriched in heavy metals. Large-scale concentration techniques 'A' (in our case classification) should only be applied as a first step if the material contains larger amounts of coarse fractions poor in heavy metals.

The realization of our decontamination process reads as follows:

January 1983: Feasibility Study TFW-00-1268-40 from LURGI UMWELT- UND CHEMOTECHNIK GMBH, Frankfurt 'Studie zur Schwermetallentgiftung von Neckarschlamm'

April 4, 1986: European Patent 0072885B1 'Process for Decontaminating Natural and Industrial Sludges'

June 1986: Recipient of the 'Philip Morris Research Prize' in the field 'Man and his Environment' for the decontamination process

December 1987: Versuchsbericht TOGV01001191087 'Verfahren zur Dekontamination von schwermetallbelasteten Schlämmen' from UHDE GMBH, Dortmund, 57p. with very positive results from experiments on a technical scale with strongly polluted dredged sediments and sewage sludge; financial support was given by the government of Baden-Württemberg within the frame of a soil-water project (PWAB)

Early 1989: The realization of the process up to this date in the hands of industry automation Sondertechnik GMBH & Co., Heidelberg, was transferred to ROM, Rudolf-Otto-Meyer, Sondertechnik, Hamburg 70. An experimental plant is under construction in the premises of ROM in Hamburg. Financial support is to be given from the German Federal Ministry of Research and Technology in Bonn.

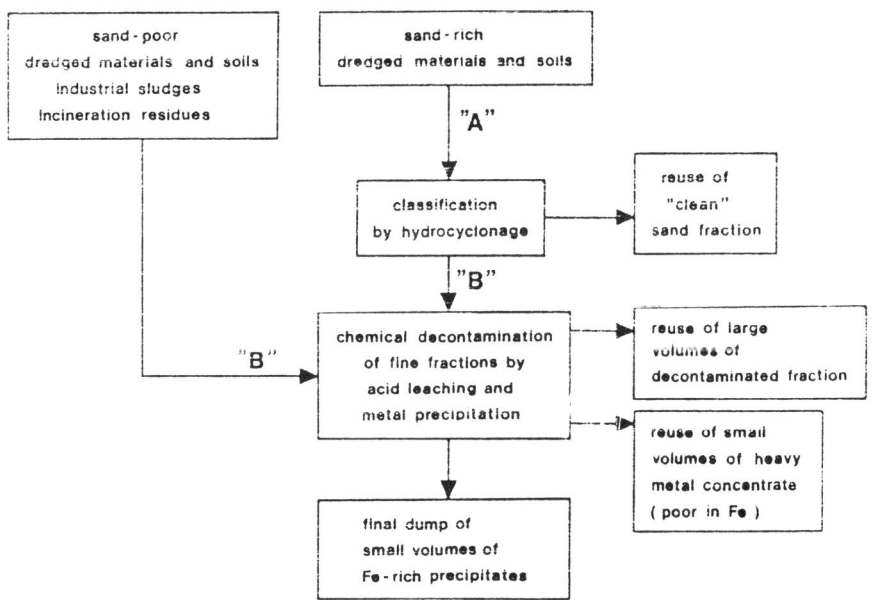

FIGURE 6. Scenario for the Chemical Decontamination of Materials Enriched in Heavy Metals.

REFERENCES

1. Van Gemert, W.J.T., J. Quakernaat, and H.J. van Veen. "Methods for the Treatment of Contaminated Dredged Sediments," in *Environmental Management of Solid Waste*, Salomons, W. and U. Förstner, Eds. (Berlin, Heidelberg, New York, London, Paris, TOKYO: Springer-Verlag 1988), pp. 44-64.

2. Christiansen, H., G. Öhlmann, and L. Tent. "Probleme im Zusammenhang mit dem Anfall von Baggergut im Hamburger Hafen," *Wasserwirtschaft* 72:385-389 (1982).

3. Werther, J. "Classification and Rewatering of Sludges," in *Environmental Management of Solid Waste*, Salomons, W. and U. Förstner, Eds. (Berlin, Heidelberg. New York, London, Paris, TOKYO: Springer-Verlag 1988), pp. 65-79.

4. Kerr, R.S. "Control of Mercury Pollution of Sediments," NTIS Rep. No. PB-213771 (1972).

5. Van Hoek, G.L., P.J. Gommers, and J.A. Overwater. "Removal of Heavy Metals from Polluted Sediments - Ion exchange, a Possible Solution," in *Proc. Int. Conf. Heavy Metals in the Environment*, Vol. 2 (Edinburgh CEP Consultants, 1983), pp. 856-859.

6. Bolt, A., M. Tels and W.J.T. Van Gemert. "Recovery of Pure Metal Salts from Mixed Heavy Metals Hydroxides Sludges," In *Recycling International*, Thome-Kozmiensky K.J., Ed. (Berlin, EF-Verlag für Energieund Umwelttechnik 1984), pp, 1025-1031.

7. Van Veen, F. "Thermische Grondreinigingstechnieken," in *Ontwikkelingen Bodemreinigingstechnieken.* Colloq. Minist. V Rom. (April 5, 1984).

8. Müller, G. and S. Riethmayer. "Chemische Entgiftung: das Alternative Konzept zur Problemlosen und endgültigen Entsorgung schwermetallbelasteter Baggerschlämme," *Chemiker-Zeitung* 106:282-292 (1982).

9. Müller, G. "Chemical Decontamination: A Concept for the Final Disposal of Dredged Materials and Sludges Contaminated by Heavy Metals," in *Proc. Int. Conf. Heavy Metals in the Environment,* Vol. 2 (Edinburgh CEP Consultants, 1983), pp. 948-951.

10. Müller, G. "Decontamination of Dredged Materials and Sludges Enriched in Heavy Metals," in *Recycling in Chemical Water and Wastewater Treatment*, Hahn, H.H., R. Klute and P. Balmer, Eds. Schriftenreihe Inst. Siedl. Wass. Wirtsch. Karlsruhe 50:237-243 (1986).

11. Kitano, Y., N. Kanamori, and R. Fujiyoshi. "Distribution of Cadmium between Calcium Carbonate and Solution (part 1)," *Geochem. J.* 12:137-145 (1978).

12. Kitano, Y., N. Kanamori, and R. Fujiyoshi. "Distribution of Cadmium between Calcium Carbonate and Solution (part 2). *Geochem. J.* 12:147-151 (1978).

DISCUSSION OF:
CHEMICAL DECONTAMINATION OF DREDGED MATERIALS, SLUDGES, COMBUSTION RESIDUES, SOILS AND OTHER MATERIALS CONTAMINATED WITH HEAVY METALS

Raymond W. Regan, Sr.
Environmental Resources Research Institute, The Pennsylvania State University, University Park, Pennsylvania, USA

INTRODUCTION

Professor Müller has identified several key items in the introductory statement that relate to the environmental impact of fine-grained sediments ('muds') in water bodies and the disposal and storage of metal-ladened dredged materials by conventional means; and to the detoxification to be achieved by a proposed new technology for the chemical decontamination of dredged materials contaminated with heavy metals. This technology is further suggested to be applicable to metal-polluted wastewater sludge; fly ash; residues from combustion and pyrolysis; and contaminated soils. However, the loading rates of materials to be processed and the costs for operating various engineered systems are presented only in general terms. The discussion to follow will not address the decontamination technology itself. Rather the emphasis will be twofold, namely (1) on aspects of the natural metal decontamination processes occurring in the aquatic environment, and (2) on a modified dredging strategy that exploits this naturally occurring process to effect metals removal. This modified strategy when used in conjunction with mechanical decontamination technologies, as represented by the Müller system, may provide an improved cost-effective approach for dredge spoil management. Data presented in the paper, supplemented with related literature for watercourses within the Federal Republic of Germany (FRG) were used for the analytical assessment to be presented.

STATEMENT OF THE PROBLEM

In order to maintain commercial navigation in shipping channels and harbors large quantities of more or less contaminated sediments are dredged on a continuous basis. As has been noted by

Müller, 'muds' may be significant sinks and sources of heavy metals and organic pollutants and may need to be decontaminated to control secondary environmental effects associated with their disposal.

Conventional means of dredging may exacerbate the clean-up requirements and the associated costs in favor of meeting the commercial objectives.

BACKGROUND

The problem definition presented for the River Rhine has been used for illustrative purposes only to describe the general environmental concerns for a heavily used waterway with a significant pollution loading due to intense utilization within the drainage basin.

The River Rhine is the busiest waterway in Europe with respect to commercial and navigation [13]. Compared to the rest of the Federal Republic of Germany, the Rhine Basin has increased densities of population, industry and agriculture. Therefore, there is a significant number of potential sources for toxic metals entering into the river [13].

The ecotoxicological significance of pollutants on European rivers, specifically the River Rhine and the Rhine-Mease estuary has been the subject of a number of research investigations, as illustrated by Breder, et al [13]. The levels of heavy metals, including Cd, Pb and Hg were determined in water and sediments (samples of sediment collected from the upper 5 cm) of the Rhine and the main tributaries from Lake Constance to the Dutch/German border (40 stations). Briefly, the following findings were noted [13]; using Bodensee as the starting point:

a) The Cd in the sediments went from a negligible amount at 160 km to an exponentially increasing 35 ppm at 850 km (Emmerich).

b) The Pb in the sediments increased at a steady rate from about 10 ppm at 160 km to 550 ppm at 590 km (Koblenz), followed by a drop to about 250 ppm then increase to 550 ppm at 850 km (Emmerich).

c) The Hg sediments followed a fluctuating increase from 3 ppm at 160 km to between 8 and 11 ppm after 590 km (Koblenz).

d) The sediments constitute the depots of the toxic metals introduced into the river ([14], as cited in [13]).

e) Hg and especially Pb have a pronounced tendency to be bound to suspended particulate matter.

f) Cd had a relatively equal distribution between the dissolved and suspended particulate phases.

It is noted that the sources of the contaminated sediments specifically found in the Rhine in this case, and generally in the rivers and harbors of other large industrialized drainage basins are the result of the natural transport of the sediments towards the sea. The development of wastewater treatment facilities specifically designed for metals removal of point sources [16] may provide some level of improvement by reducing the heavy metals loading to the river system. It is unlikely however that the building of contaminated sediments could ever be completely controlled due to non-point contaminant sources. Hence, the writer examined the general area of dredging methods.

DREDGING ISSUES

The writer emphasizes the dual concerns for harbor dredging, namely (1) the removal of sediments to construct or to maintain navigation (classically considered as paramount), and [2] the decontamination of sediments, when appropriate, followed by environmentally sound disposal or reuse practices (perhaps the wave of the future).

The greatest potential for environmental problems has been associated with construction dredging due to the persistent and significant changes in the hydraulics of the channel [15]. Following this, the disposal of contaminated sediments, when associated with maintenance dredging may become a major environmental concern.

A comparison of the concentrations of selected constituents in dredged sediments [fraction not specified] and crustal materials is presented (Table 1) [16].

Deposits of sediments which create the need for dredging of most ports can be described as two main fractions, namely, deep and surficial sediments [15]. Deep sediments, whose extraction may be associated with construction dredging, typically represents the major fraction. However, the latter fraction may be the major concern for most dredging projects because the rate of deposition of surficial materials governs the extent and frequency of maintenance dredging. It is also in these sediments that the elevated levels of contaminants are found.

The National Research Council (NRC) Marine Board included the following finding concerning the environmental problems associated with dredging and effective management practices [15].

TABLE 1.   *Concentrations of Selected Constituents in Dredged Sediments and Average Global Crustal Materials [16].*

Constituents	Dredged Materials Ranged in Moles Kg^{-1} of Sediment (except as noted)	Average Crustal Materials Range in Moles Kg^{-1}
Trace Metals		
Iron	0.02 - 0.90	0.61 - 1.03
Manganese	$(0.4 - 10) \times 10^{-3}$	$(12 - 18) \times 10^{-3}$
Zinc	$(0.5 - 8) \times 10^{-3}$	$(0.92 - 1.26) \times 10^{-3}$
Copper	$(0.8 - 9400) \times 10^{-6}$	$(460 - 1090) \times 10^{-6}$
Nickel	$(0.2 - 2.6) \times 10^{-3}$	$(0.62 - 1.69) \times 10^{-3}$
Chromium	$(0.02 - 3.8) \times 10^{-3}$	$(0.92 - 1.92) \times 10^{-3}$
Lead	$(5 - 1900) \times 10^{-6}$	$(48 - 77) \times 10^{-6}$
Cadmium	$(0.4 - 600) \times 10^{-6}$	$(0.89 - 1.6) \times 10^{-6}$
Mercury	$(1 - 10) \times 10^{-6}$	$(0.149 - 0.398) \times 10^{-6}$
Synthetic Organic Substances		
Chlorinated Pesticides		0 -10 mg kg^{-1}
Polychlorinated Biphenyl Compounds		0 - 10 mg kg^{-1}
Other Properties		
pH		6 - 9
Chemical Oxygen Demand		0.03 - 0.04
Oil and Grease		0.1 - 5 g kg^{-1}

"Finding 27

A comprehensive dredging and dredge-materials management plan should be developed for each port with a specific long-term objective being to assure that maintenance projects can be carried out on schedule while minimizing adverse environmental effects.

The plan should be based on: (1) a thorough characterization of the kinds and qualities of material to be dredged, (2) a detailed determination of the Spatial distribution of contaminants (both horizontally and vertically) within channel deposits that permits definition of the degree of homogeneity in the sediments to be dredged and to delineate prominent contaminant "hot spots," (3) a rigorous assessment of the physical and chemical behavior of these materials if placed in each of the alternative disposal environments, (4) consideration of beneficial use as an alternative, (5) an assessment of the effects resulting from each of the dredging and disposal alternatives for public health, the environment, the biota, other uses of that segment of the environment, and the relative costs, and (6) consideration of long-term continuing costs and effects associated with the plan."

and

"Finding 29

Dredged sediment should be regarded as a resource rather than as a waste.

Materials should be carefully screened to determine suitability for use as construction aggregate, sanitary landfill cover, for beach replenishment, for the creation and enhancement of wetlands, and for other uses prior to disposal."

However, to complete the perspective, the NRC [15] notes that the development of advanced dredging systems capable of reducing environmental effects are not in widespread use (outside of Japan) compared to "classic" dredging systems."

EVALUATION OF MASS BALANCE DATA

The author has presented numerical information illustrating the concentration effect of heavy metals after classification by a hydrocyclone and an eluticator for dredged materials from the port of Hamburg. The statement was also made that the "heavy metals are mainly bound to the fine particles fraction." To further support this statement based on the experimental information presented (Figure 2) the writer has related the heavy metal content to the organic material present (Table 2).

TABLE 2. Heavy Metals Associated with the Organic Fraction of Dredged Material from Hamburg Harbor (after Figure 2).

	Concentration (ppm/% Organic)			
Component	Input	Output	Input	Output
Pb	17.1	17.3	17.8	16.2
Cd	0.93	1.02	0.87	0.99
Zn	146	149	133	127
Cu	16.4	15.8	17.4	15.9

The results indicate that a fairly uniform amount of each of the heavy metals was related to the organic content of the dredgings, regardless of the process stream. As was indicated (Figure 2), the contaminated sediments represented 42.7 percent of the initial material dredged to be treated by the so-called "B" small scale decontamination technique (Figure 1). The clean fraction indicated to be 57.3 percent of the initial material dredged, was considered to be pure sand.

IN-SITU CLASSIFICATION: A CONCEPT FOR THE MINIMIZATION OF DREDGE MATERIALS TO BE DECONTAMINATED

The results presented in Figure 2 (author) and in Table 2 (discusser) indicate the bulk of the heavy metal contamination was associated with the organic matter fraction, which was only 8.6 percent of the total solids in the sludge. Therefore in-situ separation of the organic matter from the remaining sediments may offer a possibility for decreasing the amount of sludge requiring decontamination.

A comprehensive dredged-materials management plan would be utilized, as was outlined previously, to establish the quantities, spatial distribution and "hot spots" for contaminated sediments to be provided special handling prior to disposal or reuse. The "advanced" dredging procedure would be carried out in two phases, at the appropriate locations. First, where significant contaminated sediments are located, a procedure suited for in situ concentrating the organic matter (equivalent to classification and elutriation) could precede its removal. The commercial availability of equipment to effect this separation is not known. However the separation of materials by differences in specific gravity in industrial applications, including wastewater treatment facilities is well known [16]. Secondly, dredging of the remaining uncontaminated sediments would be carried out following conventional methods.

SUMMARY

The need to provide for the removal of heavy metals and other contaminants from dredged materials for environmental improvement is recognized. Professor Müller has presented a decontamination technology to achieve this objective. The cost benefits for minimizing the amounts of sludge, and concentrating the metals content in that sludge was indicated. The writer has proposed an "in-situ metals concentration concept" for further reducing the quantities to be processed for decontamination.

REFERENCES

13. Breder, R. et al, "Toxic Metal Levels in the River Rhine." Chapter 15. *Pollutants and Their Ecotoxicological Significance*, ed. H. W. Nurnberg (John Wiley and Sons. Chichester, UK 1985).

14. Forstner, U. and G. Müller, *Schwermetalle in Flussen and seen*, (Springer, Berlin, Heidelberg, New York, 1974).

15. National Research Council Marine Board Commission on Engineering and Technical Systems, *Dredging Coastal Ports---An Assessment of the Issues*, (National Academy Press, Washington, Dc 1985).

16. Patterson, J.W. *Industrial Wastewater Treatment Technology*, 2nd edition, (Butterworth, Boston 1985).

RESULTS OF BENCH-SCALE RESEARCH EFFORTS TO WASH CONTAMINATED SOILS AT BATTERY-RECYCLING FACILITIES

Judy L. Hessling
PEI Associates, Inc., Cincinnati, Ohio

M. Pat Esposito
Bruck, Hartman & Esposito, Inc., Cincinnati, Ohio

Richard P. Traver, P.E.
U.S. Environmental Protection Agency, Edison, New Jersey

Richard H. Snow, Ph.D.
IIT Research Institute, Chicago, Illinois

1. INTRODUCTION AND BACKGROUND

Under U.S. laws such as CERCLA* and the National Contingency Plan that implements it, response actions at hazardous waste sites must reduce the threat of uncontrolled wastes. Until recently, this has often meant the excavation or removal of wastes from uncontrolled situations and the movement of those wastes to permitted landfills. In 1984, Congress clearly showed its intent to minimize the volume of such wastes going to permitted landfills by passing the Hazardous and Solid Waste Act (HSWA) amendments. One effect of this legislation has been the mandate of a major change in cleanup procedures to encourage the application of waste treatment technologies prior to disposal.

The policy of the U.S. Environmental Protection Agency's (USEPA) Office of Solid Waste and Emergency Response, which is responsible for implementing the 1984 HSWA amendments, is to discourage containment-based disposal of CERCLA wastes and to encourage the use of technologies which eliminate or reduce the hazardous characteristics of the waste. On-site treatment technologies that destroy or reduce contaminant levels achieve more positive control than containment technologies. Off-site disposal in landfills will probably continue to be allowed on a more limited basis in the future, but only when destruction or treatment technologies are not available for reducing the hazards of the waste

---

* Comprehensive Environmental Response, Compensation, and Liability Act

prior to disposal. As landfill space becomes more limited and expensive, and as transportation becomes more stringently controlled, on-site waste treatment technologies will become more desireable--if they are technologically demonstrated, environmentally safe, and affordable.

Soil and debris contaminated by lead (Pb) and other heavy metals are problems at many hazardous waste sites where metal recycling and reclamation activities have been conducted. Typical examples are sites where used batteries are collected and processed by various cracking and secondary smelting operations.

Piles of spent battery casings as well as slag and dust from furnace operations are often found at such sites. Soil contamination at these sites can typically reach levels in the hundreds and thousands of parts per million (mg/kg)metals. At some sites, Pb levels as high as 10% in soil have been found. Twenty-three battery recycling sites currently appear on the United States' priority listing of contaminated sites requiring cleanup under CERCLA. Many others are known to exist which are not yet part of the priority list for remediation.

Soil washing can be an effective means of either cleansing the soil or reducing the volume of contaminated solids that ultimately must be treated or disposed. It has been under intense investigation by the USEPA for the past ten years. Recently, the USEPA and the Bureau of Mines established a Memo of Understanding for evaluating specific ore enrichment/extraction technologies with potential application to lead battery hazardous waste sites. Specific technologies centering on the use of fluosilicic acid, electrowinning, and recovery/recycle of lead-enriched soil fractions for reprocessing in secondary lead smelters are being evaluated and demonstrated.

Recently, a series of soil-washing studies sponsored by the USEPA's Risk Reduction Engineering Laboratory was completed through the collaborative efforts of a group of scientists from various research organizations. In these studies, the investigators attempted to wash samples of soil from six battery-recycling sites in the United States as well as a surrogate synthetic soil spiked with lead and other metals. The soils were subjected to a rigorous bench-scale washing cycle using either tap water or tap water plus additives (surfactant or chelate). After a 30-minute contact period, the soils were separated from the wash water and rinsed. The washed soil was separated into three distinct size fractions during the rinsing operations to study the partitioning of metals relative to particle size.

This paper presents a partial analysis of the results of these bench-scale studies. It includes a discussion of the background operations at each site that were responsible for the soil contamination problems, a description of the geophysical properties of the soil and contaminant levels at each site, an explanation of the experimental bench-scale procedures followed, and a presentation of the findings relative to total lead levels in the soils before and after treatment. Results of tests performed on EPA's synthetic soil matrix (referred to as SSM) are also presented and compared with the results for the actual site soils. The project included testing the soils for leachable Pb, but the analytical data from this portion of the study are not yet available and therefore could not be presented at this time. The leachate test results could significantly alter the initial findings and conclusions offered in this report.

2. SUMMARY OF FINDINGS

The study results available at this time indicate that soils from battery-recycling operations in general are not highly responsive to soil washing under the types of contact and washing conditions included in these experiments. Total Pb contamination was virtually unchanged in several of the soil residues after treatment, separation, and rinsing. At best, some portions of some soils showed reductions on the order of 50 to 80 percent in total Pb concentrations compared with the untreated soils; however, even with such reductions, the total amount of lead remaining in the residues was often still very high (hundreds to thousands of mg/kg). Generally, plain tap water was least effective as a washing medium. The addition of a surfactant to the water produced marginal improvement, and the addition of a chelate sowed even further promise as a washing aid, based on the increased concentrations of Pb in the spent wash waters.

These results are markedly different from those obtained when washing the synthetic soil. Lead concentrations, which were very high in the soil before treatment (>14,000 mg/kg), were substantially reduced after treatment, especially when a chelate was added to the wash water. Apparently, the Pb in the freshly spiked soil had been afforded little opportunity to weather and mineralize and was therefore more easily removed from the soil by this technology.

3. SITE PROFILES

The six sites that are the focus of this study are among the United States' highest priority sites for cleanup under CERCLA. As shown in Figure 1, these sites represent a broad range of geographic locations, climatological conditions, and native soil types. A variety of process operations and waste disposal practices over several years contributed to soil contamination at these sites.

3.1 Site A

Automotive battery-recycling and secondary lead smelting and refining operations at this 46-acre site in rural northeastern United States began in 1972 and continued for 12 years. Recycling operations consisted of cracking the batteries, draining the acid, removing the lead plates, and crushing the casings. The scrap lead was then smelted in a blast furnace or (later) rotary kiln and refined to produce soft lead or antimonial lead. Furnace gases passed through an 18-cell baghouse for particulate removal. lead-bearing wastes, including the crushed battery casings (rubber and plastic), blast furnace and kiln slag, and baghouse dust, were piled, buried, or landfilled on site. In 1980, the owner entered into an administrative Consent Order to remediate soil and ground-water contamination at the site; and in 1983, the site was listed on the Superfund National Priorities List (NPL). The interim remedial investigation/feasibility study report (January, 1989) indicates that soils in the plant area contain up to 12,700 mg/kg lead. Current activities on site are associated with closure and post-closure care of the landfill.

3.2 Site B

Lead-acid batteries were recycled at this 4.5-acre site in mideastern United States from the early 1970's until 1985. Lead and lead compounds were removed from the batteries and shipped offsite for processing. Acid was drained into onsite lagoons, and broken battery casings (primarily plastic) were shredded and stockpiled on site. During a 1986 removal action, acidic liquids were pumped from the lagoons, neutralized, and discharged to a storm sewer; sludge was excavated, blended with hydrated lime, and returned to the lagoon; and surface soils were disked with lime to a depth of 2 ft. An 800-sq-ft mound of soil mixed with battery casings remains on the site. Lead concentrations as high as 67,700 mg/kg have been measured in the soil; elevated levels of arsenic, cadmium, copper, nickel, and zinc have also been detected.

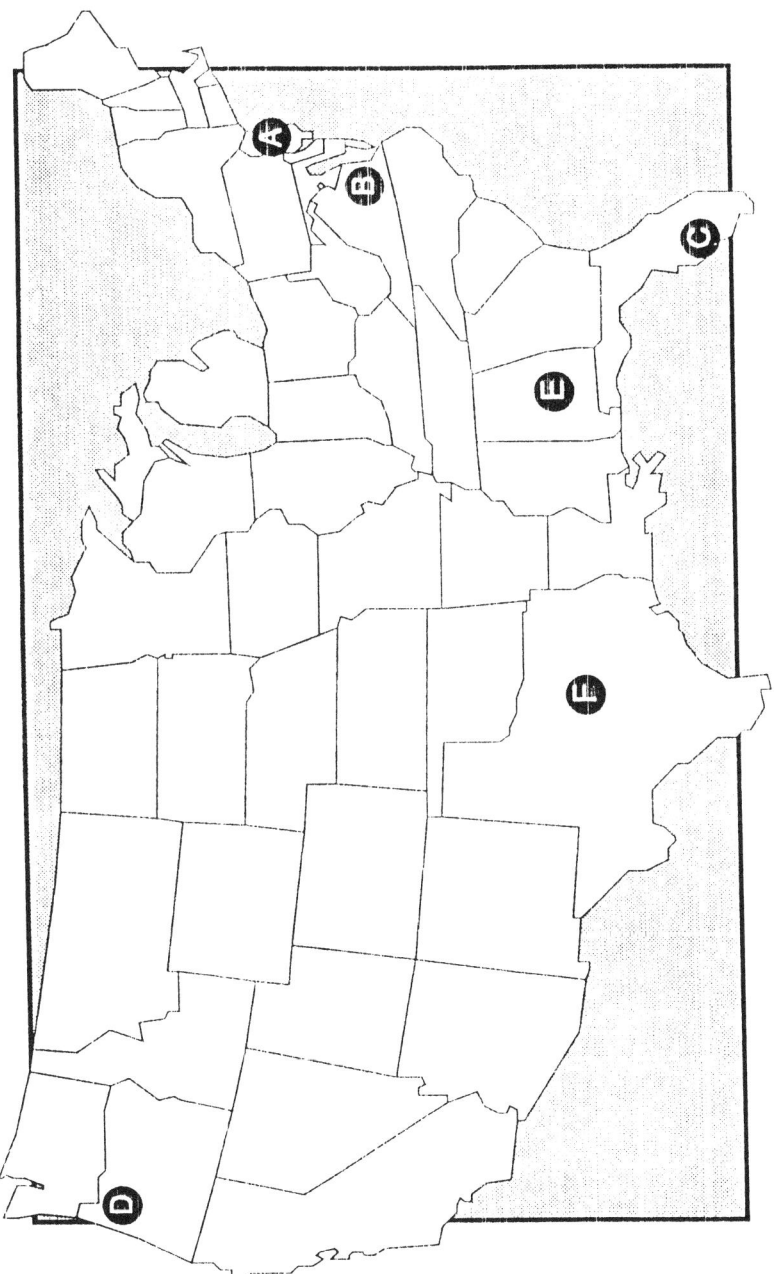

FIGURE 1. Locations of Battery-Recycling Sites.

3.3 Site C

From 1972 to 1986, a lead-acid battery reclamation facility was operated on this 17.4-acre site in a southern Atlantic State. Lead and lead oxide were removed from discarded batteries and shipped offsite for smelting. Initially, rubber battery casings were crushed and used as fill and paving near the processing area; several tons of this fill material was later excavated for recovery of additional lead. Plastic casings, which eventually replaced rubber in the manufacture of batteries, were crushed and sold to a recycler. Until 1981, sulfuric acid from the discarded batteries was treated with lime (calcium oxide) or ammonia and discharged to a 22-acre unlined holding pond; in later years, the wastewater was neutralized and discharged under permit to a publicly owned treatment works (POTW). Also, some waste acid was marketed to the phosphate industry as a processing agent. Lead concentrations in the pond sediments are generally below 500 mg/kg, but they range up to 40,000 mg/kg in the process area soils.

3.4 Site D

This Pacific Northwest site covers approximately 60 acres in a heavily industrialized area. Battery recycling, secondary lead smelting and refining, zinc alloying and casting, and cable sweating operations began in 1949; lead oxide production began in 1965. Over the 30-year operating life of the facility, 86,900 tons of waste battery casings (rubber and plastic), 11,800 tons of matte (composed of iron and lead sulfides), and 6.57 million gallons of sulfuric acid were disposed of on the site and adjacent property. Approximately 98 percent of the battery casings are buried below the surface, where they are in direct contact with the ground water. Concentrations of lead in the battery-casing wastes range up to 190,000 mg/kg; concentrations of lead in the surrounding surface soils range up to 20,000 mg/kg. An estimated 22,000 cu yd of soil requires treatment or removal. The site was listed on the NPL in 1983.

3.5 Site E

This site in southeastern United States consists of several parcels of land where lead-bearing wastes from the main lead-acid battery-recycling and secondary smelting facility were deposited as fill. Operations at the facility involved battery cracking and separation of lead-bearing soils followed by lead smelting, refining, alloying, and casting. Waste acid and rubber and plastic chips from the battery-cracking operation were shipped offsite for recycling or disposal. Slag from the smelting/refining operation was accumulated in waste piles on site. In 1986, the facility began adding calcium sulfate sludge to the blast furnace slag to immobilize the lead and then disposing of the fixed slag in the county landfill. Soil samples collected from various locations during the 1987 remedial investigation indicate that lead contamination averages more than 1000 mg/kg over most of the site and exceeds 30,000 mg/kg in some areas.

3.6 Site F

From 1979 to 1981, a nickel-cadmium battery-recycling facility operated on this 6.7-acre site in south-central United States. The batteries were charged to one of four furnaces; and cadmium, which was driven off from the process as cadmium oxide, was condensed from the exhaust gases and poured into molds. The molds were then resmelted in a ball furnace, and the cadmium was recast into 1-1/4-lb balls for shipment to various plating operations. Furnace gases were ducted through a manifold to a cyclone separator and fabric filter before being discharged to the atmosphere. Cadmium emissions from the fabric filter, along with improper storage and handling of process materials and residues, have contributed to widespread soil contamination at the site. prior to an immediate removal action in 1983, cadmium concentrations in the soil ranged up to 9000 mg/kg; although most of the contaminated materials and debris have now been removed, cadmium concentrations still range between 1000 and 5000 mg/kg over the south portion of the site. Concern over possible exposure of neighboring residents to cadmium from fugitive dust emissions prompted capping of the unpaved areas of the site. More recent sampling has shown that the soil is also contaminated with lead, copper, and nickel.

3.7 SSM

In 1986, the EPA developed a Synthetic Soil Matrix (SSM) for purposes of evaluating alternative technologies for treating hazardous wastes. The soil, which is composed of a mixture of gravel, sand, silt, clay, and topsoil, was spiked with 17 chemical contaminants (volatile organics, semivolatile organics, and metals) at concentrations typically occurring at Superfund sites. The target concentration for lead was 14,000 mg/kg. The spiked soil was prepared in 500-lb batches by blending an insoluble lead salt ($PbSO_4 \cdot PbO$) and the other contaminants with the clean soil in a 15-gal drums for subsequent testing and thus was never exposed to field conditions.

TABLE 1. Characterization of Soils from Battery-recycling Sites.

Parameter	Site A	Site B	Site C	Site D	Site E	Site F	SSM-III
Grain size distribution							
Sand and gravel, wt%	69	55	87	91	90	63	59
Silt, wt%	17	31	6	5	8	20	28
Clay, wt%	14	14	7	4	2	17	12
Predominant clay species	Illite/ kaolinite	Illite/ Smectite	Smectite	Smectite/illite	Kaolinite	Kaolinite/ smectite	aolinite/ bentonite
Moisture content, %	7.2	17.5	5.5	2.4	8.8	10.7	19.5
pH, S.U.	6.18	9.34	7.24	6.50	6.31	6.55	8.5
Cation exchange capacity, meq/100 g	36.5	36.6	40.2	23.5	5.2	13.4	133
Humic acid, %	0.34	0.04	0.76	1.21	NA[a]	NA	NA
Total organic carbon, mg/kg	16,000	7,015	14,150	5,555	3,588	14,500	32,000
Lead (total), mg/kg	57,150	75,850	3,230	27,150	3,945	302	14,318
Lead (leachable), mg/liter[b]	300	418	55.5	148	196	NA	19.9[c]
Predominant lead species	$PbCO_3$	$PB_3(CO_3)_2(OH)_2$	$PbCO_3$	$PbSO_4/PbO_2$	$Pb_4SO_4(CO_3)_2(OH)_2$	NA	$PbSO_4/PbO$

[a]NA = not analyzed.
[b]As measured by the Extraction Procedure (EP) Toxicity test, unless otherwise indicated.
[c]As measured by the Toxicity Characteristic Leaching Procedure (TCLP).
Kaolinite = $Al_2Si_2O_5(OH)_4$
Smectite = Na-Ca-Al-Si-O-H
Illite = K-Al-Si-O-H

4. SOIL CHARACTERIZATION

Samples of the raw soil from each of the six sites and SSM were characterized for physical and chemical properties, including grain size distribution, moisture content, pH, cation exchange capacity (CEC), humic acid, total organic carbon (TOC), and lead (total and leachable). These characterization data are summarized in Table 1. The predominant clay minerals and lead species, as determined by X-ray diffraction, are also indicated. Soils from Sites C, D, and E have a high percentage of sand and gravel, whereas soils from Sites A, B, and F and the SSM have a relatively high percentage of silt and clay. The moisture content of all the soils ranges from 2 to 20 percent. Soil pH is around neutral for Sites A, C, D, E, and F and slightly alkaline for Site B and the SSM. The CEC for soils from the six sites is below 40 meq/100 g, in contrast to that for the SSM, which is above 130 meq/100 g. Humic acid content for all soils measured is low (1 percent or less). Soils from Sites B, D, and E have a low TOC, and soils from Sites A, C, and F and the SSM have a high TOC. The total lead concentration in the soils ranges from a few hundred parts per million (ppm) for Site F to a few thousand ppm for Sites C and E to tens of thousand ppm for Sites A, B, and D and the SSM. Leachable lead concentrations are generally two orders of magnitude lower than total lead concentra-tions. The predominant lead species in the naturally occurring soils are cerussite (Sites A and C), hydrocerussite (Site B), hillite (Site E), and anglesite and platnerite (Site D); lead sulfate and lead oxide were used to spike the synthetic soil.

5. EXPERIMENTAL SOIL-WASHING PROCEDURES

The soil-washing procedures followed during this testing and evaluation program were based on a set of four assumptions that underlie the volume-reduction approach to washing contaminated soils. The assumptions are as follows:
 (i) A significant fraction of the contaminants in soil are either physically or chemically bound to the silt- and clay-sized particles of the soil.
 (ii) The silt and clay are attached to the sand and gravel by physical processes such as compaction or adhesion.
 (iii) physical washing (e.g., scrubbing) of the sand and gravel fractions will effectively remove the fine silt and clay materials.
 (iv) The contaminants will be removed to the same extent that the silt and clay are separated from the sand and gravel. Increasing the efficiency of the washing process will directly increase the removal rate.
 In each experiment, a 500-gram sample of soil was mixed with 5000 ml of wash water (10:1 wash water-to-soil ratio) in a 10-liter glass jar and agitated on a reciprocating shaker for 30 minutes. The

soil was then separated from the wash water by wet sieving and filtering; this operation simultaneously separated the soil into three size fractions:

>2-mm fraction	Coarse sand and gravel	No. 10 screen
0.25-to 2-mm fraction	Fine sand	No. 60 screen
<0.25-mm fraction	Silt and clay	Filtered (on 0.45μ filter paper)from the wash water that passed both screens

The solids retained on each screen were rinsed with 2000 ml of tap water and subjected to a mechanical (vibratory) dewatering device for 10 minutes. The residues and the spent wash waters were then submitted to an EPA-approved laboratory for contaminant analysis following the methods of SW-846.

Duplicate samples of each soil were washed by this method, and the analytical results for the two sample sets were averaged. The duplicate analytical results did not always match well, especially for samples with relatively high levels of lead contamination. This variability was expected, however, and must be tolerated under real field conditions where soil contamination often varies widely within small areas.

Three wash solutions were studied:
a) Tap water, pH7
b) Tap water plus anionic surfactant (0.5 percent solution)
c) Tap water plus tetrasodium ethylenediamine tetraacetate (Na_4EDTA) 3:1 molar ratio EDTA to toxic metals), pH 7-8

Figure 2 presents a schematic of the procedure followed for all experiments.

6. RESULTS AND DISCUSSION

Table 2 presents the analytical results for total Pb found in the spent wash waters. Little or no Pb was found in any of the tap-water wash solutions, which indicates that tap water alone was

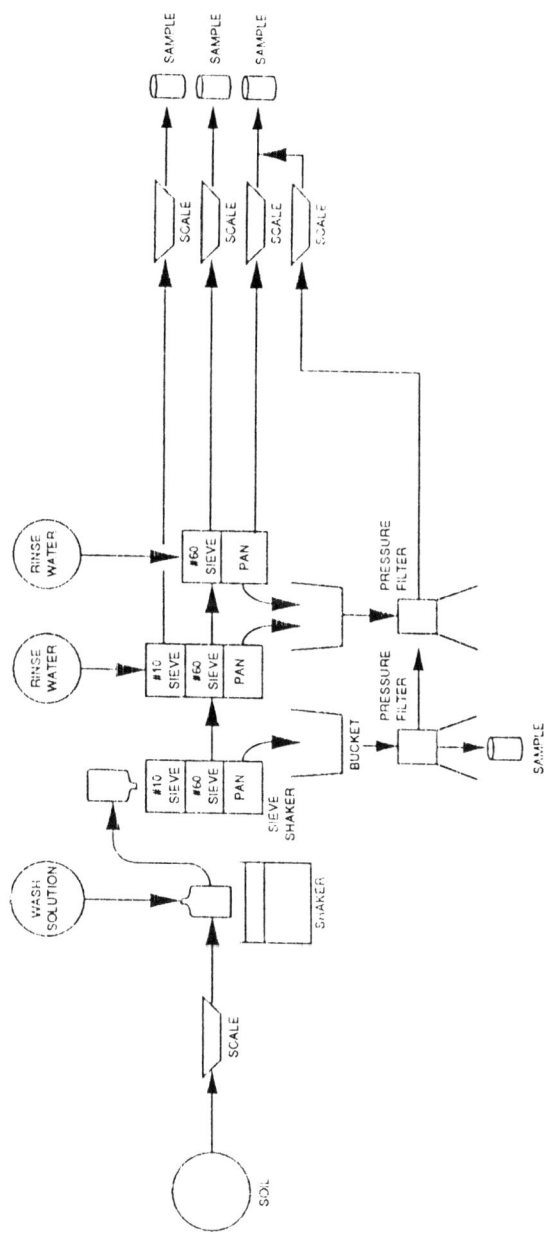

FIGURE 2.   *Bench–Scale Soil Washing Procedure.*

unable to dissolve the lead in the soil. The addition of the surfactant increased the amount of lead in the spent wash water about 100-fold, and the addition of the chelate increased the lead solubilization even more. In some case, for example soils from Sites A, B, and E and the synthetic soil, the chelate increased the amount of lead in the spent wash water about 1000-fold over the plain tap-water wash. These data indicate that most of the lead contamination in the battery-recycling-site soils is insoluble in water. They also indicate that surfactants and chelates such as EDTA offer good potential as soil washing additives for enhancing the removal of lead and possibly other metals from contaminated soils and solid debris.

TABLE 2. Total Lead Content of Spent Wash Waters. mg/liter

Site	Water Wash	Surfactant Wash	Chelate Wash
A	0.79	154	1,255
B	0.24	16.5	942
C	0.32	38.4	73.5
D	0.82	127	3324
E	0.43	Not Analyzed	456
F	<0.06	Not Analyzed	5.4
SSM	Not Analyzed	Not Analyzed	112,500

Analytical results for the soil residues from the experiments are presented in Tables 3-5. Average values or each soil size fraction recovered after washing are shown, as well as the average total contamination levels in the whole soils prior to treatment.

Overall, these data are not encouraging. No clear pattern has been identified, other than for most of the soils, the residuals were still highly contaminated with Pb, even after treatment with the chelate. Two obvious exceptions to this generalization are the soil from Site F and the SSM. Both soils
had similar grain size distributions and both had a clay content that was dominated by kaolinite. The treated fractions of all sizes for these two soils were the cleanest (had the lowest residual Pb levels) of the lot when washed with chelate.

The data collected in this study have not been statistically evaluated for trends. They have been checked for quality assurance purposes, however, and no deficiencies of abnormalities were noted.

Visual observation of the data set as a whole has revealed no apparent trend in soil or contaminant behavior relative to the type of Pb contamination (predominant lead species), type of predominant clay (percent or mineralogy), or particle size distribution. Overall, the results of this research strongly suggest that the applicability of soil washing to soils at these types of sites must be determined on a case-by-case basis.

TABLE 3. Total Lead Content of Treated Soil Recovered on Number 10 Screen (>2 MM) mg/kg.

Site	Whole soil untreated	Water Wash	Surfactant Wash	Chelate Wash
A	57,150	153,100	98,100	119,050
B	75,850	50,150	66,500	164,200
C	3,230	893	1,580	886
D	27,150	31,350	25,610	8,965
E	5,194	6,487	Not Analyzed	1,081
F	210	42	Not Analyzed	30
SSM	14,318	122	Not Analyzed	98

TABLE 4. Total Lead Content of Treated Soil Recovered on Number 60 Screen (0.25-2mm) mg/kg.

Site	Whole Soil Untreated	Water Wash	Surfactant Wash	Chelate Wash
A	57,150	22,300	32,000	24,550
B	75,850	52,200	49,670	57,350
X	3,230	2,160	1,755	2,340
S	27,150	12,800	10,960	8,670
W	5,914	2,020	Not Analyzed	1,516
F	210	312	Not Analyzed	408
SSM	14,318	491	Not Analyzed	171

TABLE 5.   Total Lead Content of Treated Soil Passing Both Number 10 and Number 60 Screen (<0.25mm) mg/kg.

Site	Whole Soil Untreated	Water Wash	Surfactant Wash	Chelate Wash
A	57,150	14,650	36,450	41,250
B	75,850	49,500	52,100	24,170
C	3,230	2,949	2,645	3,995
D	27,150	41,400	42,700	15,250
E	5,194	13,698	Not Analyzed	4,693
F	210	111	Not Analyzed	73
SSM	14,318	30,600	Not Analyzed	1,470

7.   CONCLUSIONS

It appears that soils from battery-recycling sites that have undergone years of neglect and weathering may not readily respond to soil washing as a remedial treatment technology. Also, Pb probably cannot be physically separated from the soil or concentrated into a smaller volume by particle size separation; it certainly did not partition cleanly into any of the three particle size ranges in this study.

Future efforts to interpret the results of this research effort will include results of leachate tests on the residues. These data were not yet complete at the time of this writing and unfortunately could not be included in this discussion of results. Initial leachability tests on the untreated soils (presented in Table 1) showed that all released relatively small amounts of Pb (compared with total lead levels) when exposed the mild acid extraction medium of EPA's established leaching procedures. Nevertheless, the leachate values were all substantially higher than available U.S. standards permit (5 mg/l Pb maximum). It will be interesting to see if the leachates developed from the treated soils contain reduced levels of Pb; if this is the case, it may improve the picture for soil washing as it will indicate that this technology can be used to remove the most soluble portion of the Pb, thus removing the most important negative environmental impact associated with this type of soil contamination, which is toxic metal mobility.

ACKNOWLEDGMENTS

This research was funded in its entirety by the USEPA's Office of Research and Development, Risk Reduction Engineering Laboratory. Mr Richard P. Traver was the USEPA Project Officer in charge of the work. The authors wish to thank Mr. Traver for his guidance, encouragement, and timely review of the data throughout the research effort.

The Synthetic Soil Matrix which was included in this report is available for other soil treatability research efforts. For further information on its composition, availability, and response to other treatment technologies, please contact:

>Mr. Richard P. Traver, PE
>U.S. Environmental Protection Agency
>Risk Reduction Engineering Laboratory,
>Releases Control Branch
>GSA Raritan Depot, Building 10, Woodbridge Avenue
>Edison, New Jersey, USA
>Phone: 201/321-6677

DISCUSSION OF:
RESULTS OF BENCH-SCALE RESEARCH EFFORTS TO WASH CONTAMINATED SOILS AT BATTERY-RECYCLING FACILITIES

Giovanni Tiravanti
Water Research Institute, CNR, Bari, Italy

The paper deals with preliminary results related to soil washing studies and particularly to the treatment of soils contaminated by lead, in sites where exhausted batteries were collected and treated.

The purpose of this investigation was to understand and evaluate a certain number of metal extraction procedures and technologies, from contaminated soils, with the aim to reduce the volume of solid wastes to be landfilled and to optimize lead extraction.

This study is at a preliminary stage and for this reason, the authors have not taken into full consideration the extensive literature relating the importance of surface chemistry to the fate of heavy metals, together with the role played by organic matter on the metal-binding properties.

Surface reactions are responsible for the uptake of metal ions by solid phases. Examples include clays, oxides, zeolites and other mineral precipitates, bacteria, microorganisms and humic acids as well. Functional groups on the solid surfaces react with metal ions to form metal-complexes via ion-exchange, adsorption, etc.

Most of surface functional groups exhibit acid-base properties, especially oxides, such that their net charge is pH dependent. Examples of these functional groups are represented by amphoteric hydroxides on metal oxide surfaces, which can donate or accept hydrogen ions, and carboxyl and amino groups on bacterial surfaces, with their well known acid-base characteristics. It follows that surface complex formation, or adsorption, or, generally speaking, sorption, depends on free metal ion, surface site and hydrogen ion activity, together with the other common parameters (i.e. temperature, ionic strength, etc.) [1, 2].

Sorption reaction equilibria should be investigated through adsorption isotherms by plotting adsorbed mass concentration of the ion of interest versus solution equilibrium concentration for a specified metal-solid phase system, and by correlating data to currently credited models such as Freundlich and Langmuir.

In this particular case the "desorption" isotherms of contaminated soils should also be determined, in order to assess the reversibility of the sorption reaction and to evaluate the effect of weathering processes on the release of metal ions.

For the organic matter it is important to point out that generally, modelling of metal speciation in natural systems is limited by the difficulties to account for the metal-binding properties of the organic ligands present in the soil [3].

These organic ligands include, among others, those that are classified as humic and fulvic acids, defined as heterogeneous mixtures of polymeric polyelectrolytes whose structures are poorly characterized. Generally, humic and fulvic compounds are ubiquitous and can account for a significant portion of the metal binding capacity of both surface waters and soils.

Unfortunately, their mechanisms of interaction with metals are poorly understood, because it is still impossible to assign a molecular weight to these compounds, and thus the molar concentration of the compounds, which are necessary as input data for simulation models, cannot be directly assigned. Another reason for this can be attributed to the different nature of the metal interaction by naturally occurring organic ligands which makes it difficult to define types of binding sites and approximate conditional stability constants.

Sposito [4] in 1981 suggested the use of a quasi-particle model representing simplification of the complex metal-humic interactions in waters. By this model the overall, conditional stability constants and binding capacities can be determined by use of the Scatchard function [5, 6].

Finally, more information is needed on metals speciation directly in the solid phases, by using spectroscopic techniques such as XPS-ESCA techniques, in order to evaluate the metal oxidation state, their chemical environment, the distribution of metal ions between inorganic and organic matter. This is particularly important in the kind of studies presented by the authors, when soil weathering processes have to be taken into account, with all the chemical reactions involved (i.e. dissolution, hydration, hydrolysis, oxidation, reduction, carbonation, biological activity and so on).

The decomposition of organic matter during weathering can release a number of organic compounds having the capacity to chelate or complex metal ions so that their behavior may be more important than that caused by chemical reactions alone.

The above comments underline that most significant factors affecting the uptake and the release of metal ions were not considered or substantiated by the authors in this paper.

These omissions preclude unambiguous interpretation of their experimental results.

It can be concluded that this is a difficult field of investigation and more research is needed to have a better understanding of the involved phenomena.

REFERENCES

1. Stumm, W. and P.A. Brauner. "Chemical Speciation," in *Chemical Oceanography*, 2nd ed., J.P. Riley and G. Skirrow, Eds. (Acad Press, New York, N.Y., 1975).

2. Florence, T.M. and G.E. Battey. "Chemical speciation in Natural Waters," *C.R.C. Critical Rev. Analyt. Chem.*, 9:219-296, (1980).

3. Campbell, P.G.C. and A. Tessier. "Current Status of Metal Speciation Studies," in *Metals Speciation, Separation and Recovery*, J.W. Patterson and R. Passino, Eds., (Lewis Publ. Inc., Chelsea, MI), 201-224, (1987).

4. Sposito, G. "Trace Metals in Contaminated Waters," *Environ. Sci. Technol.*, 15:396-403, (1981).

5. Scatchard, G. "The Attractions of Proteins for Small Molecules and Ions," *Ann. N.Y. Acad. Sci.*, 51:660-672, (1949).

6. Mantoura, R.F.C. and J.P. Riley. "The Use of Gel Filtration in the Study of Metal Binding by Humic Acids and Related Compounds," *Anal. Chem. Sci.*, 78:193-200, (1975).

PART VII: WASTE REDUCTION AND RECOVERY CASE STUDIES

THREE CASE STUDIES OF WASTE MINIMIZATION THROUGH USE OF METAL RECOVERY PROCESSES

M. Lynn Apel and James Bridges
U.S Environmental Protection Agency, Cincinnati, Ohio

M.F. Szabo and S.H. Ambekar
PEI Associates, Inc., Cincinnati, Ohio

1. INTRODUCTION

Metal-bearing waste streams containing heavy metals are generated through several industrial processes and are a major source of hazardous waste in the United States. These waste streams are common in the metal finishing and electronics industries. Metal finishing operations include about 46 unit [1] operations and incorporate several different industrial processes. Major waste streams consist of aqueous wastewaters, organic solvents, waste oils, and metal-containing sludges. The source of most wastewaters in the metal finishing industry is rinse waters and spent process solutions. Contaminants include metals, cyanide, and other process bath chemicals. Spent organic solvents are generated in degreasing and cleaning operations. Waste oils result from most metal fabrication operations. The solvents (usually halogenated organics) and waste oils are contaminated with metals and constitute a hazardous waste.

Passage of the 1984 Hazardous and Solid Waste Amendments (HSWA) to the Resource Conservation and Recovery Act (RCRA) of 1976 marked a strong change in U.S. policies concerning the generation of hazardous and nonhazardous wastes. In addition to authorizing very stringent treatment and disposal regulations, the Amendments also indicated (as the Nation's top waste management priority) a redirection toward "waste minimization" as a preferential strategy for encouraging improvement in environmental quality.

In 1986, the U.S. Environmental Protection Agency (EPA) responded to the Congressional request in the HSWA with a "Report to Congress on the Minimization of Hazardous Waste" [2]. In this report, the Agency defined waste minimization as:

"The reduction, to the extent feasible, of hazardous waste that is generated or subsequently treated, stored, or disposed of. It includes any source reduction or recycling activity undertaken by a generator that results in either 1) the reduction of total volume or quantity of hazardous waste or 2) the reduction of toxicity of hazardous waste, or both, so long as the reduction is consistent with the goal of minimization of present and future threats to human health and the environment."

The Report to Congress stressed that the most constructive role government could assume would be to promote voluntary waste minimization by providing information, technology transfer, and assistance to waste generators and expanding research and development activities in this area. The Agency proposed a pollution-prevention program to encourage industry to accelerate efforts to reduce the generation of wastes through implementation of process changes and the incorporation of recycling methods.

This paper presents a summary of the findings of three waste minimization research projects that illustrate source reduction and recycling practices in the electronics and metal-finishing industries. The information presented here is a composite of the research studies of Apel, Fair, and Adams in 1984 [3], Leak in 1988 [4], and Myers in 1988 [5]. The purpose of these studies was to evaluate the technical and economic effectiveness of three metal recovery operations in minimizing the volume of hazardous waste generated at each facility.

2. BACKGROUND

Wastewater is the largest source of waste in most metal-finishing facilities [6], and easily the largest in printed circuit board manufacturing facilities and radiator repair shops. Copper, nickel, lead, zinc, and tin are the contaminants most commonly found in these wastewaters. Several technologies are being developed to recover metals from the wastewaters [1,3,6-16]. The technologies involve the use of principles of extraction, ultrafiltration, ion exchange, electrochemistry, membrane separation, and others. The case studies considered herein involve: 1) spray rinse system, 2) recovery of copper and lead from metal-finishing waste streams by aluminum displacement, and 3) recovery of copper and lead from wastewaters generated in radiator repair shops by use of a nonelectrolytic process.

Nickel is one of the principal nonferrous metals of economic significance. Ranked 8th in demand in the United States, nickel is used at a rate of more than 200,000 tons/year [15]. Only about 10 percent of the nickel consumed in the United States is supplied by domestic producers. Also, only 10 percent of the nickel is currently recycled [15]. The economic incentives for developing a nickel recycling process are, therefore, self-evident. Further, with the recent land ban regulations, recovery of hazardous constituents such as lead and copper provide significant economic incentives. The case studies discussed here will, therefore, be of significance to a number of operators, especially because they involve only minor modifications to existing processes.

3. CASE STUDY NO. 1 - DESIGN AND APPLICATION OF A SPRAY RINSING SYSTEM FOR RECYCLING OF PROCESS WATERS [3]

3.1 Background

A joint research project was undertaken by a major U.S. printed circuit board manufacturer and the EPA to demonstrate the effectiveness of a multiple-spray rinsing system installed on a nickel electroplating line. The objectives of this project were 1) to determine the rinsing efficiency of the modified rinse system compared with the old system, 2) to assess the potential reduction in water usage of the modified rinse system operated under normal operating conditions, 3) to determine the quantity of nickel metal recovered for reuse, and 4) to compare the associated economic costs and benefits of the modified rinse system with those of the old system.

3.2 Printed Circuit Board Production Process

Three main production methods (additive, semiadditive, and subtractive) are used in the fabrication of printed circuit boards. The additive production method involves the deposition of plating material on the board in the pattern dictated by the circuit, rather than the removal of metal already deposited (as in the subtractive process). The semiadditive process is a compromise between the additive and substractive methods.

As a part of its daily operation, the manufacturing facility produced approximately 6000 square feet of double-sided and multilayered printed circuit boards for electronic business machines by use of the additive production process. In this process [17], fiberglass boards were initially covered with copper foil, drilled, and mechanically sanded. A rack panel plating line using electroless copper and electrolytic copper was utilized in the next step, plating through holes. In this process, palladium and tin were deposited on the surface of each board and on the inside of each hole. Electroless copper was then applied on top of the palladium and tin to establish continuity. A resistant film (usually a light sensitive polymer) was applied to the boards, followed by the application of an ultraviolet light source through a pattern onto the resistant film. After activation, the unexposed resistant film was chemically dissolved, which left the desired circuit diagram imprinted on the surface of each board. The boards were then reloaded onto racks and passed through a pattern-plating process in which electrolytic copper and electrolytic nickel were used. The copper electroplating was deposited to build the circuit to the desired thickness, whereas the nickel provided a protective coating. After the nickel electroplating had been applied, the boards were rinsed with the new innovative spray rinsing system. After the rinse, and if the application required it, a gold electroplating was deposited and the boards were rinsed again. The boards then underwent an etching process to remove the excess copper. A solder plate was then applied to each board to seal the copper circuit and to protect it from corrosion after final fabrication before the boards were subjected to a final visual and analytical inspection.

3.3 Engineering Design of the Spray Rinse System

The engineering design of the new multiple-spray rinse system required the modification of an existing Napco shuttle dip-rinse system. Construction of the new system required the installation of a polypropylene liner in the shuttle. The liner was divided into three compartments by vertically positioned baffles, as shown in Figure 1. Lower baffles were extended to the maximum height possible to permit the rack to pass through the shuttle unimpeded. Side baffles were also installed to minimize overspray between the compartments. A total of 56 Vee jet spray nozzles were vertically mounted on the sides of the shuttle liner in the second and third compartments.

After exiting the nickel bath, the loaded rack vertically entered the first compartment of the multiple-spray rinse system, where it was allowed to drip for 2 or 3 seconds. The rack then traveled

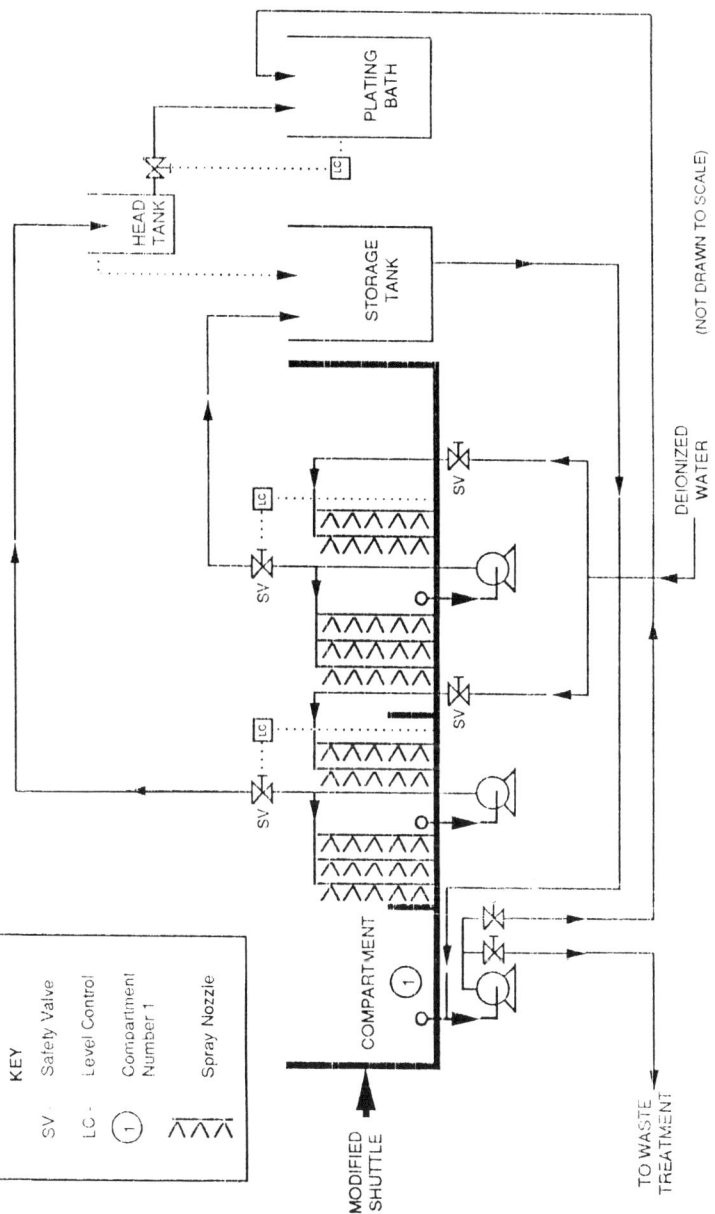

FIGURE 1. Printed Circuit Board Nickel Recycling/Recovery.

horizontally across the 30-foot shuttle at a rate of approximately 6 in./s before vertically exiting the rinse system at the far end of the third compartment. The first compartment of the rinse system (see Figure 1) was used as a drip chamber. In the second compartment, a recirculated rinse spray was applied to the boards, followed by a deionized-water spray rinse. Each nozzle directed a 2-sq.-in. area spray onto the boards. These nozzles were used to impart high water pressure to a concentrated area, which provided a mechanism for removing water and nickel from within the holes in the boards as well as from the surface of the boards. Prior development data by the facility had shown that after 10 seconds of dripping, a large proportion of the residual dragout remained in the holes and that use of jet nozzles greatly enhanced metal removal efficiency.

A minimum of three pumps were used to incorporate recycling capabilities into the design of the spray-rinse system. This provided the rinse system with not only the potential for improved rinse efficiency, but also the ability to conserve and reuse process materials. One pump was used to transfer liquid from the bottom of the first compartment back to the nickel bath. This pump could also be used to transfer liquid from the storage tank (see Figure 1) back to the bath, and to transfer liquid from the storage tank to waste treatment when necessary.

A second pump was used to recirculate rinse water from the bottom of the second compartment onto the boards. A third pump was used to recirculate rinse water in the third compartment, and to transfer excess water to the storage tank when required. When a sufficient quantity of rinse water accumulated in the storage tank, it was transferred batchwise to waste treatment, where it was treated and filtered to isolate and recover the nickel sludge.

3.4 Results and Discussion

Rinse Efficiency

The rinse efficiency of the spray-rinse system was determined and compared with that of the Napco shuttle dip-rinse system previously use. Measurements of the amount of nickel on the printed circuit board after various stages in the spray-rinsing system and the nickel concentration in the plating bath were used to calculate the dragout volume and nickel dragout per board after each rinsing sequence. Details on the sampling and analysis procedures are discussed in the completed report [3]. Metal removal at various stations in the spray-rinse system was determined, and these data are presented in Table 1.

Results showed that of the 1.29 g of nickel remaining on the board after dripping in the first compartment of the rinse system, 95.5 percent was removed by the first recycle spray, and an additional 29.5 percent was removed by the first deionized-water spray in the second compartment. Approximately 0.78 percent and 0.38 percent were removed by the third compartment recycling and deionized sprays, respectively. The rinse efficiency of the second compartment was found be 98.4 percent [(1.29 g - 0.02g)/1.29 g x 100], whereas that of the third compartment was calculated as 75 percent [(0.02g - 0.005g)/0.02g x 100].

Although these two compartments have the same rinsing pattern and flow rates, the lower efficiency of the third compartment was not expected. Metal removal in the second compartment was suspected to be easier than that in the third compartment; removal was primarily from the surface of the board and also from within the holes in the board. Because the board surface was covered with a resistant film, liquid droplets were more easily removed than was liquid from the inside surface of the drilled holes where no film was present. Partially because of these surface differences, additional metal removal in the third compartment was expected to be more difficult, as it primarily required removal of the residual material from within the holes. Removal in this compartment was further hindered by the nickel reaction occurring on the board's surface. A characteristic of this plating process was the quick drying action of the metal on the surface of the boards. Theoretically, this reaction prevents total removal of the metal by the rinse because of solubility and time limitations. This was also confirmed by the decreasing residual removal data shown in column three of Table 1.

The data provided by the printed circuit board manufacturer and reported at the bottom of Table 1 show that the previously used dip-rinse system removed 99.4 percent of the metal. In summary, the data show that the shuttle, in combination with the follow-up rinse, removed 99.8 percent of the nickel with 5 gpm of water. This compares with 99.6 percent metal removal achieved by the spray-rinse system with only 0.35 gpm of water. Overall, this equates to comparable rinsing efficiency by the spray-rinse system with a 93 percent reduction in the total amount of water used.

4. MATERIALS BALANCE

A materials balance for the spray-rinse system was performed to determine the quantity of nickel recovered for reuse in the nickel bath. Overall, the balance showed that approximately 94 percent of the nickel entering the spray-rinse system was recycled back to

TABLE 1. *Evaluation of Metal Removal Efficiency at Various Stations in the Spray-Rinse System and Napco Dip-Rinse System.*

Rinsing sequence	Nickel remaining per board after rinsing sequence	Amount of nickel removed per station		
		% of Total	% of Total	% of Accumulation
Drip only	1.29	-	-	-
Drip + first recycling spray	0.058	95.5	95.5	95.5
Drip + first recycling spray + first DI spray[a]	0.020	2.9	66.0	98.5
Drip + first recycling spray + first DI spray + second recycling spray	0.010	0.78	50.0	99.2
Drip + first recycling spray + first DI spray + second recycling spray + second DI spray	0.005	0.38	50.0	99.6
Drip + first DI spray	0.053	95.9	-	
Drip + first DI spray + second DI spray	0.015	99.8	71.7	
Napco Shuttle dip-rinse System				
3.5-gpm shuttle	0.0081	99.4	99.4	99.4
1.5-gpm overflow rinse	0.0033	0.40	59.3	99.8

[a] DI = Deionized-water spray.

the bath for reuse. Approximately 4 percent was directed to waste treatment, and 2 percent was lost through dragout and losses within the system (lines, filters, pumps, etc).

In comparison, approximately 98 percent (5.67 lb/day) of the nickel entering the rinse system was directed to waste treatment when the dip-rinse system was used. Implementation of the spray-

TABLE 2. **Material Balance** *for the Two Rinsing Systems.*[a]

Rinse System	System Inflows Drag-in Water		System Outflows Recycled to Bath			
	From Bath	Inflow				
	Nickel lb/day	Water gpd	Nickel lb/day	%	Water gpd	%
Dip Rinse System	5.77	6600	0	0	0	0
Spray Rinse System	5.77	460	5.4	94	178	39

Rinse System	System Outflows Directed to Waste Treatment				Losses (Evaporation, Dragout, etc.)			
	Nickel lb/day	%	Water gpd	%	Nickel lb/day	%	Water gpd	%
Dip Rinse System	5.67	98	6560	99	0.10	2	40	1
Spray Rinse System	0.23	4	211	46	0.14	2	71	15

[a] Based on data from 53-operating-day test period.

rinse system, therefore, resulted in a 96 percent reduction in the amount of nickel requiring treatment and/or disposal. On average, 39 percent (178/460 gpd,* Table 2) of the water inflow to the rinse system (0.35 gpm or 460 gpd) was recycled and reused for bath makeup water. Approximately 46 percent (211/460 gpd, Table 2) of the water inflow was sent to waste treatment. Based on mass balance calculations, an estimated 71 gpd was lost through evaporation and dragout from the spray rinse system. In conclusion, during the 53-operating-day period, a total of 11,206 gallons of wastewater, or an average of 211 gpd, was directed to waste treatment as a result of the implementation of the spray-rinse system.

In comparison, the dip-rinse system would have required approximately 6600 gpd of water (5.0 gpm, Table 1) over comparable time period. Assuming estimated evaporation and dragout losses of 40 gpd, approximately 6560 gpd would have been directed to waste treatment. Therefore, a 97 percent reduction in the amount of water sent to waste treatment would result from the implementation of the spray-rinse system.

5. ECONOMIC ANALYSIS

This study indicated that use of the spray-rinse system in place of the Napco shuttle dip-rinse operation reduced operating costs without sacrificing rinse efficiency. As reported earlier, water usage decreased 93 percent. Approximately 94 percent of the nickel metal and 39 percent of the water inflow to the spray-rinse system (see Table 3) was recycled back to the bath. As expected, these reductions translate into substantial economic savings; implementation of the new system resulted in an estimated payback period of less than one and a half years.

Table 3 presents an economic analysis of the spray-rinse system. All calculations were based on a work standard of 22 hours/day and 260 days of operation/year. Calculations of water-usage requirements were based on the flow rates given for each system. Water, treatment, and disposal costs were calculated based on local charges. Sludge disposal costs were determined based on a drum loading of 65 pounds and rounded to the nearest drum. Calculation of the cost of plating chemicals not recovered was based on pounds of nickel used per year. Additional costs, such as filter cartridges for the deionized-water unit, were also considered and reported as actual costs. The annual savings in each cost category

---

* Based on 22-hour workday, 260 days of operation/year.

were then determined and used in an economic analysis of the spraysystem. It is noteworthy that more than 60 percent of the annual savings resulted from the recycling of plating chemicals alone.

The installed cost of the spray rinse was $8300. This figure included all materials and in-house labor required to alter the shuttle and install the spray system. Of this amount, shuttle alteration was 12 percent; the spray system, 18 percent; the pump and recovery system, 52 percent; and the automatic panel control system, 18 percent. It is recognized that the installed cost of this system is high; however, this particular unit was built as a prototype for developing design criteria for curtain spray, full cone spray, and other rinsing systems. A unit designed specifically to isolate and recover metal in this type of paint application could be built and installed for a fraction of the cost of this prototype unit.

Annual operating costs, fixed costs, and savings were calculated based on the data presented and itemized in Table 3. Overall, installation of the new rinse system resulted in an initial positive cash flow and a net after-tax savings of over $4900, which equates to a system payback period of less than 1.5 years.

6. CONCLUSIONS

In summary, the installation this spray-rinse system has proven to be both an environmental and economic success. The simple design and very low maintenance associated with this system render it attractive to other electroplating processes as well as other rinsing processes outside the metal-finishing industry. The multiple spray-rinse system has been in operation since 1983, and no major problems have been encountered. Performance data consistent with those presented have been routinely achieved.

7. CASE STUDY NO. 2 - RECOVERY OF METALS FROM METAL-FINISHING WASTE STREAMS BY USE OF ALUMINUM DISPLACEMENT [5]

7.1 Background

Copper- and lead-bearing waste streams are common in the printed circuit industry. Copper ions in waste solutions exist with sulfate and chlorides and in complex ions with ammonia and EDTA. Lead ions generally exist with fluoborate and chloride ions. These metals-containing rinse streams result from cleaning, microetching,

etching, electroless, and electroplating processes used to fabricate printed circuits.

TABLE 3.   Economic Analysis of Spray Rinse System.

Install cost, $/yr:

Equipment (pumps, piping, control box, liner, etc).	$ 4,800
Installation	3,500
	$ 8,300

Annual operation costs ($/yr):

Maintenance labor	$ 500
Materials (filter cartridges, plating chemicals)	653
Utilities (energy, water)	440
Wastewater treatment	50
Sludge disposal	90
Plant overhead	100
	$ 1,933

Annual fixed costs, $/yr:

depreciation (10% of investment)	$ 830
Taxes and insurance (1% of investment)	83
	$ 913

Annual savings, $/yr:

Water	$ 1,990
Wastewater treatment	1,231
Sludge disposal	1.281
Plating chemicals	7,890
	$12,392

Net savings ($/yr) = [annual savings − (operating costs + fixed costs)]	$ 9,546
Net savings after taxes ($/yr), 48% tax rate	$ 4,964
Cash flow ($/yr) = (net savings after taxes + depreciation)	$ 5,794
Payback period (yr) = (investment/cash flow)	1.43

Reports of removal of copper from etching solutions by displacement with aluminum was first issued on in the mid-1960's. Because of the position of aluminum in the electromotive series, metal ions with lower oxidation potentials (such as copper and lead) are reduced to metal if brought into contact with aluminum metal. The aluminum metal is composed stoichiometrically based on the input of more noble metal ions. Several printed circuit manufacturers in California and in the Minneapolis-St.Paul area are currently using this technology for waste treatment.

The objective of this project was to study the variables that affect the recovery of copper and lead from printed circuit and metal-finishing waste streams by displacement with aluminum metal. These variables include aluminum configuration, stream composition, flow rates, contact items, solution pH, and aluminum surface activity. The project was performed by the Circuit Chemistry Corporation, Maple Plain, Minnesota [5] and was funded by the EPA's Risk Reduction Engineering Laboratory through a cooperative agreement with the Minnesota Waste Management Board and the Minnesota Technical Assistance Program.

7.2 Experimental Equipment and Procedures

Test equipment that was constructed and installed included a 15-gallon aluminum exchange reactor, an input holding tank, a metering pump, and an output holding tank.

The reactor design is illustrated in Figure 2. Agitation air was used to produce mechanical motion in the reactor to assist in removing metal particles from the aluminum surfaces as buildup occurred. Metal particles that were displaced on the aluminum and subsequently dislodged settled into the bottom of the reactor through a screen that was raised off the bottom of the tank.

Hammer-milled drill entry foil used on printed circuit board was utilized as the aluminum exchange material. It was shredded into strips 5/8 in. x 12 in. long x 0.012 in. thick. The aluminum concentration of these strips was more than 98 percent. These aluminum exchange strips were supported above the screen. The solution flowed through the reactor from a bottom distribution sparger up through the aluminum and out through a top fitting.

The filter bag (as shown in Figure 2) was not used for these tests. At the low solution flow rates through the test reactor, no significant carryover of solids occurred in the outflow.

The following common metal-containing waste streams from printed circuit manufacturer and metal finishers were selected for testing by aluminum exchange:

- Copper sulfate - sulfuric acid
- Copper ammonia chloride
- Copper - EDTA complex stream
- Hydrogen peroxide - sulfuric acid etchant
- Copper nitrate streams
- Lead fluoborate waste streams
- Solder brightener
- Nickel sulfate

Both recirculation testing and single-pass testing were conducted during the study. Recirculation through an aluminum exchange reactor has potential for shops with low-volume flow rates, where a holding tank can be used to accumulate metal-containing flows. An additional tank with a pump would then be used to produce multiple passes of the solution through the aluminum reactor until the metal reaches an acceptable level for discharge. A single pass of the waste stream through the aluminum exchange reactors, however, would be the process required for the majority of shops where holding and recirculation are not possible.

Recirculation tests were conducted on the copper sulfate/sulfuric acid and lead fluoborate streams to determine the lower limit for metal removal from these processes. Most of the testing for this project, however, was done with a single pass to evaluate the following:

- Optimum pH
- Effect of excess chloride ion as a displacement rate enhancer
- Effect of flow rate or contact time in the reactor
- Effect of pH on aluminum consumption
- Effect of air agitation on displacement efficiency

7.3 Results and Discussion

The recirculation test proved that, given sufficient contact time with the aluminum, the copper and lead streams could be controlled to very low metal concentrations (1 to 2 ppm) for compliance with publicly owned treatment works (POTW) discharge standards.

The single-pass tests showed the 1) lower flow rates provided higher metal removals; 2) a pH range of 2.0 to 3.5 was necessary for optimum metal removals of 93+ percent; 3) excess chloride ion from the use of HCl to adjust pH did not increase metal displacement,

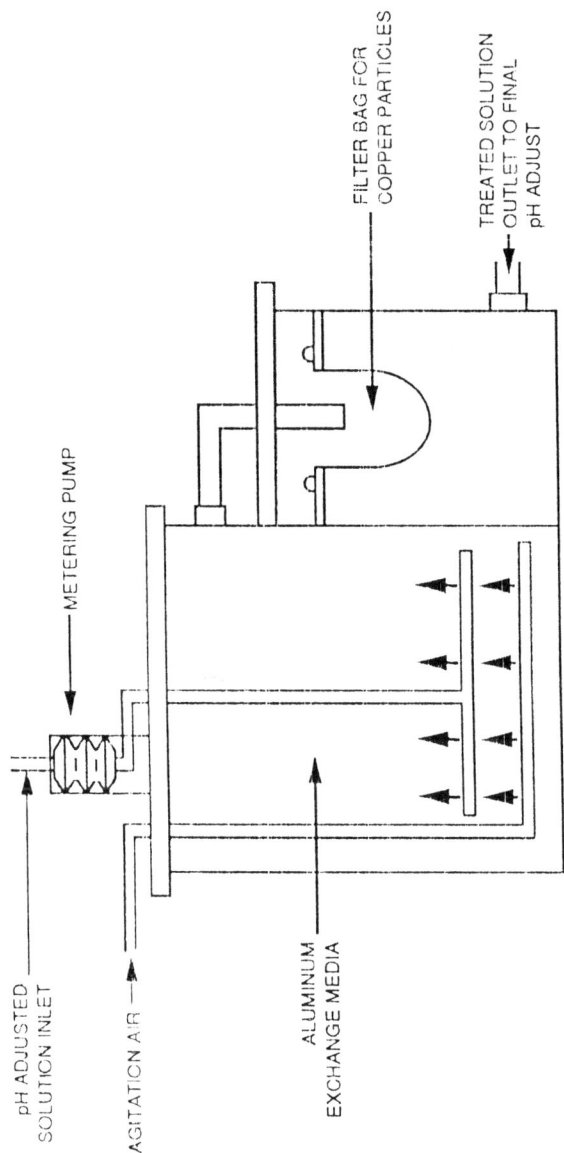

FIGURE 2. *Aluminum Exchange Reactor Test Equipment Design.*

as previously thought; 4) air agitation improves displacement efficiency; and 5) a pH of 3.5 provides the least amount of aluminum consumption. Table 4 presents a summary of the single-pass test results.

Copper in the waste streams from electroplating, alkaline etchant, and sulfuric acid microetching was removed at a rate of approximately 90 percent. Copper complexed with EDTA showed only about 50 percent removal and therefore should not be allowed to mix with the other high-efficiency displacement streams.

Copper nitrate was not displaced on the aluminum. In fact, it stripped copper from the aluminum surfaces. Nickel also was not removed by the aluminum because it is much closer to aluminum in the electromotive series than is copper. Tin from tin chloride solder brightner was removed at 85 percent efficiency.

TABLE 4. Effectiveness of Aluminum Displacement on Metal Removal (Single-Pass)[a].

Waste Stream	Metal	Percent Removed
Copper Sulfate	Copper	96
Copper Ammonia Chloride	Copper	90
Copper EDTA	Copper	51
Peroxide/Sulfuric Etchant	Copper	89
Copper Nitrate	Copper	0
Lead Fluoborate	Lead	90
Tin Chloride	Tin	85
Nickel Sulfate	Nickel	0

[a] Based on the use of 25 gallons of test solution at 0.15 gallon per minute through the reactor at a pH of 2.5.

7.4 Conclusions

The results of this project indicate that aluminum exchange systems could become a significant choice of waste treatment technology for printed circuit manufacturers.

The low equipment cost and simplicity of operation are positive attributes. The metal is recovered as metal particles and is suitable for recovery at a smelter. The metal content of the reactor was estimated to be more than 50 percent; however, the value of the metal was not determined. Aluminum material of suitable shape and purity is readily available for use in the reactors.

The aluminum exchange equipment system must be properly engineered for each unique facility to assure good results. It may be the only equipment required for some facilities, whereas it will be used as one component of a total system in other facilities.

For efficient use of the aluminum-exchange technology, the waste streams should be segregated into metal-containing and non-metal-containing streams. The flows in the metal-containing streams should be minimized by the use of dragout rinses and/or countercurrent flow rinses. The resulting low volume waste stream with high metal concentrations (up to 200 ppm) is suitable for aluminum exchange. The low-volume flow provides the retention time required for efficient displacement.

8. CASE STUDY NO.3 - METAL REMOVAL/ RECOVERY FROM RADIATOR SHOP CLEANING SOLUTIONS BY NONELECTROLYTIC METAL RECOVERY [4]

8.1 Background

Throughout the automobile radiator repair industry, the hot caustic soda boilout tank is one of the prime components of the repair line in nearly every shop, whether large or small. In the United States and Canada, more than 8000 facilities use hot caustic tanks for radiator preparation.

The automobile radiator is composed almost entirely of copper, lead, zinc, and tin alloy. Whereas most ferrous alloys are unaffected by a hot caustic bath treatment, the copper and lead alloys of an automobile radiator are dissolved to a considerable extent during the hot caustic boilout treatment. Radiator shops typically exceed the wastewater discharge standards for lead, copper, and zinc (0.7, 4.5, and 4.2 mg/L, respectively). A portion of the dissolved heavy metals are in true solution, but a much higher portion of the metals precipitate as floccules or metal hydroxide as quasi colloids. This very dilute sludge contains about 10 percent solids, and the metal levels in this sludge have been reported to be about 18,000 ppm lead, 500 ppm copper, 1000 ppm zinc, and 2500 ppm tin.

The normal repair procedure involves submerging the radiator in a caustic solution bath containing 40 to 50 percent NaOH (often referred to as the boilout tank) and heated to 140° to 200°F. The radiator is soaked in the bath for 15 to 60 minutes to soften and/or remove deposits, scale, corrosion, and other foreign materials. Newer boilout tanks may also be equipped to enable ultrasonic cleaning, a technique that significantly improves cleaning efficiency. After soaking, the radiator is withdrawn and (to varying extents) the contents are drained back into the boilout tank. Draining can range from a very limited effort on the part of the repair person to empty the radiator to a very thorough emptying that includes gravity draining of the bulk of the contents and the use of compressed air to empty the oil-cooling section of the radiator. At this point, the radiator has been surface-cleaned, but it is coated with the highly viscous contents of the boilout tank. The radiator is then rinsed with roughly 5 to 10 gallons of water per radiator in a rinse booth. The purpose of the rinsing is primarily to scour this coating from the inside and outside of the radiator.

Wastes generated by the radiator repair process include rinse wastewater from the rinse booth and sludge that must be periodically removed from the boilout tank. Some rinse booths are configured so that they discharge directly to the sewer, whereas in others, rinsing occurs in a closed-loop system that uses rinse water from a reservoir. In the latter rinsing mode, wastewater and sludge are withdrawn from the reservoir on a periodic basis and combined with the sludge removed from the boilout tank. Total sludge generated under either mode ranges from 30 to 300 gallons per year. The sludge is handled as appropriate under hazardous waste regulations.*

---

*Currently, the disposal cost in Minnesota ranges between $175 and $225 per drum. This cost includes approximately $25 for the shipping container (drum) and $50 to $75 for transportation as a hazardous material. Hazardous waste haulers consider the material primarily as a strong caustic solution and secondarily as lead-bearing material.

In addition to these wastes, most shops report the need for routine dumps of the boilout and test tanks. Dumping the boilout tank, which depends on the buildup of sludge in the bottom of the tank, may be required once a year if sludge is not otherwise removed on a more frequent basis. Test tanks may be dumped as often as once per month because of the buildup of suspended solids, which impart turbidity to the water and interfere with the leak-test process.

The objective of this study was to develop a simple concentration/precipitation/reduction technique that could function with these strong caustic solutions while minimizing large volumes of wash or dilution water during the operations. The project was conducted by the National Resources Recovery Institute at the University of Minnesota, Duluth, Minnesota [4]. Project funding was provided by EPA's Risk Reduction Engineering Laboratory through a cooperative agreement with the Minnesota Waste Management Board and the Minnesota Technical Assistance Program.

8.2 Experimental Procedures and Equipment

Research was performed on three representative automobile radiator caustic "boilout" samples. One of these samples represented the older operations, whereas the other two represented the newer technique combining an ultrasonic vibration tank with hot caustic solution. The lead content ranged from 8730 ppm to 4.8 percent.

Chemical analysis of the liquid samples was performed by atomic adsorption apparatus after suitable dilution and comparison with standard samples. The sandy/earthy conventional process sludges were much like ore/mineral samples and required a sodium peroxide fusion procedure before dissolution, dilution, and determination by atomic adsorption analysis.

Most of the bench-scale leaching tests were performed in 1-liter plastic jars with a "U.S. Stoneware" jar mill providing agitation, mixing, and blending between the leachate solution and the sludge sample. Vacuum filtration on the leach residue was accomplished with a Buechner funnel and vacuum flask. In some instances, a centrifuge was required to remove the very fine sludge sediments from the analytical samples. The centrifuge speed was not critical; 10,000 to 15,000 rpm was adequate for separation.

8.3 Results and Discussion

Leaching tests were performed to test the effectiveness of ammonium persulfate and ammonium carbonate as a copper leachate and sodium sulfide as a copper precipitant. These tests indicated that when a significant amount of copper was put into solution, the sodium sulfide would precipitate the ionic copper and leave a nearly copper-free filtrate. For example, the solubility product for the resultant copper sulfide is: Ksp (copper sulfide) = 2.4 x 10E-28 mol/liter in water. The resultant metallic ion concentration in the clear filtrate would be below detection limits of the analytical equipment, but the strong caustic soda solution interferes with the reaction equilibria.

Ammonium Persulfate and Ammonium Carbonate Leaching: Even though ammonium persulfate is known to exhibit good leaching properties for copper, it has not been particularly effective for the selective removal of copper from these caustic sludges. The resultant leach solutions are a pale blue-green color, which indicates the copper ion is present in a simple ionic state. It does not have the intense blue color of the copper-ammonium complex. Also, the copper concentration is an order of magnitude lower than the ammonium carbonate leachates. The maximum concentration of copper attained by ammonium persulfate leaching was 1280 ppm as opposed to 10, 600 ppm copper when ammonium carbonate was used under similar conditions.

Because ammonium persulfate does not seem to be an effective leaching agent under these high caustic conditions, it is not recommended for selective copper leaching from these caustic sludges.

Conversely, ammonium carbonate seems to be an effective leachate for the selective removal of copper from these caustic sludges. All leachates produced by ammonium carbonate exhibit the brilliant blue of the copper-ammonium complex, and the copper con-centrations of the leachates have reached 10,600 ppm copper while the lead levels were maintained below 100 ppm.

Figure 3 plots the copper extraction values as a function of the ammonium carbonate or ammonium persulfate leaching reagent used to remove copper from radiator sludge. Ammonium carbonate seems to function quite effectively as a copper leachate in these strongly basic solutions. Copper extraction of 80 percent can be obtained without neutralization or optimization of the leachate-copper ratio. This is an important factor for a treatment process carried out be semiskilled labor. On the other hand, the strong acid salt (ammonium persulfate) is not an effective copper leachate. This could be due to a neutralization interaction of the persulfate radical

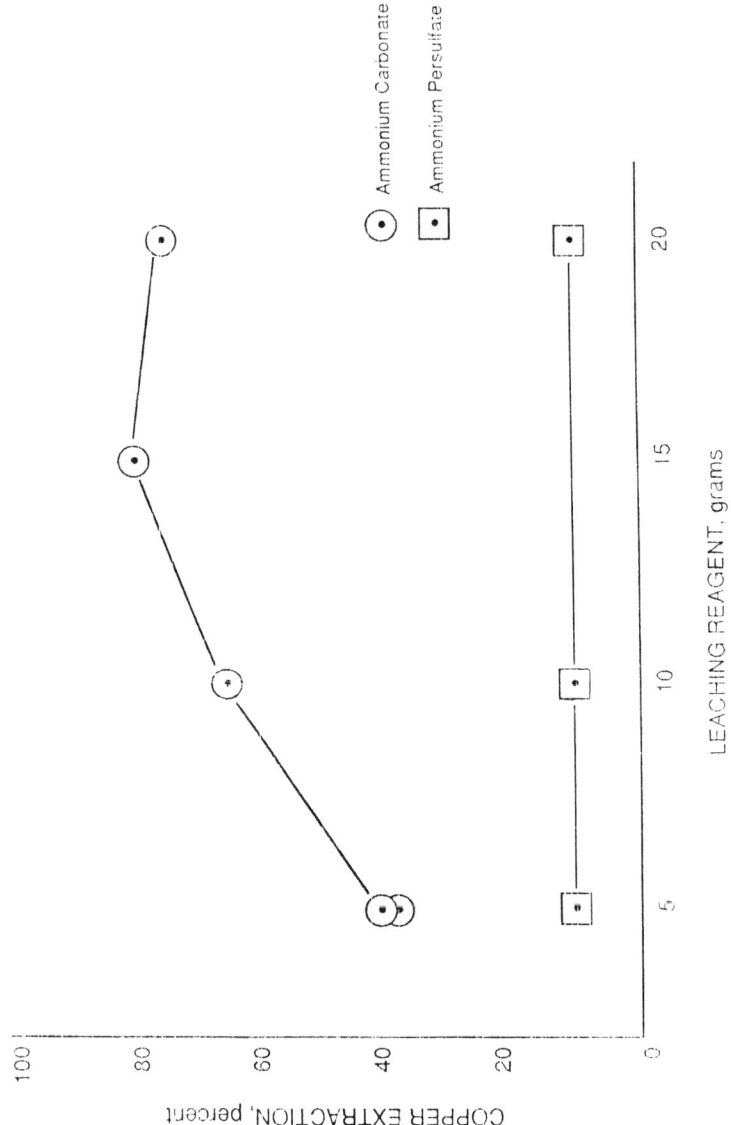

FIGURE 3. Copper Extraction from "D" Sludge with Ammonium Carbonate or Ammonium Persulfate.

and strong caustic of the boilout solution, but the mechanism has not been determined.

Sodium Sulfide Precipitation: Because of the complexity of selective leaching for a small generator, a much more simplified scheme has been developed to provide a marketable product for secondary metal recovery. Both sludges produced in ultrasonic boilout tanks had similar characteristics: flocculent precipitate, difficult to filter, difficult to settle, low-density precipitate. This type of sludge could be characteristic of all the stainless steel, ultrasonic, smaller boilout units that are coming into the industry. The cooled caustic liquid/sludge is treated with sodium sulfide (technical grade $Na_2S \cdot 9 H_2O$) in a 1 to 20 weight ratio and stirred thoroughly to dissolve the sodium sulfide and to allow the metallic sulfide precipitation to occur.

Although the very fine black precipitate is difficult to filter, it settles quite readily and the liquid may be decanted or siphoned off the precipitate. In one case, the lead in the sludge was reduced from 8730 to 11 ppm or a ratio of approximately 800:1 (99.87% efficiency of removal). The resultant stripped liquor was reduced to 14 ppm of copper, lead, and zinc combined. The tin, however, was not significantly reduce, i.e., from 2010 ppm in the original liquid to 1830 ppm in the final liquid. Apparently, the caustic sodium level is so high that the amphoteric nature of the tin as sodium stannate [$Na_2Sn(OH)_6$] overrides the formation of tin sulfide.

$$Na_2Sn(OH)_6 + S^{-2} \rightleftharpoons SnS\downarrow + SnS_2\downarrow + NaOH$$

In other words, the tin sulfide precipitation reaction is effectively inhibited by the high concentration of sodium hydroxide. Other means must be developed to reduce the tin level to the point where the operator can dispose of the resultant liquid in the sewer.

8.4 Conclusions

This preliminary test work has shown that cooling the hot caustic radiator boilout solution and then adding sodium sulfide will precipitate the metal hydroxides. The precipitates are coarse enough to settle out of the solution and leave a stripped, clarified liquid. Approximately 80 percent of the solution can then be recycled.

This means that it should be feasible to design a precipitation and clarification system that could be used by any small radiator repair shop and thus eliminate this hazardous waste stream from the environment.

The results of this test work also indicates that copper can be effectively separated from radiator boilout tank sludge by a single leaching with ammonium carbonate. Some equipment and labor would be necessary, but a large repair shop may be justified to use this procedure to provide an acceptable feed to a secondary lead refinery. Copper is the most undesirable tramp element in scrap lead materials, and it must be less than 0.02 percent in the refined lead ingots that are produced.

Whereas ammonium carbonate leaching would be the most effective on the earthy, sandy sludges produced by the traditional process, sodium sulfide treatment would be the most effective for the flocculent sludges produced in the ultrasonic boilout tanks.

Ammonium carbonate leaching of the sandy/earthy bottom sludge characteristic of the conventional radiator caustic boilout tanks is an effective way to remove copper, and it leaves a lead-rich sludge for secondary lead recovery. This process seems to involve more steps than a small/medium-sized radiator shop could handle in a cost-effective manner, however. It would be more suitable as a conditioning step for a collector/processor to use in recovering the copper and preparing the sludges for secondary smelting.

Such intermediate processing is a research and development area that should not be ignored. Much of the current effort has been directed toward reducing the volume of hazardous wastes requiring eventual hazardous waste landfill at the site of their generation.

In the area of heavy metals, it may now be the time to begin implementation of the following concepts: (a) to condition the heavy metal sludges for resource recovery at the generator level, and (b) to develop the processing technology for heavy metal resource recovery at the economics-to-scale for the intermediate collector and processor level.

In the area of municipal solid waste (MSW), resource recovery is becoming more commonplace as cleaned paper stock, high-value plastics, and glass are being separated from the MSW stream and returned to the marketplace as feedstock or as energy commodities. Heavy metal sludges should be considered from this same resource-recovery perspective.

REFERENCES

1. U.S. Environmental Protection Agency. "Development Document for Effluent Limitations and Standards for the Metal Finishing Point Source Category," EPA 440/1-83/091, Washington, D.C. (1983).

2. U.S. Environmental Protection Agency. Report to Congress, "Minimization of Hazardous Wastes." EPA/530-SW-86-033. Office of Solid Waste, Washington, D.C. (1988).

3. Apel, M.L., P.S. Fair, and J.P. Adams. "Design and Application of a Spray Rinsing System for Recycle of Process Waters," in *Proceedings of Water Reuse Symposium*, American Water Works Association, San Diego, CA, (1984).

4. Leak, V.G. "Metal Removal/Recovery from Radiator Repair Shop Cleaning Solutions Using Non-Electrolytic Metal Recovery," Report prepared by the Minnesota Waste Management Board for the Risk Reduction Engineering Laboratory, U.S. Environmental Protection Agency, Cincinnati, Ohio, (1988).

5. Meyers, S.C. "Recovery of Metals from Metal Finishing Waste Streams Using Aluminum Displacement," Report prepared by the Minnesota Waste Management Board for Risk Reduction Engineering Laboratory, U.S. Environmental Protection Agency, Cincinnati, Ohio, (1988).

6. Hunt, G.E. "Waste Reduction in the Metal Finishing Industry," *JAPCA*, 5(38):672-680 (1988).

7. Kavanaugh, D.J., and A.R. Boyce. "A Continuous Closed Loop Regeneration System for a Chronic/Sulfuric and Etchant Bath," paper presented at *42nd Purdue University Industrial Waste Conference*, West Lafayette, Indiana, (1987) p. 873.

8. Semmens, M.J., C.F. Kenfield, R. Qin, and E. L. Cussler. "The GM-IX Process: A Novel Metal Cyanide Treatment and Recovery Technique," paper presented at *42nd Purdue University Industrial Waste Conference*, West Lafayette, Indiana, (1987) p. 883.

9. Hunt, G.E., and R.W. Walters. "Cost-Effective Waste Management for Metal Finishing Facilities: Selected Case Studies," paper presented at *39th Purdue University Waste Management Conference*, West Lafayette, Indiana, (1984) p. 521.

10. Edelstein, P. "Reclaiming Platers, Nickel and Copper," *Product Finishing*, (October, 1987) p. 86.

11. Brooks, C.S. "Waste Metal Recovery Case Histories," paper presented at *41st Purdue University Industrial Waste Conference*, West Lafayette, Indiana, (1986) p. 647.

12. Ramirez, E.R., and R.C. Ropp. "Metal Finishing Wastewater Treatment Incorporating Materials Recovery and Low Temperature Evaporation," paper presented at *41st Purdue University Industrial Waste Conference*, West Lafayette, Indiana, (1986) p. 679.

13. Edwards, J.D., and J.W. Cammarn. "Design and Construction of a 200-gallon Per Minute Ultra-High Purity Water System," paper presented at *41st Purdue University Industrial Waste Conference*, West Lafayette, Indiana, (1986) p. 659.

14. Brooks, C.S. "Nickel Metal Recovery From Metal Finishing Industry Wastes," paper presented at *42nd Purdue University Industrial Waste Conference*, West Lafayette, Indiana, (1987) p. 847.

15. Ramirez, E.R., and O.F. D'Alessio. "Innovative Design and Engineering of a Wastewater Pretreatment Facility for a Metal Finishing Operation," paper presented at *39th Purdue University Industrial Waste Conference*, West Lafayette, Indiana, (1984) p. 545.

16. Brooks, C.S. "Metal Recovery From Waste Sludges," paper presented at *39th Purdue University Industrial Waste Conference*, West Lafayette, Indiana, (1984) p. 529.

17. U.S. Environmental Protection Agency. "Development Document for Proposed Existing Source Pretreatment Standards for the Electroplating Point Source Category," EPA 440/1-78-085, Washington, D.C., (1978).

DISCUSSION OF:
THREE CASE STUDIES OF WASTE MINIMIZATION THROUGH THE USE OF METAL RECOVERY PROCESSES

W. Wesley Eckenfelder, Jr.
ECKENFELDER INC., Nashville Tennessee

INTRODUCTION

Recent regulations in the United States restrict the disposal options for metallic sludges and limit liquid discharges to very low metal concentrations either through pretreatment to a municipal plant, or for direct discharge. This is a result of metal accumulation in municipal sludges and the toxic effects of metals in an aquatic bioassay. The effect of heavy metals on aquatic toxicity in an organic chemicals wastewater is shown in Table 1. It, therefore, becomes obvious that:
- Metal containing wastewater volume must be reduced to an absolute minimum.
- Metallic sludges must be severely reduced or eliminated.
- Residual metals in the wastewater must be reduced to the lowest levels possible with available cost-effective technology.

The authors have addressed the aforementioned issues with three case histories. In the first case, the wastewater volume and dragout in a printed circuit board line is reduced by 93 percent by employing a high pressure spray rinse system with recycle. Only a small portion of wastewater requires treatment.

TABLE 1.   Effect of Metal Removal on Toxicity Reduction.

	48-hr LC_{50}	% Improvement	Cu mg/l	Cr mg/l	Zn mg/l	Ni mg/l
Secondary Effluent	16	--	0.24	0.50	0.16	0.71
After Alum Flocculation	21	31	0.06	0.35	0.07	0.52
After Hydroxide Precipitation	24	50	0.03	0.25	0.03	0.49

It would appear from this example as well as others, that in many cases wastewater can be eliminated by closing the loop. Ramirez and Ropp [1] proposed a system in which counterflow dragout tanks are employed. Daily evaporation from the plating both permits the first dragout tank to be pumped into the plating tank and the last dragout tank to be filled with clean water. If operating evaporative losses are inadequate, supplementary evaporation is employed. It would appear that combining the concepts employed by the authors of the paper and that described by Ramirez and Ropp would result in zero discharge at minimum cost for most cases.

The second case history presented involves the substitution of aluminum for copper and lead. The metal is recovered as metal particles and is suitable for recovery at a smelter. This addresses the critical problem of minimizing or eliminating metallic hazardous sludges. A number of emerging technologies are becoming available for by-product recovery or immobilization of heavy metal sludges. While in the past these technologies were considered too costly, when compared to the present disposal costs of these sludges, in many cases they become cost-effective.

Sequential solvent extraction at low pH has been reported (Brooks 2) to provide selective recovery of chromium, copper, nickel and zinc in the presence of iron. Highly efficient separations were consistently obtained for copper with efficiencies ranging from good to poor for nickel, zinc and chromium, depending on the specific waste system.

Boyd and Fulk [3] reported on an electrolytic process for the removal of copper and lead. The copper concentration was reduced from 1,600 mg/l to less than 10 mg/l. There is no metallic sludge generated from this process.

In many cases, it is not possible to eliminate the generation of metallic sludges. Etzel [4] has recently shown that heavy metal sludges can be rendered non-hazardous (as defined by the EP Toxicity test) by the addition of triple super phosphate at a pH of 8.5 to 9.0. Metal concentrations after extraction with and without phosphate addition are shown below:

	W/O Phosphate	W/Phosphate
Copper	17.6 mg/l	0.007 mg/l
Lead	17.1 mg/l	0.028 mg/l
Nickel	0.7 mg/l	0.000 mg/l
Cadmium	15.5 mg/l	0.018 mg/l

Application of the approaches presented in the paper and in this discussion should result in substantial reduction in wastewater volume and metallic sludges. However, in many cases, permit levels cannot be achieved and further treatment of the liquid residuals are required. A summary of these technologies for cadmium is shown in Table 2.

In summary, the authors are to be complimented for presenting three excellent case histories of waste minimization. As we enter the next decade, source control and minimizing the generation of metallic sludges must receive top priority in any waste management program. Only minimum residual metals should receive end-of-pipe treatment.

TABLE 2. Removal of Low Levels of Cadmium.

Process	Initial Concentration (mg/l)	percent Removal
Iron Oxide Adsorption	0.056-112	76-98
Activated Carbon	20-202	27-85
Ion Exchange		
Natural Zeolite	0.216	67
Alpha Sorb	0.130	>99
Sulfide Precipitation	100	99.8
Membrane	174	78

REFERENCES

1. Ramirez, E.R. and R.C. Ropp. "Metal Finishing Wastewater Treatment Incorporating Materials Recovery with Low Temperature Evaporation," in *Proc. 41st Ind. Waste Conf.*, (Lewis Publishers, 1986) p. 67.

2. Brooks, C.S. "Waste Metal Recovery Case Histories," in *Proc. 41st Ind. Waste Conf.*, (Lewis Publishers, 1986) p. 647.

3. Boyd, D.M. and R.J. Fulk. "Treatment of Plating Wastewater Without Sludge," in *Proc. 43rd Ind. Waste Conf.*, (Lewis Publishers, 1988) p. 499.

4. Etzel, J.E. "Industrial Pretreatment Technologies for Heavy Metal Removal and Treatment of Heavy Metal Sludges to Render Them Non-Hazardous," *Symposium-Wastewater Toxics Management*, (Virginia WPCA, November, 1988).

METRO RECOVERY SYSTEMS - A CENTRALIZED METALS RECOVERY AND TREATMENT FACILITY IN TWIN CITIES, U.S.A.

J. J. Chen
Lancy International, Inc., An Alcoa Separations Technology Co., Warrendale, Pennsylvania

1. INTRODUCTION

Responding to the requirements of environmental regulations and economical necessity, a "new ethic" is created for the national waste management strategy in the United States. The enforcement of the 1976 Resource Conservation and Recovery Act (RCRA) and its subsequent Amendments has prioritized the waste management decisions in a descending order of source reduction, recycling, recovery, treatment and disposal.

In the last 40 years, most metal finishing plants have dealt with their wastewater problems using traditional end-of-pipe chemical precipitation and sedimentation. While this approach may be effective in meeting effluent discharge standards, it may not fit the new waste management strategy and may be economically undesirable due to the production of large quantities of sludge and associated high disposal costs. This is especially true for small- and medium-sized metal finishing firms which have limited resources and for whom it is economically impractical to consider any resource recovery. Alternatives to the conventional treatment are clearly needed.

One of the alternatives is the central treatment and recovery facility (CTRF) concept which, by pooling the wastes from various shops, can make the resource recovery approach feasible, benefitting from the economies-of-scale. The CTRF provides its member shops a vehicle to meet effluent requirements with less capital investment and operating and maintenance expenditures. Their exposure to environmental liability is reduced because the CTRF is managed by professional staff with qualified technical skills, and because maximization of metals recovery decreases the sludge disposal problem.

Encouraged by the authority regulating discharge to the local sewer system, a group of surface finishers and electronics manufacturers in the Twin Cities (Minneapolis and St. Paul), Minnesota area, formed a task force in 1982 to study the feasibility of CTRF. The preliminary results based on the technical, engineering, marketing and economic factors indicated that the CTRF concept was practical. Consequently, the Metropolitan Recovery Corporation was established in May, 1983 by most of the same group of firms to carry out the task of bringing the concept to reality.

A fund was established from a stipulation agreement between MRC members and the Metropolitan Waste Control Commission (MWCC), the local authority, which was used to make market evaluation, laboratory and field pilot studies, preliminary engineering, permit applications, and land purchase. In May, 1986, MRC signed a partnership agreement with Lancy Recovery, Inc. (LRI), a subsidiary of Lancy International, Inc., part of the Alcoa Separations Technology Division, to form Metro Recovery Systems (MRS) for continuing development of the project. With its technical experiences and expertise, Lancy took charge of the project, doing design, engineering, construction, project management and operation of the CTRF plant. Ground was broken on September 1, 1987, and the plant began to process wastes on July 1, 1988, after a 10-month fast-track construction schedule.

2. PROJECT EVALUATIONS

During conceptual development of the CTRF project, several technical, economical, and financial issues were considered and evaluated which may warrant some discussion.

2.1 Technical

The basic element for the CTRF concept to work in the Twin Cities is to effectively transport the wastes in concentrated and segregated forms from the customers' sites to the CTRF plant. The concentrated forms will greatly reduce the cost of transportation, and the segregation is a prerequisite to low-cost metals recovery.

A typical surface finishing shop generally produces two types of wastewaters: the concentrated batch dumps and dilute rinse waters. The batch dumps can be economically trucked to the CTRF without any special preparation since the wastes are in high concentration. The only attention which should be paid in dealing with the wastes is segregation to prevent any undesirable chemical reactions among wastes and to facilitate recovery.

It is obviously not economical to transport dilute rinse waters directly from the customer site to the CTRF without some form of transformation. During the early stage of project development, it was determined that concentration by the ion exchange process was the most desirable means to accomplish such transformation.

Several issues were addressed concerning the use of ion exchange resin as the transport medium:

Types of Resins Selected

Obviously, the resin has to be able to produce an effluent fitted to be discharged to the sewer or reused in a simple and cost effective way. Superficially, the best approach to the problem in question would be to use specialty resins selective to specific ions to be removed (chelated or nonchelated). Chelating resin will preferentially take the heavy metal cation -- the pollutants -- rather than the sodium and calcium ions, but it behaves chromatographically so that it is not practical for streams containing more than one metal. Also, it wouldn't produce deionized water for recycle. Specialty resin was considered for PCB shops -Cu only; however, it would be an operational nightmare if various types of resins have to be handled and regenerated in the CTRF at the same time. Therefore, a strong acid and a strong base resin were selected for use in this CTRF. These resins, though not selective, are applicable to all streams and are cost effective.

Cross Contamination of Resins

One of the most fearful pitfalls for resin regeneration in place is the difficulty in giving resin a complete and thorough cleaning and regeneration without physically moving out of the confined canisters. Dead zones are often encountered during regeneration in canisters. Therefore, early in design phase resin regeneration in place within canister was ruled out because it was far more economical and technically desirable to remove resin from canisters used in plating shops and regenerate batch-wise in large systems. The spent resin is to be regenerated in the batch mode with the resin bearing the same ion being regenerated at the same batch. This is another way of segregation enabling resource recovery. The problem may arise that certain sloppy shops will contaminate the resin with nasty impurities, therefore, mixing the resin with other good resins in the batch regeneration process will create a cross contamination problem, called by our technical people with a nontechnical term, the "herpes" effect. There is no good solution to this potential problem except to rely on preventive measures. It is of utmost importance to train the service technician who picks up the spent resin to have high sensitivity in detecting any trace or sign of resin abuse, both of negligence and ignorance. In addition, frequent analysis of incoming spent resin is emphasized.

Fouling of Resin

It is impossible to keep all impurities out of ion exchange resin when it is used for wastewater treatment. However, minimization of the fouling potential by protective measures is important. The resin should be protected from suspended solids and organics fouling. Cartridge filter, granular media filter and mechanical filter were considered for removing suspended solids. Cartridge filters were selected for primary consideration because of their cost advantage and without the requirement of filter backwash. They have proven adequate in most applications. For some high-solids situations, it has been necessary to use more elaborate filtration and/or load reduction through the disciplined use of slow-flowing drag-out rinses as will be discussed later. Activated carbon filter was considered for reducing organics and oil contamination of resin.

Ion Exchange Effluent Monitoring

Finding an economical means to detect the breakthrough of ions from the resin column was a hard task. Several options were studied including pH, conductivity, automatic turbidity meter, automated continuous metals analysis, on-line x-ray fluorescence analyzer, and manual colormetric method. The automated analyzers were ruled out either because they are not practical or not cost effective. The manual colormetric method was selected. A search for an economical auto-analyzer is continuing.

Water Reuse

The reuse of water after ion exchange treatment should be encouraged wherever possible. The immediate return is the reduction of city water usage and sewage discharge fees. The additional benefit is to preserve the resin exchange capacity for the target ions, not the incidental ions existing in the water supply. This is especially true for shops using hard city water. For printed circuit shops it is essential for them to practice water reuse or use softened water. The spent cation resins from printed circuit shops are regenerated with sulfuric acid. Any appreciable amount of calcium ions in the resins would foul the resins by the formation of calcium sulfate during regeneration. The water reuse also will reduce the potential for an excessive suspended solids buildup, which puts a great stress on the pretreatment filtration step, especially for the cyanide plating operation line. Unfortunately, most of the customers have taken a conservative attitude to avoid the water reuse practice, instead water softeners are used for the fear that it may impair their product integrity. Apparently, more education on the water reuse subject is needed.

Dead Rinse

It is much more economical to treat the wastewater in batch dump than via ion exchange resin. Therefore, a dead rinse was recommended before a running rinse wherever possible. This is particularly true for the cyanide plating line, not only for economic reasons, but also as a necessity to drop out the undesirable suspended solids, which will overload the pretreatment filtration unit prior to the ion exchange treatment. However, many customers were reluctant to accept the recommendation because of fund savings consideration or lack of floor space, and suffered some unpleasant surprises of having to deal with a filter plugging problem.

Sludge treatment and disposal is another interesting issue that deserves serious consideration. Although the bulk of recoverable metals are reclaimed from the waste streams, even at such a reduced sludge quantity, the ever-escalating sludge disposal cost makes this cost a major part of the CTRF operating expenditures. Therefore, every effort is made to reduce the volume of sludge by improving metals recovery efficiency and the sludge drying process to increase the solids content in the sludge cake. The land ban disposal of the metal finishing sludge (F006) has motivated the management to consider a contingent plan of providing sludge solidification and stabilization capability in the CTRF as its needed flexibility to accommodate the legislative change.

2.2 Economics

One of the most important factors to attract the potential customers to use the CTRF plant is the economical incentive. The cost of utilizing CTRF has to be less than that for them to treat the wastes in house. Some other intangible factors will also be preserving the floor space for production usage, not for waste treatment purposes, and by dedicating time, energy and money to production effort rather than operation and maintenance of complete waste treatment systems.

In order to illustrate the economical benefits for joining the CTRF, a comparison of in-house treatment cost against the CTRF usage cost was made in 1986 for two typical shops, one for an electroplating job shop (Table 1) and another for an independent printed circuit board manufacturer (Table 2) [1]. The advantage of using the CTRF alternative is very obvious. If the amortization of capital expenditure is added to the comparison, the bias to the CTRF will be even more pronounced.

TABLE 1. Example Comparison of In-Plant Cost to CTRF Cost, Electroplating Job Shop.

Cost to Treat In-House

Flowrate	40,000 – 60,000 GPD
Total Metals Concentration	50 mg/l
Total Cyanide Concentration	15 mg/l
Chromium Concentration	8 mg/l
Volume of Acid Dumped	210 GPDk
Volume of Chromate Dumped	290 GPD
Chromium Concentrate in Dump	1000 mg/l
Raw Water Alkalinity	75 mg/l
Labor Rate	$23/hr
Total Calculated Operated Cost =	$262,800

TABLE 1.   Continued.

Cost to Treat Using CTRF

Item	Units	Annual Amount	Unit Cost	Total Cost
HCl	gallons	37,000	$ .50	$20,358
Other Acids	gallons	1,300	.36	468
Chromates	gallons	75,400	.23	17,342
Alkaline (Soak, Electro)	gallons	15,600	.23	3,588
Zinc Resin	cu.ft.	1,070	22.18	23,733
Nickel resin	cu.ft.	80	39.23	3,138
Chromium Resin	cu.ft.	100	43.60	4,360
Cyanide Resin	cu.ft.	600	62.50	37,500
Mixed metal RESIN	cu.ft.	1,070	39.77	42,554
Anion Resin	cu.ft.	270	16.49	4,452
			Subtotal	$157,500
		In-House Annual Costs		14,200
		Total Calculated Operated Cost		$171,700

TABLE 2.   Example Comparison of In-Plant Cost to CTRF Cost, Independent Printed Circuit Board Manufacturer.

Cost to Treat In-House

Flowrate	50,000 - 70,000 GPD
Total Metals Concentration	30 mg/l
Volume of Acid Dumped	590 GPD
Raw Water Alkalinity	275 mg/l
Labor Rate	$23/hr
Total Calculated Operated Cost =	$232,900

TABLE 2. Continued.

Cost to Treat Using CTRF

Item	Units	Annual Amount	Unit Cost	Total Cost
HCl	gallons	8,000	$ .54	$ 4,320
Other Acids	gallons	35,900	.36	12,924
Chelated	gallons	2,000	.83	1,660
Sulfuric Acid (PC)	gallons	37,450	.40	14,980
PC Alkalis	gallons	37,700	.18	6,786
Copper Resin	cu.ft.	1,600	36.07	57,712
Nickel Resin	cu.ft.	100	39.23	3,923
Lead Resin	cu.ft.	250	36.80	9,200
Anion Resin	cu.ft.	3,400	16.49	56,066
			Subtotal	$167,600
		In-House Annual Costs		15,100
		Total Calculated Operated Cost		$182,700

Similar cost comparisons were made for 30 MRC "member" plants and 10 other potential clients. Following are the aggregated results of the computations.

30 MRC plants

Estimated in-plant waste treatment	$5,900,000/yr.
Estimated CTRF waste treatment	4,3000,000/yr.
Savings by using CTRF	$1,600,000/yr.

Samples of 10 plants not affiliated with the CTRF program

Actual in-plant waste treatment	$2,000,000/yr.
Estimated CTRF waste treatment	1,000,000/yr.
Savings by using CTRF	1,000,000/yr.

It is clearly economically attractive for the surface finishing plants to use the CTRF plant. This factor and the intangible benefits of CTRF as mentioned previously should provide sufficient incentive for individual firms to participate in the CTRF program.

2.3 Market Analysis

Information used for various market analyses came from survey results by direct mail and telephone conversations, input from regulatory database on sewage discharge, and market surveys by marketing consultants [2, 3, 4]. The analytical results for MRC pointed to a conclusion that the primary market for the CTRF in the Twin Cities area was the small- and medium-sized metal plating and printed circuit firms.

The latest market research [5] as analyzed for MRS by Touche Ross further categorized the target market into 4 tiers.

Tier I - Contract Companies

There are 20 contract firms which have signed "take or pay" contracts with MRS and are obligated to pay a minimum charge for their commitments. These 10 firms, most of which participated in the original stipulation agreement with MWCC, are called "founder" firms and have received price discounts on treatment fees for their faith and help in launching the CTRF.

The estimated annual waste volume from the Tier I market is listed in Table 3. Because of the "take or pay" contract, most of the Tier I customers undercommitted their contract waste volume, which is about 39% of the estimated waste volume. A further non-binding survey indicated that the actual revenue from
these firms should be about 62% of the estimated value through the 3-year contract period.

TABLE 3.   Estimated Annual Waste Volume for Tier I Firms.

Waste Type	Annual Estimated Waste Volume
Liquid Wastes (gallons):	
Hydrochloric Acid	714,548
Sulfuric Acid (Non PC)	120,852
Sulfuric Acid (PC)	235,776
Other Acid	211,191
Cyanide Concentrate	22,648
Cyanide Dragout Rinse	360,240
Chromate	2,026,922
Chromate Concentrate	0
Chelated	77,616
Ammoniacal	11,885
Sulfate Copper Etchant	6,177
Alkaline Cleaners	512,078
PC Alkali	87,260
Resins (ft^3):	
Zinc Resin	13,568
Copper Resin	14,705
Nickel Resin	3,798
Mixed Metal Resin	0
Chrome Resin	2,790
Cyanide Resin	2,705
Lead Resin	2,027
Anion Resin	39,039

Tier II - Other Electroplating and Printed Circuit Companies

The noncontracted firms consisted of 113 firms in the Twin Cities metropolitan area and 72 firms outside of the metropolitan area but within the CTRF 150-mile service radius. The estimated annual waste volume is summarized in Table 4.

TABLE 4.   Estimated Annual Waste Volume for Tier II Firms.

Waste Type	Annual Estimated Waste Volume
Liquid Wastes (gallons):	
Hydrochloric Acid	92,680
Sulfuric Acid (Non PC)	21,604
Sulfuric Acid (PC)	91,284
Other Acid	503,337
Cyanide Concentrate	4,247
Cyanide Dragout Rinse	273,801
Chromate	454,919
Chromate Concentrate	12,990
Chelated	159,446
Ammoniacal	9,287
Sulfate Copper Etchant	350,006
Alkaline Cleaner	132,239
PC Alkali	140,747
Resins (ft^3):	
Zinc Resin	19,599
Copper Resin	17,458
Nickel Resin	10,505
Mixed Metal Resin	943
Chrome Resin	8,878
Cyanide Resin	2,243
Lead Resin	6,121
Anion Resin	45,930

The forecast market penetration is estimated to be: 1st year - 14%, 2nd year - 23%, 3rd Year - 28%, 4th year - 33%, and 5th year - 38%.

Tier III - Other Industries Producing Inorganic Waste

Companies in other industries which generate inorganic waste may use the CTRF plant for disposal of their nasty waste sludge.

Using very limited information, four potential wastes are considered for the estimation as follows:

Hydrochloric Acid	179,430 gallons/yr.
Sulfuric acid (Non PC)	179,430 gallons/yr.
Cyanide dragout rinse	6,729 gallons/yr.
Inorganic sludge	7,543 tons/yr.

The market penetration is estimated at: 1st year - 10%, 2nd year - 15%, 3rd year - 20%, 4th year - 25%, and 5th year - 30%.

Tier IV - Future Markets

This market sector includes potential customers who may be able to use the CTRF service if minor modification is made to the facility. This may include the following market segments:
1. Regeneration of non-hazardous ion exchange resins in processes such as raw water deionization and ultra-pure water manufacturing.
2. Recovery of hexavalent chrome by chrome oxidation from such waste streams as chromic acid etchants, chromating baths, and leather tanning chrome discharge.
3. Silver waste treatment and recovery in photographic processes such as graphic arts companies, film studios, photofinishing labs and microfilm processing in banks and insurance companies.
4. Solder stripper and circuit board etchants solutions which normally contain high concentration of lead, tin and copper. The potential volume for these wastes are estimated to be solder stripper, 2,400 gallons per month and circuit board etchants, 38,000 gallons per month.

3. THE CENTRAL TREATMENT AND RECOVERY FACILITY

3.1 The In-Plant Equipment

The CTRF treatment actually starts from the customer site where the wastes are processed and prepared to be transported to the CTRF plant. This inter-link is vital to the success of the CTRF concept as preventive measures may be taken to avoid fatal problems downstream.

The concentrated batch dumps and sludges, etc., are stored in separate holding tanks to facilitate recovery and/or prevent undesirable reactions during storage. The capacity of storage normally designed is for seven days between pickups.

The dilute rinse waters are concentrated for transport using ion exchange systems. For each rinse water stream, a collection tank is provided; from there it is pumped through a cartridge filter to remove suspended solids, an activated carbon column to reduce organics and, finally, to the ion exchange canisters for metals concentration. Figure 1 depicts a typical ion exchange system at the customer site [6].

The CTRF will supply ion exchange system, it will cost customers about $10,000 to $15,000 to make it operational with all the necessary auxiliaries. For the batch holding tank, the cost is approximately $1.50 per gallon of storage capacity.

3.2 The CTRF

Figure 2 describes the overall flow of wastes through the centralized facility [6]. The spent resin canisters are transported by MRS trucks to the CTRF where the resin in each canister is first sampled and "fingerprinted" and the information stored in the computer file. After acceptance, the canisters are unloaded in the dock area then transferred to the resin regeneration station where the resin is hydraulically sluiced to the segregated storage tanks (90 ft^3 each) waiting to be regenerated. When a sufficient quantity of spent resin is collected, the resin is hydraulically transferred to the regeneration vessel where sulfuric acid, hydrochloric acid or caustic is used to drive metals or salts off the resin. After regeneration, the resin is reloaded to the canisters for the customer's reuse. The resin loaded with free or complex cyanide is regenerated in a separated 600 ft^2 airtight room. Sulfuric acid and caustic are used for regeneration purposes with sodium cyanide as the by-product.

The concentrate batch dumps are trucked to the CTRF in separate tank trucks. After "fingerprinting" and acceptance, the content in each tank truck is unloaded under a strict protocol that both the driver and plant operator have to sign off for verification. The concentrated liquids are pneumatically transferred to separate storage tanks, prepared to be processed in the next step.

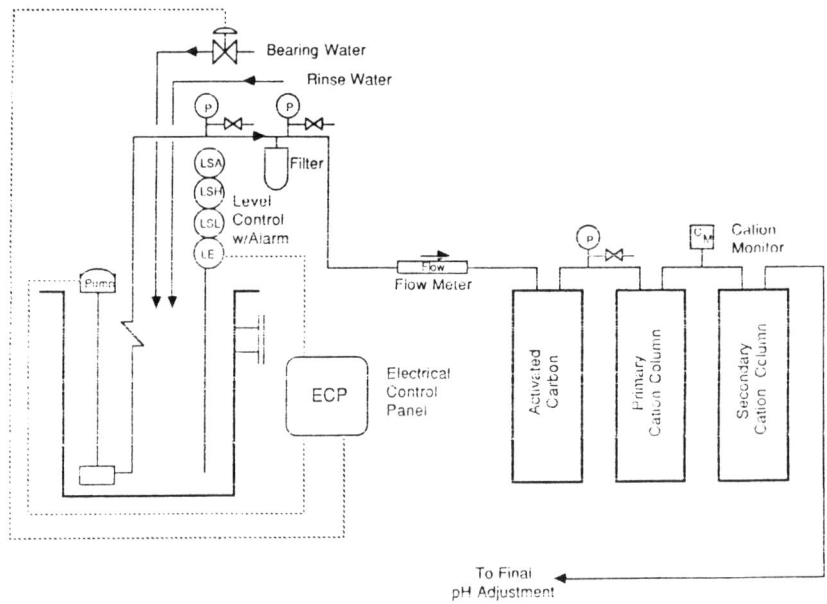

FIGURE 1. *Schematic of Customer Waste Collection Equipment.*

When the concentrated liquids and/or the resin regenerant bearing copper, nickel or zinc stored in each segregated holding tank is collected in sufficient quantity, it is subjected to one of the three electrolytic metal recovery (EMR) cells. The metal is deposited on the cathode plates from the bulk liquid; as a result, the metal concentration is greatly reduce, normally from about 20 g/l to 1 g/l. The cathodes are periodically lifted out to remove the recovered metal and sold to reclaimers.

The residual liquid and other unrecoverable liquid wastes will go through a conventional treatment system consisting of equalization, chemical treatment for metal precipitation, clarification and Ph neutralization before it is discharged to the sewer. The effluent is continuously monitored and sampled by an automatic sampler.

The resulting sludge and the sludges shipped in from the customer's site directly are pumped to a thickener and dewatered by a sludge press and a dryer. If the sludge contains sufficient metal concentration and in high purity, it may be sent to a smelter; *landfill. Another option would be to redissolve it with the incoming sulfuric acid liquid waste to facilitate the recovery.

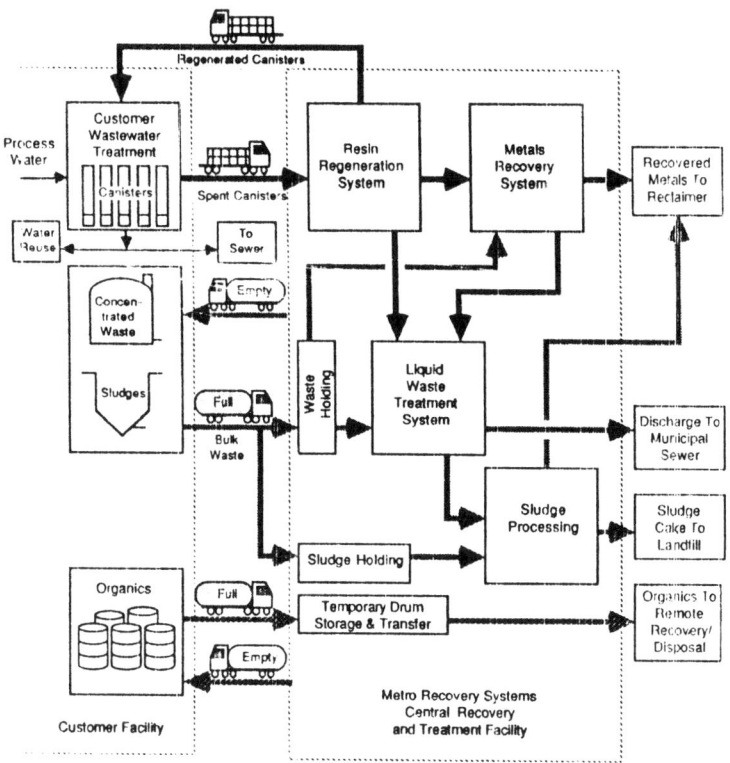

FIGURE 2. *Flow of Wastes through Centralized Facility.*

3.3 The Current Status

Since the operation started at the CTRF in July, 1988, continuous shipments of concentrated liquids and resins have been very encouraging. Table 5 shows the in-bound shipments data for August and November, 1988 [7]. At present the average daily intake to the CTRF is about 8,300 gallons of bulk liquid and 98 ft^3 of resins, while the mean daily output is 200 lbs of recovered metals, 23,000 gallons of clean water to sewer and 2.5 tons of metal hydroxide sludges (F006). The daily chemical usages are 600 gallons sulfuric and hydrochloric acids, 700 gallons sodium hydroxide, 1,300 lbs lime and 68,300 SCF natural gas.

The construction cost for the CTRF may be broken down as follows:

Land, engineering, permitting and market development	$2,200,000
Building and equipment	7,550,000
Start up	150,000
Financing cost, interest and working capital	1,100,000
Total	$11,000,000

TABLE 5.   In-Bound Streams, Month Totals.

Bulk (Gallons)	August	November
HC1	16,550	31,28
H_2SO_4	12,038	9,100
Other Acid	12,680	6,970
Cyanide	18,650	0,470
Chrome	33,830	32,565
Chelate	4,860	7,200
Ammoniacal	0	7,800
$CuSO_4$	1,800	10,100
Alkaline	5,100	34,580
	105,508	150,065

Resin (Cubic Foot)		
Zinc	446	319
Copper	365	320
Nickel	100	94
Chromium	64	67
Cyanide	13	214
Lead	39	91
Mixed Metal	872	606
Activated Carbon	0	24
	2,069	1,735

There are 20 operating staff in the CTRF organization. They are a general manager, a technical manager, a sales and marketing manager, a customer service manager, an office manager, 3 truck drivers, a customer service agent, a laboratory manager, 2 plant engineers, 2 lead operators/dock masters, 3 day shift operators, a night shift operator, a receptionist and a bookkeeper/secretary.

4. SUMMARY AND CONCLUSIONS

A new national waste management strategy is being formulated in the United States motivated by regulatory requirements and cost saving necessity. The central treatment and recovery facility (CTRF) fits the new strategy nicely.

After a thorough evaluation based on technical, economical and market factors, it was concluded that the CTRF concept is very feasible. A unique coalition of regulatory agencies, metal finishing fraternities, environmental engineering industries and the public has brought the CTRF concept to reality in the Twin Cities by formation of Metro Recovery Systems (MRS). The facility has been in continuous operation since July, 1988.

The MRS facility is obligated to provide contracted firms with the full service of transporting and processing all industrial wastes they normally generate. The concentrated liquids and sludges are transported to the CTRF by MRS for processing. The dilute rinse streams are concentrated by ion exchange resin in portable canisters at the customers' sites. The spent canisters are then trucked to the facility by MRS for regeneration. The regenerants and concentrate are processed for materials recovery and reclamation. The non-recoverable wastes are treated as required to produce acceptable effluent for discharge to sewer. All sludges generated at the customers' sites or at MRS are currently mechanically dewatered and thermally dried for ultimate disposal in classified landfills. However, the CTRF is designed with the flexibility of extracting metals for recovery from sludges using acid treatment and adding sludge fixation/stabilization steps to accommodate requirements of the land ban legislation. Organic liquids are also accepted for temporary storage for ultimate processing at an organic solvent or oil recovery facility, or disposal by incineration off site.

It is concluded that the CTRF will provide direct cost savings both capital and operation expenditures to the users. It will reduce potential liability to the customer because by means of metals recovery the sludge production is reduced, and sludge stabilization may transform the metal hydroxide sludge to nonhazardous. Ultimate disposal of sludges is the greatest liability concern to the surface finishers. The CTRF will be the first line defense to shield any allegation against the users on the liability issue. Professional management and technical expertise of the CTRF staff allow the customers to concentrate their efforts on product manufacturing and comfortably rely on the CTRF to take care of waste treatment problems.

ACKNOWLEDGMENTS

The author wishes to thank Messrs. Fred A. Steward, Greg E. Norgaard, Gary S. Hapach and Gary G. Dodge for their dedicated cowork on the project.

REFERENCES

1. PACE Laboratory. "Plating Line Survey and Cost Comparison," Unpublished report, (May, 1986).

2. Touche Ross. "MRC Waste Stream Market Analysis Survey," Unpublished report, (November, 1983).

3. Dodge, G. "MRC Loan Application to Minnesota Department of Energy and Economic Development," Unpublished document, (May, 2, 1986).

4. PACE Laboratory. "MRC Marketing Survey," Unpublished report, (April, 1985).

5. Touche Ross. "Financial Feasibility Study for Metro Recovery Systems," Unpublished report, (August 10, 1987).

6. Norgaard, G. E. "Centralized Waste Treatment Facility Opens in Twin Cities Area," *Plating and Surface Finishing*, 75(11):23-26, (1988).

7. Norgaard, G. E. Unpublished results (1988).

DISCUSSION OF:
METRO RECOVERY SYSTEMS - A CENTRALIZED METALS RECOVERY AND TREATMENT FACILITY IN TWIN CITIES U.S.A.

E. Rolle
Department of Chemistry, University "La Sapienza," Rome, Italy

A central metal recovery and treatment plant in Twin cities, Minnesota U.S.A. deals with various aspects involved in the designing and construction of a centralized treatment plant for the metal finishing industry.

The proposal drawn up by Lancy International is characterized on a technical level by the following details:

-particular attention to recovery possibilities, avoiding traditional treatment and disposal processes as far as possible;

-centralization of the treatment not only of concentrated waste but also of only slightly polluted waters for which original processes for the concentration and transport of the pollutants are devised;

-use of tried and tested treatment and recovery technologies for the centralized plant.

The centralized recovery plant has been in operation since July 1988 and the data summary of the first months has proved very satisfactory, with a progressive increase in the amount of waste dealt with by plant.

Although the initiative described in Dr. Chen's paper is directed specifically at waste treatment for the metal finishing industry, it also contains elements of a more general character which it is worth while mentioning briefly.

Despite the fact that centralized plants have been indicated as the best solution for the treatment of concentrated waste, the number of these plants constructed for recovery purposes has up to-date been more limited than would have been expected.

Various factors, even of a technical nature, may have contributed to this delay but the following are the most significant:

-the uncertainties arising legislations which are still being drawn up regarding the treatment and transport of waste and which are ambiguous with regards to the regulation of the recovery phase;

-the risk induced in using a centralized plant is considered unacceptable by same industries as an eventual closedawn of the plant would put single users who were not equipped to carry out their own waste treatment independently in serious difficulties.

With regard to the situation in Italy it can be noted that it has been possible to make use of these centralized plants only where a sufficiently reassuring solution was found to this particular problems. A typical example is that of the tanning areas made up of hundreds of small industries with similar types of manufacture which level themselves perfectly to the centralized solution for the recovery of chromium from the exhausted tanning solutions.

Despite highly favorable economic forecasts (the initial capital is recovered in less than three years of activity of the center) only one tanning area is at present using a centralized plant with fairly limited aims.

Going back to the more specific aspects of the case represented by Dr. Chen in which the above considerations are of little importance, it should be stressed how an attempt was made to give as satisfactory a reply as possible to the problems of industries, even foreseeing the treatment of the rinse waters with original solutions, while causing the minimum possible impact on the organization of the industry.

There is no doubt that the interest for a centralized solution increases in proportion with the amount of relief given to the industry from the treatment problem and with the minimum possible number of changes the production programme necessary for treatment demands. as is the case with any new initiative, the many positive aspects of the "Metro Recovery System" project, must however be verified by the actual operation of the plant.

By using a data summary, in particular, it will be necessary to confirm all those expectations which directly or indirectly affect the cost of the initiative, such as especially:

-the efficiency, the duration and the selectivity of the resin used for concentrating the diluted wastes;

-the decreased importance of the treatment operations which the industry must however in order to respect the environmental protection laws; the decrease in transport costs;

-the quality of the products recovered and how real the actual possibilities are of placing them on the market.

RECOVERY OF METALS FROM WASTE STREAMS BY HYDROMETALLURGICAL PROCESSES

C. L. van Deelen
TNO Netherlands Organization for Applied Scientific Research, Apeldoorn, the Netherlands

1. SUMMARY

This paper presents a processing route, developed by TNO and others, which has proved successful in the removal of (heavy) metals from a number of waste streams and recovery of individual metal components as salable products. The process is based on hydrometallurgical principles and can be applied to both solid and liquid waste streams.

The basic principles of the TNO process for recovery of metals from waste streams by hydrometallurgical processes are briefly outlined. The recovery of metals by solvent extraction, precipitation and electrochemical deposition is illustrated more specifically for two different hazardous wastes streams, i.e. spent hydrodesulphurizing catalyst and batteries.

2. INTRODUCTION

In the Netherlands more than one million tons of waste are produced on a yearly basis, which, according to the Dutch legislation, have to be regarded as hazardous waste. Approximately half of this amount is processed with currently available technologies. Examples thereof are the recovery of specific chlorinated hydrocarbons by means of distillation and the incineration of combustible hazardous wastes in dedicated incinerators. For the other half, however, no suitable processing methods are available as yet. Due to the severe restrictions put on the tipping of hazardous waste in the Netherlands and the decreasing export possibilities to neighboring countries, industries are faced with serious problems in finding an acceptable outlet for their hazardous waste substances.

Waste streams containing heavy metals constitute by far the majority of the 500,000 tons of "non-processable" hazardous wastes that are generated yearly. Recovery of metals from waste streams would therefore offer a good solution to the hazardous waste problem. Another argument that pleads for processing of metals containing waste streams is that they may constitute a valuable source of raw materials if the pure salts or the individual metals could be recuperated from them. It is therefore not surprising that over the years various technologies have been developed for treatment of metal containing waste streams. However, especially in the case of solid wastes and liquid waste streams with high metal contents, nearly all of these technologies are technically immature or economically not viable. With this in mind, the Netherlands Organization for Applied Scientific Research TNO and the Eindhoven University of Technology developed already in the early eighties a process for recuperation of metals from waste streams [1]. The process developed is based on hydrometallurgical principles. In section 2 of this paper an outline is given of the metal recovery process in general. The recovery of metals is illustrated more specifically in sections 3 and 4 for two different hazardous waste streams, i.e. spent hydrodesulphurizing catalyst and spent batteries. These projects (and others), which are carried out in association with companies involved in processing of hazardous wastes, have the status of contract research. The results presented therefore have a rather global character.

3. METAL RECOVERY BY HYDROMETALLURGICAL TECHNIQUES

The process for recovery of metal from metal containing waste streams developed by TNO and the Eindhoven University of Technology makes use of hydrometallurgical techniques because these are considered to have a number of advantages over other processing techniques (e.g. pyrometallurgical) when processing waste streams:

* Waste streams are often generated at quantities of some thousands of tons per year, requiring the use of flexible, small units;
* Relatively low capital investments;
* Low energy consumption;
* High recovery of metals;
* Low level of metal emissions into air and water.

The process developed can be applied for recovery of metals from both liquid and solid waste streams. When processing solid wastes the first stage of the metal recovery process involves a leaching step, possibly preceded by particle size reduction. Leaching is performed with an acid. Process conditions during leaching (e.g.

acidity of the leaching agents, temperature, residence time and liquid/solid ratio) are selected such that virtually all of the toxic metals are dissolved, leaving a residue that can be deposited as a non-hazardous waste.

After leaching and filtration the acidic solution generally contains various metal ions. Individual metal components can be removed from the leaching liquid (or from a liquid waste stream) through a number of process steps involving techniques such as solvent extraction, ion exchange, precipitation and electrochemical deposition. No general guidelines can be given as to which type of technique should be selected for removal of specific metals since this strongly depends on factors such as the type of metal to be removed, its concentration in the (leaching) liquid and the presence of the metal ions. Apart from these "technical" parameters the costs of metal recovery by a specific technique should preferably be favorable in comparison to the assets of marketable products. In this respect the type of metal compound to be recovered by a technique as well as the purity thereof are of importance.

The process for recovery of metals developed by TNO and the Eindhoven University of Technology has been tested extensively on waste streams such as hydroxide sludges and pickling baths from galvanic and surface finishing industries, various types of catalyst and fly ashes from waste incinerators [2, 3]. Metals recovered include copper, chromium, zinc, nickel, cadmium, iron, lead, manganese, cobalt, mercury and molybdenum.

The next sections illustrate the potential for recovery of metals from two different waste streams for which research projects are currently under investigation at TNO.

4. RECOVERY OF METALS FROM SPENT HYDRODESULPHURIZING CATALYST

4.1 Data on Hydrodesulphurizing Catalyst

Oil refineries make extensive use of hydrodesulphurizing (HDS) catalyst for removal of heavy hydrocarbons, carbon and other constituents of crude oil which gradually render it inactive HDS-catalyst is carried out, in situ or by specialized operations, by burning off the organics and carbon very carefully at temperatures of about 500°C. After a number of regeneration cycles activity drops to uneconomic levels and the catalyst then has to be disposed of.

In former days substantial quantities of these spent catalysts were used in steelmaking. Due to increased quality requirements this possible use is now ruled out. Lacking outlets, spent HDS-catalyst today is either dumped at hazardous waste sites at costs ranging from US $150-300, or stored at refineries, awaiting better times. The estimated amount of spent HDS-catalyst to be disposed of annually is given in Table 1. Furthermore, tens of thousands of tons of spent HDS-catalyst are stored at refineries, local storage or dump sites.

TABLE 1. Estimated Yearly Amount of Spent HDS-Catalyst.

Area	Spent HDS-Catalyst (in tons)
Western Europe	3,000
Middle East	2,000
Southern hemisphere	1,000
North America	25,000

The average composition of fresh and spent catalyst is listed in Table 2. The average metal content for spent HDS-catalyst is estimated at 2.5% wt. for cobalt or nickel and 6.5% wt. for molybdenum. The intrinsic value of spent catalyst, based on the content of valuable metals only, is calculated to be some US $700 per ton at the present, relatively low, metal prices.

In calculating this figure the following assumptions have been made:
* a ratio of Co/Mo to Ni/Mo catalyst of 70 to 30;
* recovery of cobalt and nickel as metals;
* recovery of molybdenum as MoO_3;
* metal recovery yield of 90%;
* metal sales at 80% of market prices.

4.2 Existing Processes for Recovery of Metals

In the past 25 years several processes have been developed that aim at recovery of metals from spent HDS-catalyst. These processes can be divided into two types; pyrometallurical and hydrometallurgical. At present no pyrometallurgical processes are in operation, mainly due to the relatively low prices for cobalt and molybdenum in combination with the low quality of alloys produced.

Two types of hydrometallurgical processes can be distinguished, namely alkaline and combined alkaline/acid leaching. Latter process uses a thermal pretreatment to remove carbon, hydrocarbons and sulphur from the spent catalyst.

The alkaline processes recover molybdenum and vanadium and part of or all of the alumina carrier material. Ammonia-based processes also recover cobalt and nickel as amine complexes. Metal compounds are recovered from the solution using precipitation.

The alkaline/acid processes recover molybdenum, vanadium and part of the alumina in an alkaline leach, after which the residue is submitted to an acid leach to recover cobalt and nickel. Separation and recovery is done using solvent extraction and precipitation techniques.

Nearly all literature on the above processes reports severe filtration problems, due to the poor dissolution of the alumina, producing fine alumina particles. Processes that dissolve all the alumina suffer from problems in separation and recovery of the metals from the solution containing high concentrations of alumina [4].

TABLE 2.   Average Composition (in % wt.) of Fresh and Spent HDS-Catalyst.

Component	Fresh Catalyst	Spent Catalyst
Al_2O_3	87.5 %	40 - 90 %
Co or Ni	3 %	2 - 3.5 %
Mo	9 %	4 - 9 %
C	-	0 - 20 %
C_xH_y	-	0 - 10 %
S	-	0 - 15 %
other (e.g. P, V, Si)	1 %	0 - 15 %

4.3 TNO Metal Recovery Process

Given the experience from other metal recovery processes, TNO aimed at developing a process that meets the following requirements:
* high recovery (90 % plus yield) of valuable metals (cobalt, molybdenum, nickel and possibly vanadium);
* produces a non hazardous residue;
* low level of emission to air and water;
* economic feasibility of the process at a scale of 5,000 tons/year, even if the present high dump prices are disregarded.

The research project is carried out by TNO, Chemconserve (Netherlands) and REAKT (United Kingdom) and is partly funded by the Dutch national research programme on the reuse and recycling of waste materials (NOH). The main stages of the TNO process are depicted in Figure 1 [5].

Spent HDS-catalyst arrives at the site either in drums or in bulk containers and is transferred to a feeding bunker. From there catalyst is fed into a rotary calciner to burn off carbon and hydrocarbons and to convert the metal sulphides into their oxides. In most cases the energy content of spent HDS-catalyst itself is sufficient to perform this operation, but fuel can be added (e.g. natural gas). Flue gas from the rotary calciner must be treated with lime or an alkaline solution to remove the sulphur dioxide that is formed during oxidizing of the metal sulphides. After the thermal treatment a sieving step is carried out to remove ceramic balls that have been used as support material in the desulphurizing reactor.

In the next stage of the process the treated catalyst is leached with diluted sulphuric acid at an elevated temperature, extracting cobalt, nickel, molybdenum, vanadium (if present) and part of the alumina carrier. Care is taken to prevent the bulk of the carrier material from going into solution, thus avoiding serious filtration problems. The residue from the leaching step is washed and dried and can be sold as alumina source. The residue is 60-70% of the feed stock, depending on the composition of the spent HDS-catalyst. The metal contents in the residue meet the requirements set by the Dutch law on chemical waste. The alumina in solution is partly separated from the leachate losing less than 2% of the metals to be recovered. The alumina is transformed into aluminiumsulphate that can be sold to the paper industry.

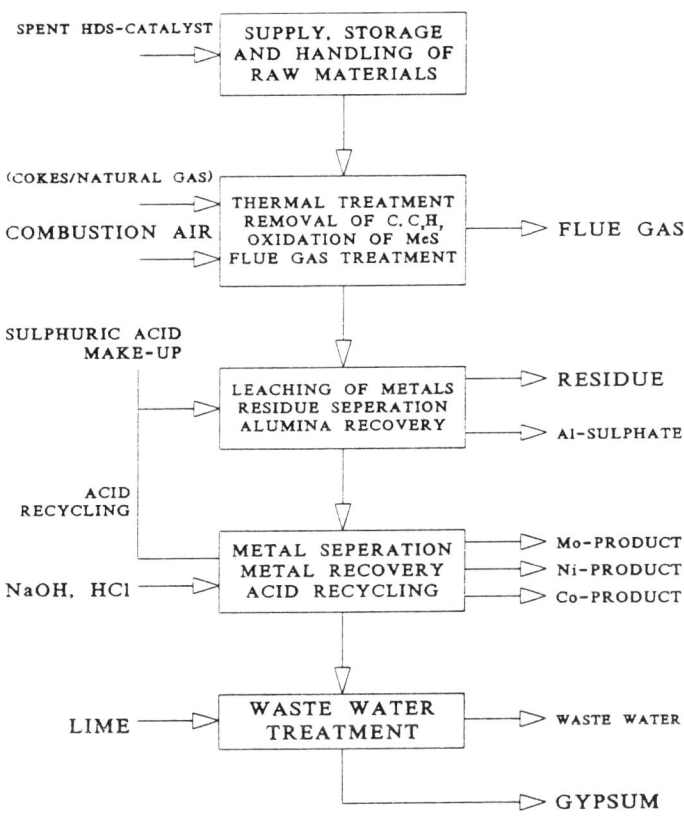

FIGURE 1.   TNO/Metrex Project for Metal Recovery from Spent HDS-Catalyst.

The remaining leachate contains all metals, including some of the alumina, at levels varying from 0.5 to 25 g/l in a sulphate environment, the acidity reaching a level of pH 0 to 2. The metals are recovered from the leachate using a combination of solvent extraction, ion exchange and electrolysis. A more detailed flow scheme on the metal separation part of the process is given in figure 2 and will be discussed below. After the removal of cobalt, nickel, molybdenum and vanadium from the leachate (in the course of which the metals ions are removed and sulphuric acid is formed) the leachate is prepared to be recycled to the leaching step. This preparation primarily consists of adding acid to lower the pH to a level that is needed in the leaching step. A small percentage (0.5-2%) of the leachate is bled off, to prevent a too high concentration of impurities. The bleed can be neutralized using lime or limestone producing a small waste gypsum stream.

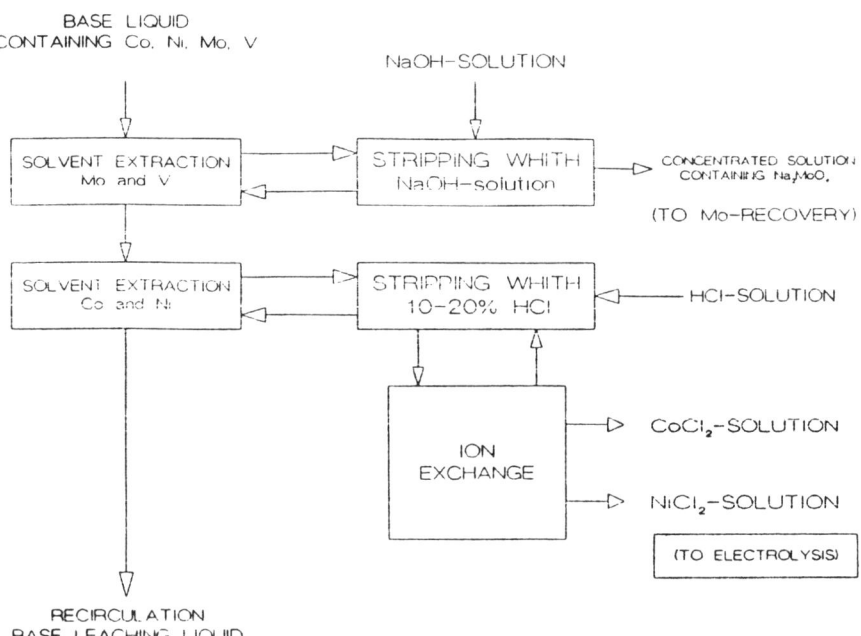

FIGURE 2.   Detail of Metal Separation in TNO/Metrex Project.

The metal separation part of the process is detailed in figure 2. Molybdenum and vanadium are co-extracted from the base liquid by using a commercially available hydroxyquinoline as solvent in a one stage extraction. Both metals extract very well, leaving less than 5 mg/l in the aqueous phase. Stripping is carried out using an alkaline solution, producing a solution of sodiummolybdate and sodiumvanadate. From this solution several products can be prepared, e.g. calciummolybdate, molybdenumsulphide or ferromolybdenum. Separation of molybdenum and vanadium can be carried out using selective precipitation and/or solvent extraction.

Cobalt and nickel are co-extracted from the base liquid using a commercially available naphthalene sulphonic acid. Extraction is carried out in three stages recovering more than 98% of both metals from the aqueous phase. Stripping is carried out using a 5-20% hydrochloric acid. Separation of cobalt and nickel can be done by ion exchange or by solvent extraction, both methods resulting in a chloride containing solution of the metal ion. These solutions can be electrolyzed directly, using membrane electrolysis to prevent chlorine gas from being formed. Another possibility is to add sulphuric acid, boil off the hydrochloric acid azeotrope (20% HCl) and produce cobalt and nickel content in the cobalt metal will be lower than 99.5%. The nickel content in the cobalt metal will be lower than 0.1%. In both cases no acid bleed off is necessary and the acid streams can be recycled directly without any loss. Other metal compounds can be produced if necessary, e.g. metal chlorides or -sulphates by using a cristallization step and metal oxides by using a precipitation method.

The process described above has been developed on a laboratory scale. Verification of the laboratory results in a pilot plant which can process 5-15 kg per hour of spent HDS-catalyst is in progress and is expected to be completed by April 1989. The pilot plant results obtained till now confirm the promising perspectives of the laboratory research. The favorable developments have led to the decision to create a legal platform for commercial realization of the process. A new company, by the name of Metrex B.V., was set up a limited company under Dutch law. Its shares are held equally between Chemconserve and Reakt which together provide for the management. A formal decision to build a commercial plant for processing spent Hydrodesulphrizing catalyst is anticipated by mid 1989. The plant will have a capacity of processing 5,000 tons of spent catalyst per year.

5. RECOVERY OF METALS FROM SPENT BATTERIES

5.1 General

In the Netherlands alone more than 100 million batteries are sold for private use each year, corresponding to a total weight of approximately 5,000 tons. In addition to this, a substantial, unknown number of batteries is used by industrial enterprises, hospitals and the like.

Batteries are cells that can supply energy by means of electrochemical reactions. They are available in a wide variety of types and dimensions, but all have in common a high metal content. Due to the fact that nearly all batteries contain cadmium or mercury, they constitute a serious environmental problem, the more so since the majority of spent batteries are discarded with municipal or industrial wastes.

The main components of the major types of batteries are given in Table 3. Presently developments are going on whereby throughout Western Europe the mercury content of alkaline batteries is gradually lowered to a level of 0.15% wt. in 1990. A further decrease is likely to come with the recent introduction of the low mercury alkalines containing 0.025% wt.

Table 4 presents an estimate of the market share of the various types of batteries in the Netherlands as per 1982. Recent figures are not available but, if anything, total sales of batteries are still increasing. Especially there is an explosive growth in the sales of rechargeable Ni-Cd batteries. In view of their high cadmium content this development gives rise to serious concern about the potential environmental impact thereof. Other developments are a gradual increase of the market share of alkalines at the cost of Leclanché, batteries and the replacement of mercury and silver oxide button cells by zinc-air cells.

TABLE 3. Composition of the Major Types of Batteries [6].

| | Percentage of Total Weight | | | | | | |
| | Type of Battery: | | | | | | |
Component	Leclanchè	Alkaline	Mercury-Oxide	Silver-Oxide	Zinc-Air	Ni-Cd Sealed	Ni-Cd Open
Zn	17	14	11	10	30	-	-
MnO_2	29	22	-	-	-	-	-
Hg	0.01	0.8	33	1	1	-	-
Ag	-	-	-	26-30	-	-	-
Ni	-	-	-	-	-	20-30	10
Cd	-	-	-	-	-	11-15	8
Steel/Plastic/Graphite	26	42	29	29	6	35-40	49
Electrolyte	5	5	9	11	7	30	33

5.2 Existing Processes for Recycling of Spent Batteries

The high content of (sometimes valuable) metals in batteries offers good perspectives for recycling. This is most of all true for the silver oxide batteries; collection of these batteries through retailers and recovery of silver by specialized companies has been common practice for many years.

Full scale commercial plants for the recovery of mercury from mercury oxide button cells have been in operation in Sweden, Germany, Belgium and Japan. All process are based on the thermal treatment of batteries in a furnace and condensation of the vaporized mercury. Due to the relatively low selling prices for mercury all of the European plants have ceased operation, although environmental problems at the sites of the recycling plants also seem to have contributed to this some cases.

TABLE 4. Estimated Number of Sold Batteries in the Netherlands in 1982 [7].

Type of Battery	Number of Batteries (in millions)
Leclanché	60
Alkaline	25
Mercury oxide	2.8
Silver oxide	4
Zinc-air	0.4
Nickel-cadmium	5

Compared to the mercury oxide cells the mercury content of Leclanché and alkaline batteries is much lower. For this reason recovery of mercury by thermal treatment is economically unattractive and proposed processes have never come beyond the research stage. All thermal processes developed aim at maximum recovery of marketable mercury. Outlet potentials of recovered products such as zinc and manganese dioxide are uncertain, and probably in most cases can only be disposed of at certain costs.

As to the recycling of nickel-cadmium batteries commercial plants are in operation in France (S.N.A.M.) and Sweden (SAB NIFE). Additionally 3 plants are presently in operation in Japan with capacities of 80-150 tons per month. These plants do not only process spent batteries but also production waste of the battery industry. All processes are based on thermal principles in which the vaporized cadmium is recovered, leaving a mixture that predominantly contains nickel, iron and char. All process seem to be able to find an outlet for this mixture; the proceeds thereof are unknown but certainly well below the price that could be obtained if pure nickel were recovered.

5.3 TNO Processes for Recovery of Metals from Spent Batteries

Already some years ago TNO's Division of Technology for Society has carried out, in cooperation with the Technical University of Twente, laboratory research on sorting techniques for mixtures of batteries and the recovery of metals from alkaline and Leclanché batteries. Very encouraging results were obtained using a sorting technique based on X-ray pattern recognition and a hydrometallurgical processing route for selective metal recovery [8]. It was estimated that processing costs were comparable with those for dumping batteries as a hazardous waste. In view of the encouraging results an extensive research project was drawn up that aimed at the development of a processing route for alkaline, Leclanché and nickel-cadmium batteries. Potential processing routes are tested on a laboratory scale with further testing of definitive process routes on a continuous, semi-technical scale. The research programme, sponsored by the European Community and the Dutch national research programme on the reuse and recycling of waste materials (NOH), is carried out by TNO in association with Leto Recycling B.V. and Technical University of Twente (all The Netherlands) and Warren Spring Laboratory (UK).

In the following a description is given of the results obtained so far from the laboratory-scale experiments on the recovery of metals from nickel-cadmium and alkaline batteries respectively.

Processing of Nickel-Cadmium Batteries

Figure 3 depicts the processing route for recovery of metals from nickel-cadmium batteries as developed by TNO.

The first step of the process involves shredding of the batteries to a size of less than 10 mm. The shredded batteries are leached with hydrochloric acid to dissolve cadmium, nickel and (inevitably) iron. Conditions are selected such that more than 99.9% of the cadmium is leached out from the shredded batteries, leaving a residue that, apart from some plastic and paper, contains 60-80% wt. of iron and nickel together. The quantity of leaching residue amounts to less than 10% of the weight of the starting material.

Following separation of the leachate and the residue by filtration, cadmium is removed by solvent extraction. Under the prevailing conditions this can be achieved with various extractants, e.g. tributyl phosphate or a tertiary amine. The selectivity of cadmium extraction is very high, provided the leachate does not contain iron (III), which easily forms a chloride complex and therefore is co-extracted with the cadmium. The loaded organic phase is stripped with demineralized water. The strip solution can be worked up further by electrolysis or precipitation, yielding metallic cadmium or cadmium hydroxide with a high purity.

The cadmium-free leachate contains high concentrations of iron and nickel. Iron can be removed completely from the leachate by oxidation of iron(II) to iron(III) with hypochlorite and adjustment of the pH to about 4. The iron hydroxide precipitate obtained contains some nickel. The nickel can be removed to a level below 0.5% wt. (Dutch limit for hazardous waste) by thoroughly washing the precipitate with a slightly acidic solution.

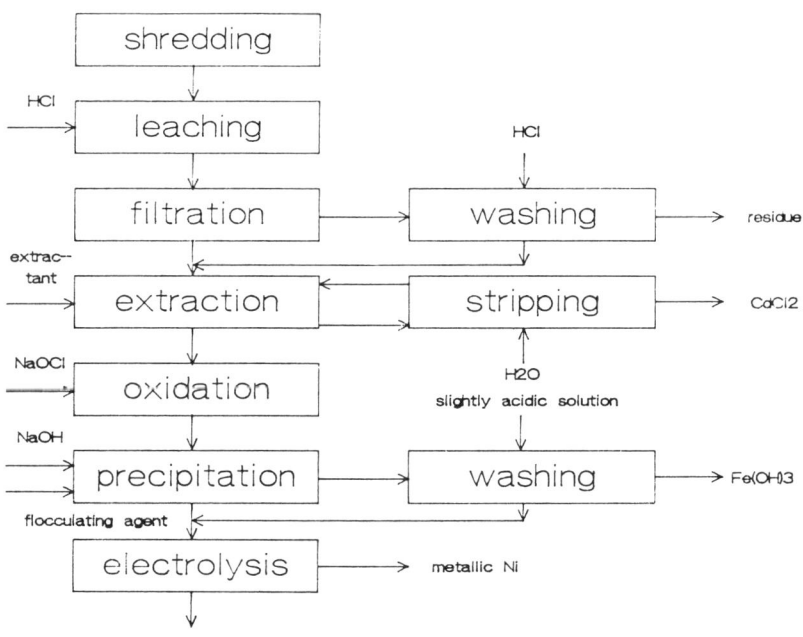

FIGURE 3.   Processing Route for Recovery of Metals from Spent Nickel-Cadmium Batteries.

Finally nickel can be recovered from the leachate by electrochemical deposition of metallic nickel or by a further increase of the pH and precipitation of nickel hydroxide.

A drawback of the above process route is that the removal of metal hydroxides by adjustment of the pH hampers the recycling of the leachate. The spent leachate contains, true enough, only trace amounts of metals, which possibly have to be removed with an ion exchanger, but the discharge to the surface water of a fairly concentrated sodium chloride solution is not really an elegant solution. Recycling of leachate can be accomplished by removal of iron and nickel by means of solvent extraction at a low pH. Extractants potentially suited for this purpose have already been tested by us.

Processing of Alkaline Batteries

As mentioned before a process route for the recovery of metals from alkaline and Leclanché batteries has been developed by TNO and tested on a laboratory scale. The process route is shown in figure 4 and the most important results of the experimental research are summarized below [8]:

* Batteries can be sorted by content on the basis of X-ray pattern recognition at a rate of 300 batteries per minute.

* Batteries can be shredded effectively in a hammer mill. After shredding it is possible to separate the battery content (carbon, zinc, mercury, manganese - the fine fraction) from the coarse fraction by sieving at a mesh of approximately 2.5 mm. The coarse fraction consists of steel casings, copper pins, paper and plastics. When processing a mixture of alkaline and Leclanché batteries the mercury content of the coarse fraction amounted to a few hundreds of ppm wt., which is well above the limit laid down for mercury in the Dutch Chemical Waste Act. Marketing of the coarse fraction is therefore doubtful, despite of its high non-ferro content. It is believed that the severe more of shredding in a hammer mill significantly contributes to the relatively high mercury content and that improvements can be obtained when another type of shredder is applied.

* Metals can be removed from the fine fraction by subsequent treatment with hydrochloric acid and sodium hypochlorite. Through filtration at pH 3 a residue is separated, mainly consisting of carbon and manganese dioxide. The residual mercury content does not exceed 50 ppm wt. which is the limit laid down in the Dutch Chemical Waste Act. A special problem occurs if, due to an

unsatisfactory performance of the sieve and/or shredder, copper pins end up in the fine fraction. The presence of copper pins leads to an incomplete dissolving of the mercury, probably due to amalgam formation. Under these circumstances a residual mercury content below 50 ppm wt. could not be obtained.

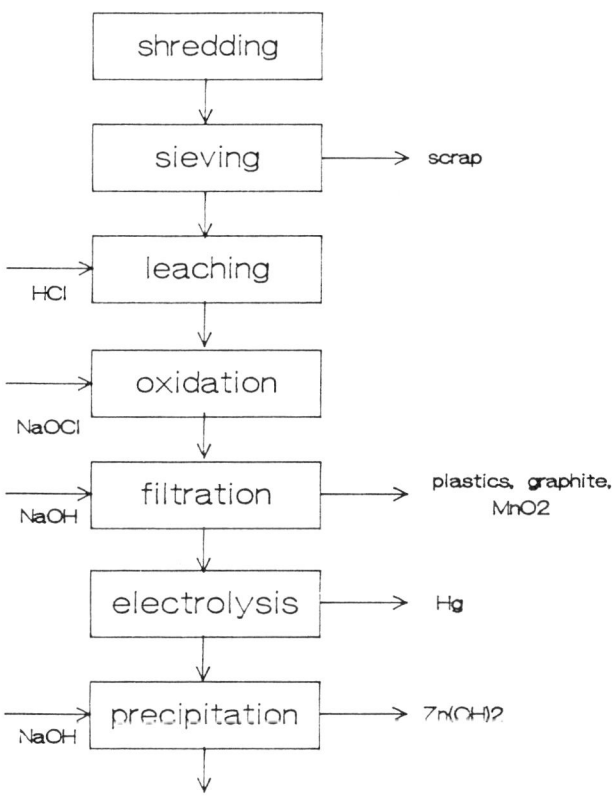

FIGURE 4.   Recovery of Metals from Spent Alkaline and Leclanché Batteries.

* By electrolytic reduction mercury can be separated selectively from the solution containing mercury and zinc. The efficiency of electrolysis is high and the residual mercury content is lower than

1 mg/l in solution. It makes sense to carry out electrolysis in two steps. The second step is meant as a polishing purification.

* By precipitation at pH 10 zinc can be separated from the solution as $Zn(OH)_2$. The residual zinc content is lower than 1 mg/l. By further treatment the metal content in the solution (Hg, Zn, Cu, Ni) can be reduced to a level of 10-100 μg/l.

A major purpose of the research project presently under investigation is to further improve the process route depicted in figure 4. In this respect the research efforts are mainly focussed on finding better marketing potentials for the scrap and the manganese dioxide/carbon mixture. This will be achieved by optimizing the shredding and sieving steps as well as the hydrometallurgical process steps.

6. CONCLUSIONS

Hazardous waste streams containing heavy metals are still being dumped, landfilled or stored as a major route of removal. Due to the potential environmental impact thereof this is not to be considered as an ideal solution. Processing of hazardous wastes with the aim to recuperate the (toxic) metals would therefore be of considerable environmental interest. In this paper it is demonstrated that recovery of metals from waste streams by hydrometallurgical processes can be an attractive option from a technical and economic point of view.

REFERENCES

1. Tels, M., A. Bolt, and W.J. Th van Gemert. European Patent 84.200.464.

2. Engelhard, W.J.F.M. and W.J.Th. van Gemert. "Development and Testing of Sequential Procedure for the Treatment of Toxic Waste Streams Containing Heavy Metals," Reports C and D, TNO reports 86-098 and 86-087, (1986).

3. Engelhard, W.J.F.M. and R.K. Bloebaum. "Processing of Metals Containing Wastes," TNO-report 87-315 (in Dutch), (1987).

4. Parkinson, G. "Recyclers Try New Ways to Process Spent Catalysts," *Chem. Eng.* 94 (2):25-29 (1987).

5. Gerritsen, R. "Metal Recovery from Spent Hydrodesulphurizing Catalyst," paper presented at conference Worldwide Spent Catalyst Disposal, Princeton (USA), September 1988.

6. "Mercury - an Environmental and Natural Resources Problem," Report (in Dutch) by Stichting Natuur en Milieu, (1983).

7. Genest, W. "Waste Problems of Spent Batteries," Part 1 (in German), *Müll und Abfall*, 17(6):194-199 (1985).

DISCUSSION OF:
RECOVERY OF METALS FROM WASTE STREAMS
BY HYDROMETALLURGICAL PROCESSES

D. Petruzzelli and A. Lopez
Istituto di Ricerca sulle Acque, Consiglio Nazionale Ricerche, Bar
Italy

INTRODUCTION

In a world of limited resources it turns out to be increasingly evident that not only lifestyle, but even life itself depend on the establishment of a conservation-oriented society.

In this context, it is imperative that human activities would be managed and planned within the needs of man itself, to provide the most rational use of resources and energy. It goes without saying that "wastefulness" spells as irrational use of more than necessary for the human "luxurious need".

In some hints it is ineluctable that production for the economical growth creates pollution; pollution, however, in a broad sense, refers to that portion of resources which is spoiled from current technology.

A resource conservation oriented technology or "non-waste technology" is the basis for the conserver-society, and this not covers only production, but the rational use of resources and energy, waste disposal management, recovery, as well as environmental protection in its largest sense.

According to a comprehensive project from 3M Company [1], the so-called "Low- and/or Non-Pollution Concept" is intended to encourage production procedures under which less, or less dangerous production wastes arises (clean technologies), and products are manufactured in such a way to produce less pollution and eventually are more cost and energy effective.

Moreover, production wastes, once originated, create a major problem for the industries in determining what to do with these noxious by-products. They have a choice, however, between destroying the contaminants and recovering the materials for reuse.

The resource-conservation aspects of "low- or non-pollution technology" also includes the use of production residues as potential resources; this concept, according to the mentioned 3M Company report [1], can be illustrated by the following simple equation:

POLLUTANTS (Industrial and social) + KNOWLEDGE (Technology) = POTENTIAL RESOURCES

In conclusion, waste reduction and recovery technologies are not only to be seen as optimum environmental solutions, but also as an effective part of innovation and plant modernization to increase profitability and competitiveness as well [2, 3]. Otherwise, rising environmental costs and liabilities together with legislative restrictions and limitations may not result in a global solution of the environmental problems.

DISCUSSION

In the ambit of low-or non-pollution-technologies, Dr. Van Deelen's paper offers two sound and valid examples of the above principles.

Based on reading the paper and personal experience on similar processes, however, the proposed laboratory investigations appear quite complicated in terms of number of steps involved; in terms of complexity of unit operations involved (thermal treatments, solvent extraction, ion exchange, filtration, leaching); in terms of chemical and/or physic-chemical operations involved (redox, complexation, precipitation reactions, electrolysis). In this context I would like very much for the author to better substantiate the economy of each step of the proposed processes even at a very preliminary level (eventually the considerations that lead the author to carry on from laboratory level to the eventual pilot plant).

Based on personal experience, strong economical and practical hindrances have been found in transferring laboratory results to full scale operations. The economical problems essentially reside in the costs of the chemicals to be used for the very peculiar chemical reactions on which the process is based; and the practical problems reside in the very complicated flow sheet of the process and in the unit operations per se' involved.

Moreover, in most cases, eventual recovered products from the operation are not in a qualitative and economic convenient condition to be recycled back to the main productive process. In the case where they are equally recycled it would eventually lead to a finished product of lower added value. This is certainly unacceptable in most industrial activities where the quality of the product is at the basis of the same production.

CONCLUSIONS

The reported processes are two typical examples of non-waste technology applied to the solution of environmental problems, giving, at the same time, allowances for recovery of valuable by-products to be recycled in the productive operations.

Reference processes, however, are actually at a stage of laboratory scale: In this context extensive studies have to be run, at least at pilot demonstrative level, to evaluate an eventual scale effect as well as to allow for a detailed economic evaluation of the processes and see if costs would be comparable with simple landfilling of the solid and/or land disposal of the liquid wastes.

Needless to say the more complex the process the higher the costs and the less competitive the process would be from an applicative point of view.

REFERENCES

1. Ling. J.T. "Low- or Non-Pollution Technology Through Pollution Prevention," An Overview, 3M Company Rep. to the United Nations Environment Programme, Office of Industry and Environment, (1982).

2. Overcash, M.R. *Techniques for Industrial Pollution Prevention,* (Lewis Pub. Co. Chelsea Michigan, 1986).

3. Royston, M.G. "Making Pollution Prevention Pay," *Harvard Business Review*, Nov.-Dec. (1980).

CHROMIUM RECOVERY FROM TANNERY SLUDGE BY INCINERATION AND ACID EXTRACTION

M. Beccari, L. Campanella, E. Cardarelli, M. Majone, and E. Rolle
Department of Chemistry, University "La Sapienza", Rome, Italy

1. INTRODUCTION

In Italy, it frequently happens that wastewaters from the various tanning operations are not kept separate. The treatment of these unsegregated wastewaters results in very large quantities of sludges (about 100,000 metric tons per year, as dry solids) in which Cr (III) is present in concentrations ranging from 1 - 5% on a dry solid basis [1].

A general survey of the possible alternatives of sludge disposal (with or without recovery) is shown in Figure 1.

As a general rule the alternatives calling for the reuse of the substances contained in the sludges are to be preferred since they are consistent with the general trend towards a maximum recovery of resources. Alternatives involving the use in agriculture of the organic and nutrient matter contained in the sludges (spreading on farmland; composting together with urban solid waste; incineration combined with flash drying and use of the dried material) are not feasible owing to the presence of chromium, which can affect plant metabolism and can also have cumulative effects; chromium also hinders the reuse of tannery sludges in animal feed manufacturing.

Pyrolysis (i.e. high temperature cracking of the organic matter contained in the sludges in absence or shortage of oxygen) has only been implemented in a small number of demonstration plants treating combined municipal refuse and sewage sludge; therefore, no reliable data are available on which to base technical-economic projections for tannery wastes.

The alternative based on the acid treatment of sludges followed by mechanical separation of the acid extract (subsequently subjected to selective chromium separation for recycling in the tanning process) from solid (for use in agriculture or in animal feed manufacturing) is of great interest. Theoretically, numerous separation methods are available, although none have yet been exhaustively tested: electrolytic oxidation, selective precipitation, membrane processes.

Promising results have been obtained [2] by using cupferron as a precipitating agent for the Al and Fe interfering in the tanning process. However, an economic limitation to this method is its inability to completely recover cupferron, a very expensive reagent, from the precipitate. Further studies to improve reagent recovery yield (for example, through reagent immobilization on inert matrixes) have been planned.

Nevertheless, disposal methods without recovery, such as mechanical dewatering and transportation to a supercontrolled landfill (with or without intermediate incineration), continue to be widely used. Possible options in this direction have been taken into account in a previous work [3] on the basis of well-defined criteria for designing treatment units and evaluating capital and operating costs.

This analysis showed that if high post-combustion temperatures are not required, the incineration of the sludges (dewatered up to 35 - 40% solid content) before transportation to disposal sites represents the most economical solution.

Furthermore, there is a much smaller quantity of material to be disposed of and this, as well as making it easier to find a suitable disposal site, can in some cases, make incineration a compulsory choice.

However, landfilling with ashes from tannery sludge incineration can pose an environmental hazard as possible oxidation of Cr (III) to Cr(VI) during incineration renders the metal more toxic and more leachable [4].

Some patent methods for preventing chromium oxidation or minimizing leaching by heating in a reducing atmosphere [5,6] or by adding chemicals [7,8], have also been proposed.

Conversely, owing to the increased leachability of Cr in its hexavalent form, an easier and more selective separation of chromium (for recycling in the tanning process) could be performed by extraction from the ashes rather than from sludges if, during incineration, most of the Cr(III) content were oxidized to Cr(VI). In this regard, the high calcium content of sludges is a positive factor for increasing the percentage of chromium oxidation during incineration, in analogy with a pyrometallurgical type of process. However, deeper knowledge of chromium behavior during the incineration of tannery sludges will have to be gained in order to evaluate both the environmental impact of the process and possible metal recovery from the ashes.

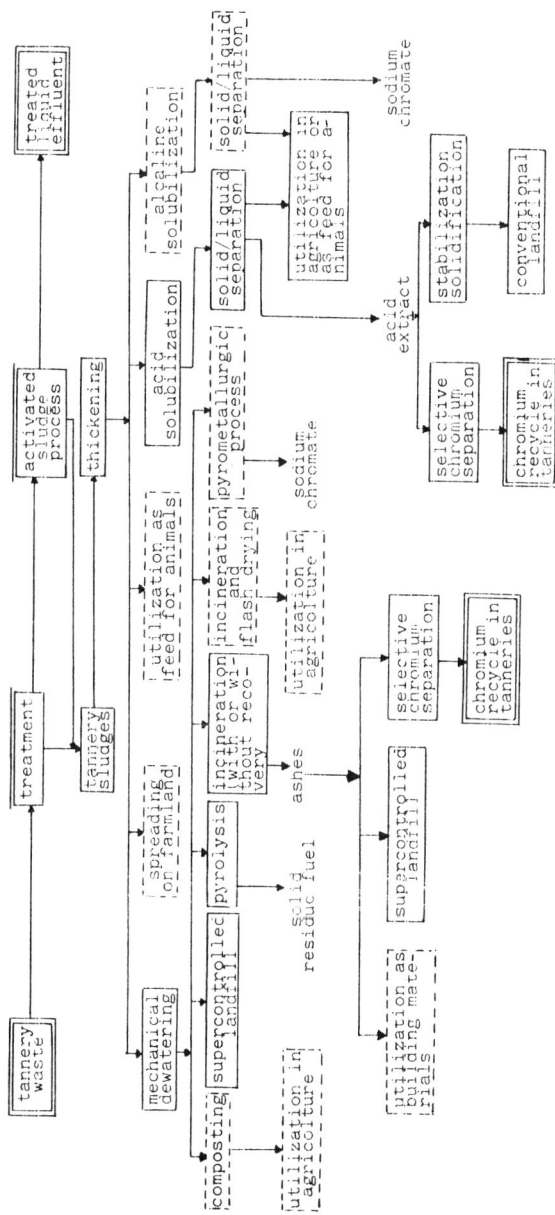

FIGURE 1. Treatment and Disposal Alternatives of Tannery Wastes.

2. DEFINITION OF OBJECTIVES

From the state of the art as described above, it appears that a dearth of knowledge exists with regard to the most practicable and economical alternatives of tannery sludge disposal.

Among the alternatives with chromium recovery the one based on incineration followed by acid extraction, chromium reduction and final stabilization of residual ashes seems worth examining more closely.

In the present paper, particular emphasis has been laid on the following:

a) effect of experimental conditions (temperature, time and oxygen excess) and Cr/Ca ratio in sludges on Cr(VI) formation during incineration;

b) optimization of the experimental parameters (e.g. ashes/acid solution ratio, pH) in chromium extraction from ashes with regard to degree of chromium recovery, concentration of chromium and interfering substances, and consumption of acid used as extracting agent;

c) environmental impact of the process, in particular, chromium leachability from residual ashes (after acid extraction).

3. EXPERIMENTAL

3.1 Sludge Characterization

Sludge was supplied by the wastewater treatment consortium of Arzignano (Vi), Italy. The sludge was produced in the primary sedimentation unit without any flocculating agent and was sampled before conditioning and filtration. To determine Fe, Ca and (total) Cr contents the sludge was treated with HNO_3 and H_2O_2 at boiling temperature for at least 2 hours according to [9]. The resulting solution was analyzed after filtration on a 0.45 μm filter by flame atomic absorption spectrometry using a Varian instrument, model SpectrAA10.

For the purpose of distinguishing between the two Cr oxidation states, the Cr(VI) content was also determined. This procedure is described in the following table.

Sludge characterization is shown in Table 1. For reference purposes, also another sludge has been characterized (from the wastewater treatment consortium of S. Croce (Pi), Italy), which had been used in a previous work, the main difference being the Ca content.

TABLE 1. Sludge Characteristics (Mean Values); (1) Previous Work - (2) This Work.

	S. Croce Sludge (1)	Arzignano Sludge (2)
pH	7.3	7.3
Water Content (%)	94.8	96.4
on Dry Basis (%)		
Weight Lost at 600°C	48.8	46.4
Weight Lost at 800°C	56.6	50.0
Ca	15.7	10.2
Fe	2.5	0.7
Cr(III)	1.8	3.1
Cr(VI)	absent	absent

3.2 Incineration Procedure

The thickened sludge was dewatered at 105°C, then carefully ground and passed through a 80 mesh sieve. Samples for each incineration experiment were taken from the same aliquot of this sifted sludge after careful mixing. In order to check that sampling was correct, several total Cr determinations were made on different samples and the mean value taken. The standard deviation was found to be less than 3%.

Also modified sludges were used. In order to increase or decrease sludge Cr/Ca ratio, aliquots of dewatered sludge were added respectively to distilled water and to saturated $Ca(OH)_2$ solution in different ratios. After separation of the liquid, the solids were dried again.

Incineration experiments were performed using a Mazzali (Italy) Nikro model muffle. Accurately weighed 10 - 20g samples were treated at different temperatures (400 - 800°C) in the presence of a stream of air. The overall air volume passing through the ashing room during the experiment was at least 6 Nm^3/VSS kg. Samples were introduced into the muffle at room temperature. The pre-set temperature was attained in one hour and maintained for 4 hours.

As the muffle was not suitable for short term tests designed to investigate the effect of incineration time, sludge incineration was also performed using a Du Pont, model 951, thermogravimetric device. In this case the apparatus was brought to the final test temperature before introducing a 10 mg sample and maintained for a pre-set time (0.5 - 60.0 minutes) with no current of air. Introduction time was taken as initial test time.

3.3 Ash Characterization and Extraction Procedure

For the purpose of determining Cr(VI) content in the ash, a previously standardized [10] Cr(VI) extraction procedure was used. This consisted in treating about 100 mg of ash (about 10 mg in the case of kinetic tests) for one hour under continuous magnetic stirring with 30 mL of dilute H_2SO_4 at pH3. The pH was kept constant by adding more H_2SO_4 (1:100 or 1:25) by means of an automatic pH stat system. The solution was separated from residual solids by filtration on a 0.45 μm filter and the filter cake washed with distilled water. Both filtrate and washing aliquots were mixed and the Cr(VI) content of the resulting solution determined by the diphenylcarbazide spectrophotometric method [11] using a Beckmann instrument, model 24. In the following, this procedure is reported as standard extraction.

Cr extraction tests under different operating conditions were also performed by changing the pH or the ash/solution w/w ratio with reference to the standard extraction, or by adding an Na_2SO_3 solution as a Cr(VI) reducing agent (100% excess with respect to the stoichiometric ratio). In these tests Ca, Fe and (total) Cr were determined in the extracted solution as described above. Total acid consumption was also recorded: for tests performed at pH greater than 3, more acid was added to the separated solution to bring the pH to 3 and corresponding consumption accounted for.

Leaching tests using 0.5 N acetic acid according to [9] were performed on residual solids from the extractions and both Cr and Cr (VI) were determined as previously described.

4. EXPERIMENTAL RESULTS

Preliminary tests [10] carried out on sludges supplied by the waste treatment consortium of S. Croce and with a high Ca content (15.7%) showed that:
- Cr(VI) oxidation during incineration took place with a high yield (80 - 100%) over a wide range of temperatures (400 - 800 °C) without any need to add reagents or for a large excess of oxygen;

TABLE 2.   Cr(VI) Content and Residual Ashes vs/Incineration Temperature.

Temperature (°C)	Residual Ashes (Weight %)	Cr(VI) (%)
400	72.4	2.5
450	57.3	67.2
500	55.6	77.1
600	53.7	79.1
700	50.9	82.7
800	50.0	81.4

- in almost all cases no chromium loss through gaseous emission was found;
- at pH 3 a large aliquot of total chromium was extracted from the ashes in the form of Cr(VI).

However, the above-mentioned favorable aspects were accompanied by a consumption of acid (20 - 30 kg of 100% H_2SO_4 per kg of chromium recovered) high enough to have a considerable effect on the economics of the process. This high consumption was required to neutralize the high alkalinity due to the large Ca content of the ashes (about 30%).

Tests were therefore continued using sludges supplied by the wastewater treatment consortium of Arzignano, the calcium content of which was due only to hide liming. The aim of the experiment was to ascertain whether the lower calcium content of these sludges (10.2%) would in any case allow a satisfactory percentage oxidation of Cr(VI) to be obtained, as well as an acid consumption possibly lower than that obtained with the sludges used in the preliminary test.

4.1 Incineration Tests

Table 2 reports both the Cr(VI) content of the ashes (expressed as a percentage of total chromium contained in the treated sludges) and the residual weights (expressed as a percentage of the weight of treated sludges) as a function of the temperature maintained during incineration (carried out for 4 h in excess oxygen conditions).

TABLE 3.   *Reproducibility of the Incineration Tests.*

	Temperature (°C)	residual Ashes (Weight %)	Cr(VI) (%)
Run 1	800	49.8	83.5
Run 2		50.7	83.9
Run 3		49.9	83.3
Run 4		50.1	78.6
Run 5		49.4	77.6
Mean Value % Standard		50.0	81.4
Deviation		1.0	3.7
Run 1	600	54.0	87.0
Run 2		54.7	83.4
Run 3		53.1	75.5
Run 4		53.1	70.6
Mean Value		53.7	79.1
% standard deviation		1.5	9.4

Table 2 shows that Cr(III) oxidation is negligible below 400°C. Most of the oxidation occurs over the 400 - 450°C range and remains practically constant (about 80%) between 600 and 800°C. The similar trend in the loss of weight seems to indicate that a good proportion of the oxidation can be achieved only if the heat treatment ensures the near-total removal of the organic matter contained in the sludges.

Table 3 shows how the oxidation yield is highly reproducible, above all at higher temperatures (600 - 800°C) no significant increase in percentage oxidation is observed after one hour's incineration. Conversion of Cr(III) into Cr(VI) is negligible during the first minute of incineration corresponding to the combustion of the organic matter, which removes the oxygen required to oxidize the chromium.

Comparison between Tables 2 and 4 reveals an apparent slight increase in percentage oxidation with decreasing quantities of sludge used, all other factors being equal. This is probably due to an effect of scale which will have to be further investigated in later research.

Table 5 shows both Cr(VI) content in the ash (expressed as a percentage of total chromium contained in the treated sludges) and the residual weights (expressed as a percentage of the residue remaining at the end of the tests) as function of incineration time

TABLE 4. Influence of Incineration Time on Cr(VI) Content and on Residual Ashes at Various Temperatures.

Temperature				Time (min)				
(°C)	0.5	1.0	2.0	4.0	8.0	15.0	30.0	60.0
450	0.0	0.2	6.9	12.8	23.8	33.7	46.7	55.1
500	0.0	0.1	6.4	24.1	35.0	45.2	51.2	60.7
550	0.1	0.1	5.3	37.0	49.0	58.0	69.1	77.4
600	0.0	0.0	14.9	43.2	58.0	70.0	78.3	86.2
700	0.0	0.0	39.5	66.9	87.8	89.1	98.1	89.5
800	0.0	0.0	80.5	91.1	87.1	82.9	84.6	88.6

Temperature				Cr(VI) (%) Time (min)				
(°C)	0.5	1.0	2.0	4.0	8.0	15.0	30.0	60.0
450	n.a.	66.5	60.8	60.8	59.1	57.7	56.8	56.4
500	67.0	64.0	58.8	58.1	57.1	56.8	56.4	56.2
550	66.5	62.6	57.4	56.7	56.0	55.3	55.3	54.1
600	63.7	59.2	55.7	54.9	54.2	53.9	53.8	53.0
700	60.6	56.4	53.2	52.5	50.1	49.9	49.1	49.1
800	57.3	51.3	49.7	49.4	49.0	48.7	48.5	47.4

and of the Cr/Ca weight ratio in the sludges to be treated. This ratio was changed from its original value of 0.31 (see Table 1) by washing the matrix with either distilled water or a saturated solution of calcium hydroxide. Table 5 indicates a decrease in kinetics (which cannot be explained in terms of a different trend in the combustion of the organic substance), as well as a decrease in final

TABLE 5. *Influence of the Ratio Cr/Ca in the Sludge on the System Behavior During the Incineration. (at 600°C).*

Cr/Ca Ratio (w/w)	0.5	1.0	2.0	4.0	8.0	15.0	30.0	60.0
0.11	0.57	1.07	70.76	81.10	95.87	102.42	110.14	109.32
0.16	0	0	57.41	77.28	86.96	98.40	94.08	101.43
0.31	0	0	14.93	43.15	58.03	70	78.32	86.18
0.36	1.11	1.27	0.96	52.57	62.73	63.52	74.94	81.56
0.41	0	0	2.05	30.69	39.82	48.31	54.10	57.46

Cr (VI) (%)

Cr/Ca Ratio (w/w)	\multicolumn{8}{c}{Time (min)}
	0.5	1.0	2.0	4.0	8.0	15.0	3.0	60.0
0.11	68.97	77.20	84.25	84.98	96.02	88.03	91.44	100
0.16	74.78	83.68	95.24	97.97	96.75	98.55	96.89	100
0.31	77.32	86.81	94.36	95.91	97.36	98.17	98.38	100
0.36	73.59	80.61	80.26	94.88	95.60	95.92	97.84	100
0.41	72.14	77.37	88.89	96.48	97.29	98.35	98.45	100

Residual ashes (weight %)

TABLE 6. *Influence of the Incineration Temperature on Acid Consumption (Standard Extraction).*

Temperature (°C)	Acid Consumption	
	Acid/Ash (w/w)	Acid/Recovered Cr (w/w)
450	0.38	10.8
500	0.41	9.8
600	0.42	8.8
700	0.40	7.5
800	0.38	7.3

Cr(VI) content in the ashes, with increasing Cr/Ca ratio. The calcium content in the Arzignano sludges (produced by liming treatment alone) actually represents a critical value below which is not economical to descend without decreasing the efficiency of Cr(VI) oxidation.

The results of the incineration tests reported in this section were found to be only negligibly affected by oxygen excess (defined as the ratio between the quantity of O_2 actually available during the test and the stoichiometric quantity of O_2), at least in the range of values tested for this parameter (1.5 - 3).

Lastly, the incineration tests carried out using undewatered sludges gave Cr(III) to Cr(VI) conversion rates comparable to those obtained using dewatered sludges, all other factors being equal.

4.2 Extraction Tests

Table 6 reports acid consumption (with reference to standard extraction conditions) as a function of incineration temperature (from 450 - 800°C). Below 600°C, consumption tends to increase, albeit only slowly. Acid consumption is, in any case, lower than for sludges with higher Ca content [10].

On the basis of the results reported in Tables 2 and 6, the extraction conditions were investigated in particular for ashes obtained at 600 and 800°C.

Table 7 presents the extraction data (expressed as a percentage of the initial content in the treated sludge and/or as a concentration in the extracted solution) as a function of pH and of the weight ratio between ashes (resulting from an incineration treatment carried out at 800°C for 4 h in excess oxygen conditions) and the final quantity of extracting solution.

The results in Table 7 indicate a considerable increase in extraction yield as pH decreases down to 6. Any further decrease in pH has a negligible effect on Cr(VI) accounts for more than 90% of the chromium extracted and is recovered almost completely from the ashes. The ashes/extracting solution ratio does not appear to have any appreciable effect on the yields obtained, at least for ratio values not greater than 6 - 7%, thus allowing chromium concentrations to be achieved in the extract (3 - 4 g/L) such as to be of interest to direct recycling inside the tannery. Iron concentration in the extract is generally less than 1 mg/L. Any Fe interference during possible recycling can therefore be excluded.

Calcium concentration remains rather high despite the use of H_2SO_4 as extractor and could perhaps entail further separation of the calcium from the chromate.

Lastly, acid consumption (5 - 10 kg of H_2SO_4 per kg of chromium) does not seem to be significantly affected by extraction conditions.

TABLE 7. Cr, Ca and Fe Extraction and Acid Consumption vs Ash/Solvent Weight Ratio at Various pH (* at the Natural pH of Ashes), Incineration Temperature 800°C.

Ash Solvent (w/w %)	pH	Recovered Cr		Extracted Cr(VI)/Cr (%)	Concentration in the Extract			Acid Consumption	
		Cr (%)	Cr(VI) (%)		Ca (%)	Cr (mg/L)	Ca (mg/L)	Fe (mg/L)	Acid/Ash (w/w)	Acid/Recov'd Cr (w/w)
0.3	3	86.1	81.4	94.5	81.7	178.6	273.6	0.28	0.38	7.3
1.2		90.8	84.2	92.7	74.8	675.2	911.6	0.45	0.33	9.1
3.9		91.5	82.5	90.2	61.7	2271.2	2467.4	0.63	0.33	5.8
6.5		85.6	81.5	95.2	33.5	3504.8	2212.1	0.71	0.33	6.2
8.3		74.1	69.0	93.0	15.5	3935.8	1297.9	0.93	0.37	8.1
0.3	6	81.9	83.5	101.9	68.8	172.5	237.3	0	0.43	8.6
1.3		86.5	84.0	97.1	69.1	685.8	987.1	0.49	0.32	5.9
4.1		77.2	80.7	104.4	46.4	2010.1	1946.7	0.25	0.31	6.5
7.4		82.3	79.4	96.5	30.7	3857.0	2319.9	0.42	0.29	5.7
9.2		65.2	66.3	101.8	23.2	3874.1	2175.3	0.87	0.38	9.1
0.3	8	64.2	59.1	92.1	55.0	134.7	184.1	0.13	0.37	9.2
11.8		52.6	54.5	103.5	26.2	3986.5	3124.4	1.03	0.33	9.7
0.3	>10 *	53.6	57.7	107.5	25.5	112.9	87.9	0	0.12	3.8
1.7		58.5	54.7	93.5	28.9	608.5	493.1	0	0.08	2.1
8.3		35.1	38.4	109.5	19.5	1974.9	1645.8	0.61	0.06	2.8

Table 8 reports the extraction data referring to the ashes produced by incineration treatment carried out at 600°C. No substantial differences with respect to the results in Table 7 were found.

4.3 Evaluation of Environmental Impact

Table 9 reports the data referring to chromium release from residual solids after chromium extraction (residual solids derived from extraction tests carried out for each pH at the higher ashes/solvent ratio set out in Tables 7 and 8). It can be seen that Cr(VI) concentration in the leachate, although corresponding to a very low percentage of the initial chromium content, is, under all conditions, much higher than the maximum allowable values. Therefore, in the light of these results, it is necessary to subject the residue to stabilization treatment before discharge. Preliminary tests of chromium release were also carried out on residual ashes after extraction treatment (at pH 3 and with a high ashes/extracting solution ratio) combined with simultaneous reduction of the extrac-

TABLE 8. *Cr, Ca and Fe Extraction and Acid Consumption vs Ash/Solvent Weight Ratio at Various pH (* at the Natural pH of Ashes), Incineration Temperature 600°C.*

Ash Solvent (w/w %)	pH	Recovered Cr		Cr(VI) /Cr (%)	Extracted Ca (%)	Concentration in the Extract			Acid Consumption	
		Cr (%)	Cr(VI) (%)			Cr (mg/L)	Ca (mg/L)	Fe (mg/L)	Acid/ Ash (w/w)	Acid/ Recov'd Cr (w/w)
0.3	3	80.3	79.1	96.58	91.8	155.9	307.6	0.18	0.42	8.8
1.3		78.9	75.5	95.71	62.9	593.1	810.5	0.25	0.36	7.8
4.0		77.1	76.5	99.29	51.2	1810.1	2063.3	0.71	0.37	8.3
6.6		75.7	75.5	99.72	22.2	2952.8	1481.6	0.71	0.38	8.6
8.4		61.4	66.2	107.85	17.9	3036.6	1518.0	0.94	0.38	9.6
0.4	6	76.7	75.4	98.24	64.8	158.9	230.3	0.10	0.29	6.5
1.5		75.0	74.5	99.32	61.0	652.0	910.8	0.23	0.19	4.3
4.5		75.4	74.9	99.44	64.9	1992.8	2946.1	0.80	0.22	4.8
8.4		72.6	72.9	100.36	32.3	3591.6	2739.1	0.98	0.19	4.5
8.4		65.1	71.9	110.47	33.2	3221.1	2812.9	1	0.17	4.4
11.6		59.6	62.6	105.12	23.0	4082.9	2705.7	1	0.17	4.4
0.4	8	76.1	73.6	96.58	56.8	157.5	201.7	0.10	0.24	5.3
13.3		57.8	61.8	106.97	26.9	4543.4	3628.6	0.97	0.11	3.0
0.4	>10	70.9	68.4	96.42	45.9	155.3	172.3	0.10	0.10	2.8
8.3	*	52.5	54.0	102.76	24.9	2592.1	2108.6	0.78	0.05	1.5

ted Cr(VI) (sulphite excess: 100%): Table 10 shows that Cr(VI) is absent in the leachate; under such conditions, however, the extraction yield is slightly lower than in the absence of simultaneous reducing treatment and the total chromium content exceeds the allowable values.

TABLE 9. *Leaching Test on Residual Solids After Cr Extraction (* at the Natural pH of Ashes).*

Incineration Temperature (°C)	Extraction Test (on Ashes) pH	Cr (mg/L)	Cr(VI) (mg/L)	Leaching Test (on residual Solids) Cr (%)	Cr(VI) (%)
600	3	136.0	107.4	2.7	2.1
	6	245.4	245.5	4.6	4.4
	8	358.1	338.9	6.4	5.8
	> 10 *	815.7	785.0	13.8	12.8
800	3	80.7	69.3	1.4	1.3
	6	367.4	364.7	6.9	6.8
	8	710.2	671.0	12.2	11.3
	> 10 *	734.0	698.5	12.1	11.1

Lastly, no chromium was found in the gases given off during the preliminary incineration tests. In any case, further research is necessary on this aspect of the problem and experimental apparatus is currently being developed for the purpose of enhancing detection sensitivity.

5. CONCLUSIONS

Investigation of the state of the art of treatment/disposal of tannery sludges has shown that an interesting method for selectively recovering chromium from the sludges in order to allow it to be recycled in the tannery consists of oxidative incineration followed by chromate extraction, reduction to Cr (III), residue stabilization and disposal of the inert ashes in a non-special controlled discharge.

Ad hoc experiments have been carried out to evaluate process feasibility.

Incineration tests have shown that in the 600 - 800°C temperature range, Cr(III) oxidation takes place with high yields (about 80%) without the need for additional reagents (provided that the Cr/Ca weight ratio in the sludges to be treated is $\leq$ 0.30 - 0.35, as in most tannery sludges) or with high oxygen excess.

A contact time of 1 h is sufficient, at pH 3 - 6, to extract a very large aliquot in the form of Cr(VI) of the total chromium originally contained in the sludge even for comparatively high ashes/extracting solution ratios (up to 6 - 7%). The corresponding acid consumption has been found to be quite low (5 - 10 kg of 100% H_2SO_4 per kg of chromium actually extracted).

TABLE 10.   Extraction/Reduction Test and Leaching Test on Residual Solids.

Incineration Temperature (°C°)	Recovered Cr		Leached Cr		Leached Cr	
	Cr (%)	Cr(VI) (%)	Cr (mg/L)	Cr(VI) (mg/L)	Cr (%)	Cr(VI) (%)
800	66.8	0.11	35.6	0	0.67	0
600	46.6	0.28	48.9	0	1.01	0

The results obtained in preliminary extraction tests combined with a simultaneous reduction of the Cr(VI) extracted are promising in the light of the environmental hazards associated with the leachability of Cr(VI) from the residual ashes. On the basis of the favorable results described in the present paper, an adequately parametrized model will be developed for the purpose of predicting performance as a function of a given set of adopted conditions and which will, therefore, allow a cost/benefits analysis to be made of the process under optimal operating conditions, also taking environmental constraints into account.

Research funded by a Ministry of Education grant

REFERENCES

1. Simoncini, A. "Quantita', Caratteristiche e Distribuzione Geografica dei Fanghi da Impianti di Depurazione Delle Concerie Italiane," *Acqua Aria* 1:55-57(Italian 1982).

2. Majone, M. "Aluminium and Iron Separation from Chromium Solutions by Precipitation with Cupferron," *Enviromental Technology Letters* 7:531-538 (1986).

3. Beccari, M., M. Majone, R. Passino, and E. Rolle. "Confronto Tecnico Economico fra Processi Alternativi di Trattamento e Smaltimento dei Fang hi di Concerial," *Ingegneria Sanitaria*, Italian, 34 (3):137-145 (1987).

4. U S Environmental Protection Agency, Ed. *Enviromental Effects of Pollutants III Chromium*, (Springfield VA: NTIS, 1978) p. 6-8.

5. Matsumoto, Y. Jpn. Kokai Tokkyo Koho 79,128,158 (Cl.C02C3/00), (October 4, 1979).

6. Aritoshi, H., E. Tamura, H. Nagamatsu, and S. Usada Jpn. Kokai Tokkyo Koho 79,139,261 (Cl.CO2C3/00), (October 29, 1979).

7. Sugawara, T. Jpn. Kokai Tokkyo Koho 81,15,883 (Cl.BO9B3/00), (February 16,1981).

8. Ebara-Infilco Co LTD Jpn. Kokai Tokkyo Koho 82,22,000 (Cl.CO2F11/14), (December 18, 1981).

9. Istituto di Ricerca sulle Acque, Ed. *Quaderno 64 Metodi analitici peri fanghi*, Vol III, (Rome, Italy: CNR, 1985), in Italian.

10. Beccari, M., M.R. Coretti, M. Majone, and E. Rolle "Chromium Recovery from Tannery Sludge by Incineration," *Trib. de Cedbedeau*, 40:29-34 (1987).

11. Istituto di Ricerca sulle Acque, Ed. Quaderno 11 *Metodi analitici perle acque*, Vol II, (Rome, Italy: CNR, 1972), in Italian.

DISCUSSION OF:
CHROMIUM RECOVERY FROM TANNERY SLUDGE BY INCINERATION AND ACID EXTRACTION

Chriso Petropoulou
Illinois Institute of Technology, Chicago, Illinois

INTRODUCTION

As the disposal costs of the metal bearing sludges increases and the value of the metals also increases, the potential of metal recovery and reuse seems to be a promising waste management option. In general large quantities of sludge are produced in the precipitation of metal hydroxides from industrial wastes. The specific choice of base for pH adjustment will also influence the sludge production, with lime yielding roughly three times more sludge compared to soda ash (Na_2CO_3) or caustic soda (NaOH) due to the formation of calcium carbonate and sulphate [1].

The handling and disposal of tannery sludges together with the chromium recovery, will become a more important and costly aspect of waste treatment as tanneries upgrade their wastewater treatment systems to comply with waste discharge limitations [2]. There are a number of appropriate adaptations that can be made of extractive metallurgy practices (i.e. Acid or Microbial leaching) to recover metals from industrial sludges. The incentives for metal recovery from metal wastes rest primarily in obtaining reductions in waste disposal costs and improving the ease of meeting pollution standards [3].

The authors are to be commended for their contribution in the research for the treatment of sludges that contain heavy metals such as chromium, through the in-depth investigation of the incineration process. The ultimate purpose of this research is to elucidate the possibility for the recovery and reuse of the chromium from the tannery sludges.

METHODS FOR THE REUSE AND RECOVERY OF CHROMIUM

Chromium is found in the tannery waste streams from chromium tanning, rinsing after chromium tanning, blue stock wringing, retanning, rinsing after retanning, coloring, fatliquoring, and drying.

Three general methods have been used to reuse or recover chromium from process wastewaters [2]:

1) Isolating the spent chromium tanning solutions for reuse as "pretannage" in the pickle liquor, as chromium tanning liquor (after refortification) or as retanning solution;

2) Concentrating the effluent chromium by chemical precipitation for use in formulating new chromium tanning liquors;

3) Incinerating the chromium bearing sludges and recovering hexavalent chromium from the incinerator ash.

The amount of chromium recovered varies with each of the three methods and depends on the amount of chromium sent to the recovery system and the efficiency with which chromium is recovered from the treated wastes.

The chromium recovery systems have all been designed to recover only the chromium in the spent tanning solutions. However, the washwaters associated with chromium tanning also contain a significant amount of waste chromium [2].

The volume of washwater is usually so large that the chromium in the wastewater cannot be recovered in these reuse systems because the washwater has greatly diluted the spent chromium liquors and the resulting volume would be more than is needed for pickling or tanning.

To solve this problem, chromium recovery systems designed to treat the spent tanning liquors and rinse waters employ chemical precipitation to concentrate the chromium. Previous work [4], showed that Cr could be solubilized almost quantitatively from unsegregated sludges by acid extraction. In order to separate interfering metals, such as Al and Fe, selective precipitation using cupferron was successfully used.

These systems can generally recover more chromium than the direct reuse systems; but on the other hand more equipment, greater capital investment, and more sophisticated operating controls are required.

The most comprehensive chromium recovery system that has been proposed involves incineration of the chromium-bearing sludges and recovery of chromium from the ashes. Successful operation of the system depends on low temperature, alkaline incineration to oxidize trivalent chromium to the hexavalent state. These conditions are necessary to retain the chromium in the ash.

DISCUSSION OF THE RESULTS

One of the experimental parameters that have been evaluated in this research is the Cr/Ca ratio in the sludge. Table 5 in the paper summarizes the results. As the Cr/Ca ratio increases the % Cr(VI) concentration diminishes. Therefore it appears that the high calcium content of the sludges is a positive factor for increasing the percentage of chromium oxidation during incineration.

From the literature review there are some findings that can be used to postulate the reasoning behind this effect: In general the heavy metals in the residual phase are usually in the state of the oxides, which are stable at ultimate disposal. But, the chromium (III) compounds are converted to the chromium (VI) compounds in the oxygen-rich atmosphere of the incineration process. Kashiwara indicated that this conversion is accelerated in the presence of calcium compounds through the following reactions [5]:

(Below 300 °C) $4Cr(OH)_3 + 4Ca(OH)_2 + 3O_2 = 4CaCrO_4 + 10H_2O$ (1)

(Above 300 °C) $2CrO_3 + 4CaO + 3O_2 = 4CaCrO_4$ (2)

Majima, et al. [5] conducted detailed experiments in the temperature range (600-800 °C) and clarified that the following reaction took place in addition to reaction (1).

(600-800 °C) $12CaCrO_4 + 6CaO + 7/2O_2 = 18CaO.10CrO_4.Cr_2O_3$ (3)

It is also reported [6] that chromite, $FeCr_2O_4$, which is the chief mineral (with Chromium in the third oxidation state and iron in the second oxidation state), is treated with molten NaOH (M.P=318.4 °C) and O_2 to convert Cr(III) to CrO_4^{2-}.

The similarities between the two chromium oxidation processes are the presence of alkalinity sources ($Ca(OH)_2$ and NaOH), the high temperature, and the oxygen requirement. Hence, it also seems interesting to further examine the effect of alkalinity on the oxidation rate. The importance of the alkalinity for the oxidation of chromium during the incineration process has already been pointed out by other researchers [2].

The authors also studied the effect of the incineration Temperature on the formation of Cr(VI). Most of the oxidation process was found to occur over the 400 - 450 °C range and remains practically constant (about 80 %) between 600 and 800 °C. The importance of the Temperature is related to the fact that it determines the loss of chromium through vaporization. It is reported in the paper that in almost all cases no chromium loss through gaseous emissions was found. This is in accordance with the results of other researchers. Takeda, et al. [7] report that pyrolysis of sewage sludge at low temperature minimized the vaporization of heavy metals contained in the sludge without sacrificing the reduction of bulk density. The analysis of heavy metals caught in the particulates and scrubbed water suggested that only some of the metals will be caught in a dust collector, but much will remain unrecovered as fumes. Cadmium and lead are lost to a great extent in the pyrolysis and/or incineration process. On the other hand, for manganese, copper, zinc and chromium, the residual ratios are high (Figure 1). Finally, an important finding of this research that is worth mentioning here, is the chromium release from the residual solids after chromium extraction (residual solids derive from extraction). It can be seen from Table 9 in the paper that Cr(VI) concentration in the leachate, is under all conditions much higher than the maximum allowable values. Therefore, it is necessary to subject the residue to stabilization treatment before discharge.

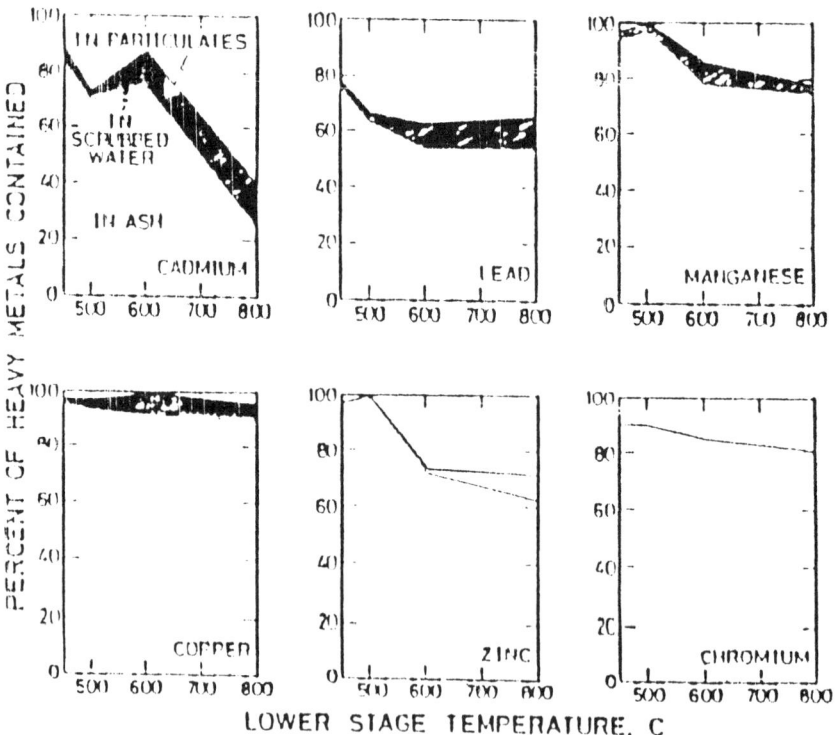

FIGURE 1. Distribution of Heavy Metals (Neither Zinc in Particulates Nor Chromium in Particulates and Scrubbed Water Was Measured) [7].

In general hexavalent chromium is recovered from the ash by a-two stage process. The 1st stage, in which most of the chromium is recovered, involves leaching of chromate from the ash using a filter to separate the leachate and ash residual. Two different methods have been proposed for the second stage. The first method entails mixing the residual ash from the first stage in a reducing solution to convert the chromium from the hexavalent state to the third state. The pH of the resulting solution is then adjusted to approximately 8.5 to form a precipitate of chromium hydroxide, which is recovered by using a dissolved air flotation unit.

The second method involves transferring the residual ash from the first stage to a pressure filter and pumping a reducing solution through the ash cake in the filter to reduce the hexavalent chromium in the ash to insoluble oxide. The ash, excluding the chromium oxide is then landfilled.

CLOSING REMARKS

From the available literature review, it appears that there are no full-scale incineration - recovery systems for tannery sludges, currently operating. Therefore the long-term operating characteristics and problems associated with continuous operation are not known. However, the system has several apparent advantages:

1. Almost all chromium discharged from a tannery may be recovered and eliminated from the wastewater stream;

2. Scrap leather pieces and buffing dust can also be incinerated, and the chromium in them can be recovered; scraps and buffing dust are believed to contain approximately 8% of the chromium purchased by a leather tannery; and

3. The problem of tannery sludge disposal is reduced because the volume and weight of sludge are substantially decreased by incineration.

REFERENCES

1. Patterson, J.W. *Industrial Wastewater Treatment Technology*, (Stoneham, MA:Butterworth Publishers, 1985), pp.53-87.

2. U.S. Environmental Protection Agency, Ed. *Development Document for Effluent Limitations Guidelines and Standards for Leather Tanning and Finishing*, (EPA 440/1-82/016, 1972) pp.175-184.

3. Brooks, C. S., "Metal Recovery from Industrial Wastes," *Journal of Metals*, (July, 1986), pp. 50-57.

4. Majone, M., "Aluminum and Iron Separation from Chromium Solutions by Precipitation with Cupferron," *Environmental Technology Letters* 7: 531-538 (1986).

5. Pojasek, D., Ed. *Toxic and Hazardous Waste Disposal*, Chapter 11, Vol.3., (1980)

6. Cotton, F.A. and G. Wilkinson. *Basic Inorganic Chemistry* pp.391-395.

7. Takeda, N. and M. Hiraoka, "Combined Pyrolysis and Combustion of Sludge Disposal," *Environmental Science & Technology* 10:1147-1150 (1976).

RESEARCH NEEDS

RESEARCH NEEDS IN METALS SPECIATION, SEPARATION AND RECOVERY

J.W.Patterson and C.Petropoulou
Illinois Institute of Technology, Chicago, Illinois

At the conclusion of the second international symposium on metals speciation, separation and recovery, a research needs workshop was held. Participants were divided into four discussion groups, with the following assigned topics:

A. Analysis and Speciation Chemistry
B. Solid-Liquid Interface Chemistry
C. Selective Separation Technology
D. Metal Sludges Management

The following sections present the research needs identified by the four working groups.

GROUP A: ANALYSIS AND SPECIATION CHEMISTRY

F. J. Millero (Chairman)
M. Achilli
M. Baldi
M. Betti
L. Campanella
S. Capri
J. Diaz
K. Karstensen
G. Macchi
A. Mathews
R. Minear
P. Papoff
M. Pettine

In order to understand the chemical properties of aquatic systems we need to characterize different systems in terms of their dominant variables-major ions, oxidation-reduction status, acid-base components, minor ions, complexing components, and adsorbing surfaces. No single method presently available permits unequivocal identification of a species. Usually, the evidence for a particular form of occurrence is circumstantial and is based on complementary evaluations together with kinetic and thermodynamic considerations [Stumm & Morgan, 1983]. It is well recognized that the particular behavior of trace metals is determined by their specific form rather by their total concentration [Förstner, 1987]. Such experience led to the introduction of the term "speciation", which is used in a vague manner both for operational procedure for determining metal species in the environmental samples and for describing the distribution and transformation of such species in various media.

RESEARCH NEEDS

1. Identification of Minor Species

The identification of minor species has proved to be very difficult, as such species occur at concentrations smaller than 10^{-6} M in the presence of larger excess of substances that often interfere with specific in situ sensing methods. Because of this, investigations using synthetic solutions in which the variables are known and can be controlled may frequently provide valuable clues to the types of species that exist in real water.

2. Size Fractionation

The concept of chemical species as "dissolved" or "particulate" as defined by the pore size of a membrane filter can no longer be considered adequate. Among the size fractionation methods that seem to be particularly promising for characterizing the molecule-size distribution of soluble organic macromolecular material and of the metal ions associated with it is gel filtration (filtration and elution from columns containing gels of dextran, silica, or other molecular sieves). So far, it has been used mostly for the fractionation of humic compounds and other colored components of natural waters. Additional research is required for the wider application of this and similar innovative size fractionation techniques.

3. Factors Affecting Metal Speciation

It is interesting to examine how metal-ion speciations vary as a result of the addition of complex forming organic matter and what is the effect of pH change (e.g. resulting from acid discharge) on the free metal-ion concentration. In addition, chemical speciation under anoxic conditions needs further investigation.

4. Sequestration of Metals by Particles

Particles-through their surfaces-are scavengers of reactive elements. However, there are questions that need to be answered: What effect do particles, capable of adsorbing metal ions, have on metal-ion speciation? What is the competition between particles and soluble ligands for the metal ion?

5. Model Calculations

Additional data are needed both for the thermodynamics and for the kinetics of the chemical processes in waters and wastewaters. As has been pointed out in the literature [Morel, 1983]: " Aquatic chemical kinetics is an extremely complicated and largely undeveloped discipline which is still awaiting systematic treatment". In modelling and design, a number of reaction types many of them heterogeneous (e.g., precipitation, adsorption, reductive dissolution, surface photochemistry) must be considered. Water chemistry models for predicting metals removal should increasingly draw upon available kinetic information for those chemical processes which appear potentially to exert kinetic control on metals behavior in the system. Another important task is species "discovery" as well as the determination of thermodynamic properties for species and the working out of reliable analytical methods to measure the various species of metals.

GROUP B: SOLID-LIQUID INTERFACE CHEMISTRY

W. Van Driel (Chairman)
A. Bowers
C. Haraldsson
A. Kuiters

B. Luo
D. Marani
D. Petruzzelli
M. Pollak
A. Sengupta
W. Van Riemsdijk

Most chemical reactions that occur in natural waters take place at surface discontinuities; that is the atmosphere-hydrosphere and the lithosphere-hydrosphere interface. The discussion of this group focused on the solid-liquid interface. This is still a very broad topic; even in simple and reasonably well-defined systems the interfacial and colloidal properties involve so many variables and represent such complicated physical and chemical interactions that is difficult to describe such systems as a unified subject.

RESEARCH NEEDS

1. Complex Systems

There is a need to study experimentally and to model more complex systems: Both with respect to solution phase composition: multi-metal, multi-ligand (anion) mixtures, and with respect to more complex solid phase composition. Apart from that, our basic understanding of the processes occurring at interfaces should be increased.

2. Surface Phenomena

Kinetic studies are needed in order to characterize mechanisms of surface phenomena. These studies can be based on techniques such as autoradiography, X-Ray microprobe, ESCA, XPS, and so on. Information would be finalized for simulation models, and for interpretation of ionic migration in natural and synthetic porous materials.

3. Precipitation Process

Research is needed on the optimization of the metals precipitation process in terms of precipitate size distribution. A well designed and controlled precipitation process (for instance fluidized bed) should yield large dense particles, which do not require subsequent coagulation/flocculation post treatment. This leads also to a reduction in sludge volume.

GROUP C: SELECTIVE SEPARATION TECHNOLOGY

G. Tiravanti (Chairman)
V. Cunningham
K. Fytianos
R. Haldorsen
G. MÜller
P. Nwogu
C. Ostroff
G. Petruzzelli
R. Ramadori
A. Zouboulis
A. Grappelli

There are a number of separation and concentration processes potentially applicable to metal recovery from industrial wastes. The two principal classifications of processes are those applicable to soluble and insoluble metals. Existing and prospective metals control technologies can be classified into two groups; metals (including salts) concentration (tends to be non-selective), and selective extraction. Selective recovery is considered to be the most promising area of future development. However, such development and application will require a far more sophisticated understanding and utilization of chemical speciation and kinetic practice than exists at the present time (See Groups A and B).

RESEARCH NEEDS

1. Fundamental Studies on Factors Governing the Release of Metals from Solid Matrixes

The mobility of metals from solid matrixes or sediments is a function of many factors. The degree of sediment-water interaction and the conditions under which it occurs will determine to a great extent whether constituents will migrate from the sediment into the water, be adsorbed from the water onto the sediment or be released into the water only to be precipitated or altered quickly. Factors that affect the release of heavy metals from sediments and need additional investigation include:

1.1 Conductivity (Ionic Strength)

It is realized that conductivity changes can alter the sorption-desorption tendencies of chemical constituents by varying the competition for solid exchange sites.

1.2 pH of the Leachate

Although it is known that pH changes can alter the solubility of some salts as well as the species distribution of inorganic and organic complexes in water, the mechanisms by which leachate pH affects the metal release are not yet readily apparent.

1.3 Agitation

It has been reported [Brannon et.al, 1980] that leaching conducted under quiescent conditions may represent somewhat less than the actual amounts of exchange that would occur in the natural environment. However, completely mixed systems may overestimate release in natural environment because the same degree of mixing in the laboratory and the natural environment is rarely achieved. It is interesting to examine not only the rate of release under stirred and quiescent conditions but also the release patterns that occur [Pomeroy et. al., 1965].

1.4 Time of Contact

Time as a factor influencing long-term release from sediments depends on several physical factors. Types of chemical reactions and reaction rates become more important than initial release in determining net mass release as the contact time between sediments and water increases. In general, time could exert a large effect for constituents that must undergo some type of hydrolysis reaction before release.

2. Geochemical Factors and Mobility of Metals in the Environment

2.1 Soil factors

Acidity: The effects of acidification on release of toxic metals from soils is an area that is currently receiving increased research effort, due to the ecological dangers associated with acid rain. Effects have been postulated, but the picture is far from clear. All soils are not equally susceptible to acidification. The buffering capacity of soils depends on mineral content, texture, structure, pH, salt content and soil permeability. Biological processes in the soil necessary for plant growth could also potentially be affected by soil acidification, including nitrogen fixation [Bubenick, 1984]. However there are few data available in this area.

Ion exchange capacity: Few comprehensive studies have been conducted to assess the potential impacts of acidic precipitation on cation exchange. Those that have been conducted tend to concur that increases in acidification of rainfall leads to reduced cation exchange capacity and increased rates of mineral loss [Bubenick, 1984].

Receiving Waters

Metals in natural waters may be suspended, colloidal or soluble. Due to physical, chemical and biological reactions within the water, there may be dynamic interactions among the various particle sizes and chemical forms [Rubin, 1974]. There are several types of potential interactions that can take place between incoming metals and the water bodies they enter that will affect the concentration and distribution of metals in the water, and need additional research:

1).The effect of pH and Eh of the receiving water on the solubilization or agglomeration and therefore subsequent sedimentation of the metal species.

2).Another group of interactions of metals and receiving waters that has not been extensively studied, can be classed as biochemical, or reactions involving living organisms. The different mechanisms by which aquatic organisms assimilate metals and need further investigation, include [Martin, 1970]: i).Particulate ingestion of matter containing metals suspended in the water, ii).Ingestion of food, iii).Solubilization and assimilation through secretion of biological chelating or complexing agents, iv).Incorporation into physiological systems, and vi).Ion exchange and sorption on tissue and membrane surfaces.

3. Separation Processes

3.1 Selective Adsorption

This technology could employ biological materials (inactivated or nonviable microorganisms), mineral oxides, or polymeric resins. One interesting recent trend in this technology is to attempt to coat coarse or porous media such as sand with an adsorbing surface [Edwards,M. and M.M.Benjamin, 1988]. However, additional research is needed in order to promote progress in the development of selective adsorption methods.

3.2 Differential Solids Extraction

Extractive metallurgy, in various modifications, appears promising for recovery of metals from wastewater treatment sludges. Its potential will be enhanced by application to relatively "clean" sludges as contrasted with complex sludges typically produced by end-of-pipe treatment of combined wastewaters [Patterson, 1987]. There are two approaches to the application of extractive metallurgy; Pyrometallurgy and Hydrometallurgy (extensively covered in the discussion of group D).

3.3 Selective Solvent Extraction

Non-selective removal of metals from mixtures of metals in aqueous solution can be achieved by solvent extraction with a variety of organic reactants. Some examples of such agents are 8-hydroxyquinoline for cobalt, molybdenum, copper, iron, nickel, cadmium, zinc, vanadium, manganese, lead and titanium [Baes, 1962, Seeton, 1964, Clevenger & Novak, 1983].In some instances highly specific extractants can be selected, such as dimethyl glyoxime for nickel, where only palladium seriously interferes. Experimental investigation would provide more data for metal specific solvents. Solvent extraction can be advantageously combined with electrodeposition, ion exchange or precipitation to provide selective separation and concentration of individual metals.

3.4 Differential Precipitation

The most promising differential precipitation strategy is the sequential, step-wise precipitation of relatively pure solid phases. Although attempts have been made to define selective precipitation schemes for metals based upon thermodynamic models [Jenke and Diebold, 1983], these have been largely unsuccessful due to incomplete understanding of thermodynamics with regard to coprecipitation. Kinetic limitations are also cited as reasons for lack of conformity of results with predictions[Karra et al., 1983]. Further investigation of hydrolysis and precipitation kinetics may help to design selective precipitation processes based on kinetic rather than thermodynamic control.

3.5 Recovery of Metal Hydroxides of Colloidal Particle Size

There are techniques under experimental evaluation with a promising potential of recovering insoluble metal hydroxides in the colloidal particle size range. Insoluble hydroxides in the small particle size range can be transferred from an aqueous phase to a hydrocarbon phase by surfactants [Brooks, 1986]. Transfer is non-selective but has a useful potential for efficient separation and dewatering of fine particle sludges.

GROUP D: METAL SLUDGES MANAGEMENT

W. W. Eckenfelder (Chairman)
M. Beccari
K. Giesler
M. Loizidou
J. Patterson
A. Pereira
C. Petropoulou
M. Santori
M. Szabo
C. Van Deelen

The treatment of metallic sludges provides an alternative to the present extensive land disposal of these materials, and most significantly, it obtains value or credit from the recovery of the metals. The incentives for metal recovery from metal wastes rest primarily in obtaining reductions in waste disposal costs and

improving the ease of meeting pollution standards. These factors also combine with conservation of metals of economic significance and strategic value. Principal economic factors and technical aspects that need to be considered include:
- Process Complexity
- Commercial Proven Technology
- Sufficient Market Demand for the Facility
- Potential for Reincorporation into Manufactured Product
- Disposal Costs
- Value of Recovered Materials
- Markets for Recovered Materials
- Minimal Environmental Liability
- Sludge Variability

RESEARCH NEEDS

1. Improved Methods to Assess the Immobilization Efficiency of Heavy Metals

In order to evaluate and improve physical and chemical stabilization options that are already available and to design new ones, in-depth investigations of the mechanisms involved in the fixation process need to be performed.

2. Metals Recovery from Wastewater Sludges

There are a number of appropriate adaptations that can be made of extractive metallurgy practices to recover metals from solid wastes, whether low grade ores, mine tailing, industrial sludge or municipal wastes [Brooks, 1986].

Pyrometallurgy: In some instances, pyrometallurgical practice has been used to react sulfide or oxide ores with SO_2/SO_3 to produce water soluble sulfates of metals such as copper, nickel, and cobalt. Ammonium carbonate proves effective in leaching the sulfated metals. Although, many wastewater treatment plant sludges are richer in target metals than the source ores from which the metals were initially extracted, the sludge matrices are often not acceptable for direct pyrometallurgical extraction in existing smelters. Research is needed so that applications of pyrometallurgy would be possible for all the metals and not only for precious metals where it is restricted today.

Hydrometallurgy: Hydrometallurgical processes such as acid or microbiological leaching in combination with electrowinning, solvent extraction, flotation, selective precipitation and magnetic separations continue to be evaluated for metals present as oxides or sulfides.Since many applications of the hydrometallurgy for metals recovery are still in the development stage additional experimental data are needed in order for this recovery process to become more effective. The newer technology of biohydrometallurgy is still in its formative stages, but moving rapidly toward full-scale application for metal extraction from ores [Patterson, 1987]. A further, investigation of the biological step, which employs organisms characteristic of those indigenous to acid mine drainage, is necessary in order to better control the microbial leaching process, for metals from wastewater sludges.

REFERENCES

Baes, C. "The Extraction of Metallic Species by Dialkyl Phosphoric Acid," *J. Inorganic Nuclear Chemistry*, 24, (1962), p.107.

Brannon et. al. "Long-Term Release of Heavy Metals from Sediments," in *Contaminants and Sediments*, Baker, R. Ed. Vol.2, (Ann Arbor Science, Michigan, 1980).

Brooks, C. S. "Metal Recovery from Industrial Wastes," *Journal of Metals*, (July, 1986), pp.50-57.

Bubenick, D. V. *Acid Rain Information Book*, Edition 2, Noyes Publications, New Jersey, (1984).

Clevenger, T.and J. Novak. "Recovery of Metals from Electroplating Wastes Using Liquid-Liquid Extraction," *J. Water Pollution Control Federation*, 55, (1983), pp.984-989.

Edwards, M. and M.M. Benjamin. "Adsorptive Filtration Using Coated Sand: A New Approach for Treatment of Metal-Bearing Wastes," presented at WPCF Annual Conference, Dallas, Texas, (1988).

Förstner, U. "Changes in Metal Mobilities in Aquatic and Terrestrial cycles," in *Metals Speciation Separation and Recovery*, Patterson and Passino Eds., (Lewis Publishers, Inc., Chelsea, Michigan, 1987).

Jenke, D.R. and F.E. Diebold. "Recovery of Valuable Metals from Acid Mine Drainage by Selective Titration," *Water Research*, 17:11:1585, (1983).

Karra, S.B., C.N. Haas, V. Tare and H.E. Allen. "Kinetic Limitations in the Selective Precipitation Treatment of Electronic Wastes," *Air, Water and Soil Pollution*, (1984).

Martin, J.H. "The Possible Transport of Trace Metals via Moulted Copepod Exoskeletons," *Limnol. Oceanog.*, 15, (1970), p.756.

Morel, F.M.M. *Principles of Aquatic Chemistry*, (Wiley, New York, 1983).

Patterson, J.W. "Metals Separation and Recovery," in *Metals Speciation, Separation, and Recovery*, Patterson and Passino Eds., (Lewis Publishers Inc., Chelsea, Michigan, 1987).

Pomeroy, L., Smith and C. Grant. "The Exchange of Phosphate Between Estuarine Water and Sediments," *Limnol. Oceanog.* 10:167-172 (1965).

Rubin, A. *Aqueous Environmental Chemistry of Metals*, (Ann Arbor Science, Ann Arbor, Michigan, 1974).

Seeton, F. "Solvent Extraction Recovers Vanadium from Waste Stream," Chemical Engineering, (1964), pp.112-114.

Stumm, W. and J. Morgan. *Aquatic Chemistry*, Wiley Interscience Publications, (1981).

INDEX

Acetylacetone 150-152, 440
Acid 32-34, 36, 144, 150, 237, 283, 284, 286, 289, 360, 406, 452, 483, 549, 551-554, 556-560,
    distribution 617
    extraction 50, 144, 165, 242, 267, 279, 381, 389, 440, 481, 498, 500, 502-505, 510, 512, 563, 568-569, 571-575, 578, 579, 580, 587, 590-593, 597, 603, 604
    redissolution 177
Activated Carbon 203-206, 212, 216, 220, 222, 230, 231, 232, 417, 544, 550, 559, 562
Adsorption 37, 96, 97, 108-110, 199, 242, 248, 286, 311, 312, 394, 395,
    activated carbon 204, 231, 232, 417, 544, 550
    Freundlich isotherm 219, 220, 512
    isotherm 329, 331, 333, 330, 334, 339, 342, 343, 349, 356, 512
    kinetics 45, 216, 254
    metal ion 184, 331, 333, 341, 342, 343, 345, 356, 357, 359, 373, 452, 512
    proton 329, 331, 335, 336, 339, 341, 342, 356, 357
Affinity 222, 255, 286, 330, 331, 339, 342-344, 348, 349, 355-357, 362, 367, 407, 408
Aging Process 181, 193
Alkaline Precipitation 92, 117, 169, 170, 193, 194
Alkaline Batteries 576-578, 580
Alkyllead Compounds 4, 5
Aluminum 36, 97, 167, 422, 424, 438
    displacement 518, 527, 529-533
    displacement - recirculation and single pass testing 530
    displacement - reactor design 529
    Nuclear Magnetic Resonance 247
    speciation in geothermal brines 237-258
Ammonium Persulfate 536, 537
Ammonium Carbonate
    copper leaching efficiency 536, 537, 539, 620
Anions 93, 94 97, 126, 134, 135, 166, 177, 422, 429
    Chloride 170, 171, 208
    Hydroxide 93, 116, 178, 208, 222, 275, 482
    Nitrate 86, 170, 175, 178
    Sulfate 97, 170, 171, 199, 530

Atmospheric
    fluxes 3, 7-11, 15, 85, 187
    deposition 8, 9, 14-18
Auger Emission 48
Basic Cupric Salts 170, 171, 189
Batteries 498, 500, 502, 503
    alkaline 576, 580-582
    beryllium 73, 76, 79, 80, 81, 83
    nickel-cadmium 567, 576, 577, 578-580
    spent 567, 568, 570, 572, 575, 576, 577, 578, 581
Bidentate Complexes 329, 345-349
Biotic Indicators 303
Cadmium 34-36, 46, 47, 50, 57, 59-61, 63, 64, 73, 91, 93, 94, 97-101, 106, 114-117, 418, 425, 433
    adsorption 45, 46, 50, 92, 96
    carbonate 35, 120, 463, 464, 466, 483
    environment 576
    precipitation 3, 45, 47, 93, 94, 114, 483, 484, 567, 569
    recovery 29, 30, 34, 440, 457, 503, 547, 558, 569, 570, 575, 577, 578
    solubility 93, 116
    speciation 69, 304, 306, 313
Cadmium Hydroxide 47, 64, 91, 100, 114-117, 120, 579
    complexation 115, 441
    minimum solubility pH 117, 120
    total soluble concentration 94, 101
Canister 549, 559
Chemical 46-49, 53, 60, 69-71, 308
    decontamination 407, 477, 478, 481, 486-487, 490-496, 503-504
    precipitation 45, 92, 174, 190, 200, 203-206, 208, 222, 227, 393-395, 406, 481, 483, 547, 560, 567, 569, 585, 604
    speciation 34, 69, 73, 78, 79, 83, 142, 314-315, 324, 325, 377, 382, 406, 613, 615
Chloride 79, 81, 86, 99, 119, 120, 136, 171, 174, 208, 219, 227, 277, 421, 429, 480, 481, 483, 527, 530, 532, 575, 579, 580
Chromium 91, 142, 399, 482, 493, 552, 553, 562, 606
    recovery 32, 33, 400, 405, 406, 408, 409, 418, 435, 438, 440, 543, 547, 558, 566, 567, 569, 570, 575, 581, 587-601, 603, 607
Chromatography 311
    size-exclusion 292, 294

Citric Acid
    adsorption 45, 92, 96, 288
    distribution 97, 106-110
    inhibition 91-93, 108, 109

Colloidal Copper Hydroxide 170, 209
Colloidal Stability 169
Combined waste 208, 220, 221
Combustion residues 477, 490
Complexes
    bidentate 329, 345-349
    monodentate 329, 343-345
Concentration 8-10, 12-15, 17-19, 29-33, 50, 53, 64, 70, 76, 79-81, 83, 85, 98-100, 106, 107, 109, 114, 115, 117, 119, 120, 122, 130, 133, 134, 136-138, 143, 151-158, 160, 174-178, 181-184, 186, 188-190, 193, 194, 200, 204, 205, 208-212, 227, 237, 239, 241-243, 245, 252-254, 267, 276-279, 284, 285, 301, 319, 324, 344, 348, 362, 375, 378, 379, 386, 388, 390, 393, 395, 399-405, 406, 407, 409-411, 417-419, 421, 422, 424, 425, 435, 442, 446, 447, 449, 455, 458, 463, 468, 469, 473, 474, 478, 480, 485, 494-496, 504, 512, 522, 529, 535, 536, 538, 543, 544, 549, 552, 553, 558-560, 565, 569, 574, 590, 597-599, 605, 606, 612, 613, 615, 617, 618
Construction cost 562
Copper pthalocyanine 204
Copper 15, 16, 30, 35, 36, 38, 50, 71, 169-194, 203-213, 215-223, 286-295, 312, 468, 493
    free 125, 175, 176, 183, 184, 187, 193, 207, 208, 222, 226, 286
    leaching tests 536, 537
    precipitant tests 174,
    minimum solubility pH 208, 209
    ion electrode 204
    recovery 36, 203, 218, 222, 406, 457-459, 518-520, 527, 529, 532, 547, 558, 569, 606
Cost savings 563
Cost comparison 564
Cross contamination 549
Crystal growth 91-93, 95-97, 103, 106, 108, 110, 114, 115
Cu(II) 137, 169-194, 360, 362-367, 369, 371, 373, 375
Cupric ion electrode 204, 206, 207
Cyanide 117, 118, 121, 122, 457, 517, 551-553, 556-559, 562
Dead rinse 551

Distribution 46, 48, 65, 69, 76, 80, 91, 93, 96, 99, 100, 103, 142, 143, 159, 175, 180, 190, 191, 238, 263, 264, 275-279, 329-331, 334, 342, 378, 388, 464, 466, 467, 469, 470, 474, 479, 480, 483, 492, 494, 495, 504, 509, 513, 529, 606, 612, 614, 616, 617
Double layer models 329, 331, 334, 335, 342
Dryer 560
Economics 539, 552, 593
Electrochemistry 518
Environment 3, 4, 27, 38, 46, 47, 73, 85, 91, 142, 263, 264, 275, 276, 302, 390, 417, 433, 452, 454, 468, 486, 487, 490, 494, 513, 518, 539, 573, 586, 616, 617
Experimental procedure 206, 207, 420
Extraction 31, 32, 34, 36-38, 45, 92, 144, 149-152, 158-162, 165-167, 224, 232, 237, 238, 242, 250, 251, 256, 311, 315, 381-383, 385-389, 406, 437, 438, 440, 441, 443, 418, 449, 463, 464, 469, 473, 474, 480-482, 492, 498, 504, 510, 512, 518, 536, 537, 543, 567, 569, 571, 573-575, 579, 580, 585, 587, 588, 590, 592, 596-601, 603, 604, 606, 615, 618, 620-622
Filtration
    gel 243, 261, 286, 289
    ultra 33-34, 237, 239, 242, 249, 250, 362, 518
Fingerprinting 559
Forest soil 283, 284, 291, 292, 293, 295
Freundlich isotherm 206, 219, 220
Fulvic acids 294, 296, 330, 334-336, 345, 354, 464, 466, 467, 513
Future market 557, 558
Hazardous waste streams 437-449, 452-459, 517, 518, 534, 539, 567-582
Hazardous waste 31, 73, 74-78, 82, 497, 498
    incineration 73, 82, 587, 591, 593, 598-601
Heavy metal uptake 283
Heavy metals 45, 69, 91, 203-205, 224, 225, 295, 284, 302, 304, 309, 370, 377-380, 382, 388, 390, 393-396, 408, 417, 418, 429, 433, 434, 435, 452, 463-467, 469-470, 473, 477, 479-485, 487, 490-492, 494-496, 498, 512, 517, 534, 539, 542, 549, 567, 603, 605, 606, 616, 620, 621
Heterogeneity 330, 339, 357
Humic acids 312, 329, 330, 334-336, 345, 381, 512, 514
Hydrolysis precipitation 170, 177
Hydrometallurgical processes 149, 150, 165, 407, 567, 571, 584, 621
Hydroxide precipitation 91-93, 101, 114, 177, 208-211, 222, 227, 482, 485

Incineration 563
    combustion residues 73, 477, 490
    flyash particulates 73
    metal emissions 74, 76, 86, 588
    solid and hazardous waste 74-78, 82, 567, 587, 588, 598-601, 603-607
Industrial wastewaters 93, 169, 170, 225
Information propagation 308
Ion exchange 28, 29, 31-34, 45, 92, 243, 252, 256, 359, 417-422, 424, 427, 433, 438, 452, 456, 458, 569, 573, 575
    cross contamination 549
    dead rinse 551
    monitoring 550
    resin 549-551, 558, 559, 563
    resin regeneration 400-405, 406, 459, 549-551, 558, 559, 563
    regeneration process 399-405, 550, 559
    water reuse 550
Iron 36, 125, 143, 149-153, 155, 160, 161, 165-167, 224, 227, 242, 309, 314, 320, 362, 418, 440, 457, 469, 482, 483, 493, 502, 543, 544, 569, 578-580, 597, 601, 605, 618
Kinetics 57, 77, 91-93, 95-97, 100, 101, 175
Lead 12, 14, 15, 27, 30, 31, 35-38, 60, 70, 71, 73, 76, 77, 79, 81, 85-87, 91, 304, 314, 315, 336, 339, 342, 407, 418, 425, 426, 427, 435, 457, 474, 493, 498-500, 502-506, 508-510, 512, 518, 519, 527-529, 530, 533-534, 535, 537, 539, 540, 569, 606, 616
Leaching 27, 29, 37, 149, 165, 283-285, 289, 292, 293, 298, 464, 466, 479, 481, 504, 510, 535, 536, 538, 539, 554, 556-558, 562, 568, 569, 571-574, 578, 585, 588, 592, 600, 601, 603, 607, 616, 620, 621
Leaf-litter 285, 289, 298
Litter leachates 283-285, 287, 289, 290, 291-294
Manganese 314, 468, 569, 580, 618
Market
    analysis 45, 69, 70, 548, 555
    future 557, 558
    penetration 557-558
    segments 558
Mass balance 479, 480, 494, 526
Maximum acceptable concentration 305, 306
Membrane separation 29, 518
Mercury 27, 46, 47, 49, 50, 52, 53, 57, 59-61, 63, 64, 70, 73, 224, 312, 313, 407, 457, 479, 487, 493, 569, 576, 577, 580, 581
    metallic 59

Metal 3, 4, 7, 10, 11, 14-18, 21, 45-47, 49, 53, 57, 59, 60, 64, 69-71, 74-76, 93, 114
    Complexing abilities 93, 108, 283-289, 291-299, 329, 481
    complexing capacity 283, 298-299
    contaminants 399, 481, 477, 490, 496, 498, 508, 517
    desorption 184, 291, 393
    finishing 91, 452, 547, 563
    finishing operations 517, 547
    finishing industries 203, 417, 517, 518, 527-533, 547-559
    finishing waste streams 518, 529
    ion adsorption 184, 302, 329, 341-349, 356, 357
    ions 51, 93, 165, 175, 183, 283, 284, 285, 295, 298, 301, 302, 304, 308, 317, 318, 319, 329, 354-357, 359-360, 362-367, 369, 373-375, 380, 421, 437, 438, 441, 448, 452-459, 481, 512-513, 569, 575
    maximum acceptable concentration 306
    mobilization 284, 299, 380, 394, 464, 473, 510
    recovery 29, 30, 31-32, 33, 34-35, 37, 38, 149-150, 203, 362, 405, 406, 437-438, 457, 459, 518, 519, 527, 533, 542-544, 607
    solubilization 94, 99, 106, 169, 175, 176, 183, 267, 284, 291, 295, 318, 380, 456, 498, 547-563, 565-566, 567-582
    toxicity reduction 302-308, 418, 542, 569
Metals speciation 69, 75, 80, 85, 87, 125, 167, 301, 318, 324, 325, 377, 393, 412, 513, 514, 574, 611, 621, 622
MEXICO process 46, 69
Minimum solubility pH 208, 209, 211
Mobility 99, 101, 102, 142, 175, 268, 270, 271, 278, 393-395, 615, 617
Monitoring 263, 285, 287, 309, 324, 444, 446, 447, 453, 455, 458, 550
Monodentate complexes 329, 343-345
Natural waters 25, 69, 71, 125, 126, 131, 142, 149, 261, 275, 277, 279, 301, 302, 304, 312, 318, 319, 324, 325, 354, 514, 612, 614, 617
Nickel 12, 14, 36, 378
    recovery 29, 30, 33, 34, 406, 407, 437, 440, 518, 519, 527, 533, 543, 547, 553, 554, 556, 557, 558, 560, 562, 569-573,
    cadmium batteries 503, 577-580
Nitrate 49, 86, 97, 169-171, 175, 177, 178, 181, 183, 185, 189-194, 198-200, 421, 429, 482, 532, 539
Nucleation rate 91, 96, 97, 105, 106, 110
Organics 27, 33
    acids 38, 211-213, 216

INDEX 629

Organo-
    metal complexes 226, 227, 284, 315, 437
    metallic dye 203, 216
Organotin
    aquatic environment 263, 264, 275, 276
    biocidal applications 275
    compounds 263, 276, 315
Oxidation 27-29
    metals 28, 31, 46-48, 60, 69, 71, 125-142, 149-158, 160,
        165-167, 283, 284, 287, 305, 312, 317, 381, 408, 444,
        480, 481, 513, 529, 558, 579, 587, 588, 590, 592-594,
        597, 600, 605, 611
    nonmetals 125, 126, 232, 275, 312
Persistence 264, 276
Phenolic acids 283, 284, 286, 294, 295, 298
Phthalic acid 329-331, 343, 345-347, 349, 355, 356
Pilot studies 203, 223, 548
Plating industry 404, 408, 417, 438, 440, 454, 555, 556-557
Polyelectrolyte 364, 365, 373
Polyphenols 283, 284, 294, 298
Polysaccharides 283, 295, 298, 360, 369, 370
Precipitation 27-29, 31, 32, 34-37, 45-47, 49, 53, 57, 59, 60, 64,
    69, 71, 91-95, 97, 100-104, 106-110, 149, 151, 152, 156,
    198-200, 203-206, 208-212, 309, 310, 373
    carbonate 177, 438, 483, 486
    alkaline 36, 92, 117, 149, 169, 170, 174, 193, 194, 212, 417,
        532
    chemical 92, 174, 203, 205, 208, 227, 393, 406, 481, 535,
        547, 560, 567, 585, 604
    hydroxide 91, 92, 93, 114, 116, 171, 174, 178, 193, 208,
        222, 224, 359, 438, 482, 483, 486, 534, 542, 575, 580,
        603
    hydrolysis 92, 97, 169, 170, 194, 198
    sulfide 211, 220, 222, 228, 229, 231, 539, 544, 575
Printed circuit board 30, 36, 453, 518, 519, 521-523, 529, 542, 552,
    553, 556
Protocatechuic acid 286, 288
Proton adsorption 329, 331, 334-336, 339, 341-343, 349
    isotherm 329, 331, 333, 334, 335, 339, 342, 349
Pyrocatechol Violet (PCV)
    colorimetric method 241, 246, 256

Radiator
    boilout tanks 533, 536, 540
    repair process 533-535
    shop cleaning operations 533-536
    wastewater 518, 528, 534
Ratios 57, 59, 60, 65, 70, 91, 100, 110
    monomer/polymer 237, 250
    supersaturation 91, 94
Raw pig manure 377, 381, 388, 393

Recovery 28, 36, 38, 46, 47, 69
    of metals from waste streams 144, 150, 165-167, 304, 377, 405, 406, 407, 409, 410, 437, 438, 457-459, 487, 498, 502, 514, 517-519, 521, 527, 529, 533, 535, 538, 542-544, 547-563, 567-582, 603-607, 611, 615, 618-622
Recycling 85, 377, 406, 409, 410, 438, 485, 487, 488, 497-504, 508, 510, 512, 518, 519, 521-524, 527, 547, 572, 577, 578, 580
Regeneration process 399, 402, 404, 405, 422, 433, 435, 550
Residence time 74, 77, 97, 100-102, 192, 569
Resin 32, 177
    regeneration 34, 243, 252, 400-404, 406-410, 411, 440, 456, 480, 538, 549, 551, 558, 559, 563, 566
Salicylic acid 329, 345, 346, 349, 355-357
Sediment core 265, 267, 269, 271, 277-279
Segragated stream treatment 548-550
Segregation 30, 548-550,
Sludge 27, 29, 38, 69, 85, 92, 477, 479, 484-486, 500, 503
    biological 377-380, 393-396, 513
    disposal 86, 380, 438, 526, 528, 551, 563, 587, 590, 600, 603
    economics 539, 551, 552, 603
    press 560
    solidification 551
    stabilization 551, 563
    tannery 165, 362, 587, 588, 590, 600
    wastewater 30, 35, 37, 46, 50, 60, 64, 92, 377, 438, 440, 490, 517, 518, 528, 534-539, 543, 545, 590, 591, 593, 603-607
Smelter 533, 543, 560
Soil 377-388, 393-395, 479, 498-500, 502-511, 512, 513
    batch experiment 14, 284, 291, 292, 298, 377-380, 498
    equilibrium solution 284, 291, 292, 293, 294, 295
    forest 283, 284, 291-292, 293, 295, 298, 299
    solution 283, 284, 383
Solidification 551
Source reduction 29, 30, 518, 547

Speciation
    aluminum 237-258, 261
    chemical 3, 190, 314-318, 377, 382, 390, 408
    of metals 28, 34, 36, 45, 47, 59, 69, 71, 73-75, 85, 86, 106, 108, 125, 126, 166, 265-267, 275, 276, 279, 284, 301, 342, 464, 474, 513
    trace metals 74, 125
Spent batteries 567-568, 570, 572, 575, 580
Stability 49, 52, 93, 108, 133, 136, 137, 141, 169, 193, 198, 276, 277, 279, 295, 360, 441, 513
Stabilization 47, 60, 275, 551, 563, 590, 599, 600, 620
Starch Xanthate 45, 49, 50
    polymer structure 53
    reactions 57
    sludges characterization 49
Sulfate 86, 97, 170, 171, 174, 287, 437, 446, 454, 458, 503, 505, 527, 530, 532, 551, 556, 557
Sulfide precipitation 211, 220, 222, 227-229, 231, 538, 544
Surfaces 46, 95
    charge 330, 331, 333, 334, 342, 344, 345
    potential 332-334, 336, 342, 344, 348, 349
Tannery sludge 165, 362, 587, 588, 590, 600
Tin 263-273, 275, 276, 279, 315, 518, 520, 532-534, 538, 558
Toxicity reduction 542
Trace metals 3, 4, 7, 14, 17, 19, 20, 69, 73, 74, 304, 313, 324-325, 468, 612, 622
    Antimony 73
    Arsenic 500
    Beryllium 73, 76, 85-87
    Cadmium 15, 35, 46, 50, 57, 64, 73, 93, 98, 304, 313, 379, 399, 418, 438, 440, 457, 484, 493, 500, 503, 543, 569, 579, 606, 618
    Copper 125, 500, 503, 606
    Lead 15, 27, 35, 36, 38, 73, 76, 81, 86, 304, 457, 474, 493, 500, 503, 504, 505, 518, 527, 530, 532, 533, 538, 543, 569, 606, 618
    Manganese 313, 468, 493, 569, 580, 606, 618
    Mercury 50, 407, 493, 569, 576, 580
    Nickel 30, 36, 313, 379, 386-388, 390, 399, 418, 422, 429, 438, 440, 482-484, 493, 500, 503, 518, 519, 524, 530, 532, 543, 560, 618, 620
    Selenium 142

Treatment 45, 49, 69, 377, 479-483
    combined waste 204, 220, 221
    comparison 222, 237, 250, 293, 378, 547, 552, 594
    recovery 29, 203, 417, 418, 420, 429, 502, 512, 547-558, 565, 566, 577,
    recycling 85, 377, 406, 438, 500, 518, 519, 527, 533, 547
    segragated stream 30, 203, 222, 548-550, 604
    source reduction 518, 547
    technologies 27-29, 38, 497-498, 510, 533
Tributyltin 263, 270, 275, 315
Twin Cities 547, 548, 555, 556, 563, 564, 565
Ultra filtration 33-34, 237, 239, 242, 249, 250, 362, 518
Variably charged surfaces 330
Waste 69
    hazardous 45, 73-83, 85-87, 203, 204, 215, 437-449, 477, 487, 497-502, 503, 517, 518, 534, 539, 542-544, 565, 567-570, 572, 574, 577-580, 584-586, 587, 592, 603, 604
    minimization 29-34, 36, 518, 550
Wastewater 35-37, 57, 63, 65, 69
    industrial 27, 28, 33-34, 37, 45, 93, 169, 170, 223, 377, 399, 401, 404, 405, 437, 452, 517-535, 547, 550, 551, 590, 593, 603-604, 607
    pollutants - copper, lead, zinc, tin 57, 399, 438, 452, 490, 517, 543
    sludge 29, 30, 59, 440, 490, 500, 526, 528, 537, 551, 590, 591
    treatment 30, 93, 95, 223, 477, 492, 496, 502, 526, 547, 550, 551, 566, 568, 589
Water reuse 550, 551
Water-soluble 45, 283, 294, 298, 380
X-ray photoelectron spectroscopy 46, 47, 69
    XPS Data Analysis 51, 59
Zinc 16, 30, 35, 36, 304, 313, 378, 386-388, 390, 399, 406, 418, 438, 440, 484, 493, 500, 502, 518, 533, 534, 538, 543, 553, 556, 557, 560, 562, 569, 576, 577, 580, 606, 618